Developments in Soil Science - Volume 30

Development of Pedotransfer Functions in Soil Hydrology

Developments in Soil Science

Series Editors: A.E. Hartemink and A.B. McBratney

Developments in Soil Science - Volume 30

Development of Pedotransfer Functions in Soil Hydrology

Edited by

Ya. Pachepsky
*USDA-ARS Environmental Microbial Safety Laboratory
Beltsville, MD, USA*

W.J. Rawls
*USDA-ARS Hydrology and Remote Sensing Laboratory
Beltsville, MD, USA*

2004

ELSEVIER

Amsterdam • Boston • Heidelberg • London • New York • Oxford • Paris
San Diego • San Francisco • Singapore • Sydney • Tokyo

ELSEVIER B.V. ELSEVIER Inc. ELSEVIER Ltd ELSEVIER Ltd
Sara Burgerhartstraat 25 525 B Street, Suite 1900 The Boulevard, Langford Lane 84 Theobalds Road
P.O. Box 211, 1000 AE Amsterdam San Diego, CA 92101-4495 Kidlington, Oxford OX5 1GB London WC1X 8RR
The Netherlands USA UK UK

© 2004 Elsevier B.V. All rights reserved.

This work is protected under copyright by Elsevier B.V., and the following terms and conditions apply to its use:

Photocopying
Single photocopies of single chapters may be made for personal use as allowed by national copyright laws. Permission of the Publisher and payment of a fee is required for all other photocopying, including multiple or systematic copying, copying for advertising or promotional purposes, resale, and all forms of document delivery. Special rates are available for educational institutions that wish to make photocopies for non-profit educational classroom use.

Permissions may be sought directly from Elsevier's Rights Department in Oxford, UK: phone (+44) 1865 843830, fax (+44) 1865 853333, e-mail: permissions@elsevier.com. Requests may also be completed on-line via the Elsevier homepage (http://www.elsevier.com/locate/permissions).

In the USA, users may clear permissions and make payments through the Copyright Clearance Center, Inc., 222 Rosewood Drive, Danvers, MA 01923, USA; phone: (+1) (978) 7508400, fax: (+1) (978) 7504744, and in the UK through the Copyright Licensing Agency Rapid Clearance Service (CLARCS), 90 Tottenham Court Road, London W1P 0LP, UK; phone: (+44) 20 7631 5555; fax: (+44) 20 7631 5500. Other countries may have a local reprographic rights agency for payments.

Derivative Works
Tables of contents may be reproduced for internal circulation, but permission of the Publisher is required for external resale or distribution of such material. Permission of the Publisher is required for all other derivative works, including compilations and translations.

Electronic Storage or Usage
Permission of the Publisher is required to store or use electronically any material contained in this work, including any chapter or part of a chapter.

Except as outlined above, no part of this work may be reproduced, stored in a retrieval system or transmitted in any form or by any means, electronic, mechanical, photocopying, recording or otherwise, without prior written permission of the Publisher. Address permissions requests to: Elsevier's Rights Department, at the fax and e-mail addresses noted above.

Notice
No responsibility is assumed by the Publisher for any injury and/or damage to persons or property as a matter of products liability, negligence or otherwise, or from any use or operation of any methods, products, instructions or ideas contained in the material herein. Because of rapid advances in the medical sciences, in particular, independent verification of diagnoses and drug dosages should be made.

First edition 2004

Library of Congress Cataloging in Publication Data
A catalog record is available from the Library of Congress.

British Library Cataloguing in Publication Data
A catalogue record is available from the British Library.

ISBN: 0 444 51705 7
ISSN: 0166-2481 (Series)

∞ The paper used in this publication meets the requirements of ANSI/NISO Z39.48-1992 (Permanence of Paper). Printed in The Netherlands.

Working together to grow
libraries in developing countries

www.elsevier.com | www.bookaid.org | www.sabre.org

ELSEVIER   BOOK AID International   Sabre Foundation

FOREWORD

The reason to introduce the term *pedotransfer function* by Bouma and van Lanen in 1987 and later, in a more accessible form, by Bouma in 1989, was to emphasize the possible link between soil survey ("pedology") and soil hydrology. Characteristics, such as texture, bulk density and organic matter content obtain a broader meaning when they are directly related to natural soil bodies in the field, as defined by soil survey. Textures provide information on the geologic origin of sediments or weathering products and they vary in characteristic patterns in the landscape. Bulk densities usually increase with depth in the soil unless compaction occurs in surface soil. Some soils have by origin significantly lower bulk densities than others. Organic matter is usually concentrated in surface soil but it decreases in characteristically different patterns with depth in different soil types, landscapes and climate types. Soil databases, based on soil surveys, have been established in many countries and provide therefore an attractive source of soil data to be used in *pedotransfer functions*. Such data don't stand apart but have a structural link with soils in the field, thereby providing a direct connection with uses of land that are geographically defined. The link of *pedotransfer functions* with soil survey was emphasized by Bouma and van Lanen (1986) as they defined both *continuous* and *class pedotransfer functions*. The latter were based on soil horizon designations or even on entire soil types, to which particular soil information was attached (e.g. Wosten et al., 1985). This has not been much further developed because soil classifications and soil horizon designations have always retained a somewhat arbitrary descriptive character which contrasts with a clearcut quantitative regression equation. Still, I believe that more attention should be paid in future to using such *class pedotransfer functions*.

The *pedotransfer* concept is not restricted to water movement in soil. An early example was presented by Breeuwsma et al. (1986) who related the phosphate adsorption capacity (PAC) of soils to their Fe and Al content as presented in soil survey reports. The rather cumbersome method of measuring PAC could be replaced by a rapid calculation method.

Soil information is increasingly used for designing innovative agricultural production systems and for spatial planning using participatory approaches where stakeholders and policy makers are much more involved right from the start than was the case in earlier times (e.g. Bouma, 2003). In this context, earlier soil survey interpretations and land evaluations are not satisfactory anymore as they only provide qualitative, descriptive information that certainly is of value but does not have the required quantitative focus that is needed for modern scenario analysis. Here, modern simulation models are increasingly being applied. Such models, however, have a high data demand. We all know examples where sophisticated models are being run, requiring, perhaps, 80 types of input data of which only 20 have really been measured and where the rest is "estimated". When results are spectacularly presented using modern GIS technology, the impression of a triumph of science and technology may in reality represent a hoax.

Obviously, *pedotransfer functions* are ideal to at least partly satisfy the data demand. Without proper *pedotransfer functions* many modeling applications – indispensable for modern work – could not have been realized, simply because funds are often lacking to measure all required parameters. Of course, one should never forget that *pedotransfer functions* are based on state-of-the-art measurements of particular parameters, including, of course, measurements made elsewhere. *Pedotransfer functions* can only be derived when a sufficient number of measurements have been made, but no more than what may be considered to be "sufficient". When deriving such functions it is therefore important to continuously check the error involved with predictions. As more measurements become available, the error usually tends to decrease. Once this error is considered acceptable, further measurements of the particular parameter can be omitted and scarce funds can be applied elsewhere. This particular error should always be presented as an essential element of the particular *pedotransfer function*.

One concern when using *pedotransfer functions* is what may be called the "sorcerer's apprentice" dilemma. User-friendly models, fed with *pedotransfer functions* can be used by anyone who knows how to handle a computer keyboard, which by now defines the entire population. There is a clear risk here. Some soil scientists have expressed their concern about deriving *pedotransfer functions* for exactly this reason. "Any engineer or consultant can use our data and run with it". The answer is, of course, a proactive approach emphasizing quality control. One major advantage of *pedotransfer functions* is the structural link with soil survey and pedology. Results should always be checked: do results make sense considering the particular properties of soils in the area being considered: do soils exhibit preferential flow? Are they hydrophobic? Is there lateral subsurface flow etc. etc. The recent establishment of the Hydropedology working group of the Soil Science Society of America is an excellent initiative also to ensure proper use of models and *pedotransfer functions*.

We may further emphasize the latter aspect by considering use of *pedotransfer functions* in the context of "knowledge chains". They range from soil data to soil information, all the way to knowledge and wisdom. Information is data to which a certain meaning has been attached. Knowledge is internalized information that has become one's own expertise. Wisdom represents the ability to apply knowledge in a selective manner, which includes not applying it at all. A final wish, therefore, can be that *pedotransfer functions* will be wisely used. If so, they can contribute significantly to increasing our understanding of the dynamic functioning of our land which is essential when designing the sustainable land use systems of the future.

Johan Bouma
Emeritus Professor of Soil Science
Wageningen University and Research Center
Wageningen, The Netherlands

PREFACE: STATUS OF PEDOTRANSFER FUNCTIONS

Agricultural and environmental modeling and assessment have many uses for soil parameters governing retention and transport of water and chemicals in soils. These properties are notorious for the difficulties and high labor costs involved in measuring them. Often, there is a need to resort to estimating modeling-related soil parameters from other readily available data. Modeling in a wide range of scales, from general circulation models to the fine scale precision agriculture decision support, has the need in such estimations.

Following Briggs and Shantz (1912), who were the pioneers in the field, generations of researchers quantified and interpreted relationships between soil properties. Terms "prediction of" or "predicting" soil properties, "estimation of" or "estimating" soil properties, "correlation of" or "correlating" soil properties, were used interchangeably to name contents, procedures and results of such studies (Carter and Bentley, 1991; Rawls et al., 1991; Van Genuchten and Leij, 1992; Timlin et al., 1996a; Pachepsky et al., 1999; McBratney et al., 2002). Recently, statistical regression equations expressing relationships between soil properties were proposed to be called "transfer functions" (Bouma and Lanen, 1987) and later "pedotransfer functions" (Bouma, 1989). Estimating soil hydraulic property dominates the research field, although soil chemical and biological parameters are also being estimated. Several reviews on PTF development and use have been published (e.g., Rawls et al., 1991; Van Genuchten and Leij, 1992; Timlin et al., 1996a; Pachepsky et al., 1999; Wösten et al., 2001). Large databases, such as UNSODA (Leij et al., 1996), HYPRES (Lilly, 1997; Wösten et al., 1999), WISE (Batjes, 1996), and NRCS pedon database (USDA Natural Resource Conservation Service, 1994), are established that can be used for the purposes of the PTF development.

The apparent easiness of developing pedotransfer functions (PTFs) by applying statistical regressions should not overshadow several basic questions about PTFs that need to be answered by hydrologists and soil scientists. Why do PTFs exist? How to assess the reliability of PTFs? How to quantify the accuracy and reliability of PTFs? Will a grouping of soils by some criterion enhance both the accuracy and the reliability of PTFs? Is there a limit of accuracy and reliability of PTFs and what does this limit depend on? What are the most appropriate techniques to evaluate a PTF? What input variables are more preferable or necessary to be included in a PTF? An inventory of existing PTFs seems to be in order to begin answering these questions in a systematic fashion. Such inventory is the purpose of this book.

The book encompasses the international experience in the field. The authors present the spectrum of the ideas behind PTF development. Multiple PTFs are presented that have been successfully used in applications. The general objective has been to assemble the first-of-a-kind treatise that would be useful for both developers and end users of PTFs. This section outlines the book.

Part I of the book comprises the methods that have been used to-date to develop PTFs. In Chapter 1, H. Vereecken and M. Herbst describe statistical regression that has been a traditional tool of PTF development for decades. The advantage of the regression technique is the possibility to obtain rigorous estimates of statistics for predicted values and coefficients in the equations. However, as the authors show, a blind use of the statistical computer package without a preliminary and a *posteriori* data analysis can have a deleterious effect on the results of a regression. Model misspecification, presence of outliers, multicollinearity, and other drawbacks are examined, and techniques to deal with them are outlined. The authors emphasize that the PTF building with regression is an iterative procedure that may require many iteration steps.

Applying the statistical regression to predict soil properties that are difficult to measure requires (a) deciding which properties are to be used as predictors and (b) which regression equation to use. Those decisions are not straightforward, especially when the databases contain many potential predictors and the relationships between soil properties may be different in different parts of the databases. For that reason, PTF development has recently employed data mining and exploration methods to introduce algorithms that automate predictor and equation selections. Ya. Pachepsky and M. Schaap describe three of the most popular such methods in Chapter 2. Artificial neural networks (Section 1) are becoming a common tool for modeling complex "input-output" dependencies because of their ability to mimic the behavior of complex systems. Being powerful approximators, the artificial neural networks, however, lack a built-in methodology to sieve out the less or the least important input variables. Therefore, PTF development may benefit from methods that would not assume a type of dependence and which still would be able to eliminate some predictors. One of them, the group method of data handling, is presented. The important predictors may also be different in different parts of the database. Then the database has to be quite large to apply the artificial neural network or group method of data handling. A data exploratory technique aiming on uncovering structure in data can be useful in such cases. The regression tree modeling is such a technique that gives transparent results and can be applied not only to numerical but also to categorical soil data. A common disadvantage of all techniques in this chapter as compared to statistical regression is the heuristic element involved so that the rigorous statistical judgment of results is hard to make.

Evaluation of PTFs is an essential element of their development and use. Pachepsky et al. (1999) broadly defined the accuracy of a PTF as a correspondence between measured and estimated data for the data set *from which a PTF has been developed*. The reliability of PTF was assessed in terms of the correspondence between measured and estimated data for the data set(s) *other than the one used to develop a PTF*. Finally, the utility of PTF in modeling was viewed as a correspondence between *measured and simulated environmental variables*. The concept of the PTF uncertainty (Schaap and Leij, 1998) encompasses the ambiguity in PTF predictions and parameters caused by the input data variability and uneven representation of soils with different properties in the database. In Chapter 3, M. Schaap shows that the PTF accuracy in parameters may be dependent on the optimization criteria, and reviews established techniques available to assess the uncertainty in PTF estimates.

Part II of the book presents a panoramic view on the research to estimate soil water retention and soil hydraulic conductivity. Particle size distribution and its parameters are used in almost any pedotransfer function. In Chapter 4, A. Nemes and W. Rawls present a comprehensive review of the use textural parameters in PTF. A special attention is paid to the problem of PTF portability caused by the differences in particle size classes in various national and international classifications. The authors show it is very important to use the boundaries of sand and silt textural classes from the classification used during PTF development rather than from a national classification. That may require interpolation of particle size distributions that can introduce an error, although interpolation techniques suggested by the authors seem to mitigate such errors.

Water retention PTFs generally become more accurate if some experimental information about water retention is used as the PTF input. Chapter 5, written by D. Timlin, W. Williams, L. Ahuja, and G. Heathman, shows that, for PTF development, it is beneficial to combine experimental information about water retention with scaling laws. The nonlinearity in soil hydraulic properties creates substantial difficulties for their pedotransfer estimation. The authors explored logarithm-based linearizations of soil hydraulic properties for the USDA textural classes and found strong correlations between slopes and intercepts for both water retention and hydraulic conductivity across the classes. One parameter remains to be estimated to use those correlations as PTFs.

Soil organic carbon content and composition affect both soil structure and adsorption properties and therefore soil bulk density, water retention and hydraulic conductivity may be affected by soil organic carbon. However, literature on the effect of soil organic carbon on soil hydraulic properties is contradictory. In Chapter 6, W. Rawls, A. Nemes and Ya. Pachepsky show that that this effect depends on proportions of textural components and amount of organic carbon. The authors use the massive US National Soil Characterization Database to develop their pedotransfer functions for bulk density, water retention and hydraulic conductivity.

Using soil morphological attributes and soil structure in pedotransfer functions is extremely important because of dominating role of soil structure in soil hydrology. It is also very difficult because of semi-quantitative nature of soil morphology and structure descriptors. Chapter 7, written by A. Lilly and H. Lin, shows that it is possible to develop pedotransfer relationships to predict soil hydraulic properties and soil hydrological functioning from soil morphology data including soil structure. These relationships can be expressed as either pedotransfer functions or pedotransfer rules. A considerable volume of soil morphological data residing in national and regional databases can be of great use in providing information on soil hydrology. However, while there is a scope for further developments with existing data, future progress may require an acceptance of standardized methodologies both to describe and quantify soil morphological attributes and to measure soil hydraulic properties in the field.

Soil aggregate composition is often assumed to be related to soil structure and as such has been expected to affect soil water retention. In Chapter 8, A. Guber and co-authors show that it may indeed be the case if the quantiles of distributions are used instead of contents of individual aggregate fractions. Parameters of van Genuchten equations have had those quantiles as leading predictors for soils in the database dominated by Mollisols.

Soil mineralogical and chemical properties are conceptually important factors of soil ability to retain and transmit water. Chapter 8, written by A. Bruand, shows that much could be gained in understanding and estimating soil hydraulic properties when mineralogical and chemical data have been coupled with soil physical information. This chapter also shows that such coupling has not often happen in the past, and there is both a room for development and urgent need in regions where soil salinity and alkalinity constitute environmental and management problem.

Grouping soils appeared very early as a way to increase the reliability and applicability of PTFs. A. Bruand have authored Chapter 10 in which a comprehensive outlook of the grouping strategies is presented. The author concludes that texture, soil bulk density, and the type of parent material appear to be the most efficient criteria to improve accuracy of PTFs.

Part III of the book deals with PTF that estimate parameters of processes closely related to soil water transport and retention. Soil erosion modeling has developed a substantial body of pedotransfer research. Evolution of PTFs in this field is described by D. Flanagan in Chapter 11. The author stresses that, for wide-spread applicability, soil erosion prediction models often rely upon a variety of soil pedotransfer functions that most often estimate soil erodibility and soil infiltration parameters. The chapter highlights the importance of existing long-term historical erosion plot data (e.g. the USLE database) that may provide sufficient information to create erosion model pedotransfer equations.

PTFs for solute and gas retention and transport parameters are reviewed in Chapter 12 written by B. Minasny and E. Perfect. Their review suggests that input parameters required in solute adsorption and transport models can be predicted reasonably well from basic soil properties although caution should be exercised because of possible scale effects, i.e., the effect of scale of solute transport on the solute retardation. The unified diffusivity model can use existing PTFs to estimate solute and gas diffusivities as a function of soil water content. Based on current knowledge, pore-size distribution appears to be an acceptable dispersivity predictor, and water retention PTFs can be helpful in developing dispersiity PTFs. The lack of coupled data on solute dispersivity, soil hydraulic properties and other soil parameters presents the current knowledge gap hampering the dispersivity estimation.

The shrink–swell dynamics of soils requires quantification in many regions of the World. Chapter 13 by E. Braudeau, R. Mohtar, and N. Chahinian describes one concerted attempt to estimate parameters of soil shrinkage. Several characteristic points on the shrinkage curve have been related to texture, the coefficient of linear extensibility, and two points on soil water retention curves approximately corresponding to water contents at field capacity and wilting point.

Chapter 14 describes an attempt to relate soil water retention and soil rheology. The authors, E. Shein, A. Guber and A. Dembovetsky exploit a method to define several characteristic points on the water retention curve that separate segments of the water retention curve at which the forces causing water retention have distinctly different physical nature. Water contents in characteristic points are termed key water contents and are related to the soil matric potential. The authors maintain that key water contents can be estimated from the liquid limit, the plastic limit and data on evaporation from soil. As soon as they are estimated, their relationships with soil matric potential provide four points on water retention curve which may be sufficient for some applications. It is interesting to note that, in this and previous chapters, rheology-related PTFs operate with gravimetric water contents whereas the majority of PTFs for soil hydrologic properties include the volumetric water contents.

Part IV of the book takes a peek in the Pandora box of issues related to the application of PTFs in spatial context. PTFs are built from small point samples, and yet are to be used for large spatial units. Results of applications will depend on soil basic data availability within those units, on soil variability within those units, on the capability of PTFs to reproduce this variability, and on our ability to modify PTFs according the scale of the application. Chapter 15 by K. Smettem and colleagues provides a framework for that.

The authors provide an elaborated discussion of the general role of PTFs in simulating and understanding hydrologic functioning of catchments. The issue of soil basic data paucity is specifically addressed. The proposed strategy is to use as simple as possible PTF that would use input soil basic information that could be estimated from remote sensing data.

Soil properties are known to be related to landscape position, geomorphic information has long been routinely used in soil mapping, and geomorphometry was proposed as a data source to predict soil basic properties. In particular, soil texture, organic matter content, and bulk density are known to reflect both landscape position and land surface shape. Because these soil properties are most often included in pedotransfer functions, one can hypothesize that soil hydraulic properties should have some relationship to landscape position and land surface shape. Chapter 16, written by N. Romano and G. Chirico summarizes available approaches to relate terrain attributes to soil basic and hydraulic properties. Overall, the reported studies and the related discussion confirm the usefulness of topographic attributes as ancillary data to indirectly estimate soil hydraulic properties. The authors rightly note that the extent of such utility is site-specific and relies on integration of local environmental conditions in certain primary and secondary terrain variables.

N. Romano in Chapter 17 continues the exploration of the spatial component in PTF development. Is the spatial variability of PTF estimates the same as the spatial variability of measured hydraulic properties? Is it possible to "calibrate" an existing PTF by using auxiliary information about spatial variation in environmental variables that may affect soil properties? Examples given in the chapter show the research framework is appearing to take shape to answer these questions in a satisfactory manner.

Part V of this book explores user-oriented techniques and software. A PTF user has to able to make an informed choice of PTF and to have a convenient tool to apply the PTF technology. Chapter 18 written by A. McBratney and B. Minasny gives an excellent example of packaging an extensive research results in a multipurpose tool. The authors use the term "soil inference system" for their technology. A soil inference system takes measurements known with a given level of (un)certainty, and infers data that are not known with minimal inaccuracy, by means of logically conjoined pedotransfer functions. Uncertainty is quantified in terms of the model (PTFs) uncertainty and input uncertainty. The user defines the distribution for each input variable and their correlations. If the sample is outside the PTF training set, the uncertainty is increased using fuzzy logic methods. The authors also provide a compendium of PTF software available via Internet. More details about the PTF software are presented by M. Schaap in his Chapter 19 by M. Schaap who concentrates on graphic user interfaces currently in use with PTF. The use of models depends on the easiness to operate them, and PTFs are not an exception.

How to discriminate between existing PTFs? Chapter 20 describes the most popular approaches and elaborates on one technique developed by authors. Section 1 of this chapter written by M. Donatelli, H. Wösten and G. Belocchi decribes the general framework of PTF evaluation. Authors show that that PTF evaluation can be based on evaluating uncertainty in equations and data sets, on comparing estimates and measurements, and on sensitivity analysis of a specific model with respect to PTFs in question. Section 2 of Chapter 20 presents a set of statistics and derived integrated indices for pedotransfer function evaluation. M. Donatelli, M. Acutis, A. Nemes, and H. Wösten stress that no single statistic can provide a portrayal of the PTF performance. Various aspects of PTF performance are gauged by separate statistics that are later combined in unique indices using the fuzzy logic methods. An example of using this technique to access the reliability of 10 PTFs is presented. Finally, Section 3 of Chapter 20 written by H. Wösten, A. Nemes, and M. Acutis, considers the evaluation of PTFs on the base of their utility. The functional evaluation of PTFs uses criteria directly related to applications rather than statistics to characterize the accuracy. This section shows that the PTF accuracy may be not an issue as at least four factors affect the performance of a PTF in simulations. Those are the accuracy of basic soil data used as inputs in PTFs, the accuracy of PTF itself, specific features of the simulation model, and the output used in functional criteria. This section also shows the scarcity of published reports on PTF functional evaluation. Further PTF development needs information about ways and results of PTF use.

Part VI of the book presents examples of regional PTF development. Chapter 21 summarizes results of a unique work undertaken by J. Tomasella and M. Hodnett and their colleagues. The authors documented inadequacy of temperate region PTFs for tropics, explained reasons for that, collected a large database of hydraulic properties for Brazilian soils, and applied modern data mining techniques to obtain PTFs specifically for tropical soils.

Chapter 22 by H. Wösten and A. Nemes encapsulates results of another large project on creating European database on soil hydraulic properties and deriving pedotransfer functions on a continental scale. The chapter shows problems in constructing this large international database and reports on methods to solve those problems. The chapter also describes the study to identify the relevance of PTF from international data collections for individual countries that contributed to the database or for countries located in areas with comparable climatic conditions.

Chapter 23 by W. Rawls contains pedotransfer functions developed from the large database on soil hydraulic properties containing thousands of samples collected in several regional and national projects in USA. The methodology of the projects has been coordinated, and that has allowed the author to develop PTFs for water retention at specific soil water potentials and for parameters of most common water retention equations. The chapter also contains unique PTFs to estimate saturated hydraulic conductivity.

Chapter 24 summarizes results of the country-wide PTF development in Poland. The authors, R. Walczak, B. Witkowska-Walczak, and C. Sławiński, describe pedotransfer functions for soil water retention hysteresis, for organic soils and for unsaturated hydraulic conductivity. Such PTFs have not been developed in other regions, they reflect needs of soil water management specific for the country, and may be of interest in other regions with similar soil cover and land use. Finally, Chapter 25 by V. Štekauerová and J. Šútor presents results of estimating water retention for a small, agriculturally important region in Slovakia. The authors have used the independent data set to evaluate the PTF reliability and found an excellent correlation between different PTF accuracy indices.

Each chapter of the book contains an extensive bibliography. Editors have decided to include also the additional bibliography that contains important information on regional pedotransfer function development in USA, Canada, Sweden, Australia, Japan, and other countries. Some of the references point to the valuable data sources. Other references pertain to emerging issues of the effect of soil management on hydraulic properties, of using non-traditional inputs for PTFs, and of using physical models of soil structure to make PTFs more robust.

The PTF research is a fast-developing field driven by the high demand in hydraulic and related parameters existing in environmental modeling. Recently the volume of PTF applications increased due to development of GIS-based regional modeling. Relying on PTFs presumes a level of the PTF accuracy and reliability sufficient to obtain acceptable uncertainty in results of modeling. As many PTFs have been developed and are under development, the users face the problem of PTF selection. Discussions of reasoning and consequences of such selection are currently underrepresented in literature. One criterion of PTF selection is its reliability. Although no general conclusion could be derived from the review of the PTF reliability studies, some general observations can be made. PTFs developed from regional databases give good results in regions with similar soil and landscape history. Water retention PTFs developed in Belgium (Verekeen et al., 1989) were the most accurate as compared with 13 others for the data base of the Northern Germany (Tietje and Tapkenhinrichs, 1993). Water retention PTFs developed for the Hungarian Plain (Pachepsky et al., 1982) were applicable for the Caucasian Piedmont Plain (Nikolaeva et al., 1986). PTFs developed in Australia were more accurate for the Mississippi Delta as compared with other regional PTFs (Timlin et al., 1996b). It remains to be seen whether this observation holds for other cases, and which soil and landscape features have to be similar in two regions to assure the mutual reliability of the PTFs developed.

The database size and measurement methods are among important factors affecting PTF accuracy and reliability. The number of samples has to be large enough to develop both accurate and reliable PTF (Pachepsky et al., 1999). PTFs developed from the USA database by Rawls et al. (1982) are more robust than PTFs developed from regional databases. When the reliability of the all-USA PTFs has been compared with the accuracy of several other regional databases, and the PTFs were ranked by their reliability, the all-USA PTFs usually had one of the highest rankings (Tietje and Tapkenhinrichs, 1993; Kern, 1995; Timlin et al., 1996b). Collection and analysis of data suitable for the PTF development is important step in providing essential parameters for the environmental and agricultural modeling.

There may exist a limit in accuracy and reliability of PTFs caused by temporal variation of soil properties related to the changes in vegetation and soil management. Incorporation of organic matter content (Felton and Ali, 1992), soil erosion (Fahnenstok et al., 1995), tillage practice (Azooz et al., 1996) can cause variations in hydraulic properties that are comparable with variations within regional data bases. Pachepsky et al. (1992) have reported 20% changes in water retention at -1 kPa and 5% changes in water retention at -30 kPa in soils under wheat during growing season in three different climatic zones. The amount of published data on temporal variations of soils hydraulic properties remains small. A temporal component may be required in PTFs to improve their reliability.

A relatively new promising direction in PTF development is the use of spatially dense physical information related to the soil cover. First steps in using remote sensing data and topographic information have been discussed in this book. Geophysical techniques, such as ground-penetrating radar, penetrometers, electric conductivity meters, etc., and remote sensing provide spatial coverage that shows a potential to be included in PTFs.

The reliability of a PTF is not directly related to its utility (Pachepsky et al., 1999). The latter is affected by the sensitivity of the model to PTF predictions, and also by the uncertainty in other model inputs (Leenhardt, 1995). When the PTF uncertainty is factored in a modeling effort, the variation in predictions of different PTFs has to be considered along with uncertainty of individual PTF predictions. The procedures to do that are yet to be developed. Using weighted-average predictions of several different PTF instead of predictions of a single PTF may be a viable option. As the hydrological model calibration technology develops, more opportunities appear to compare calibrated and PTF-estimated hydraulic properties (Wang et al., 2003) and also to use PTF predictions as initial estimates for model calibration (Jacques et al., 2002).

An emerging challenge is the upscaling of PTF estimates, i.e., determination of equivalent hydraulic parameters for large spatial units using PTF estimates and information on their variations in space. Scale dependencies in soil hydraulic properties has become recognized (Bork and Diekrügger, 1990; Feddes et al., 1993). Currently, these dependencies are ignored and may limit PTF reliability. Scale dependencies of hydraulic properties need to be included in PTFs. As the scale becomes coarser, hydrologic models change and so do the hydrologic parameters. Soil field capacity presents an example of such parameter that cannot be predicted as water retention at some fixed value of soil matric potential. PTF predictions of water retention curves have to be supplemented with a scale correction to become utilizable at coarser scales.

The PTF research demonstrates the need in and value of integrative studies. Systematic coupling measurements of various soil properties adds value to all of those measurements as pedotransfer relationships can be derived to be used in the variety of applications. The PTF development contributes to understanding of soil functioning and enhances role of soil information in multiple applications.

Ya. Pachepsky
W.J. Rawls

REFERENCES

Azooz, R.H., Arshad, M.A., Franzluebbers, A.J., 1996. Pore size distribution and hydraulic conductivity affected by tillage in Northwestern Canada. Soil Sci. Soc. Am. J. 60, 1197-1201.
Batjes, N.H., 1996. Development of a world data set of soil water retention properties using pedotransfer rules. Geoderma 71, 31-52.
Bork, H.-R., Diekrügger, B., 1990. Scale and regionalization problems in soil science. Trans. 14 Int. Congr. Soil Sci. 1, 178-181.
Bouma, J., van Lanen, H.A.J., 1987. Transfer functions and threshold values: from soil characterisctics to land qualities. *In*: Proc. of the Int. Workshop on Quantified Land Evaluation Procedures, 27/04–2/05/1986, Washington, DC, USA, pp. 106–110.
Bouma, J., 1989. Using soil survey data for quantitative land evaluation, 1989. Adv. Soil Sci. 9, 177-213.
Briggs, L.J., Shantz, H.L., 1912. The wilting coefficient and its indirect measurement. Bot. Gazette 53, 20-37.
Carter, M., Bentley, S.P., 1991. Correlations of Soil Properties. Pentech Press, London.
Fahnenstok, P., Lal, R., Hall, G.F., 1995. Land use and erosional effects on two Ohio alfisols, 1. Soil properties. J. Sustainable Agric. 7, 63-84.

Feddes, R.A., Menenti, M., Kabat, P., Bastiaansen, W.G.M., 1993. Is large scale inverse modeling of unsaturated flow with areal average evaporation and surface soil moisture as estimated from remote sensing feasible? J. Hydrol. 143, 125-152.

Felton, G.K., Ali, M., 1992. Hydraulic parameter response to incorporated organic matter in the B horizon. Trans. ASAE 35, 1153-1160.

Jacques, D., Simunek, J., Timmerman, A., Feyen, J., 2002. Calibration of Richards and convection–dispersion equations to field-scale water flow and solute transport under rainfall conditions. J. Hydrol. 259, 15-31.

Kern, J.S., 1995. Evaluation of soil water retention models based on basic soil physical properties. Soil Sci. Soc. Am. J. 59, 1134-1141.

Leenhardt, D., 1995. Errors in estimation of soil water properties and their propagation through a hydrological model. Soil Use Manag. 11, 15-21.

Leij, F., Alves, W.J., van Genuchten, M.Th., Williams, J.R., 1996. The UNSODA unsaturated soil hydraulic database, User's manual version 1.0. EPA/600/R-96/095. National Risk Management Laboratory, Office of Research and Development, Cincinnati, OH.

Lilly, A., 1997. A description of the HYPRES database (HYdraulic Properties of European Soils). The use of pedotransfer functions in soil hydrology research. Proc. Second workshop of the project Using existing soil data to derive hydraulic parameters for simulation modeling in environmental studies and in land use planning. Orleans, France, 10–12/10/1996, p. 161–184, INRA, Orleans and EC/JRC Ispra.

McBratney, A.B., Minasny, B., Cattle, S.R., Vervoort, R.W., 2002. From pedotransfer functions to soil inference systems. Geoderma 109, 41-73.

Nikolaeva, S.A., Pachepsky, Ya., Shcherbakov, R.A., Shcheglov, A.I., 1986. Modeling moisture regime of ordinary Chernozems. Pochvovedenie 6, 52-59, in Russian.

Pachepsky, Ya., Rawls, W.J., Timlin, D.J., 1999. The current status of pedotransfer functions: their accuracy, reliability, and utility in field- and regional-scale modeling. *In*: Corwin, D.L., Loague, K., Ellsworth, T.R. (Eds.), Assessment of non-point source pollution in the vadose zone, Geophysical monograph 108. American Geophysical Union, Washington, DC, pp. 223-234.

Pachepsky, Ya., Shcherbakov, R.A., Varallyay, G., Rajkai, K., 1982. Soil water retention as related to other soil physical properties. Pochvovedenie 2, 42-52, in Russian.

Pachepsky, Ya., Rawls, W.J., Timlin, D.J., 1999. The current status of pedotransfer functions: their accuracy, reliability, and utility in field- and regional-scale modeling. *In*: Corwin, D.L., Loague, K., Ellsworth, T.R. (Eds.), Assessment of non-point source pollution in the vadose zone, Geophysical monograph 108. American Geophysical Union, Washington, DC, pp. 223-234.

Pachepsky, Ya., Mironenko, E.V., Scherbakov, R.A., 1992. Prediction and use of soil hydraulic properties. *In*: van Genuchten, M.Th., Leij, F.J., Lund, L.J. (Eds.), Indirect Methods for Estimating the Hydraulic Properties of Unsaturated Soils. University of California, Riverside, CA, pp. 203-212.

Rawls, W.J., Brakensiek, D.L., Saxton, K.E., 1982. Estimation of soil water properties. Trans. ASAE 25, 1316-1320.

Rawls, W.J., Gish, T.J., Brakensiek, D.L., 1991. Estimating soil water retention from soil physical properties and characteristics. Adv. Soil Sci. 16, 213-234.

Schaap, M.G., Leij, F.J., 1998. Database-related accuracy and uncertainty of pedotransfer functions. Soil Sci. 163, 765-779.

Tietje, O., Tapkenhinrichs, M., 1993. Evaluation of pedo-transfer functions. Soil Sci. Soc. Am. J. 57, 1088-1095.

Timlin, D.J., Ahuja, L.R., Williams, R.D., 1996a. Methods to estimate soil hydraulic parameters for regional scale applications of mechanistic models, Application of GIS to the modeling of non-point source pollutants in the vadose zone. SSSA Special publication 48. American Society of Agronomy, Madison, WI.

Timlin, D.J., Pachepsky, Ya., Acock, B., Whisler, F., 1996b. Indirect estimation of soil hydraulic properties to predict soybean yield using GLYCIM. Agric. Syst. 52, 331-353.

USDA Natural Resource Conservation Service, 1994. National soil pedon database. Lincoln, NE.

Van Genuchten, M.Th., Leij, F., 1992. On estimating the hydraulic properties of unsaturated soils. *In*: van Genuchten, M. Th., Leij, F.J., Lund, L.J. (Eds.), Indirect Methods for Estimating the Hydraulic Properties of Unsaturated Soils. University of California, Riverside, CA, pp. 1-14.

Verekeen, H., Maes, J., Feyen, J., Darius, P., 1989. Estimating the soil moisture retention characteristics from texture, bulk density and carbon content. Soil Sci. 148, 389-403.

Wang, V., Neuman, S.P., Yao, T.-M., Wierenga, P.J., 2003. Simulation of large-scale field infiltration experiments using a hierarchy of models based on public, Generic, and Site Data. Vadose Zone J. 2, 297-312.

Wösten, J.H.M., Lilly, A., Nemes, A., Le Bas, C., 1999. Development and use of a database of hydraulic properties of European soils. Geoderma 90, 169-185.

Wösten, J.H.M., Pachepsky, Ya., Rawls, W.J., 2001. Pedotransfer functions: bridging the gap between available basic soil data and missing soil hydraulic characteristics. J. Hydrol. 251, 123-150.

CONTRIBUTORS

Ahuja L.R. USDA-ARSL, Great Plains Systems Research Laboratory, 2150 Centre Ave., Bldg. D, Suite 200, Fort Collins, CO 80256, USA

Belocchi G. ISCI (Research Institute for Industrial Crops), Via di Corticella, 133, 40128 Bologna, Italy

Braudeau E. IRD/PRAM, Quartier Petit Morne, BP 214, 97285 Le Lamentin Cedex 2, Martinique

Bruand A. Institut des Sciences de la Terre d'Orléans (ISTO), Bâtiment Géosciences, Université d'Orléans, BP 6759 45067, Orléans Cedex 2, France

Chahinian N. IRD/PRAM, Quartier Petit Morne, BP 214, 97285 Le Lamentin Cedex 2, Martinique

Chirico G.B. Department of Agricultural Engineering, Division for Land and Water Resources Management, University of Naples "Federico II", Via Università, 100, 80055 Portici (Naples), Italy

Dembovetsky A. 119992 Moscow, Leninskie Gory, Moscow State University, Soil Science Faculty, Russia

Donatelli M. ISCI (Research Institute for Industrial Crops), Via di Corticella, 133, 40128 Bologna, Italy

Flanagan D. USDA-Agricultural Research Service, National Soil Erosion Research Laboratory, 1196 Building SOIL, Purdue University, 275 S. Russell Street, West Lafayette, IN 47907, USA

Guber A. Environmental Microbial Safety Laboratory, USDA-ARS-BA-ANRI-EMSL, Bldg. 173, Rm. 203, BARC-EAST, Powder Mill Road, Beltsville, MD 20705, USA

Harper R. Centre for Water Research, The University of Western Australia, MO15, 35 Stirling Highway, Crawley, Western Australia, Australia, 6009

Heathman G.C. National Soil Erosion Research Laboratory, 275 S. Russell Street, West Lafayette, IN 47907, USA

Herbst M. Institut Agrosphäre, ICG-IV, Forschungszentrum Jülich GmbH, Leo Brandt Straβe 52425, Jülich, Germany

Hodnett M. Centre for Ecology and Hydrology, Wallingford, OX10 8BB, United Kingdom

Lilly A. Macaulay Land Use Research Institute, Craigiebuckler, Aberdeen AB15 8QH, Scotland, United Kingdom

Lin H. Department of Crop and Soil Sciences, 116 A.S.I. Building, The Pennsylvania State University, University Park, PA 16802, USA

McBratney A.B. Faculty of Agriculture, Food & Natural Resources, The University of Sydney, JRA McMillan Building A05, NSW 2006, Australia

Minasny B. Faculty of Agriculture, Food & Natural Resources, The University of Sydney, JRA McMillan Building A05, NSW 2006, Australia

Mohtar R.H. Department of Agricultural and Biological Engineering, 1146 ABE Building, Purdue University, West Lafayette, IN 47907, USA

Nemes A. USDA-ARS Hydrology & Remote Sensing Lab, Bldg. 007, Rm. 104, BARC-W, Beltsville, MD 20705-2350, USA

Oliver Y. Centre for Water Research, The University of Western Australia, MO15, 35 Stirling Highway, Crawley, Western Australia, Australia, 6009

Pachepsky Ya. Environmental Microbial Safety Laboratory, USDA-ARS-BA-ANRI-EMSL, Bldg. 173, Rm. 203, BARC-EAST, Powder Mill Road, Beltsville, MD 20705, USA

Perfect E. Department of Earth and Planetary Sciences, 1412 Circle Drive, University of Tennessee, Knoxville, TN 37996-1410, USA

Pracilio G. Centre for Water Research, The University of Western Australia, MO15, 35 Stirling Highway, Crawley, Western Australia, Australia, 6009

Rawls W.J. USDA-ARS Hydrology & Remote Sensing Lab, Bldg. 007, Rm. 104, BARC-W, Beltsville, MD 20705-2350, USA

Romano N. Department of Agricultural Engineering, Division for Land and Water Resources Management, University of Naples "Federico II", Via Università, 100, 80055 Portici (Naples), Italy

Schaap M.G. George E. Brown, Jr. Salinity Laboratory (USDA/ARS), 450 Big Springs Road, Riverside, CA 92507, USA

Shein E. 119992 Moscow, Leninskie Gory, Moscow State University, Soil Science Faculty, Russia

Smettem K. Centre for Water Research, The University of Western Australia, MO15, 35 Stirling Highway, Crawley, Western Australia, Australia, 6009

Sławiński C. Institute of Agrophysics, Polish Academy of Sciences, Doswiadczalna 4, 20-290 Lublin 27, PO Box 201, Poland

Štekauerová V. Institute of Hydrology, Slovak Academy of Sciences, Račianska 75 83102 Bratislava, Slovak Republic

Šútor J. Institute of Hydrology, Slovak Academy of Sciences, Račianska 75 83102 Bratislava, Slovak Republic

Timlin D.J. USDA-ARS Alternate Crops and Systems Laboratory, Bldg 001, Rm 342 BARC-W, 10300 Baltimore Ave., Beltsville, MD 20705, USA

Tomasella J. INPE/CPTEC, Rod. Presidente Dutra km. 39, 12630-000 Cachoeira Paulista/SP, Brazil

Vereecken H. Institut Agrosphäre, ICG-IV, Forschungszentrum Jülich GmbH, Leo Brandt Straβe 52425, Jülich, Germany

Walczak R. Institute of Agrophysics, Polish Academy of Sciences,

Doswiadczalna 4, 20-290 Lublin 27, PO Box 201, Poland

Williams R.D. USDA-ARS Alternate Crops and Systems Laboratory, Bldg 001, Rm 342 BARC-W, 10300 Baltimore Ave., Beltsville, MD 20705, USA

Witkowska-Walczak B. Institute of Agrophysics, Polish Academy of Sciences, Doswiadczalna 4, 20-290 Lublin 27, PO Box 201, Poland

Wösten J.H.M. Alterra, Droevendaalsesteeg 3, 6700 AA Wageningen, The Netherlands

CONTENTS

Foreword	v
Preface	vii
Contributors	xvii

Part I. Methods to Develop Pedotransfer Functions

Chapter 1. Statistical Regression
H. Vereecken and M. Herbst — 3

1. Objectives of Statistical Regressions	3
2. Preliminary Analysis of Soil Data	4
2.1. Simple data analysis	4
2.2. Multivariate methods	7
3. Model Building	9
3.1. Model fit	12
3.2. Poor model specification	14
3.3. Confidence intervals on estimated soil properties values	15
3.4. Outlier detection	15
4. Validation of Regression Models	16
5. Summary	17
References	18

Chapter 2. Data Mining and Exploration Techniques
Ya. Pachepsky and M.G. Schaap — 21

1. Artificial Neural Networks	21
2. Group Method of Data Handling	24
3. Regression Trees	26
4. Cross-Validation Procedures	29
5. Concluding Remarks	29
References	30

Chapter 3. Accuracy and Uncertainty in PTF Predictions
M.G. Schaap — 33

1. Optimization Criteria	33
2. Criteria for Evaluating the Accuracy of PTFs	36
3. Evaluating the Uncertainty of PTF Predictions	39
References	41

Part II. Soil Hydraulic Properties: Water Retention and Hydraulic Conductivity

Chapter 4. Soil Texture and Particle-Size Distribution as Input to Estimate Soil Hydraulic Properties
A. Nemes and W.J. Rawls — 47

   1. Introduction — 47
   2. Particle-Size and Soil Texture Class Systems — 47
   3. Soil Texture Data in Pedotransfer Functions — 49
      3.1. The use of texture class information in pedotransfer functions — 49
      3.2. The use of particle-size distribution data in pedotransfer functions — 50
      3.3. Pedotransfer functions based solely on texture or particle-size distribution information — 52
   4. Interpolations to Fill in Missing Particle-Size Data — 53
   5. Evaluation of Different Representations of Particle-Size Distribution — 55
      5.1. Soil data — 55
      5.2. Methods — 57
      5.3. Results — 57
   6. Summary — 63
      References — 64

Chapter 5. Simple Parametric Methods to Estimate Soil Water Retention and Hydraulic Conductivity
D.J. Timlin, R.D. Williams, L.R. Ahuja and G.C. Heathman — 71

   1. Introduction — 71
   2. Estimating Soil Water Contents and Soil Water Retention — 72
      2.1. A scaling method to estimate soil water retention curves — 72
      2.2. The One-Parameter Gregson–Hector–McGovan (GHM) Model — 73
      2.3. Air-entry potential and saturated water content and the GHM Model — 78
      2.4. The GHM One-Parameter Model with generalized parameters — 79
         2.4.1. Implementation of the GHM One-Parameter Model with generalized parameters — 81
      2.5. Use of available water capacity with the GHM One-Parameter Model — 83
   3. Hydraulic Conductivity — 83
      3.1. Determining saturated hydraulic conductivity, K_{sat} — 83
         3.1.1. Predicting saturated conductivity from effective porosity — 84
      3.2. Relationships for unsaturated hydraulic conductivity — 86
         3.2.1. Extending the One-Parameter Model to unsaturated hydraulic conductivity–matric potential relationships — 86
   4. Applications of Pedotransfer Functions for Simulation Models — 88
   5. Summary — 90
      References — 91

Chapter 6. Effect of Soil Organic Carbon on Soil Hydraulic Properties
W.J. Rawls, A. Nemes and Ya. Pachepsky — 95

   1. Introduction — 95
   2. Bulk Density/Porosity — 95

3. Soil Water Retention	97
3.1. Data	98
3.2. Methods to quantify the effect of organic carbon content on water retention	99
3.3. Regression Trees	100
3.3.1. Predictors: soil texture class and organic carbon content	100
3.3.2. Predictors: soil texture class, soil taxonomic order and organic carbon content	101
3.3.3. Predictors: soil taxonomic order and organic carbon content	102
3.3.4. Predictors: sand, silt, clay and organic carbon content	103
3.3.5. Summary	103
3.4. Group method of data handling	104
3.4.1. No split of the data	106
3.4.2. Split by taxonomic order	106
3.4.3. Split by texture classes	106
3.5. Pedotransfer models	107
3.6. Summary	109
4. Saturated Hydraulic Conductivity	110
5. Conclusions	111
References	111

Chapter 7. Using Soil Morphological Attributes and Soil Structure in Pedotransfer Functions
A. Lilly and H. Lin   115

1. Introduction	115
2. Using Soil Morphology and Structure in Estimating Soil Hydraulic Properties	117
2.1. Qualitative or semi-quantitative approaches	117
2.1.1. Predictions of hydraulic conductivity	117
2.1.2. Predictions of moisture retention	120
2.1.3. Grouping and classification of soil hydrological functions and pedotransfer rules (PTRs)	122
2.2. Quantitative approaches	126
2.2.1. Quantitative calculations of hydraulic conductivity and moisture retention using micromorphometric data	126
2.2.2. Quantification of macromorphological attributes in developing PTFs	127
2.2.3. Other quantitative uses of qualitative morphological attributes in PTFs	130
3. Future Improvements	133
3.1. Standardization of soil morphology descriptions and hydraulic measurements	133
3.2. Quantification of soil morphology including soil structure	134
3.3. Derivation of PTFs for soils with unusual characteristics	135
3.4. Grouping soils based on terrain and geomorphology	135
4. Summary	135
References	136

Chapter 8. Soil Aggregates and Water Retention
A. Guber, Ya. Pachepsky, E. Shein and W.J. Rawls   143

1. Introduction	143
2. Soil Database	144

3. Regression Tree Modeling	145
4. Discussion and Conclusion	148
References	150

Chapter 9. Utilizing Mineralogical and Chemical Information in PTFs
A. Bruand   153

1. Mineralogical Composition of the Clay Fraction	153
2. Cation Exchange Capacity	154
3. Soil Chemical Properties	155
4. Concluding Remarks	156
References	157

Chapter 10. Preliminary Grouping of Soils
A. Bruand   159

1. Origin of the Variability and Grouping Strategy	159
2. Grouping Criteria	160
2.1. Genetic grouping	160
2.2. Horizon-based grouping	161
2.3. Texture grouping	163
2.4. Grouping based on structure and bulk density	165
2.5. Parent material grouping	167
2.6. Consecutive grouping	167
3. Grouping Decreases the Number of Predictors	167
4. Comparison of Groupings and Improvement of Prediction after Grouping	168
5. Conclusion	171
References	172

Part III. Hydrological and Physical Parameters

Chapter 11. Pedotransfer Functions for Soil Erosion Models
D. Flanagan   177

1. Introduction	177
2. History of Early U.S. Erosion Research	177
3. The Universal Soil Loss Equation	179
4. Parameterization of Erosion Prediction Models	180
4.1. Erosion prediction models	180
4.2. Sediment particle fractions and particle composition	181
4.3. WEPP infiltration parameterization	183
4.4. WEPP erodibility parameterization	185
5. Procedures to Develop Erosion Model Pedotransfer Functions	186
5.1. Experimental techniques	187
5.2. Interrill erodibility	188
5.3. Rill erodibility and critical shear stress	188
5.4. Effective hydraulic conductivity	189
6. Summary	190
References	191

Chapter 12. Solute Adsorption and Transport Parameters
B. Minasny and E. Perfect   195

    1. Introduction   195
    2. Solute Adsorption   196
    3. Diffusive Solute Transport   201
    4. Convective-Dispersive Solute Transport   204
       4.1. Convection dispersion equation (CDE)   204
       4.2. Mobile–immobile model (MIM)   207
       4.3. Other physico-empirical models   213
    5. Upscaling Pedotransfer Function Predictions   213
    6. Conclusions and Future Directions   216
       References   217

Chapter 13. Estimating Soil Shrinkage Parameters
E. Braudeau, R.H. Mohtar and N. Chahinian   225

    1. Importance of Shrink–Swell Properties   225
    2. Soil–Water Medium Functional Model   225
       2.1. Soil–water medium hierarchy and functionality   225
       2.2. Characterization of the pedostructure using shrinkage curve   227
    3. Seeking pedotransfer Functions for the SC using the
Pedostructure Characterization   229
       3.1. The required parameters for crossing scales from laboratory
to the field   230
       3.2. Significance of the SC parameters and its
corresponding approximation   231
       3.3. Construction of the SC from primary data of soil   235
    4. Application Example   235
       4.1. Pedotransfer functions for calculating FC and PWP (W_D and W_B)   235
       4.2. Values of LS_{mod} for the four types of soil   236
       4.3. Value of K_{bs} as a function of texture   237
       4.4. Equations used to build the shrinkage curve   237
    5. Conclusion   238
       Appendix A. List of Parameters and Abbreviations Used   238
       References   239

Chapter 14. Key Soil Water Contents
E. Shein, A. Guber and A. Dembovetsky   241

    1. Introduction   241
    2. Materials and Methods   243
    3. Estimating Soil Water Contents at Field Capacity   243
    4. Selection of Key Water Contents to Estimate Van Genuchten's
Parameters   245
    5. Concluding Remarks   248
       References   248

Part IV. Spatial Component in PTF Development

Chapter 15. Data Availability and Scale in Hydrologic Applications
K. Smettem, G. Pracilio, Y. Oliver and R. Harper   253

1. Introduction	253
2. Describing One-Dimensional Flow	254
3. Some Issues in Extrapolating from Point-Based Soil Water Balance	255
3.1. Background of a simple physico-empirical pedotransfer function	257
3.2. Difficulties with estimation of the "air entry" point	260
3.3. An intercomparison of three simple PTFs	260
3.4. Estimating the hydraulic conductivity "matching point" in the Brooks–Corey $K(h)$ or $K(\theta)$ relation	263
4. Spatial Mapping of Clay Content Using Ancillary Data	264
4.1. Gamma radiometric techniques	264
4.2. High resolution airborne radiometric systems	265
5. Redundancy of soil textural classes and the interrelation with climate	267
6. Concluding remarks	267
References	268

Chapter 16. The Role of Terrain Analysis in Using and Developing Pedotransfer Functions
N. Romano and G.B. Chirico   273

1. Introduction	273
2. Terrain Analysis for Landscape Description	275
2.1. Primary terrain attributes	279
2.2. Secondary terrain attributes	280
3. Terrain Attributes as Auxiliary Data for Interpolating Soil Properties	280
4. Terrain Attributes as Input Parameters in PTFs	283
5. Concluding Remarks and Future Developments	288
References	290

Chapter 17. Spatial Structure of PTF Estimates
N. Romano   295

1. Background and Justification	295
2. Soil Hydraulic Property Variations and the Role of Simplified Predictive Methods	298
3. Case Study and Discussion	303
3.1. Potential and limitations of using PTF estimates to capture the spatial structure of soil hydraulic parameters	304
3.2. Assessment of soil hydraulic spatial variability using ANNs and terrain attributes	313
4. Concluding remarks with an eye on scale issues	315
References	317

Part V. User-Oriented Techniques and Software

Chapter 18. Soil Inference Systems
A.B. McBratney and B. Minasny — 323

1. Software for Pedotransfer Functions — 323
2. Soil Inference Systems — 324
3. A Scheme for Defining Uncertainties of Data Inside/Outside the Training Set — 327
4. Example of SINFERS — 328
5. General Discussion and Conclusions — 344
   References — 345

Chapter 19. Graphic User Interfaces for Pedotransfer Functions
M.G. Schaap — 349

1. Soil Water Characteristics from Texture — 349
2. SOILPAR — 350
3. ROSETTA — 351
4. NEUROPACK — 353
   References — 355

Chapter 20. Methods to Evaluate Pedotransfer Functions — 357

1. Evaluation of Pedotransfer Functions
   M. Donatelli, H. Wösten and G. Belocchi — 357
   1.1. Evaluating uncertainty in equations and data sets — 358
   1.2. Comparing estimates and measurements — 358
   1.3. Pedotransfer as inputs for simulation models: sensitivity analysis — 362
2. Integrated Indices for Pedotransfer Function Evaluation
   M. Donatelli, M. Acutis, A. Nemes and Wösten — 363
   2.1. Integrated indices to evaluate PTFs for soil water retention — 363
      2.1.1. The "Accuracy" module — 364
      2.1.2. The "Correlation" module — 366
      2.1.3. The "Pattern" module — 366
      2.1.4. Aggregation of modules — 368
      2.1.5. The soil data set — 369
      2.1.6. The pedotransfer functions evaluated — 370
   2.2. Evaluating pedotransfer functions using the integrated indices — 377
      2.2.1. The index I_{PTFSW} — 377
      2.2.2. The index I_{PTFRC} — 388
      2.2.3. Final remarks about integrated indices for pedotransfer function evaluation — 388
3. Functional Evaluation of Pedotransfer Functions
   H. Wösten, A. Nemes and M. Acutis — 390
   3.1. Example of functional evaluation of PTF uncertainty — 390
   3.2. Application of PTFs in a functional context — 391

Appendix A: Numerical Indices and Test Statistics for Model Evaluation
G. Belocchi 394
A.1. List of abbreviations 394
A.2. Difference-based statistics 394
A.3. Correlation-based statistics 398
Appendix B: Fuzzy Expert Systems
G. Fila and G. Belocchi 400
References 405

Part VI. Pedotransfer Functions Developed for Different Regions of the World

Chapter 21. Pedotransfer Functions for Tropical Soils
J. Tomasella and M. Hodnett 415

1. Introduction 415
2. Materials and Methods 417
3. Results 420
4. Discussion 420
5. Conclusions 424
   Appendix: PTFs Used to Estimate Volumetric Water Content at Selected Potentials and Available Water Capacity 425
   References 428

Chapter 22. Pedotransfer Functions for Europe
J.H.M. Wösten and A. Nemes 431
   References 435

Chapter 23. Pedotransfer Functions for the United States
W.J. Rawls 437

1. Introduction 437
2. Soil Water Retention 437
   2.1. Pedotransfer functions for specific water potentials on the soil water retention curve 437
   2.2. Estimation of soil water retention model parameters 441
3. Saturated Hydraulic Conductivity 446
   References 446

Chapter 24. Pedotransfer Studies in Poland
R. Walczak, B. Witkowska-Walczak and C. Sławiński 449

1. Water Retention 449
   1.1. Importance of various soil solid phase parameters 449
   1.2. Pedotransfer functions for mineral soils 450
   1.3. Comparison of selected pedotransfer function models 452
   1.4. Approach to pedotransfer functions for organic soils 453

2. Hydraulic Conductivity	455
2.1. Saturated hydraulic conductivity	455
2.2. Unsaturated hydraulic conductivity	456
References	462

Chapter 25. Pedotransfer Functions of the Rye Island – Southwest Slovakia
V. Štekauerová and J. Šútor   465

1. Area Description	466
2. Methods	468
3. Results and Discussion	468
4. Conclusion	471
References	472

Additional Bibliography   475

Index   497

PART I

METHODS TO DEVELOP PEDOTRANSFER FUNCTIONS

Chapter 1

STATISTICAL REGRESSION

H. Vereecken[*] and M. Herbst

Institut Agrosphäre, ICG-IV, Forschungszentrum Jülich GmbH, Leo Brandt Straβe, 52425 Jülich, Germany

[*]Corresponding author: Tel.: +49-2461-61-6392; fax: +49-2461-61-2518

Many of the available and well-established PTFs for the prediction of soil hydraulic properties from continuous soil properties are based on statistical regressions (Pachepsky et al., 1982; Cosby et al., 1984; Rawls and Brakensiek, 1985, 1989; Puckett et al., 1985; Vereecken et al., 1989, 1990; Wösten et al., 1997, 1999; Scheinost et al., 1997). Statistical regression is concerned with the analysis and construction of dependence structures between dependent (response) variables like parameters describing the moisture retention curve and independent (predictor) variables, e.g., bulk density or textural information. Depending upon the objectives, the process of regression will differ, but it is possible to construct a general modeling approach as was proposed by Draper and Smith (1966). They distinguish the following three phases:
- The first phase encompasses the planning stage. At this level, the problem is defined and the objectives are specified, the *a priori* knowledge is screened and eventually existing data gathered. It includes a preliminary data analysis.
- The second phase is the genuine model building with the development of the regression models and their testing against the objectives in an iterative way.
- The third phase is the validation of the obtained models, including the stability of parameters, prediction over the sample space and evaluation of the model adequacy.

1. OBJECTIVES OF STATISTICAL REGRESSIONS

In general, three main objectives can be distinguished when using statistical regressions to model relations between two sets of variables:
(1) prediction;
(2) model specification;
(3) parameter estimation.

In using regression analysis for prediction purposes, the concern is mainly to obtain the best possible estimation of the response variable. Correct model specification and parameter accuracy are of secondary importance. In model specification, one is mainly interested in the relative importance of individual predictor variables on the predicted responses. This implies that all variables should be available in the database and that the model contains the correct functional form of the predictor variables. Only then can the predictor variables be correctly assessed (Gunst and Mason, 1980). Application of regression methodology, in order to estimate parameters, requires that the model is correctly specified, the predictions are accurate and that the data allow a good estimation. Limitations of the database and the inability to measure all relevant predictors constrain the estimation of the parameters.

The choice of the objective criterion or criteria to evaluate the developed model is determined by the objectives. The objective function (criterion) should be the quantitative expression of the modeling objective. A widely accepted objective criterion is the coefficient of determination (R^2), which evaluates the performance of the model to explain the variation in the data.

Most of the statistically based PTFs are either multiple linear regression equations or polynomials of nth order. Multiple linear regression is a common statistical tool used for the prediction of the response variable y from a number of n predictor variables x_i. A multiple linear regression equation can be written as (Herbst and Diekkrüger, 2002):

$$y = a + \sum_{i=1}^{n} b_i x_i + \varepsilon \qquad i = 1, ..., n \tag{1}$$

with the constant a (intercept), the regression coefficients b_i and the error ε. A nonlinear regression equation based on a second-order polynomial has the following form:

$$y = a + \sum_{i=1}^{n} (b_i x_i + c_i x_i^2) + \varepsilon \qquad i = 1, ..., n. \tag{2}$$

where besides from the intercept a for every predictor variable x_i, two regression coefficients b_i and c_i have to be determined (Rawls and Brakensiek, 1985).

2. PRELIMINARY ANALYSIS OF SOIL DATA

Different techniques are available to analyze the available soil data ranging from simple descriptive statistics (first and second moment of distribution, range) to graphical techniques (scatter plots) to multivariate statistical methods.

2.1. Simple data analysis

Scatter plots, plotting different soil properties of interest against selected response variables, are extremely useful in detecting trends and extreme measurements The latter can also be done by means of biplots. One should be cautious, however, to delete at this stage so-called outliers, except when it is clear that these observations are incorrectly specified or measured. Scatter plots give information with regard to the linear or nonlinear behavior of variables and about the kind of transformation to be performed to eliminate nonlinearity. Transformations of the response variable are to be considered when different possible predictor variables show the same nonlinear behavior with respect to the response.

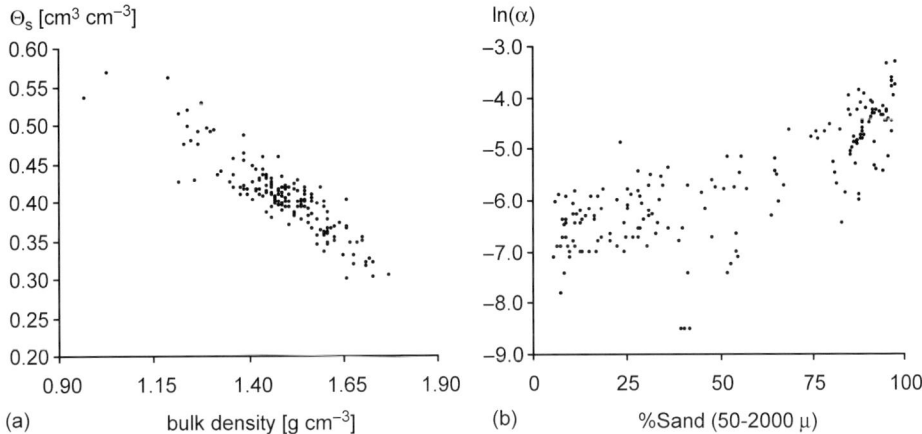

Figure 1. Scatter plots of (a) the saturated water content versus bulk density and (b) the log-transformed van Genuchten's parameter α versus sand content from the data set used by Vereecken et al. (1989).

Figure 1 gives two examples of scatter plots (Vereecken, 1988). Figure 1a reveals the linear relationship between the saturated water content θ_s (response) and bulk density (predictor), while Figure 1b exhibits the positive correlation between the sand content and log-transformed van Genuchten's α (Van Genuchten, 1980).

With statistical inferences and hypothesis testing in mind, it is interesting to examine the distributions of response variables. Examinations of sample distributions give information about transformations in order to obtain distributions more similar to the normal distribution, which is a precondition for the statistical regression techniques explained in Section 3. Frequently used numerical tests to evaluate the normality of a distribution are the Kolmogorov–Smirnov and Shapiro–Wilk statistics. The Kolmogorov–Smirnov statistic is usually used for data sets including more than 50 observations, while in other cases the Shapiro–Wilk statistic is used. The one-sample Kolmogorov–Smirnov test calculates the D-value, which is the absolute maximum difference between the cumulative sample distribution and the cumulative distribution of a normal population. A two-sample Kolmogorov–Smirnov is used to test whether two samples are drawn from the same population. Rather than measuring differences of means and variances of the populations, the Kolmogorov–Smirnov statistic measures differences in shapes.

The test statistic is a function of the sample size and can be either one- or two-tailed, testing the null hypothesis that the sample data are random samples from a normal distribution. Critical values for a specific level of significance can be looked up in special tables in order to decide whether or not the null hypothesis is to be rejected.

For sample sets smaller than 50 observations the Shapiro–Wilk statistics should be applied. This kind of statistics is the ratio of the best estimator of the variance to the usual corrected sum of squares estimator of the variance. The value of W ranges from 0 to 1, with small values rejecting the null hypothesis that the sample is drawn from a normally distributed population.

Information about the shape of distribution can be obtained from the third and fourth moments of the distribution. The third moment measures the skewness of the distribution S:

$$S = \left[\frac{n}{(n-1)(n-2)} \right] \sum_{i=1}^{n} \frac{(X_i - \bar{X})^3}{s^3} \qquad (3)$$

where s is the standard deviation, X_i is the observation value, $\bar{X}$ is the mean value and n is the number of observations. A population is said to be positive or rightly skewed, if the tail occurs in the smallest values. A negatively or left skewed population has the tail in the larger values (Figure 2). For normally distributed data, the skewness equals zero.

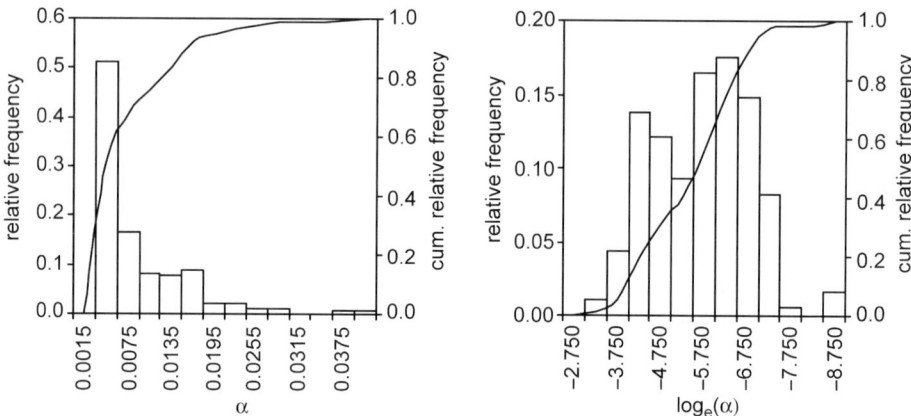

Figure 2. Histogram (bars) and cumulative relative frequency (solid line) of van Genuchten's α (cm^{-1}) and the log-transformed α from the data set used by Vereecken et al. (1989), $n = 182$ observations.

The fourth moment of the distribution measures the extent of heaviness of the tails of a distribution. A heavily tailed distribution has a positive kurtosis, a flat distribution with short tail has a negative kurtosis:

$$K = \left[\frac{n(n+1)}{(n-1)(n-2)(n-3)} \right] \sum_{i=1}^{n} \frac{(X_i - \bar{X})^4}{s^4} - \frac{3(n-1)^2}{(n-2)(n-3)} \qquad (4)$$

The measure of kurtosis K is zero for a normally distributed population.

An alternative to the numerical tests concerning the distribution of response variables is the use of graphical aids. Basically there are three types of data plots:
(1) the stem and leaf plot;
(2) the box or schematic plot;
(3) the normal probability plot.

They often give a better overall view of the extent of non-normal behavior of the response variables. The examination of the distribution of variables is mainly based on the analysis of the plots.

The example in Table 1 reveals the strong increase in W-value for the log-transformed α and n values compared to the original variables (see also Figure 2). The distribution of θ_r

Table 1
The W-value, the skewness and kurtosis of the parameters of the moisture retention characteristic of the data set used by Vereecken et al. (1989)

	W-value	Skewness	Kurtosis
θ_r	0.88	1.57	4.41
θ_s	0.94	0.61	1.76
α	0.76	2.03	5.15
n	0.87	1.35	1.75
$\ln(\alpha)$	0.96	0.02	−0.58
$\ln(n)$	0.97	0.24	−0.26

Soil hydraulic properties are based on the Van Genuchten model $(\theta - \theta_r)/(\theta_s - \theta_r) = (1 + |\alpha h|^n)^{-m}$ with residual water content θ_r, saturated water content θ_s, inverse of the bubbling pressure α, shape parameters n and m and pressure head h (Van Genuchten, 1980) with $m = 1$, $N = 182$ observations.

resembles the normal distribution to the least extent having the lowest W-value and high values of skewness and kurtosis. Various transformations, however, did not result in a much higher W-value. The high probabilities of rejecting the null hypothesis are likely to be a result of the sensitivity of the Wilks test towards observations belonging to the tails of distributions.

2.2. Multivariate methods

The principal component analysis is a powerful tool in analyzing the structure of data matrices and the interdependence between variables. Success depends upon the existence of correlations among at least some of the original variables. New variables, called principal components, are formed that they are orthogonal to each other and uncorrelated. The first principal component explains the largest part of the variation. The second one explains a part of the remaining variation, and so on until all variations have been accounted for. In total, the number of components is the same as the total number of variables. Principal component analysis is mainly applied on the standardized data matrix X_s or on the correlation matrix R, which can be written as:

$$X_d = \left(X - \bar{I}\bar{X}_m^T \right)$$
$$X_s = X_d D^{-1/2}$$
$$C = \frac{X_d^T X_d}{N - 1} \tag{5}$$
$$R = D^{-1/2} C D^{-1/2}$$

where X is the data matrix, D is the diagonal matrix of the variances, $\bar{I}$ is the unit column vector, $\bar{X}_m^T$ is the row vector of means of the different variables, X_d the matrix of mean corrected scores, C is the variance–covariance matrix and N is the number of observations. Choice between the use of the X_s or R matrix depends upon the units the variables are measured in. Typically, when the variables have been measured in the same units, the variance–covariance matrix is preferred. Even then, some authors (Jollife, 1986) prefer working with the correlation matrix because of the opportunity of direct comparison between the analysis results obtained from different sets of variables. Davis (1973), however, states that in geologic studies on granulometric composition of materials, where the relative magnitudes of variables are important, it is better to use the variance–covariance matrix.

Derivation of the principal components is based on the solution of the eigenvalue problem of the correlation matrix or the X_s matrix subjected to different constraints. For the correlation matrix, this can be expressed mathematically as:

$$(R - \lambda_j I)\bar{X}_j = 0 \tag{6}$$

with $|R - \lambda_j I| = 0$ for non-zero $\bar{X}_j$ vectors and $\bar{X}_j^T \bar{X}_j = 1$. In matrix notation this becomes:

$$(R - D_e I)U = 0 \tag{7}$$

where the columns of U contain the eigenvectors $\bar{X}_j$ of R and D_e is a diagonal matrix containing the eigenvalues of R that are equal to the variances of the respective principal components, such that:

$$R = U D_e U^{-1} \tag{8}$$

Because R is a symmetric matrix, $U = U^{-1}$ and the previous equation becomes:

$$R = U D_e U \tag{9}$$

The orientation of the different principal component axes can be found from the columns of the rotation matrix U. The non-standardized principal component scores Z are calculated as:

$$Z = X_s U \tag{10}$$

Standardizing these scores, so that each component explains the same amount of variability is done as follows:

$$Z_s = X_s U D_e^{-\frac{1}{2}} \tag{11}$$

Principal component analysis resulting in standardized scores is sometimes given the name principal factor analysis. The component loading matrix, containing the correlation of the original variables with the principal components, is calculated as:

$$F = UD_e^{\frac{1}{2}} \qquad (12)$$

Reproduction of the original matrix of standardized scores can be obtained from:

$$X_s = Z_s F^T = Z_s(Z_s^T X_s). \qquad (13)$$

Using equation (13), it is possible to reconstruct the original variables in a plot of which the x- and the y-axes are given by the first and second principal factor and the factor scores. The length of these reconstructed original variables is a measure for the success of reconstruction. The position and length of the reconstructed original variables is a measure of their correlation. These graphs, called biplots, enable the user to analyze the relations existing between objects described in the axes generated by the principal components and the original variables.

Figure 3 is a biplot of the principal component analysis carried out for the data set of Vereecken et al. (1989). This is a plot of the reconstructed original variables and the individual observations on the first two principal factors of the variables exhibiting, e.g., that the vectors representing the clay percentage and θ_r are very close to one another, confirming their positive correlation (see also Table 2). Values of $\ln(\alpha)$, $\ln(n)$ and the sand fraction are positively correlated indicating that the larger the amount of sand in the soil, the higher $\ln(\alpha)$ and $\ln(n)$ become. The values of θ_s and the bulk density, pointing in the opposite direction but laying in the same line, are strongly negatively correlated. Carbon content seems to be not strongly correlated with any of the parameters. The first two principal factors explain together 64% of the variability.

3. MODEL BUILDING

The regression models are constructed in an iterative way. The most important step is the selection of variables to enter the equation. This can be done either using *a priori* knowledge and hypothetical reasoning or by trial and error using special regression techniques. Routinely applied techniques to select variables are (Gunst and Mason, 1980):
(1) all possible regression equations;
(2) backward elimination;
(3) forward selection;
(4) stepwise method;
(5) stagewise method.

Discussion on each of these techniques can be found in many statistics books. Whenever using these methods to construct regression models, care should be taken not to eliminate potential predictor variables. This is one of the reasons why it is important to check the mathematical relationship between the potential predictor variables with respect to the response in the data analysis (e.g., exponential or square root dependencies).

Once an acceptable model is obtained, it is examined. General examination of a regression model incorporates the following topics:
(1) verification of the error assumptions;
(2) assessing goodness-of-fit of the equation;
(3) examination of model misspecification;

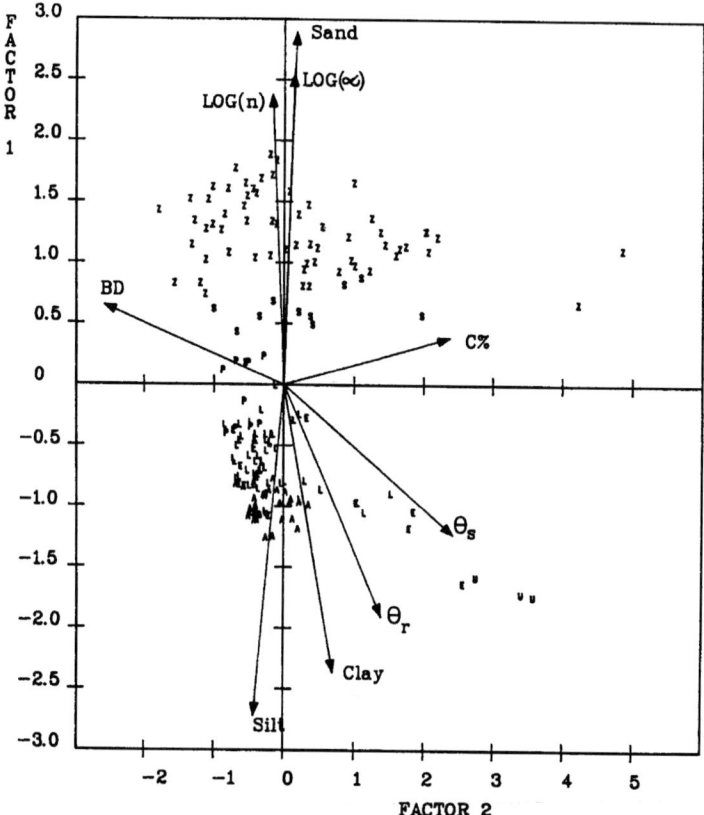

Figure 3. Plot of the reconstructed original variables and the individual observations on the first two principal factors. The length of the reconstructed variables is tripled for clearness. $log(\alpha) = log_e(\alpha)$, $log(n) = log_e(n)$, BD = dry bulk density, C% = percent organic carbon. Data points are represented by the following symbols: U = heavy clay, E = clay, A = clay silt loam, L = sandy silt loam, P = light sand loam, S = loamy sand, Z = sand according to the belgian textural classes (Verheye and Ameryckx, 1984).

(4) determination of confidence intervals on estimated regression coefficients;
(5) determination of confidence intervals on estimated response variables;
(6) detection of outliers.

A special problem related to the estimation of parameters and their confidence intervals is the multicollinearity. Multicollinearity is the problem of redundant information in the predictor set, and is defined as an approximate linear dependence between predictor variables. An extreme form of multicollinearity is exact linear dependence. Multicollinearity can be examined pairwise by means of a correlation analysis while multivariable collinearities can be detected by examination of the eigenvalues and eigenvectors of the $X_s^T X_s$ matrix (Section 2.2). Small eigenvalues reveal multicollinearity while the large scores of the variables for the corresponding eigenvectors identify the

Table 2
Correlation matrix of predictor and response variables and the corresponding significance levels of the data set used by Vereecken et al. (1989), soil hydraulic properties are based on the Van Genuchten model with $m = 1$, $n = 182$ observations

	θ_s	θ_r	$\log(\alpha)$	$\log(n)$	Clay	Sand	Silt	BD	C%
θ_s	1 0	0.58 0.0001	−0.25 0.0006	−0.43 0.0001	0.48 0.0001	−0.32 0.0001	0.24 0.000	−0.88 0.0001	0.52 0.0001
θ_r	0.58 0.0001	1 0	−0.42 0.0001	−0.46 0.0001	0.78 0.0001	−0.51 0.0001	0.37 0.000	−0.43 0.0001	0.19 0.0084
$\log(\alpha)$	−0.25 0.0006	−0.42 0.0001	1 0	0.77 0.0001	0.59 0.0001	0.78 0.0001	−0.75 0.0001	0.07 0.2973	0.00 0.901
$\log(n)$	−0.39 0.0001	−0.46 0.0001	0.77 0.0001	1 0	−0.62 0.0001	0.70 0.0001	−0.64 0.0001	0.19 0.011	0.05 0.4406
Clay	0.48 0.0001	0.78 0.0001	0.59 0.0001	−0.62 0.0001	1 0	−0.69 0.0001	0.51 0.0001	−0.28 0.0001	−0.10 0.1405
Sand	−0.32 0.0001	−0.51 0.0001	0.78 0.0001	0.70 0.0001	−0.69 0.0001	1 0	−0.97 0.0001	0.16 0.0262	0.14 0.0414
Silt	0.24 0.000	0.37 0.000	−0.75 0.0001	−0.64 0.0001	0.51 0.0001	−0.97 0.0001	1 0	−0.10 0.151	−0.13 0.0478
BD	−0.88 0.0001	−0.43 0.0001	0.07 0.2973	0.19 0.011	−0.28 0.0001	0.16 0.0262	−0.10 0.151	1 0	−0.64 0.0001
C%	0.52 0.0001	0.19 0.0084	0.00 0.901	0.05 0.4406	−0.10 0.1405	0.14 0.0414	−0.13 0.0478	−0.64 0.0001	1 0

Clay(<2 μ), Sand(2000–50 μ), Silt(50–2 μ), BD is the dry bulk density [g cm^{-3}] and C% the carbon content.

variables involved. Multicollinearity has a deleterious effect on the least squares parameter estimation and regression methodology is very sensitive to this problem. Gunst and Mason (1980) sum up four different fields of the regression analysis that are adversely affected by multicollinearity:
(1) the numerical values of estimated coefficients;
(2) the variance–covariance matrix (variance inflation);
(3) test statistics;
(4) predicted responses.

Different strategies are available to handle multicollinearity. The first one is to delete one of the involved parameters. The difficulty is often to decide which parameter to delete, especially when a good prediction is important. Another possibility is to use biased regression estimators for multicollinearity through deletion of those eigenvectors defining multicollinearity. Principal component regression is based on the elimination of eigenvectors with small eigenvalues, resulting in a biased estimate. Another type of principal component regression is given by Freund and Littell (1986), where all predictor variables are transformed to principal components. Because they are all uncorrelated, no need exists to use some kind of variable selection technique.

Eigenvalues regression is conceptually the same as the first type of principal component analysis, except that eigenvalues and eigenvectors are differently extracted and the criteria are different for deleting multicollinear variables. Yet one more alternative is the ridge regression where the effect of the eigenvectors defining multicollinearity is strongly reduced by addition of a small constant K to diagonal elements of the $X_s^T X_s$ matrix. The key problem of this alternative is to determine the value of this ridge parameter.

A correlation matrix helps to choose the predictor variables that should be used for the regression. Table 2 shows a correlation matrix of the Van Genuchten parameters (response variables) with parameter $m = 1$, textural information, bulk density and carbon content (predictor variables). A significant correlation among the response variables $\ln(\alpha)$ and $\ln(n)$ can be found, which can be explained by the fact that soils having a clearly defined air entry value like coarsely textured soils and thus having a relative large $\ln(\alpha)$ are in general characterized by a narrow pore size distribution represented by a large $\ln(n)$ value. The opposite is true for the finer textured soils. Among the predictor variables, a clear correlation exists, e.g., between the silt and the sand content, indicating that these parameters should not be combined in a regression equation, whereas the high correlation coefficient between θ_s and bulk density indicates the use of bulk density for the prediction of θ_s.

3.1. Model fit

Evaluation of the goodness-of-fit of a model is based on the partitioning of the total sum of squares (TSS) representing the total variability in the database, expressed in an analysis of variance table. The TSS can be partitioned into three components:

$$TSS = SSM + SSR + SSE \tag{14}$$

where SSR is the sum of squares attributable to the model and TSS is equal to:

$$TSS = \bar{Y}^T \bar{Y} \tag{15}$$

where $\bar{Y}$ is the column vector representing the dependent variable and $\bar{Y}^T$ is the transposed column vector of the dependent variable. SSM is the sum of squares attributable to the mean:

$$\text{SSM} = n^{-1}(\bar{I}^T\bar{Y})^2 \tag{16}$$

where $\bar{I}^T$ is the transposed unit column vector. SSE the residual sum of squares is written as:

$$\text{SSE} = \bar{e}^T\bar{e} \tag{17}$$

where $\bar{e}$ is the error vector and $\bar{e}^T$ is the transposed error vector.

Parallel with the TSS the total degrees of freedom (d_f) can be partitioned as: $\text{SSM} = 1d_f$, $\text{SSR} = n - p - 1d_f$, $\text{SSE} = pd_f$ and $\text{TSS} = nd_f$. Most analysis packages give the corrected TSS in the form of an ANOVA table expressed as:

$$\text{CSS} = \text{TSS} - \text{SSM} \tag{18}$$

Dividing each of the right-hand sided terms by its degrees of freedom results in their respective mean squares. Ratios of mean squares, called F-ratios, can be used to test the hypothesis regarding the model parameters. An important F-test is:

$$F_r = \frac{\text{MSR}}{\text{MSE}} \tag{19}$$

F_r is used to check whether the model is capable of explaining the variation in the data. The F-test is designed to perform simultaneous hypothesis tests on model parameters, while the t-test is used for individual hypothesis tests on the parameters. An important value to evaluate the goodness-of-fit is the coefficient of determination R^2. It is a measure for the amount of variability explained by the model. A possible way to calculate the term is:

$$R^2 = \frac{\text{SSR}}{\text{SSR} + \text{SSE}} = \frac{\text{SSR}}{\text{TSS} - \text{SSM}} \tag{20}$$

According to Kvålseth (1985), there is no unanimous agreement on the mathematical description of R^2 and at least eight different equations are in use. For some nonlinear models, certain expressions for the coefficient of determination yield values greater than one, as it is the case for the equation mentioned above. To account for the number of variables in the model, an adjusted coefficient of determination is used, which can be written as:

$$R^2_{\text{adj}} = 1 - (1 - R^2)\left(\frac{n-1}{n-m-1}\right) \tag{21}$$

where n is the number of observations and m is the number of variables.

3.2. Poor model specification

Two main reasons can cause the poor specification of a model: the omission of variables and/or the poor formulation of the functional expression of some or all variables in the model. Preliminary indications concerning the latter type or misspecification can be found from scatter plots, relating model variables (response, predictor) one to another (Figure 1). At the beginning of the statistical model building, necessary transformations of variables should be performed, otherwise important predictors can be lost. Once a model is built, misspecification of both types can be deduced from two types of residual plots: plots of residuals versus predictor variables (RVP plots) and partial residual plots (PR plots). Raw residuals, studentized residuals or standardized residuals can be used, but most frequently the plots involve the raw residuals. RVP plots provide information regarding the functional form of the predictor variables and the need for extra variables such as cross products or quadratic terms. PR plots give also information about the correct functional form of predictors, but assess in addition the nonlinearity in a predictor variable and the importance of a predictor variable in presence of others. These plots can only be used when the model is linear in the parameters.

A numerical analytical technique to assess model inadequacy is the lack of fit test (Draper and Smith, 1966). This test can be performed if repeated measurements are available. The availability of repeated measurements enables to calculate the sum of squared errors due to pure error and to partition the total sum of squared errors into two components.

$$SSE = SSE_p + SSE_l, \tag{22}$$

where SSE_p is the pure error component and SSE_l the component due to lack of fit. The test whether this lack of fit is significant or not, is performed by the following ratio:

$$F = \frac{MSE_l}{MSE_p} \tag{23}$$

which is an F-test, with MSE_p equal to:

$$MSE_p = \frac{\sum_{i=1}^{k} \sum_{j=1}^{n_i} (Y_{ij} - \bar{Y}_i)^2}{\sum_{i=1}^{k} n_i - k} \tag{24}$$

where k is the number of different points in the prediction range where measurements have been made; n_i the number of repeated observations in a given point of the prediction range.

The MSE (mean squared error) is obtained from the regression analysis. The difference between the MSE and the MSE_p gives the MSE due to lack of fit. The ratio given in equation (23) is then compared against the $100(1 - \alpha)\%$ point of an F-distribution. If this ratio is not significant, then there is no reason to doubt about the adequacy of the model. In the opposite case, there is a considerable bias term and attempts should be made to discover where and how the inadequacy is generated. Model misspecification due to the exclusion of important variables may not be detected with this technique, owing to the fact that both MSE_p and MSE_l are biased. In order to be able to use the lack of fit test, repeated measurements should be available. Although often no real repeated measurements (identical predictor value set) are available, the error component can be estimated by assuming that the hydraulic parameters for the same site and horizons are considered as repeated measurements.

3.3. Confidence intervals on estimated soil properties values

Two types of intervals related to the response variables can be distinguished:
(1) confidence interval on the expected responses; and
(2) prediction interval for future responses.
For the first case, the confidence interval can be written as:

$$\hat{Y} \pm t\left[\left(v, 1 - \frac{1}{2\alpha}\right)\right] s \sqrt{\bar{u}_o^T S \bar{u}_o} \tag{25}$$

where $\bar{u}_o$ is the column vector for a specific set of values of predictor variables, S the variance–covariance matrix, s the mean squared residual, v is equal to the degrees of freedom ($n - p - 1$) and $1 - 1/2\alpha$ is the confidence level, with α being the level of significance. The limits for the prediction interval for a future observation are:

$$\hat{Y} \pm t\left[\left(v, 1 - \frac{1}{2\alpha}\right)\right] s \sqrt{1 + \bar{u}_o^T S \bar{u}_o} \tag{26}$$

3.4. Outlier detection

The purpose of the detection of outliers is to identify observations having extremely large residuals, which do not fit in the pattern of the remaining data. Identification of outliers remains, even with the above definition, a very subjective and risky business and asks for careful examination. To detect outliers, the majority of plots and tests make use of studentized residuals. These studentized residuals behave more like standard normal deviates than either raw or standardized residuals, because they are divided by their standard error. Even then, raw and standardized residuals are still frequently used. The best is to combine all three types of residuals in an analysis. Commonly used graphical methods are scatter plots of the original response variables against predictor variables, relative frequency histograms of studentized residuals, plots of residuals against predictors or fitted responses and plots of residuals against deleted residuals detected as outliers. A deleted residual is defined as:

$$r_{(-i)} = Y_i - \bar{u}_i^T \hat{\beta}_{(-i)} \tag{27}$$

where $\bar{u}_i^T$ is the 1st row of the data matrix and $\hat{\beta}_{(-i)}$ the estimated regression coefficient vector for $n-1$ observations with the ith case removed. The distribution of the residuals is equal to that of a student t-test with $N-p-2$ degrees of freedom. Its value can be obtained from a transformation of the raw residuals.

Most of the statistical measures developed to detect outliers are based on the evaluation of the effect of the deletion of an observation on the estimated regression coefficient vector. These statistical tests are a measure for the distance between the estimated regression coefficient vector from the null observation set and the coefficient vector with one observation deleted. The concept is based on the multivariate confidence regions for the regression coefficient vector written as:

$$\frac{(\hat{\beta} - \bar{\beta})^T X^T X (\hat{\beta} - \bar{\beta})}{(p+1)\text{MSE}} \leq F_{(1-\alpha)} \quad \text{with } (p+1, n-p-1) \text{ degrees of freedom} \quad (28)$$

A measure for the distance or closeness between $(\hat{\beta} - \bar{\beta})$ is:

$$D_i = \frac{(\hat{\beta} - \hat{\beta}_{(-i)}) X' X (\hat{\beta} - \bar{\beta})}{(p+1)\text{MSE}} \quad (29)$$

The equation above defines the size of the confidence region containing $\hat{\beta}_{(-i)}$ by finding the $F_{1-\alpha(p+1,n-p-1)}$ value corresponding to the D_1 value and evaluating the $(1-\alpha)$ confidence level. Practically, $\hat{\beta}_{(-i)}$ should stay close to β. If that is not true, the observation should be rejected. Different simplifications have been introduced in the previous equation:

$$D_i = \frac{h_{ii} r_{(-i)}^2}{(p+1)\text{MSE}} \quad \text{or} \quad D_i = \frac{t^2/p+1}{h_{ii}/1-h_{ii}} \quad (30)$$

with h_{ii} the diagonal elements of $X(X^T X)^{-1} X^T$, measuring the influence of each observation, t_i the ith studentized residual and $r_{(-i)}$ the deleted residual. Separate examination of t_i, $h_{ii}/(1-h_{ii})$ and D_i are useful whether or not an observation is to be deleted. The ratio $h_{ii}/(1-h_{ii})$ denotes the influence that the associated response Y_i has on the determination of $\hat{\beta}$. Even with all these possible techniques and graphs, one has to be careful in deleting observations because this often leads to forcing a linear relationship through data exhibiting a nonlinear behavior or showing interaction effects. This is especially true when not enough data are available over the complete range of interest. Using "forced" equations for prediction purposes can lead to serious mispredictions.

4. VALIDATION OF REGRESSION MODELS

Validation of PTFs, developed by means of regression analysis is frequently overlooked. This is especially the case when variable selection techniques have been used to construct the model. These techniques display a penchant towards capitalizing on chance variation (Green and Caroll, 1978) for specific data sets, introducing too many independent variables. Absence of any validation is mainly due to the lack of additional data to perform the validation on.

Basically, the regression models can be validated in two ways. The model can be statistically evaluated, e.g., with the estimated regression coefficients or by cross-validation. In addition, models can be practically evaluated towards the further use of the estimated responses as input in other models. Only the statistical validation will be considered here. A commonly applied method is the double cross-validation as designed by Green and Caroll (1978). The advantages of this method are that no additional data are needed and its simplicity. The method is designed to evaluate the stability of the estimated regression coefficients and the prediction level of the equation.

In a first step, the complete set of observations is randomly split in two halves. On each of the halves, a regression analysis is performed using the variables retained on the complete set of observations, as found in the modeling procedure. Then the regression model obtained from the first half is applied to the second half and vice versa. For each case, the simple correlation between observed and estimated values is calculated.

Table 3 gives the results of a stepwise regression analysis to predict θ_s from bulk density and clay content (Vereecken et al., 1989). This regression model is adequate, because the

Table 3
Results of the stepwise regression analysis and cross validation for the water content at saturation θ_s (response variable).

θ_s	Retained variables and estimated regression coefficients at a 5%	Partial R^2_{adj}	Model R^2_{adj}	F-value for lack of fit	F-value at 95%
	0.838 − 0.283 BD + 0.0013	0.789			
		0.059	0.849	0.46	1.5
	0.845 − 0.303 BD + 0.0013 (+)		0.810		
	0.788 − 0.260 BD + 0.0014 (+)		0.870		

BD is the dry bulk density (g cm^{-3}), Clay the clay content [%] and (+) double cross-validation.

F-value of the lack of fit test is clearly smaller than the critical F-value at 95%. The double cross-validation procedure shows a maximum coefficient of determination of 87%.

A comparable procedure to the technique described above is the jackknife method (Pachepsky and Rawls, 2003), where a secondary data set is necessary. It is used to validate the model developed from the prediction data set by predicting the values of the independent data set. For both cross-validation approaches the mean absolute error, mean square error, root mean square error and graphical plots of predicted versus measured values are used to assess the quality of the predictions (Müller et al., 2001).

5. SUMMARY

We described the use of statistical regression techniques to develop pedotransfer functions (PTFs) estimating hydraulic properties from basic soil properties. In this section, PTFs are considered as regression models with soil data as predictor variables and hydraulic properties as response variables. Three basic steps are presented that are usually applied in an iterative procedure: Analysis of the soil data, the model building step and the model validation. Different methods are presented on how to analyze the soil data. They range from simple scatter plots providing, e.g., information on the type of relationship between variables and the existence of outliers to multivariate statistical analyses allowing to examine the soil data in a holistic manner. Analysis of the statistical distribution of predictor and response variables may provide important information for the model-building step. Extremely useful is the application of principal component analysis to examine the linear dependence between variables. Due to inherent correlation between predictor variables used in PTFs, care needs to be taken to avoid the problem of multicollinearity. This can be avoided by transforming predictor variables to independent variables using principal component analysis. The second step in developing PTF consists in the building of the regression model. Available methods include, e.g., backward and forward regression techniques, stepwise and stagewise methods. Once a first acceptable model is identified, six basic topics need to be checked: verification of the error assumption, goodness-of-fit of the model, identification of model misspecification, examination of confidence intervals on estimated regression coefficients and response variables and finally outlier detection. The model-building step is an iterative procedure requiring many iteration steps to find the best PTF model. The last step consists in the validation of the regression model or PTF. This step is usually overlooked but it is essential in establishing confidence in the developed model. Two basic but completely different methods are available: functional validation of the models and statistical validation. Functional validation aims at examining the variability of the outcome of simulation model (e.g., water balance, solute transport in soils) for a specific application caused by the uncertainty in PTF. In this chapter, we focused on statistical validation aiming at checking the validity of the prediction level (e.g., coefficient of determination) and the stability of the estimated regression coefficients using the double cross-validation technique.

REFERENCES

Cosby, B.J., Hornberger, G.M., Clapp, R.B., Ginn, T.R., 1984. A statistical exploration of the relationships of soil moisture characteristics to the physical properties of soil. Water Resour. Res. 20, 682-690.
Davis, J.C., 1973. Statistics and Data Analysis in Geology. John Wiley and Sons, New York.
Draper, N.R., Smith, H., 1966. Applied Regression Analysis. John Wiley, New York.
Freund, R.J., Littell, R., 1986. SAS System for Linear Models. SAS Institute Inc., Cary, NC.
Green, P.E., Caroll, J.D., 1978. Analyzing Multivariate Data. John Wiley, New York.
Gunst, F.R., Mason, R.L., 1980. Regression Analysis and Its Applications. A Data Oriented Approach. Marcel Dekker Inc., New York.
Herbst, M., Diekkrüger, B., 2002. The influence of the spatial structure of soil properties on water balance modeling in a microscale catchment. Physics and Chemistry of the Earth, Part B 27, 701-710.
Jollife, I.T., 1986. Principal Component Analysis. Springer Verlag, New York.
Kvålseth, T.O., 1985. Cautionary note about R^2. The American Statistician 39, 279-285.

Müller, T.G., Pierce, F.J., Schabenberger, O., Warncke, D.D., 2001. Map quality for site-specific management. Soil Sci. Soc. Am. J. 65, 1547-1558.

Pachepsky, Y., Rawls, W.J., 2003. Soil structure and pedotransfer functions. Eur. J. Soil Sci. 54, 443 451.

Pachepsky, Y., Shcherbakov, R.A., Varallyay, G., Raijkai, K., 1982. Statistical analysis of water retention relations with other physical properties of soils. Pochvovedenie 2, 42-52, (in Russian, English Abstr.).

Puckett, W.E., Dane, J.H., Hajek, B.F., 1985. Physical and mineralogical data to determine soil hydraulic properties. Soil Sci. Soc. Am. J. 49, 831-836.

Rawls, W.J., Brakensiek, D.L., 1985. Prediction of soil water properties for hydrologic modelling. In: Jones, E.B., Ward, T.J. (Eds.), Proceedings of the Symposium of Watershed Management in the Eighties. April 30–May 1, 1985, Denver, CO. Am. Soc. Civil Engng, New York, NY, pp. 293-299.

Rawls, W.J., Brakensiek, D.L., 1989. Estimation of soil water retention and hydraulic properties. In: Morel-Seytoux, H.J. (Ed.), Unsaturated Flow in Hydrological Modeling: Theory and Practice. Kluwer Academic Publishers, Dordrecht, pp. 275-300.

Scheinost, A.C., Sinowski, W., Auerswald, K., 1997. Regionalization of soil water retention curves in a highly variable soilscape. I. Developing a new pedotransfer function. Geoderma 78, 129-143.

Van Genuchten, M.T., 1980. A closed form equation for predicting the hydraulic conductivity of unsaturated soils. Soil Sci. Soc. Am. J. 44, 892-898.

Vereecken, H., 1988. Pedotransfer functions for the generation of hydraulic properties for belgian soils. Thesis, Doctoraatsproefschrift Nr. 171 aan de Fakulteit der Landbouwwetenschappen van de Katholieke Universiteit te Leuven, pp. 254.

Vereecken, H., Feyen, J., Maes, J., Darius, P., 1989. Estimating the soil moisture retention characteristic from texture, bulk density and carbon content. Soil Sci. 148, 389-403.

Vereecken, H., Maes, J., Feyen, J., 1990. Estimating unsaturated hydraulic conductivity from easily measured soil properties. Soil Sci. 149, 1-12.

Verheye, W., Ameryckx, J., 1984. Mineral fractions and classification of soil texture. Pedologie 2, 215-225.

Wösten, J.H.M., 1997. Pedotransfer functions to evaluate soil quality. In: Gregorich, E.G., Carter, M.R. (Eds.), Soil Quality for Crop Production and Ecosystem Health. Developments in Soils Science, Vol. 25. Elsevier, Amsterdam, pp. 211-245.

Wösten, J.H.M., Lilly, A., Nemes, A., Le Bas, C., 1999. Development and use of a database of hydraulic properties of European soils. Geoderma 90, 169-185.

Chapter 2

DATA MINING AND EXPLORATION TECHNIQUES

Ya. Pachepsky[1] and M. G. Schaap[2,*]

[1]Environmental Microbial Safety Laboratory, USDA-ARS-BA-ANRI-EMSL, Bldg. 173, Rm. 203, BARC-EAST, Powder Mill Road, Beltsville, MD 20705, USA

[2]George E. Brown, Jr. Salinity Laboratory (USDA/ARS), 450 Big Springs Road, Riverside, CA 92507, USA

*Corresponding author: Tel.: +1-(909)-369-4844; E-mail:mschaap@ussl.ars.usda.gov

Applying statistical regression techniques to predict soil properties that are difficult to measure requires deciding which properties are to be used as predictors and which regression equation to use. Those decisions are not straightforward, especially when the databases contain many potential predictors and the relationships between soil properties may be different in different parts of the databases. The data mining and exploration methods introduce algorithms that automate predictor and equation selections. This chapter describes three such methods that have recently been used in the PTF development.

1. ARTIFICIAL NEURAL NETWORKS

Artificial neural networks are becoming a common tool for modeling complex "input–output" dependencies (Maren et al., 1990; McCord and Illingworth, 1990). The advantage of ANN is their ability to mimic the behavior of complex systems by varying the strength of the influence of network components to each other and by varying the structure of the interconnections among components. After establishing network structure and finding coefficients to express the strength of influence of the network components to each other, an artificial neural network becomes a complex formula of special type, relating input values with output values (Alexander and Morton, 1990). This formula can be used like a regression formula.

Artificial neural networks are sometimes described as "universal function approximators" that can approximate any continuous function to any desired degree of accuracy (Hecht-Nielsen, 1990; Haykin, 1994). This property potentially makes ANNs very well suited for developing pedotransfer functions (PTFs). The distinct advantage of choosing ANNs over traditional regression approaches is that no *a priori* model concepts are needed. Instead, the function approximating capabilities in ANNs take care that near optimal relations between basic soil properties and hydraulic properties are found in an iterative calibration procedure. There are considerable advantages of using ANNs for developing PTFs. However, this section will also discuss some of the pitfalls of ANNs.

ANNs were first used for PTFs by Pachepsky et al. (1996), Schaap and Bouten (1996) and Tamari et al. (1996); subsequent publications by Schaap et al. (1998), Koekkoek and Bootlink (1997), Minasny et al. (1999) and Pachepsky et al. (1999) further pursued this topic. Although many types of ANN exist (Haykin, 1994), most of these publications dealt with so-called feed-forward back-propagation ANNs (or perceptrons). Tamari et al. (1996) used a different type of ANN of the "radial basis function" type.

Feed-forward ANNs are most often used to map input–output relationships (Maren et al., 1990). A diagram of a multilayer feed-forward ANN is given in Figure 1. This figure shows that feed-forward ANNs are organized in layers that contain neurons (also called nodes). The number of neurons in input and output layers corresponds to the number of input and output variables, J and L, respectively. The number of "hidden" neurons, K, can be chosen freely, and determines the complexity that can be modeled. The J inputs X_1 through X_J normally do not undergo any operation and are all connected to the K hidden neurons. At each hidden neuron $k = 1$ through K each input X_j is weighted, summed and biased (with a value b) to produce a single value, S_k:

$$S_k = \sum_{j=0}^{J} (w_{jk} X_j) + b_k \qquad (1)$$

Each neuron has its own values of w_i and b, the coefficients in each layer are situated in the matrix and vector elements w_{jk} and b_k, respectively. Similar operations occur at the output nodes 1 through L, with matrices and vectors with elements w_{kl} and b_l, respectively. The values S_k in hidden and output nodes are operated on by an *activation* or *transfer* function, $\varphi()$. The activation function is usually a monotonic function that can be easily

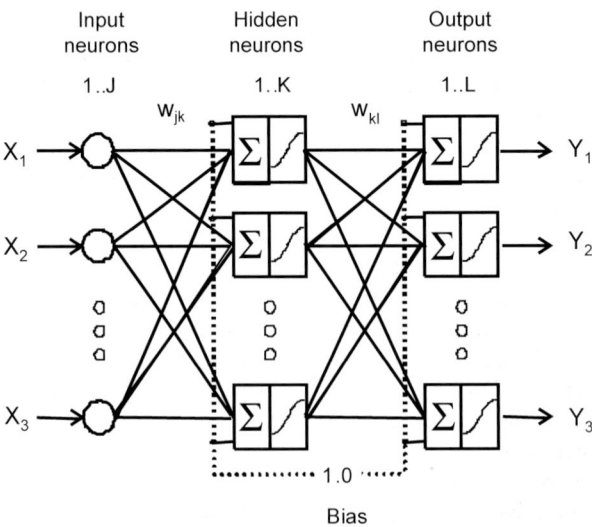

Figure 1. General overview of a feed-forward artificial neural network. See text for explanation.

evaluated (Hecht-Nielsen, 1990). Most commonly the sigmoidal function

$$\varphi(S_k) = \frac{1}{1 + \exp(-S_k)} \qquad (2)$$

is used most often for neurons in hidden and output layers, thus limiting the output Y_1 though Y_L between 0 and 1. However, linear output functions like

$$H(S_k) = S_k \qquad (3)$$

are sometimes used for output layers to extend the output range to, theoretically, plus and minus infinity. Other activation functions such as hyperbolic tangents can also be used (Demuth and Beale, 1992).

The neural network model is thus specified by the set of coefficients w_{jk}, w_{kl}, b_k, b_l and the activation functions used in the hidden and output layer. The values for the matrices and vectors, which can total more than 50 coefficients are obtained in an iterative calibration procedure. Originally, the backpropagation algorithm was used for this purpose (Rummelhart et al., 1986); however, faster alternatives are now available such as the Levenberg–Marquardt algorithm (Demuth and Beale, 1992). The most common objective function minimizes the squared residuals:

$$O(w_{jk}, b_k, w_{kl}, b_l) = \sum_n^N \sum_m^M (t_{n,m}(w_{jk}, b_k, w_{kl}, b_l) - t'_{n,m})^2 \qquad (4)$$

where N is the number of samples, M the number of parameters (with n and m as index) and t and t' are the observed and predicted variables (e.g., water contents at specific pressure heads or hydraulic parameters). Note it is possible here to predict a set of hydraulic parameters with one model, whereas separate equations are needed to predict each parameter with regression PTFs.

Although it has been shown that ANNs are able to perform better than regression-based PTFs (Schaap and Bouten, 1996; Schaap et al., 1998, Tamari et al., 1996), there are several issues to be aware. Neural networks are black-box models that allow little insight into the relations that are used to predict hydraulic variables. Except for trivially small networks, it is almost impossible to describe the models in short and easy-to-understand equations, thus making the practical implementation of ANNs difficult. At the same time, some existing "traditional" PTFs have similar problems, such as the regression PTF by Rawls and Brakensiek (1985). Despite the black-box nature of ANNs, something can be still learned from neural networks. Because properly calibrated neural networks can extract a maximum amount of information from the data, they can serve as benchmarks that indicate in how far other empirical or semi-physical models can potentially be improved. Further, it is possible to carry out sensitivity analyses with neural networks to indicate the influence of different input variables on the model's results (Schaap and Bouten, 1996).

The "universal function approximator" capabilities of many ANNs can also cause problems. A common issue is the problem of "overfitting" that is caused by the usually large number of free coefficients inside an ANN. The iterative minimization of the objective function will first proceed rapidly after which a slower decrease will take place (Figure 2). In many cases, however, the decrease in sum of squares will never entirely stop.

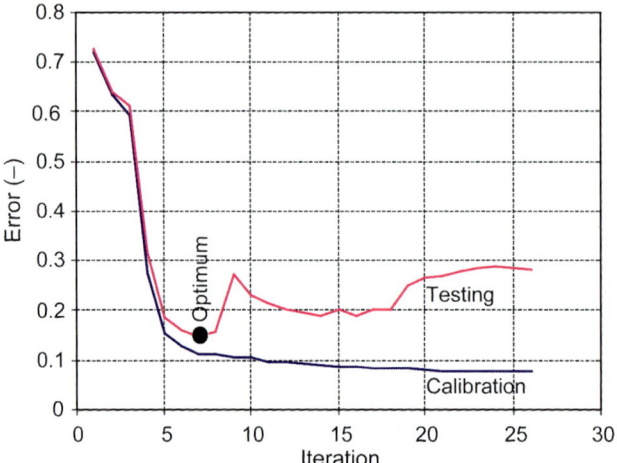

Figure 2. Evolution of calibration and testing errors with the number of iterations.

A different pattern arises when an independent dataset is evaluated simultaneously. After a similar steep descent during the first few iterations, the error starts to rise again, indicating a worsening performance of the ANN. Clearly, the optimization of the ANN is going awry after seven or eight iterations, probably because the optimization is trying to minimize the objective function on noise and artifacts in the calibration data. This adversely affects the generality of the model being implemented by the ANN and leads to an increasing error for the testing data set. Although the testing data set likely contains noise and artifacts also, it is clear that the ANN model at seven iterations is the most applicable to independent data. It is therefore prudent not to execute too many iterations while calibrating an ANN and limit the allowable number to a preset maximum. A better approach is the one depicted in the figure where calibration and testing data sets are evaluated simultaneously. The set of ANN coefficients where the testing error is the lowest is taken as the most optimal and used for practical applications. This approach can be efficiently combined with the Bootstrap Method (Efron and Tibshirani, 1993), such as explained in Chapter 3. Other approaches for preventing overfitting can be found in Demuth and Beale (1992).

2. GROUP METHOD OF DATA HANDLING

After the original set of predictors for PTFs has been selected, the subsequent PTF development may: (a) retain all selected predictors in the PTF; (b) eliminate part of them based on statistical tests; (c) eliminate some predictors and define the relative importance of the remaining predictors. The regression method (Chapter 1) can eliminate insignificant predictors, but regression equations presume *a priori* knowledge about the type of dependence that should exist in PTFs. The neural network methods (Section 1) do not assume any particular type of dependence as they can mimic any reasonable relationship between dependent and independent variables. However, they lack a built-in methodology to sieve out the less or the least important input variables. PTF development would therefore benefit from methods that would not assume a type of dependence and that still would be able to eliminate some predictors.

The group method of data handling (GMDH) combines advantages of regression analysis and artificial neural networks (Hecht-Nielsen, 1990). The GMDH constructs a flexible neural network-type equation to relate the inputs to outputs, and at the same time has a built-in algorithm to retain only essential input variables (Farlow, 1984). The GMDH has been recently used to develop PTFs (Pachepsky and Rawls, 1999; Gimènez et al., 2001).

The GMDH algorithm can be understood from the following example. Let the original data contain one column of observed values of y, and N columns containing observed values of the independent variables $x_1, x_2, \ldots, x_N$. The algorithm works by iterating over three steps. Step 1 consists of obtaining preliminary estimates of y using quadratic regressions:

$$z = A + Bu + Cv + Du^2 + Euv + Fv^2 \tag{5}$$

where A, B, C, D, E, and F are regression coefficients. All independent variables $x_1, x_2, \ldots, x_N$ are taken two at a time to become u and v in this equation, and regressions are found so that values of z best fit the dependent variable y. The resulting columns of z_m values, $m = 1, 2, N(N-1)/2$, contain estimates of y from each polynomial, and are interpreted as new variables that may have better predictive capability than the original $x_1, x_2, \ldots, x_N$. Step 2 consists of screening out the least effective new variables using a statistical selection criteria (Farlow, 1984). The list of input variables is modified at the end of step 2. Step 3 consists in testing whether the set of equations can be further improved. The smallest value of the selection criterion obtained from this iteration is compared with the smallest value obtained from the previous iteration. If an improvement is achieved, steps 1 and 2 are repeated, otherwise the iterations stop and the network is built.

The general GMDH algorithm is a heuristic one, and there are several versions that differ in the number of input variables that can be considered at a time to obtain preliminary estimates, and in the criteria to screen out the least efficient variables. The versions have been called learning networks, abductive networks, polynomial networks, multilayered iterational algorithms (MIA), and combinatorial algorithms (COMBI) among others. To-date, the version of the algorithm used in soil research is the one coded in the software ModelExpert (Pachepsky et al., 1998; Ungaro and Calzolari, 2001; Tomasella et al., 2003). This software uses either linear combinations of input variables or cubic polynomials of two or three independent variables. The number of variables to be retained in the input list is limited. Both original input variables and the output variable are normalized to have zero mean and unit variance, and the normalized variables participate in network building. The selection criteria, known as Barron criteria (Barron, 1984), is based on "punishment for complexity". The predicted square error (PSE) of the model is minimized. It is given by the sum of two components: the sum of squared errors resulting from the use of the model FSE and the complexity penalty, CP

$$CP = C\frac{2P}{N}s_{ap}^2 \tag{6}$$

where P is the total number of regression coefficient in the network, N is the number of observation in the data set, s_{ap}^2 is an *a priori* estimation of the model error and the constant C is the complexity penalty multiplier.

Examples of the GMDH applications can be found in Chapters 4, 6 and 21.

3. REGRESSION TREES

PTF dependencies can be very different in different parts of the data base, and using the same PTF equation for the whole data base may be misleading. It may be beneficial to subdivide the database into more homogeneous parts and then to build essentially different PTFs for the different parts.

Regression tree modeling is an exploratory technique based on uncovering structure in data (Clark and Pregibon, 1992). The resulting model partitions data first into two groups, then into four groups, and so on, providing groups as homogeneous as possible at each of the levels of partitioning. Each partitioning can be viewed as a branching, and the final fit of model to data looks like a tree with two branches originating in each node. Regression trees first became popular in environmental sciences (Baker, 1993; Lees and Ritman, 1991), and were later used in studies on land quality assessment and soil properties estimation (Van Lanen et al., 1992; McKenzie and Jacquier, 1997; McKenzie and Ryan, 1999). Regression trees can use both categorical and numerical variables as predictors (Breiman et al., 1993). Regression trees can be developed with the software SPLUS (Mathsoft, 1999), and also with the SAS software.

The regression tree algorithm works as follows. Suppose that a database is organized as a table with columns x_1, x_2, x_3,..., x_N representing predictor variables and the column y representing the response variable. The minimum number of samples or database lines before a partitioning, M_{split}, and the minimum number of samples in a group after partitioning, M_{group}, have to be set first. Then all possible partitions are formed for each of the predictor variables.

The method of forming a partition depends on the type of the variable. If the variable is numerical, then the whole database table is sorted by the column of this variable and then split up into two parts, each having the number of samples greater than M_{group}. An example of such partitioning is shown in Table 1 where raw data are shown in columns 1–4; variable x_1 is numerical, variable x_2 is categorical with three possible levels a, b, and c, and y is the variable of interest to be predicted. Columns 4–8 show the same database sorted by the numerical variable x_1. Let the value of M_{group} be equal to 3. Then possible partitions of the database by the variable x_1 are:

(1) either $x_1 < 1.6$, samples 1, 3, 6, or $x_1 > 1.6$, samples 11, 5, 2, 9, 8, 10, 7, 4;
(2) either $x_1 < 2.15$, samples 1, 3, 6, 11, or $x_1 > 2.15$, samples 5, 2, 9, 8, 10, 7, 4;......
(7) either $x_1 < 3.1$, samples 1, 3, 6, 11, 11, 5, 2, 9, 8, or $x_1 > 3.1$, samples 10, 7, 4.

If the variable is categorical, then the partitions are formed after dividing the database into subsets having the same value of this variable and splitting it by all possible combinations of subsets. In the example of Table 1, the subsets are shown in columns 9–13. Then possible partitions are:

(1) a|bc, i.e, either $x_2 = a$, samples 5, 6, 9, or $x_2 = b$ or $x_2 = c$, samples 1, 3, 4, 10, 11, 2, 7, 8.
(2) ab|c, i.e, either ($x_2 = a$ or $x_2 = b$), samples 5, 6, 9, 1, 3, 4, 10, 11, or $x_2 = c$, samples 2, 7, 8.

Table 1
Synthetic database to explain the regression tree algorithm.

Raw data				Data sorted by the variable x_1					Data sorted by the variable x_2			
Sample	x_1	x_2	y	Sample	x_1	x_2	y	ΔD	Sample	x_1	x_2	y
1	0.8	b	7.9	1	0.8	b	7.9	ND	5	2.4	a	3.2
2	2.7	c	4.8	6	1.1	a	3.5	ND	6	1.1	a	3.5
3	1.3	b	6.1	3	1.3	b	6.1	ND	9	2.8	a	3
4	4.2	b	2.5	11	1.9	b	5.5	14.1	1	0.8	b	7.9
5	2.4	a	3.2	5	2.4	a	3.2	19.7	3	1.3	b	6.1
6	1.1	a	3.5	2	2.7	c	4.8	14.5	4	4.2	b	2.5
7	3.6	c	0.6	9	2.8	a	3	18.5	10	3.3	b	1.7
8	2.9	c	5	8	2.9	c	5	14.7	11	1.9	b	5.5
9	2.8	a	3	10	3.3	b	1.7	23.4	2	2.7	c	4.8
10	3.3	b	1.7	7	3.6	c	0.6	ND	7	3.6	c	0.6
11	1.9	b	5.5	4	4.2	b	2.5	ND	8	2.9	c	5

(3) ac|b, i.e, either ($x_2 = $ a or $x_2 = $ c), samples 5, 6, 9, 2, 7,8, or $x_2 = $ b, samples 1, 3, 4, 10, 11.

The total possible number of partitions with a categorical variable is $2^{k-1} - 1$ where k is the number of levels. For example, for a variable having four levels a, b, c, d, there are $2^{4-1} - 1 = 7$ possible partitions (a|bcd, ab|cd, abc|d, ac|bd, ad|bc, abd|c, acd|b).

The first subset of samples in a partition is called left branch and the second is called right branch. For example, in partition (1) of the database in Table 1 by the variable x_1, the left branch will consist of samples where $x_1 < 1.6$ (samples 1, 3, 6), and the right branch will consist of samples where $x_1 > 1.6$ (samples 11, 5, 2, 9, 8, 10, 7, 4).

All partitions by all variables are compared by the reduction in non-homogeneity that they provide. The non-homogeneity in a group of samples is measured by computing deviances and defined for a group of observed values y as

$$D = \sum_i (y_i - \bar{y})^2 \qquad (7)$$

Here $\bar{y}$ is the mean value across all observations y_i. Each partition generates left $D_L = \sum_L (y_i - \bar{y})^2$ and right $D_R = \sum_R (y_i - \bar{y})^2$ deviance values where subscripts "L" and "R" indicate collections of numbers of samples in branches in a partition. The partition that maximizes the change in deviance

$$\Delta D = D - D_L - D_R \qquad (8)$$

is the partition to choose. From Table 1, values of ΔD for the partitioning by the variable x_1 are shown in column 9. The largest change in deviance $\Delta D = 23.4$ is achieved after partitioning "either $x_1 < 3.1$ or $x_1 > 3.1$". Values of ΔD obtained after partitioning by the categorical variable x_2 are $\Delta D = 2.3$ for a|bc, $\Delta D = 1.1$ for ab|c, and $\Delta D = 5.6$ for ac|b. Therefore, the partitioning "either $x_1 < 3.1$ or $x_1 > 3.1$" is the one to choose in this example.

Each of the branches obtained after partitioning is partitioned again according to the limitations imposed by values of M_{split} and M_{group}. In Table 1 as example, the group of samples where $x_1 > 3.1$ cannot be partitioned anymore while the group with $x_1 < 3.1$ can be. Attempts to partition the latter group by the variable x_1 lead to the best values of $\Delta D = 6.1$ whereas partitioning "either $x_2 = a$ or ($x_2 = b$ or $x_2 = c$)" gives the value $\Delta D = 12.9$, and becomes the partitioning to choose. Further partition is not possible because the minimum size of a group is achieved, and this node of the tree is a terminal node. The final regression tree for this example is shown in Figure 3. Here the number of a terminal node is shown in brackets, and the average value of the variable y for this node, standard deviations of the variable y in parentheses, and the count of samples pertaining to this node are shown beneath the terminal node numbers. The average value is the value predicted for the whole group of samples forming the terminal node. Regression trees do not have much in common with classical statistical regression described in Chapter 1. The term "regression" is used because they can be used as predicting tools in the same sense as statistical regression equations.

In the limit, the recursive partitioning of a large database may produce a tree with a very large number of terminal nodes. There is a chance that the predictive ability of such a large

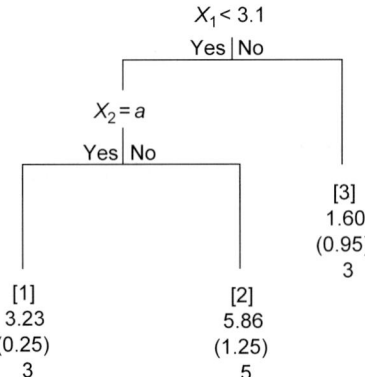

Figure 3. Regression tree for the test data set in Table 1. The node number in brackets, the average value of the dependent variable for the group, the standard deviation of the dependent variable within groups in parentheses, and the number of samples in the group are shown beneath terminal nodes.

tree will be limited, because the later branches will show intricacies of small groups of samples specific for the database. To avoid such "over-fitting," a tree has to be pruned to be useful for predictions. The regression tree methodology has variations regarding tree pruning (Bell, 1999). To snip off the least important partitions, the software SPLUS (Mathsoft, 1999), for example, applies the cost-complexity measure D_K:

$$D_K(T) = D(T) + KN_T \tag{9}$$

Here $D(T)$ is the deviance of the subtree (T), N_T is the number of terminal nodes, and K is the cost-complexity parameter.

Examples of the regression tree applications for PTF building can be found in Chapters 6 and 8 of this book.

4. CROSS-VALIDATION PROCEDURES

Both the regression trees and the GMDH are iteratively building models of progressively increasing complexity. The processes have to be stopped to prevent overfitting, otherwise the predictive capability of the resulting models with respect to new data will be deplorable. Practically, this means that the dataset has to be split into development and testing subsets, and the CP multiplier in Equation (2) or the cost-complexity parameter K in Equation (5) have to be varied to provide the level of complexity that minimizes average model error for the testing datasets. This procedure is known as cross-validation. Four techniques of cross-validation are in general use (Good, 1999):

1. *K-fold* in which the data are subdivided into K roughly equal-sized parts, then repeated the modeling process K times leaving one section out each time for validation purposes.
2. *Leave-one-out,* an extreme example of *K*-fold, in which the data are subdivided into as many parts as there are observations. One observation is left out of our regression procedure and the remaining $n - 1$ observations are used as a development set. Repeating this procedure n times, omitting a different observation each time, one arrives to the estimate of accuracy.
3. *Jackknife,* an obvious generalization of the leave-one-out approach, where the number left out can range from one observation to half the sample.
4. *Delete – d,* where a random percentage d of the observations is set aside for validation purposes, and the remaining $100 - d\%$ are used as a development set, then average over 100–200 such independent random samples.

The cross-validation is a heuristic approach to balance complexity and accuracy. The jackknife cross-validation with development – testing ratio equal to 9:1 has been used in PTF development by Pachepsky and Rawls (1999) and Rawls and Pachepsky (2002a, 2002b). The bootstrap (Chapter 3) can complement the cross-validation.

Examples of the application of the cross-validation procedure can be found in Chapters 6 and 8 of this book.

5. CONCLUDING REMARKS

The main reason in using data mining and database exploration techniques in PTF development is probably due to the complexity of relationships between soil properties. Other data base exploratory techniques have been successfully used for this purpose, e.g., numerical classification (Williams et al., 1983). Each of the methods in this chapter has its advantages and disadvantages. For example, the advantage of the regression trees is in the transparency of results, whereas the advantage of the neural networks is the ability to mimic practically any relationship. The disadvantage of all techniques in this chapter as compared to statistical regression is the heuristic element involved so that the rigorous statistical judgement is hard to make. One example of comparison of artificial neural networks, GMDH, and regression trees is given in Figure 4 (Wösten et al., 2001). As can be seen, the accuracy of the three methods is quite comparable. In all three cases, the root-mean square error is around 3.4 vol.%, the R^2 value of the regression is around 0.9, the slope of the regression is close to 1 and the intercept is about 0.001 vol.%. The three

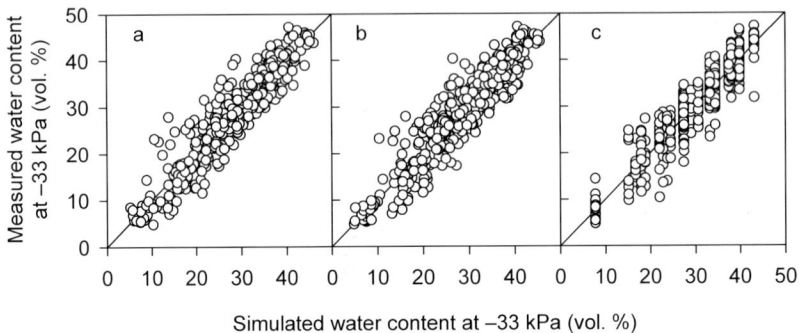

Figure 4. One-to-one diagrams showing accuracy of predicting water content at −33 kPa pressure head with three methods, (a) backpropagation artificial neural network, (b) GMDH, (c) classification and regression trees (after Wösten et al., 2001).

techniques produced practically identical PTF accuracy. The database exploration has to be viewed as a useful step that may generate PTFs that are either sufficient for the intended application or may suggest further applications of more rigorous or more flexible PTF-building techniques.

REFERENCES

Alexander, I., Morton, H., 1990. An Introduction to Neural Computing. Chapman and Hall, London.
Baker, F.A., 1993. Classification and regression tree analysis for assessing hazard of pine mortality caused by Heterobasidion annosum. Plant Dis. 77, 136-139.
Barron, A.R., 1984. Predicted Square Error: a Criterion for Automatic Model Selection. *In*: Farlow, S.J. (Ed.), Self-Organizing Methods in Modelling: GMDH-type Algorithms. M. Dekker, New York, NY, pp. 87-104.
Bell, J.F., 1999. Tree-based methods. *In*: Fielding, A.H. (Ed.), Machine Learning Methods for Ecological Applications. Kluwer Academic Publishers, Boston, pp. 84-111.
Breiman, L., Friedman, J.H., Olshen, R.A., Stone, C.J., 1993. Regression Trees. Chapman and Hall/CRC, Boca Raton, FL.
Clark, L.A., Pregibon, D., 1992. Tree-based models. *In*: Hastie, T.J. (Ed.), Statistical Models. S. Wadsworth, Pacific Grove, California, pp. 377-419.
Demuth, H., Beale, M., 1992. Neural Network Toolbox Manual. MathWorks Inc., Natick, Massachussetts.

Efron, B., Tibshirani, R.J., 1993. An Introduction to the Bootstrap. Monographs on Statistics and Applied Probability. Chapman and Hall, New York, NY.

Farlow, S.J., 1984. The GMDH algorithm. *In*: Farlow, S.J. (Ed.), Self-Organizing Methods in Modeling: GMDH-Type Algorithms. M. Dekker, New York, pp. 1-24.

Gimènez, D., Rawls, W.J., Pachepsky, Y., Watt, J.P.C., 2001. Prediction of a pore distribution factor from soil textural and mechanical parameters. Soil Sci. 166, 79-88.

Good, P.I., 1999. Resampling Methods: A Practical Guide to Data Analysis. Birkhäuser, Boston.

Haykin, S., 1994. Neural Networks, a Comprehensive Foundation, 1st Ed. Macmillan College Publishing Company, New York, NY.

Hecht-Nielsen, R., 1990. Neurocomputing. Addison-Wesley Publishing Company, Reading, MA.

Koekkoek, E., Bootlink, H., 1997. Development of a neural network model to predict soil water retention. Eur. J. Soil Sci. 50, 489-495.

Lees, B.G., Ritman, K., 1991. Decision-tree and rule-induction approach to integration of remotely sensed and GIS data in mapping vegetation in disturbed or hilly environments. Environ. Manage. 15, 823-831.

Maren, A.J, Harston, C.T., Pap, R.M., 1990. Handbook of Neural Computing Applications. Academic Press, San Diego.

Mathsoft, 1999. SPLUS 2000 Professional. User's Manual.

McCord, N.M., Illingworth, W.T., 1990. A Practical Guide to Neural Nets. Addison-Wesley, Reading, Mass.

McKenzie, N.J., Jacquier, D.W., 1997. Improving the Field Estimation of Saturated Hydraulic Conductivity in Soil Survey. Aust. J. Soil Res. 35, 803-825.

McKenzie, N.J., Ryan, P.J., 1999. Spatial prediction of soil properties using environmental correlation. Geoderma 89, 67-94.

Minasny, B., McBratney, A.B., Bristow, K.L., 1999. Comparison of different approaches to the development of pedotransfer functions for water-retention curves. Geoderma 93, 225-253.

Pachepsky, Ya.A., Rawls, W.J., 1999. Accuracy and reliability of pedotransfer functions as affected by grouping soils. Soil Sci. Soc. Am. J. 63, 1748-1757.

Pachepsky, Ya.A., Timlin, D., Varallyay, G., 1996. Artificial neural networks to estimate soil water retention from easily measurable data. Soil Sci. Soc. Am. J. 60, 727-733.

Pachepsky, Ya.A., Rawls, W., Gimenez, D., Watt, J.P.C., 1998. Use of soil penetration resistance and group method of data handling to improve soil water retention estimates. Soil and Tillage Research 49, 117-128.

Pachepsky, Ya., Timlin, D., Ahuja, L., 1999. Estimating saturated soil hydraulic conductivity using water retention data and neural networks. Soil Sci. 164, 552-560.

Rawls, W.J., Brakensiek, D.L., 1985. Prediction of soil water properties for hydrologic modeling. *In*: Jones, E.B., Ward, T.J. (Eds.), Proceedings of the Symposium of Watershed Management in the Eighties. April 30–May 1, 1985, Denver, CO. Am. Soc. Civil Eng., New York, NY, pp. 293-299.

Rawls, W.J., Pachepsky, Ya.A., 2002a. Soil consistence and structure as predictors of water retention. Soil Sci. Soc. Am. J. 66, 1115-1126.

Rawls, W.J., Pachepsky, Ya.A., 2002b. Using field topographic descriptors to estimate soil water retention. Soil Sci. 167, 235-243.

Rummelhart, D.E, Hinton, G.E., Williams, R.J., 1986. Learning internal representations by error propagation. *In*: Rummelhart, D.E., McClelland, J.L. (Eds.), Parallel Distributed Processing: Explorations in the Microstructure of Cognition, Vol. I. MIT Press, Cambridge, MA, pp. 318-362.

Schaap, M.G., Bouten, W., 1996. Modeling water retention curves of sandy soils using neural networks. Water Resour. Res. 32, 3033-3040.

Schaap, M.G., Leij, F.L., van Genuchten, M.Th., 1998. Neural network analysis for hierarchical prediction of soil hydraulic properties. Soil Sci. Soc. Am. J. 62, 847-855.

Tamari, S., Wösten, J.H.M., Ruiz-Suárez, J.C., 1996. Testing an artificial neural network for predicting soil hydraulic conductivity. Soil Sci. Soc. Am. J. 60, 1732-1741.

Tomasella, J., Pachepsky, Y.A., Crestana, S., Rawls, W.J., 2003. Comparison of two approximation techniques to develop pedotransfer functions of water retention of Brazilian soils. Soil Sci. Soc. Am. J. 67, 1085-1092.

Ungaro, F., Calzolari, C., 2001. Using existing soil databases for estimating water-retention properties for soils of the Pianura Padano-Veneta region of North Italy. Geoderma 99, 99-121.

Van Lanen, H.A.J., van Diepen, C.A.J., Reinds, G.J., de Koning, G.H.J., 1992. A comparison of qualitative and quantitative physical land evaluations, using an assessment of the potential for sugar-beet growth in the European Community. Soil Use Manag. 8, 80-89.

Williams, J., Prebble, R.E., Williams, W.T., Hignett, C.T., 1983. The influence of texture, structure and clay mineralogy on the soil moisture characteristic. Aust. J. Soil Res. 21, 15-32.

Wösten, J.H.M., Pachepsky, Y.A., Rawls, W.J., 2001. Pedotransfer functions: bridging the gap between available basic soil data and missing soil hydraulic characteristics. J. Hydrol. 251, 123-150.

Chapter 3

ACCURACY AND UNCERTAINTY IN PTF PREDICTIONS

M. G. Schaap

George E. Brown, Jr. Salinity Laboratory (USDA/ARS), 450 Big Springs Road, Riverside, CA 92507, USA
Tel.: +1-(909)-369-4844; E-mail:mschaap@ussl.ars.usda.gov

Several statistical criteria are available to evaluate and quantify the potential usefulness of pedotransfer functions (PTFs) in making accurate and reliable estimates of soil hydraulic properties. In this section we will deal with three closely related terms: optimization, accuracy, and uncertainty. Optimization criteria are the objective functions used to calibrate empirical parameters in PTFs. Accuracy criteria relate to the statistical methods used to test the performance of indirect methods, preferably using data distinct from those used for calibration. Uncertainty estimates provide information about the probability distribution of estimated hydraulic quantities. Thus, a PTF can make accurate estimates (i.e., it produces the correct values *on average*), but it may not be reliable when, for example, real-world variability is larger than estimated with the PTF. Alternatively, a PTF can be deemed inaccurate when it produces estimates that, on average, differ systematically from observations. Optimization criteria are the objective functions used to calibrate PTFs. Although optimization criteria are in a strict sense not measures to independently evaluate PTFs, their discussion in this chapter is warranted. Because PTF calibration is normally done before evaluation, we will discuss optimization criteria first.

1. OPTIMIZATION CRITERIA

All indirect methods, whether physically based or empirical, contain parameters that must be optimized using databases containing both independent data (predictors) and dependent data (hydraulic properties or parameters in hydraulic functions). Virtually all indirect methods are optimized by least-squares methods, and define the objective function as

$$SSQ = \sum_N (\varsigma - \varsigma(\mathbf{b})')^2 \tag{1}$$

where N is the number of data points and $\mathbf{b}$ is a parameter vector containing the empirical PTF coefficients. For point-based methods ς and ς' represent data points for measured and estimated water retention or hydraulic conductivity, respectively. In the case of parametric methods, ς and ς' represent parameters of hydraulic functions, such as those in the Brooks–Corey (1964, BC) or van Genuchten (1980, VG) relationships. It is often necessary to transform some of the hydraulic parameters before the optimization of Equation (1) is carried out. For parametric methods it is common to deal with log-transformed values of α, n and λ in the BC or VG equations (e.g., Rawls and Brakensiek (1985) and Wösten et al. (1999), among many others). The general rationale for these transformations is to convert the distribution of the parameters into a more statistically normal distribution (Carsel and Parrish, 1988). For both point and parametric PTFs, log-transforms can also be used in conjunction with Equation (1) to remove a bias towards large conductivity values. For example, depending upon soil type, the saturated conductivity (K_s) can vary by several orders of magnitude. Optimization according to Equation (1) without a log-transform would yield then a PTF that is generally more accurate for highly permeable soils than for low permeability soils.

The optimization of point-based PTFs according to Equation (1) is relatively straightforward. For parametric methods the situation is more complicated since *two* distinct optimizations are necessary. First the parameters in the BC, VG or other hydraulic functions must be optimized using observed water retention and/or conductivity points by minimizing the objective function given by Equation (1). This optimization is only possible if the number of observations exceeds the number of free parameters in the hydraulic function. For this reason, most parametric methods require rather detailed retention or unsaturated conductivity data sets. The actual PTFs are constructed in the second optimization by choosing appropriate functions and minimizing the sum of squares between estimated and previously fitted hydraulic parameters, again using an objective function of the type given by Equation (1).

Given a calibration data set, subtle differences may arise when point- or parametric PTFs are optimized. Being optimized directly to observed – but possibly transformed – water retention or conductivity points, point-PTFs yield optimal results in terms of those observed quantities. Parametric PTFs on the other hand estimate derived quantities (i.e., hydraulic parameters) and not quantities that are directly observed such as retention points. Because the hydraulic functions often do not perfectly fit the hydraulic characteristics, parametric PTFs typically perform slightly worse than point PTFs (Schaap and Bouten, 1996). This fact, combined with the strong non-linearity of the water retention and hydraulic conductivity equations, means that an optimum SSQ in terms of *parameter values* does not guarantee an optimum SSQ in terms of estimated water contents or conductivities.

A solution to this problem is to optimize parametric PTFs directly in terms of the observable quantities rather than optimizing the parameter values. This, in effect, will replace the two-step optimization approach for parametric PTFs with a one-step optimization method. We illustrate this approach using the work of Scheinost et al. (1997) and Minasny et al. (1999) who developed PTFs to estimate VG water retention parameters using "extended non-linear regression." Three steps are necessary in this approach. Firstly, four simple equations (f_1 through f_4) are selected that express the four VG retention parameters (θ_r, θ_s, α, n) in terms of predictors (texture parameters, porosity, organic matter content). Secondly, these equations are substituted in a modified van Genuchten (1980) equation, thus yielding (Scheinost et al., 1997)

$$\theta(h) = f_1(\text{clay}, C) + \frac{f_2(\text{clay}, \phi) - f_1(\text{clay}, C)}{[1 - \{f_3(d_g)h\}^{f_4(\sigma_g)}]} \tag{2}$$

The functions f_1 through f_4 are given by

$$f_1 : \theta_r = r_1 \text{clay} + r_2 C \tag{3}$$

$$f_2 : \theta_s = s_1 \phi + s_2 \text{clay} \tag{4}$$

$$f_3 : \alpha = a_1 + a_2 d_g \tag{5}$$

$$f_4 : n = n_1 + n_2/\sigma_g \tag{6}$$

where $\{r_1, r_2\}$, $\{s_1, s_2\}$, $\{a_1, a_2\}$, and $\{n_1, n_2\}$ are eight unknown parameters associated with the four VG retention parameters. The predictors d_g and σ_g are texture parameters (according to Shirazi et al., 1988), C is organic carbon content and ϕ is the porosity. In the third step, the eight unknown parameters are optimized according to Equation (1).

The extended non-linear regression approach has several advantages. Firstly, no fitted retention parameters are required because f_1 through f_4 are not optimized directly. Therefore, this optimization method can be applied also to databases containing fewer than four points for each retention characteristic. Secondly, this method still estimates VG parameters (f_1 through f_4) thus keeping the flexibility of parametric methods. Thirdly, the method allows the optimization of a parametric PTF directly on observed water contents. A potential drawback is that the number of free parameters must be kept to a minimum otherwise the eight-parameter system is difficult to optimize and may even yield non-unique solutions. This requires that f_1 through f_4 be kept relatively simple, possibly leaving these inadequate for certain conditions. However, Minasny et al. (1999) showed that this is not a major problem as this method provided similar results as a neural network approach. Minasny and McBratney (2002) demonstrated that a similar approach can be used for artificial neural network-based PTFs.

Although Equation (1) is used for the calibration of almost all PTFs, it is possible to use alternative objective functions. When calibrating PTFs, one is often confronted with databases that contain "outlier" points that deviate from the general trend. Outlier points can be caused by several factors such as undocumented measurement errors, or the presence of a few soils that have different characteristics from the larger population (e.g., some fine-textured soils in a larger dataset of coarse textured soils). Such outlier data points may have a large contribution to the sum of squares, and may lead to poor calibrations. Short of deleting outlier points one may use Robust Estimation (Press et al., 1988), which is relatively insensitive to outliers. Instead of minimizing the sum of squares (Eq. (1)), some variants of Robust Estimation minimize the sum of the absolute value of the difference between estimated and observed quantities. We refer to Press et al. (1988) for more information about robust parameter estimation.

2. CRITERIA FOR EVALUATING THE ACCURACY OF PTFs

The accuracy of PTFs is commonly tested using independent data sets. That is, the real value of a PTF can only be evaluated using data that were not used for the calibration of the PTF in question. When a PTF is developed for a particular data set, this PTF should perform well on that data set. However, success on other data sets is not guaranteed as illustrated by Schaap and Leij (1998) who showed that the performance of a PTF depends on the data set used for calibration *and* the one used for evaluation. In many cases, soils databases are organized along local (e.g., case studies), regional, or national boundaries and may have a bias to agricultural soils. Also, current international databases have a serious bias towards soils from temperate climates (Imam et al., 1999). For example, very little data is available for tropical soils (Epebinu and Nwadialo, 1993; Tomasella and Hodnett, 1997). Because it is difficult to define truly representative soils databases, the development, calibration and evaluation of PTFs may lead to somewhat arbitrary results. It is, therefore up to the user of PTFs to decide whether or not a particular PTF – or group of PTFs – is suitable for a particular application. This means that users should somehow obtain a data set and test the PTF using accuracy criteria.

Several studies can be found in the literature that compared and evaluated PTFs (Tietje and Tapkenhinrichs, 1993; Tietje and Hennings, 1996; Kern, 1995). These studies have used different criteria for evaluating the accuracy of indirect methods. Most commonly, criteria such as correlation coefficients, mean errors and root mean square errors are used.

Correlation coefficients (R) or coefficients of determination (R^2) are relatively simple statistics that provide insight in how well data sets of estimated and measured (fitted) hydraulic points (parameters) are related. Although the statistical significance of the value of R or R^2 depends on the size of the data set being tested they give a quick normalized impression of how much of the variance between measured and estimated hydraulic data is explained by the PTF.

Mean errors (ME) and root mean square errors (RMSE) are often used to measure the match of estimated hydraulic properties to such measurable quantities as water contents and hydraulic conductivities. General expressions for ME and RMSE are

$$\mathrm{ME} = \frac{1}{N} \sum_N (\varsigma' - \varsigma) \tag{7}$$

$$\mathrm{RMSE} = \sqrt{\frac{1}{N} \sum_N (\varsigma' - \varsigma)^2} \tag{8}$$

respectively, where N is the number of datapoints, and – for example – ς' and ς are the estimated and measured water contents or conductivities, respectively. When parametric PTFs are used the estimated parameters are first converted to conductivities or water contents at the pressure head *points* for which measurements are available. Similar to the optimization criteria in the previous section, ς' and ς may pertain to transformed variables. We note that in the case of log-transformations, the resulting RMSE and ME values will have dimensionless units (Tietje and Hennings, 1996; Schaap and Leij, 2000). When no transformations are applied the RMSE and ME values will have the same units as the measured data.

The ME and RMSE criteria serve several purposes. For example, mean errors quantify systematic errors as they reflect the average deviation between estimated and measured hydraulic data. RMSE values quantify the root of the average bivariate variance between estimated and measured data. RMSE values can thus be interpreted as a random error (not unlike a standard deviation). For a truly well-behaved PTF, both ME and RMSE should be as low as possible. A negative ME indicates that the PTF underestimates the quantity being evaluated, whereas positive values indicate an overestimation. We note, however, that the ME pertains to an *average* over N data points. It may very well be that while the ME is zero the PTF overestimates hydraulic quantities for fine-textured soils and produces overestimates for coarse-textured soils. Likewise in the case of water retention characteristics, it may be that a PTF overestimates water contents near saturation, but underestimates water contents in the dry range (Figure 1). To partially resolve this problem, absolute mean errors (AME) may be computed:

$$\text{AME} = \frac{1}{N} \sum_N |s' - s| \tag{9}$$

AME are always equal to or greater than zero but do not provide information about the sign of the systematic errors.

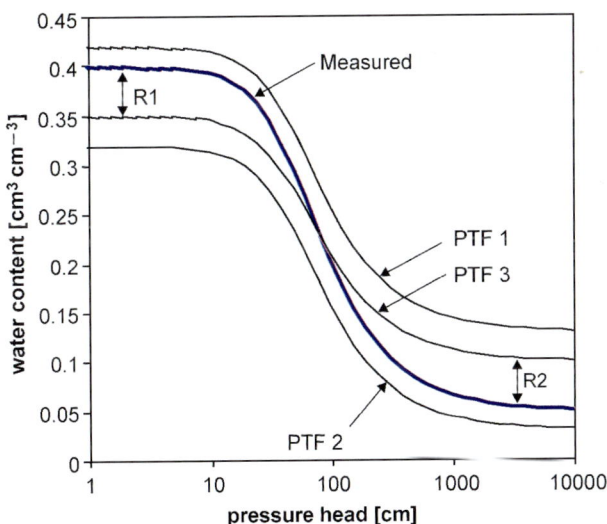

Figure 1. Examples of systematic and relative errors for a typical water retention characteristic. Shown are "measured" data (thick line) and estimates with three hypothetical PTFs (PTF 1, PTF 2, PTF 3). PTF 1 shows an overall overestimation, PTF 2 exhibits an overall underestimation, while PTF3 shows an underestimation near saturation and an overestimation in the dry range. Although the errors produced by PTF 3 at the points R1 and R2 are of similar size (about 0.05 cm^3 cm^{-3}), the relative errors at these points are different: 12.5% at R1 and 100% at R2.

Occasionally it is useful to provide an impression about the relative size of the systematic errors. For example, a PTF that produces a systematic error of 0.05 cm^3 cm^{-3} for a measured saturated water content of 0.40 cm^3 cm^{-3} leads to a relative error of 12.5% (R1 in Figure 2). However, a PTF that produces the same error for a measured retention point at 0.05 cm^3 cm^{-3} yields a relative error of 100% (R2 in Figure 2). Such errors can be quantified using the relative mean errors (RME) (Williams et al., 1992; Kern, 1995)

$$\text{RME} = \frac{1}{N} \sum_N \frac{s' - s}{s} \tag{10}$$

If mean errors are present then the RMSE given by Equation (8) includes both systematic and random errors. To better separate the systematic from the random errors one may compute unbiased root mean square errors (Tietje and Hennings, 1996)

$$\text{URMSE} = \sqrt{\frac{1}{N} \sum_N (s' - s - \text{ME})^2} \tag{11}$$

where the ME is derived from Eq. (7). Similar as ME, URMSE may still include systematic errors for particular parts of the data set.

The error measures discussed thus far are well suited for comparison of parametric PTFs, i.e., PTFs that estimate hydraulic parameters that can be converted to water retention or conductivity points for which measurements are available. A more complicated situation arises when point-PTFs are evaluated and compared because these usually estimate different numbers of water retention points at different pressure heads (Rawls et al., 1982, 1983; Ahuja et al., 1985; Minasny et al., 1999). This may thus lead to the situation where no measured water retention points are available for a particular PTF, or vice versa, no PTF estimate is available for pressure heads for which measured data are available. The available data is then not used in an optimal way. Also, the error criteria discussed previously above will be evaluated on a different number of points for each PTF (3–12 points are estimated by Rawls et al. (1982, 1983), Ahuja et al. (1985), and Minasny et al. (1999)). Results for different point PTFs may thus be difficult to compare.

Two approaches can be followed to facilitate a more objective evaluation and comparison of point PTFs. Saxton et al. (1986) fitted Brooks–Corey (1964) water retention curves to the estimates of the point-PTFs of Gupta and Larson (1979) and Rawls et al. (1982). Such an approach then allows one to evaluate the PTF as a parametric PTF by computing water contents at pressure heads for which measurements are available. The drawback of this approach is that curve fitting point PTF estimates invariably introduces some random errors and perhaps systematic errors as well. Furthermore, this approach is not suitable for PTFs that estimate a limited number of retention points since curve fitting such data is not possible or leads to unreliable results.

Tietje and Tapkenhinrichs (1993) followed a somewhat related approach in their evaluation of point and parametric PTFs. Instead of curve fitting the estimated retention points, they calculated the estimated water contents at the desired pressure heads by linear interpolation between the estimated retention points (using log-transformed pressure heads). Subsequently, they defined the mean errors and root mean square errors by integration over a certain pressure head range as follows

$$\text{IME} = \frac{1}{b-a} \int_a^b (s' - s) dh \qquad (12)$$

$$\text{IRMSE} = \sqrt{\frac{1}{b-a} \int_a^b (s' - s)^2 dh} \qquad (13)$$

where the integration interval runs form $a = \log(1 \text{ cm})$ and $b = \log(15,000 \text{ cm})$ and h is the pressure head. The integration was carried out numerically. Instead of evaluating ME and RMSE at pressures for which measurements are available, Equations (12) and (13) compute the normalized area between interpolated measured characteristics and interpolated estimated characteristics. Tietje and Tapkenhinrichs (1993) claim three advantages of their approach: (i) the method is applicable to all PTFs, thus making it easier to compare different PTFs objectively, (ii) no specific function is assumed for the water retention characteristic, and (iii) the method is also possible for a limited measured retention points. We note, however, that although log-linear interpolation is a generally very appropriate choice it *does* make an assumption about the underlying shape of the water retention characteristic, especially when only a few measured or estimated points are available. Secondly, though likewise appropriate the integration interval of Equations (12) and (13) is still somewhat arbitrarily chosen. Thirdly, the amount of information available is limited to the number of measured retention points. Interpolation or integration will never extend this information. It seems that interpolation of estimated points to pressure heads for which measured water contents are available yields similar information that is easier to access since no integration is required.

3. EVALUATING THE UNCERTAINTY OF PTF PREDICTIONS

PTFs generally provide estimates of hydraulic properties with only a modest level of accuracy. Overconfidence in PTF estimates should be avoided because the random or systematic errors made by PTFs may for example lead to inaccurate estimates of contaminant transport through soils and sediments. In some cases it would be worthwhile to use a probabilistic approach that associates an uncertainty to a PTF estimate, such as a confidence interval. Three general approaches are possible.

The first approach is the one followed by Wösten and van Genuchten (1988) who estimated water contents and conductivities at 12 pressure heads for coarse, medium and fine-textured soils. Using measured data they computed RMSE values (standard deviations) at each pressure for the three soil groups, and plotted the 90% confidence interval together with measurements and PTF estimates. The measured and estimated characteristics were generally situated within the confidence intervals, which were between 0.05 and 0.20 $cm^3 cm^{-3}$ for water retention, and between one or two orders of magnitude for the unsaturated conductivity. The advantage of this approach is that it can easily be implemented for any PTF. Disadvantages are that measured hydraulic data must be available and – to obtain reliable statistics – that the data must be grouped into classes (e.g., according to texture, or any other relevant subdivision). Because this approach requires measured data, different data sets may lead to different confidence intervals. We note that confidence intervals with this approach are computed from error criteria defined in Section 2.

The second approach is to quantify parameter uncertainty and co-variance in the PTFs during the calibration process. Many statistical optimization methods allow the analytical or numerical quantification of the variance and co-variance matrices of the optimized parameters (Press et al., 1988). Assuming the underlying model is correct, this type of parameter uncertainty relates to random measurement errors in the actual calibration database. Uncertainty in the calibration parameters subsequently leads to uncertainty in the PTF estimates. Unfortunately, this type of parameter uncertainty is often not reported for PTFs.

A third related approach quantifies parameter uncertainty or the parameter confidence intervals associated with sampling effects in the calibration database. In general, the available calibration database is assumed to be a representative sample of the natural population of soils. In reality, however, the calibration data set is just one sample of this somewhat hard to define population. Calibration of a PTF on another calibration database taken from the same population, will likely lead to a slight variation in the calibration parameters and hence translate into uncertainty in PTF estimates. In most cases, however, only one calibration database is available, making it impossible to directly compute parameter confidence intervals. However, by using Monte Carlo analysis (Press et al., 1988) or the Bootstrap Method (Efron and Tibshirani, 1993) one may gain insight into the parameter confidence intervals without having to collect large quantities of additional data. By using these methods it is also possible to provide uncertainty estimates for each individual PTF estimate (Schaap et al., 2001).

Monte Carlo approaches involve the generation of many synthetic data sets from a calibrated model. The model is re-calibrated on each synthetic data set and the distributions of the resulting parameters are converted to co-variance matrices. Carsel and Parrish (1988) developed joint probability distributions of saturated conductivity and van Genuchten (1980) water retention parameters for the 12 USDA soil textural classes. Besides providing uncertainty estimates for the VG parameters for each textural class, Carsel and Parrish (1988) also accounted for the joint probability distributions among the hydraulic parameters. Results of this type can subsequently be used to generate sets of correlated hydraulic parameters for use in, for example, Monte-Carlo studies for developing uncertainty estimates of solute transport (Carsel and Parrish, 1988) or Bayesian analyses (Meyer et al., 1999). Minasny et al. (1999) also used a Monte Carlo approach to study the effects of uncertainty in textural data on estimated hydraulic parameters. Depending on how texture was determined (either from imprecise field estimates or more detailed lab measurements) different effects of input data uncertainty were found.

The Bootstrap method is a non-parametric approach that assumes that multiple alternative realizations of a population can be *simulated* from a single calibration data set. Thus, contrary to using synthetic data such as the Monte Carlo method, real data are used that is re-ordered in new datasets. The Bootstrap method accomplishes this by repeated random resampling with replacement of the original data set of size N to obtain B alternative data sets, also of size N. The PTF is calibrated B times and, because of slight variations in the alternative data sets, the bootstrap method, therefore leads to B alternative

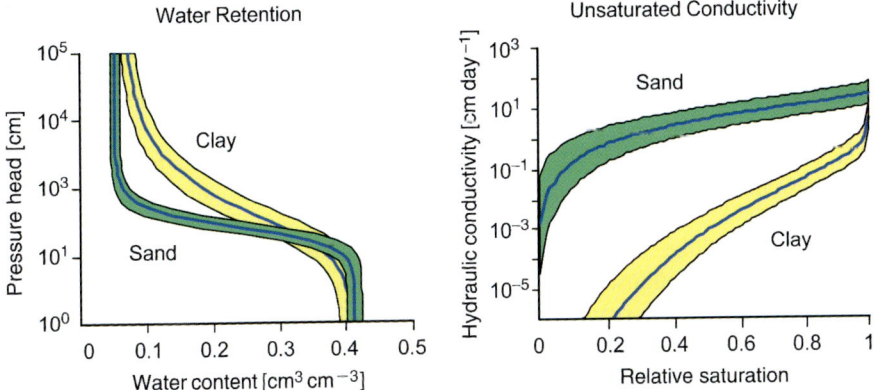

Figure 2. Examples of confidence intervals as generated with the Bootstrap method for water retention (left) and the unsaturated hydraulic conductivity (right). Shown are estimates and 90% confidence intervals for a sand (95% sand, 2.5% clay) and a clay (25% sand, 60% clay); both soils have a bulk density of 1.4 g cm^{-3}.

PTFs, resulting in a distribution of estimated hydraulic parameters. These distributions can be further manipulated for confidence intervals and co-variances. The non-parametric nature of the Bootstrap method allows it to be used for a wider range of PTFs than Monte Carlo analysis, such as neural networks which do not have easily identifiable calibration parameters.

Figure 2 shows confidence intervals as generated with the Bootstrap method. Shown are estimated retention and unsaturated hydraulic conductivity characteristics for a coarse textured soil (a sand with 95% sand, 2.5% clay, bulk density 1.4 g cm^{-3}) and a fine textured soil (a clay soil with 25% sand, 60% clay, bulk density 1.4 g cm^{-3}). The estimated characteristics are surrounded by an area that delineates the 90% confidence intervals. For water retention the 90% confidence area for the sand is smaller than that for the clay, indicating that the PTF estimate for the sand is more reliable (0.03–0.05 cm^3 cm^{-3} for the sand and 0.05–0.10 cm^3 cm^{-3} for the clay). For unsaturated hydraulic conductivity there is little difference between the confidence intervals for the clay and the sand; in both cases the confidence interval is approximately one to two orders of magnitude. We note that these confidence intervals can be generated without external data and are specific for each individual estimate (i.e., the confidence intervals change with the estimated parameters).

REFERENCES

Ahuja, L.R., Naney, J.W., Williams, R.D., 1985. Estimating soil water characteristics from simpler properties and limited data. Soil Sci. Soc. Am. J. 49, 1100-1105.

Brooks, R.H., Corey, A.T., 1964. Hydraulic Properties of Porous Media. Hydrol. Paper 3. Colorado State University, Fort Collins, CO.

Carsel, R.F., Parrish, R.S., 1988. Developing joint probability distributions of soil water retention characteristics. Water Resour. Res. 24, 755-769.

Efron, B., Tibshirani, R.J., 1993. An introduction to the bootstrap. Monographs on Statistics and Applied Probability. Chapman and Hall, New York, NY.

Epebinu, O., Nwadialo, B., 1993. Predicting soil water availability from texture and organic matter content for Nigerian soils. Commun. Soil Sci. Plant Anal. 24, 633-640.

Gupta, S.C., Larson, W.E., 1979. Estimating soil water characteristics from particle size distribution, organic matter percent, and bulk density. Water Resour. Res. 15, 1633–1635.

Imam, B., Sorooshian, S., Mayr, T., Schaap, M.G., Wösten, H., Scholes, B., 1999. Comparison of pedotransfer functions to compute water holding capacity using the van Genuchten model in inorganic soils. Report to the IGBP-DIS soils data tasks, IGBP-DIS working paper #22. IGBP-DIS office, CNRM, 42 avenue G. Coriolis, 31057 Toulouse Cedex, France.

Kern, J.S., 1995. Evaluation of soil water retention models based on basic soil physical properties. Soil Sci. Soc. Am. J. 59, 1134-1141.

Meyer, P.D., Gee, G.W., Rockhold, M.L., Schaap, M.G., 1999. Characterization of soil hydraulic parameter uncertainty. In: van Genuchten, M.Th., Leij, F.J., Wu, L. (Eds.), Proceedings of International Workshop, Characterization and Measurements of the Hydraulic Properties of Unsaturated Porous Media. University of California, Riverside, CA, pp. 1439-1451.

Minasny, B., McBratney, A.B., 2002. The neuro-m method for fitting neural network parametric pedotransfer functions. Soil Sci. Soc. Am. J. 66, 352-362.

Minasny, B., McBratney, A.B., Bristow, K.L., 1999. Comparison of different approaches to the development of pedotransfer functions of water retention curves. Geoderma 93, 225-253.

Press, W.H., Flannery, B.P., Teukolsky, S.A., Vetterling, W.T., 1988. Numerical recipes. The Art of Scientific Computing. Cambridge University Press, Cambridge.

Rawls, W.J., Brakensiek, D.L., 1985. Prediction of soil water properties for hydrologic modeling. In: Jones, E.B., Ward, T.J. (Eds.), Watershed management in the eighties. Proc. Irrig. Drain. Div., ASCE, Denver, CO. April 30–May 1, 1985, pp. 293-299.

Rawls, W.J., Brakensiek, D.L., Saxton, K.E., 1982. Estimation of soil water properties. Trans. ASAE 25, 1316-1320.

Rawls, W.J., Brakensiek, D.L., Soni, B., 1983. Agricultural management effects on soil water processes, part I: soil water retention and Green and Ampt infiltration parameters. Trans. ASAE 26, 1747-1752.

Saxton, K.E., Rawls, W.J., Romberger, J.S., Papendick, R.I., 1986. Estimating generalized soil-water characteristics from texture. Soil Sci. Soc. Am. J. 50, 1031-1036.

Schaap, M.G., Bouten, W., 1996. Modeling water retention curves of sandy soils using neural networks. Water Resour. Res. 32, 3033-3040.

Schaap, M.G., Leij, F.J., 1998. Database related accuracy and uncertainty of pedotransfer functions. Soil Sci. 163, 765-779.

Schaap, M.G., Leij, F.J., 2000. Improved prediction of unsaturated hydraulic conductivity with the Mualem–van Genuchten model. Soil Sci. Soc. Am. J. 64, 843-851.

Schaap, M.G., Leij, F.J., van Genuchten, M.T., 2001. ROSETTA: a computer program for estimating soil hydraulic parameters with hierarchical pedotransfer functions. J. Hydrol. 251, 163-176.

Scheinost, A.C., Sinowski, W., Auerswald, K., 1997. Regionalization of soil water retention curves in a highly variable landscape, I. Developing a new pedotransfer function. Geoderma 78, 129-143.

Shirazi, M.A., Boersma, L., Hart, J.W., 1988. A unifying quantitative analysis of soil texture: improvement of precision and extension of scale. Soil Sci. Soc. Am. J. 52, 181-190.

Tietje, O., Hennings, V., 1996. Accuracy of the saturated hydraulic conductivity prediction by pedo-transfer functions compared to the variability within FAO textural classes. Geoderma 69, 71-84.

Tietje, O., Tapkenhinrichs, M., 1993. Evaluation of pedotransfer functions. Soil Sci. Soc. Am. J. 57, 1088-1095.

Tomasella, J., Hodnett, M.G., 1997. Estimating unsaturated hydraulic conductivity of Brazilian soils using soil–water retention data. Soil Sci. 162, 703-712.

van Genuchten, M.Th., 1980. A closed-form equation for predicting the hydraulic conductivity of unsaturated soils. Soil Sci. Am. J. 44, 892-898.

Williams, R.D., Ahuja, L.R., Naney, J.W., 1992. Comparison of methods to estimate soil water characteristics from limited texture, bulk density, and limited data. Soil Sci. 153, 172-184.

Wösten, J.H.M., van Genuchten, M.Th., 1988. Using texture and other soil properties to predict the unsaturated soil hydraulic functions. Soil Sci. Soc. Am. J. 52, 1762-1770.

Wösten, J.H.M., Lilly, A., Nemes, A., Le Bas, C., 1999. Development and use of a database of hydraulic properties of European soils. Geoderma 90, 169-185.

PART II

SOIL HYDRAULIC PROPERTIES: WATER RETENTION AND HYDRAULIC CONDUCTIVITY

Chapter 4

SOIL TEXTURE AND PARTICLE-SIZE DISTRIBUTION AS INPUT TO ESTIMATE SOIL HYDRAULIC PROPERTIES

A. Nemes[*] and W.J. Rawls

USDA-ARS Hydrology & Remote Sensing Lab, Bldg. 007, Rm. 104, BARC-W, Beltsville, MD 20705-2350, USA

[*]Corresponding author: Tel.: +1-301-504-5707; fax: +1-301-504-8931

1. INTRODUCTION

Soil texture represents the relative proportion of soil particles with different sizes, which is a fundamental physical property of soils, correlated to just any other soil property. Soil hydraulic properties are difficult to determine, especially for large areas of land. Predictive models to estimate soil hydraulic properties – mostly termed pedotransfer functions (PTFs) – usually use soil texture and/or particle-size data as their most basic input. Most commonly, the particle-size distribution (PSD) is represented in a texture diagram based on the sand, silt and clay content of a soil. However, different standards exist and are in use to characterize and describe soil texture as well as soil PSD. In the following, we give an overview on how soil texture can be characterized and described, how those data are considered in different PTFs and what methods are known that can be used to fill in missing data required by some PTFs. We also show a study that compares different representations of soil PSD in estimating soil water retention.

2. PARTICLE-SIZE AND SOIL TEXTURE CLASS SYSTEMS

Textural classification of soils is based on particle-size analyses in most parts of the world. A number of standards exist internationally for the description of soil PSD and particle-size classes. Most national systems follow one of the internationally recognized standards. FAO (1990) and USDA (Soil Survey Staff, 1951) define clay as the particle-size fraction <2 μm, silt as the fraction between 2 and 50 μm and sand as the fraction between 50 and 2000 μm. Nemes et al. (1999), however, gave examples for countries where soils are classified in a different manner. In most of those cases, the silt/sand boundary is defined differently – at 20 μm – as adopted by the International Society of Soil Science (ISSS, 1929). That system was first proposed by Atterberg (1905) and is termed today as the "International system". Katschinski's system (Katschinski, 1956) – which was widely applied in the Central and Eastern regions of Europe – defines clay as the mass fraction of particles smaller than 1 μm and defines the upper cutoff limit for sand at 3000 μm. An example of frequently used cutoff limits in a number of countries is given in Figure 1.

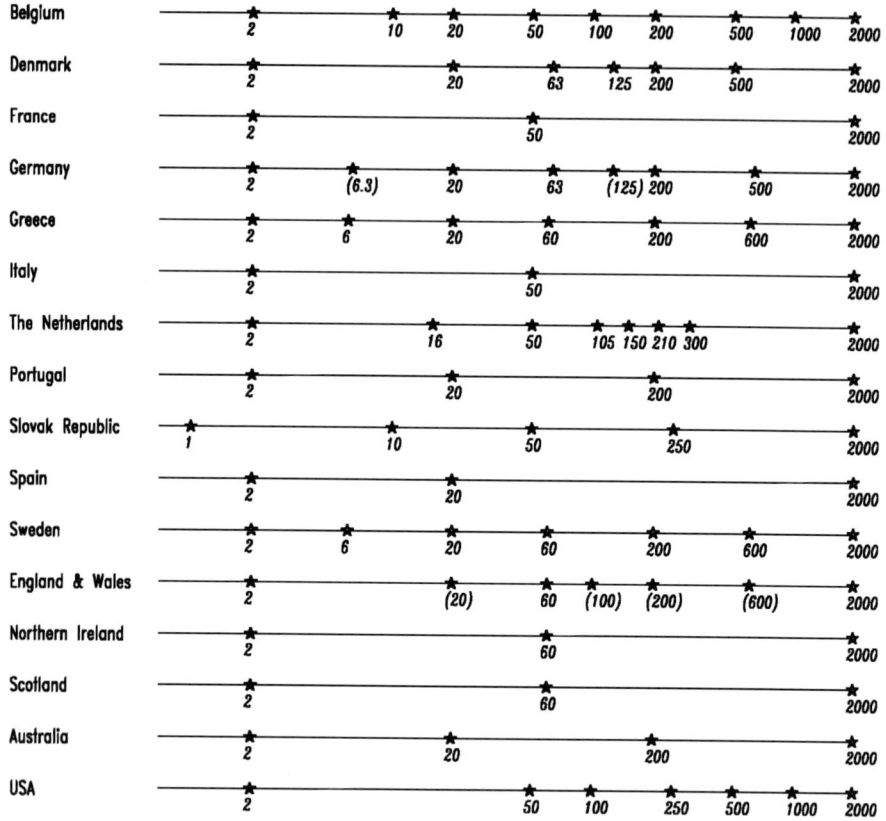

Figure 1. Particle-size limits (μm) used in some European countries, Australia and the US (after Nemes et al., 1999; Minasny and McBratney, 2001).

Filep and Ferencz (1999) summarized the most widely used particle-size categorization systems in soil research. There are different other standards that are observed primarily in geography and engineering (Soil Survey Staff, Soil Conservation Service, 1991; Vanoni, 1980).

Soil texture classes are defined based on sand, silt and clay content limits and are usually displayed in soil texture diagrams. Texture class systems that are based on different particle-size class systems are apparently different from each other. Even when the same particle-size class system is used, texture class definitions may still differ. An example for that is the difference between the USDA texture class system (Soil Survey Staff, 1951) and the system used by the FAO Georeferenced Soil Map of Europe (European Soil Bureau, 1998) (Figure 2). Examples for other textural diagrams and class definitions are those adapted in England, Germany, France and Australia (Hodgson, 1974; Arbeits Gemeinschaft Bodenkunde, 1982; Baize, 1993; McDonald et al., 1990).

Other soil textural systems are/were also in use that used different, but particle-size related soil information. An example is given by Filep and Ferencz (1999) for a

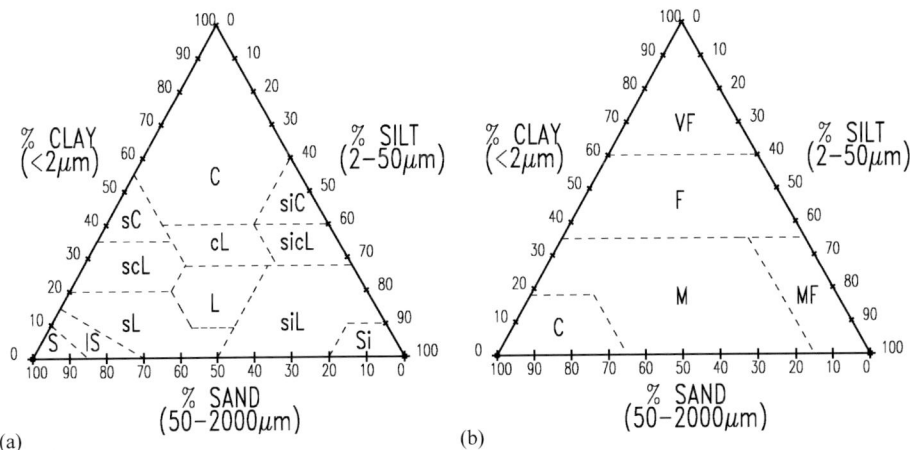

Figure 2. Textural classification systems according to: (a) USDA (Soil Survey Staff, 1951) and (b) as defined by the FAO Soil Map of Europe (European Soil Bureau, 1998). (USDA classes: S, sand; lS, loamy sand; sL, sandy loam; scL, sandy clay loam; sC, sandy clay; L, loam; cL, clay loam; C, clay; siL, silt loam; sicL, silty clay loam; siC, silty clay; Si, silt; FAO classes: C, coarse; M, medium; MF, medium fine; F, fine; VF, very fine).

system that is based on soil hygroscopic properties and the so-called "sticky point index" of the soil.

3. SOIL TEXTURE DATA IN PEDOTRANSFER FUNCTIONS

The use of soil mechanical/textural properties in predictive functions dates back to the early 20th century. Briggs and McLane (1907) were probably the first ones who attempted to make predictions of the "moisture equivalent" from the mechanical composition of soils. Later, Briggs and Shantz (1912) estimated the wilting coefficient as a function of the mechanical composition. Later examples of the use of textural components in predictive functions include the study of Nielsen and Shaw (1958), who presented a parabolic relationship between clay content and the permanent wilting point, and the study of Salter and Williams (1969) who described the relationships between soil texture classes and available water capacity and developed functions relating PSD and available water capacity.

3.1. The use of texture class information in pedotransfer functions

Soil texture class information is primarily used as input to "class" PTFs. A class PTF does not use detailed particle-size data as input, but gives calculated – usually averaged – values of the output variable in tabular format for each soil class/unit (Clapp and Hornberger, 1978; De Jong, 1982; Williams et al., 1983; Carsel and Parrish, 1988; Wösten et al., 2001). Hall et al. (1977) established relationships between textural classes and numerous soil–water properties, such as field capacity, permanent wilting point and available water content (AWC), for soils of England and Wales. Wösten et al. (1999)

developed class PTFs for top- and subsoils of 13 European countries using the texture classes recognized by FAO. Schaap et al. (1998, 2001) included in their system of PTFs a model that uses USDA soil texture class as the only input to estimate soil hydraulic properties. The user of such PTF will determine the texture of the soil according to the classification used by the PTF and will typically associate the soil with a response value looked up in a table.

Categorical type information – such as soil texture class – can also be used as input to regression tree models that can uncover structure in data. The analysis results in a multi-level grouping of data. The final fit of the model to data looks like a tree with a response value at each branch. The analysis provides a multi-level look-up table that is similar in nature to a class PTF. McKenzie and Jacquier (1997) presented a regression tree estimating saturated hydraulic conductivity (K_s) from field morphology classes – among other texture classes. Rawls and Pachepsky (2002) used this technique to estimate water retention from soil texture, structure and consistence data.

Another common use of soil texture class information is when it is used as a grouping factor prior to developing PTFs. The rationale of this approach is to define more homogeneous subgroups of soils for which PTFs developed separately may provide better accuracy. Pachepsky and Rawls (1999) used soil texture information to group soils prior to the development of PTFs, from other soil properties. Wösten et al. (2001) defined different sets of "continuous" PTFs for soils with coarse and with finer texture. Zhuang et al. (2001) used two different functional relationships – depending on soil texture – to obtain detailed PSD information, which then would be used to estimate soil water retention.

When class PTFs developed according to a certain classification system are to be used, often the original classification of the application data is according to a different system. If the original classification is based on the same particle-size classes as the system of the PTF, a simple conversion is needed, which can be done based on the texture diagram. This is the case, for example, when conversion is needed between, e.g., the FAO and the USDA classification systems, which both rely on the use of the 2, 50 and 2000 μm particle-size cutoff limits. The task is more complicated when the original and the new texture classification systems use different particle-size cutoff limits. This is the case, for instance, when conversion is required between the USDA/FAO system and the ISSS classification-based systems, latter of which uses 20 μm as silt/sand boundary. If measured data are not available that conforms both classification systems, some conversion/interpolation technique is required to obtain an estimate of the needed fraction. Examples for such techniques are described and discussed in a later section of the chapter.

3.2. The use of particle-size distribution data in pedotransfer functions

Detailed PSD and/or parameters derived from it are used in almost all PTF. As particle-size classes differ in different classification systems, so differs the number and the size of classes used in PTFs. Using sand, silt and clay contents in PTFs is the most common approach. Most internationally published PTFs adhere to the FAO/USDA system. Some authors, however, took advantage of the presence of finer textural classes in their data sets (Jamison and Kroth, 1958; MacLean and Yager, 1972; Pachepsky et al., 1982; Vereecken et al., 1990; Rajkai and Várallyay, 1992; Williams et al., 1992; Shein et al., 1999; Tomasella et al., 2003). Hartge (1969) suggested an empirical procedure to convert PSDs into water retention curves (WRCs) using several critical points on the distribution graphs and total porosity. Puckett et al. (1985) used fine sand as a separate PTF input along with the total sand. Proportions of particles in the 0.25–0.5 and 0.1–0.25 mm range controlled water release properties in horticultural growing media (Handreck, 1983).

Correlations between contents of textural components are often observed in regional databases (Nielsen and Shaw, 1958), and therefore a small number of parameters may be sufficient to characterize the texture for PTF development. To characterize the PSDs, the median diameter was found useful first in sandy media (Bedinger, 1961), and then in soils with a wide range of textures (Bloemen, 1980; Campbell, 1985). Minasny et al. (1999) and Scheinost et al. (1997) used the geometric mean diameter (GMD) along with its standard deviation to estimate soil water retention; Mishra et al. (1989) did the same to estimate K_s. Jarvis et al. (2002) estimated near-saturated hydraulic conductivity of soils using the GMD. Schaap and Bouten (1996) used a logistic equation to fit PSD functions and extracted the median diameter and a spread parameter from the results of fitting. Jonasson (1992) suggested using particle diameters at 25 and 75% quartiles of the PSD.

A certain group of water retention PTFs is based on the additivity hypothesis, i.e., composing soil water retention from water retention of soil constituents (Gupta et al., 1977; Arya and Paris, 1981; Haverkamp and Parlange, 1986; Ambroise et al., 1992; Serra-Witting et al., 1996; Zeiliguer et al., 2000). The hypothesis is that soil water retention can be approximated by summing up water retention of pore subspaces related to the soil components. The subspaces are defined from masses of soil textural components. The model of Arya and Paris (1981) is often referred to. They proposed a physico-empirical model that first translates PSD into pore-size distribution, which in turn is related to the distribution of water contents and associated pressure heads. Such methods also use information on the bulk density and particle density of the soil. They reportedly work well in sandy soils. In soils with a broader PSD and hierarchical structure, fine material hidden between large particles takes over the space assumed to be reserved for water retention of larger particles. Several authors reported lower prediction accuracy for loamy and clayey soils using modified versions of the Arya–Paris model (Haverkamp and Parlange, 1986; Tyler and Wheatcraft, 1989). Haverkamp and Parlange (1986) used an adopted version of van Genuchten's equation (van Genuchten, 1980) to fit the cumulative PSD. Smettem and Gregory (1996) fitted a sigmoidal function to the PSD and WRC of some predominantly sandy soils in Australia. They found good correlation between the slopes of the two curves but reported problems with the asymptotic minima of the two curves for soils with larger clay contents. A power-law distribution function for the PSD and the exponent of it as a predictor of water retention were also used by Paydar and Cresswell (1996). A number of other studies that relate PSD to soil hydraulic properties also applied their own techniques to obtain detailed PSD information. Zhuang et al. (2001) used two functional relationships depending on soil texture. They used the van Genuchten equation (van Genuchten, 1980) for sand and loamy sand soils and a logarithmic function for soils with other texture. Mishra et al. (1989), Wu et al. (1990) and Hwang and Powers (2003a) used a simple lognormal distribution function to obtain detailed PSD information. Other studies that present similar models include Campbell (1985); Schuh et al. (1988).

Tyler and Wheatcraft (1989) interpreted the α parameter of the Arya–Paris model (Arya and Paris, 1981) as the fractal dimension of a tortuous pore. Since then there has been considerable interest in using fractals to estimate hydraulic properties from PSD (Rieu and Sposito, 1991; Chang and Uehara, 1992; Tyler and Wheatcraft, 1992; Shepard, 1993; Kravchenko and Zhang, 1998). Arya et al. (1999a) refined the Arya–Paris model by introducing a variable α parameter. Arya et al. (1999b) derived a model that relates unsaturated hydraulic conductivity directly to the PSD assuming similarity in their shape. Unlike other models in literature, their model does not use measured (or estimated) WRC or K_s. Hwang and Powers (2003b) compared different models to describe PSD and examined their effect on the accuracy at which the models of Arya and Paris (1981) and Arya et al. (1999a,b) estimate water retention and hydraulic conductivity from the derived PSD. Four models – that use one to four parameters – were compared and the use of the one parameter model of Jáky (1944) was found to result in better $\theta(h)$ and $K(h)$ estimations than the other three models with a greater number of fitting parameters.

3.3. Pedotransfer functions based solely on texture or particle-size distribution information

An overview of all PTFs that used soil texture, PSD and/or derived information as input would virtually include the entire literature on soil hydraulic PTFs. However, there are several authors who used soil textural components as the only input parameters to their predictive equations (Husz, 1967; Rivers and Shipp, 1978; Cosby et al., 1984; Campbell, 1985; Saxton et al., 1986; Dane and Puckett, 1992; Campbell and Shiozawa, 1992; Smettem and Gregory, 1996). Tietje and Tapkenhinrichs (1993) tested various PTFs on their data. Three PTFs that used only particle-size data as input (Husz, 1967; Cosby et al., 1984; Saxton et al., 1986) yielded root-mean-squared residuals (RMSR) of 0.071, 0.067 and 0.079 $m^3 m^{-3}$ of volumetric water content, respectively, over the range of applicable pressure. Kern (1995) included the models of Cosby et al. (1984) and Saxton et al. (1986) in his comparisons and concluded that the model of Saxton et al. is adequate for estimating the water holding capacity of soils. Tomasella and Hodnett (1998) estimated soil water retention for tropical soils of Brazil using soil texture, and found an average accuracy of 0.062 $m^3 m^{-3}$ along the WRC. Smettem and Bristow (1999) used clay content to estimate K_s for Australian soils. Texture alone was reported to be a good predictor of K_s in sandy soils (Aronovici, 1946; Jaynes and Tyler, 1984; El-Kadi, 1985). Clay content was a leading texture parameter to correlate with K_s in soil databases including other than sandy soils (Puckett et al., 1985). It is good to note that most likely soil structure would be a very important input – along with soil texture – to estimate properties like K_s. However, soil structure is very difficult to quantify.

There are studies in which the authors opted to develop PTFs using different levels of input complexity using the same methodology and the same data sets. Schaap et al. (1998) named this approach the "hierarchical approach" to develop PTFs. They used texture class information as input to the most simple model (TXT) and sand, silt and clay content was used at the next level (model SSC). They also grouped soils by texture classes. They reported slight improvement from the TXT to the SSC models for water retention estimates (RMSR = 0.107 and 0.104 $m^3 m^{-3}$, respectively) and for K_s estimates (RMSR = 0.627 and 0.602 log(cm/d), respectively). They also recalculated RMSRs by texture classes. Using the TXT model RMSR was between 0.095 and 0.123 $m^3 m^{-3}$; using the SSC model they found RMSR between 0.092 and 0.124 $m^3 m^{-3}$. They also tested the model of Cosby et al. (1984), which also uses PSD as its sole input and obtained RMSR between 0.090 and 0.134 $m^3 m^{-3}$. For the estimates of K_s, they obtained RMSR between 0.510 and 0.724 log(cm/d) using the TXT model, 0.433 and 0.711 log(cm/d) using the SSC model and 0.545 and 0.980 log(cm/d) using the model of Cosby et al. (1984). Schaap et al. (2001) presented the ROSETTA program that provides estimates of soil hydraulic properties in a similar (hierarchical) approach. Using detailed particle-size data instead of soil texture class information yielded slight but consistent improvement in water retention estimation (RMSR = 0.076 vs. 0.078 m^{3}m^{-3}), in the estimation of K_s (RMSR = 0.717 vs. 0.739 log(cm/d)) and in the estimation of unsaturated hydraulic conductivity (RMSR = 1.020 vs. 1.060 log(cm/d)). To obtain estimates of unsaturated hydraulic conductivity they first estimated water retention parameters from the basic soil information which were then used in the Mualem–van Genuchten model (van Genuchten, 1980) with the assumptions of $K_0 = K_s$ and $L = 0.5$. Minasny and McBratney (2002) used only particle-size information in one of their neural network models to test different approaches of PTF development. They also tested the ROSETTA program of Schaap et al. (2001) against independent data sets. Depending on the testing data sets and on the models, they obtained RMSRs between 0.050 and 0.085 m^{3}m^{-3}. Nemes et al. (2003) used solely particle-size information in one of their models, using different databases to develop PTFs. Depending on the data set used for PTF development, they obtained RMSR between 0.058 and 0.087 m^{3}m^{-3}.

The accuracy of different PTFs obtained from different sources of literature may show considerable dependence on the size and origin of the development and validation data sets, on the method used to develop the PTFs and on the actually estimated parameters/properties (e.g., point vs. parameter estimations, c.f. Pachepsky et al. (1996)). Therefore, direct comparisons of such PTFs may not show the true difference caused by different inputs.

Comparisons between different representations of the PSD in PTFs are rare. Most authors use PSD as the first and most basic input parameter to their PTF, thus comparisons are mostly made to evaluate the usefulness of additional input parameters. McKenzie et al. (1991) found substantial improvement in their PTF to estimate available water capacity when silt was defined in the range from 2 to 60 μm instead of 2–20 μm. The studies of Schaap et al. (1998, 2001) described above provide comparisons between models that use texture class or sand, silt and clay content as input.

4. INTERPOLATIONS TO FILL IN MISSING PARTICLE-SIZE DATA

A common problem with the application of PTFs is when data for the application area does not match the classification, type or detail of data needed to apply a particular PTF. The range and detail of available measured PSD data greatly depend on the capability and feasibility of the applied measurement technique(s) and, of course, on the needs of a particular study. To use PSD, it has been found desirable to have functions suitable to approximate the whole distribution or to approximate the PSD within a large diameter range, or at a certain point. This is needed mostly when the PSD is characterized with small number of fractions (Zeiliguer et al., 2000), or when data come from different sources with different fraction diameter ranges (Nemes et al., 1999). Several studies suggest that PSD in soils show an approximately lognormal distribution (Campbell, 1985; Shirazi and Boersma, 1984). However, soils with bimodal PSD also exist (Walker and Chittleborough, 1986). Buchan (1989) described the applicability of lognormal models of PSD and found that these are only applicable for soil in some of the USDA soil texture classes (Soil Survey Staff, 1975). The same is true for using other parameterizations employing two parameters (Dapples, 1975; Terzagi and Peck, 1967; Shirazi et al., 1988). Buchan (1989) also discussed the effects of the number of particle-size fractions that are measured on the shape of the cumulative PSD curve. The more complex the cumulative distribution is, the greater the number of required model parameters is. Rousseva (1987) applied two different techniques (graph and polynomial fit) to transform PSD from Katschinski's texture scheme (Katschinski, 1956) to the scheme used by the United States Department of Agriculture (USDA) (Soil Survey Staff, 1975). She concluded that polynomial fits do not convert soil texture data adequately and that use of graphs is better, even though it is time- and labor-consuming and subjective. Rousseva (1997) defined closed-form models of exponential and power law. She investigated the suitability of these models to fit cumulative PSD of different shapes and with varying numbers of measured points. Suitability of the models appeared to be influenced by texture type (coarse or fine textured soils) rather than by measured size ranges. Shirazi et al. (1988) established connections between texture classifications adopted by the USDA (Soil Survey Staff, 1975), the International Society of Soil Science (ISSS, 1929) and the American Society of Civil Engineers (Vanoni, 1980). This work was based on a description of the clay, silt and sand fractions by the geometric mean and the geometric standard deviation of their size ranges. To allow such calculations for the clay content, the lower limit of the particle-size and a corresponding percent mass have to be specified. Shirazi et al. (1988) used 0.05 μm as lower size limit with 0.01% as the corresponding cumulative mass percent. Buchan et al. (1993) compared five different models that were derived from the lognormal distribution function to describe soil PSD. All five models accounted for more than 90% of the variance in the PSD of most of the examined soils. However, the algorithm did not converge for about 10% of the soils in their study. The logistic curve was successfully fitted to PSDs in several studies (Haverkamp and Parlange, 1986; Schaap and Bouten, 1996; Rajkai and Várallyay, 1992). Rajkai et al. (1996) applied logistic distribution functions to parameterize the PSD of Swedish soils, which was considered successful for 88% of the dataset. They also used a parameter estimation technique to obtain distribution parameters directly from measured particle-size data, without invoking a particular distribution function. Several studies fitted power-law functions with a single fractal dimension to experimental PSD data (Matsushita, 1985; Turcotte, 1986; Tyler and Wheatcraft, 1992; Medina et al., 2002). Some authors, however, reported that a single fractal dimension is not sufficient to describe PSD in soils (Wu et al., 1993; Grout et al., 1998; Bittelli et al., 1999). Kozak et al. (1996) evaluated the applicability of the fractal power-law distribution and found it applicable only in 20% of cases in their database. Wu et al. (1993) and Bittelli et al. (1999) identified three domains with three power exponents. Bittelli et al. (1999) associated the power exponent in each domain with fractal dimensions defining scaling in the clay, silt and sand domains. Posadas et al. (2001) showed that multifractal scaling provides additional information to characterize soil PSD. They found that models based on a single fractal are limited to PSDs with clay contents of $<10\%$. Baumer et al. (1994) presented typical PSDs for soil of all textural classes and various clay contents within those classes. Nemes et al. (1999) compared four methods to interpolate PSD. They concluded that – depending on the number and the position of the measured points – either fitting a non-parametric spline, or applying the so-called "similarity procedure" – a nearest neighbor type technique – may offer the best solution. The procedure does not rely on mathematical interpolation but involves finding similar PSD curves in a sufficiently large external data set. Minasny et al. (1999) and Minasny and McBratney (2001) developed empirical models to convert the 2–20–2000 μm fraction scheme, primarily to the 2–50–2000 μm scheme to enable the testing of existing PTFs on

Australian soil data. Skaggs et al. (2001) suggested and tested a generalized logistic model to estimate PSD from only clay, silt and fine plus very fine sand contents of the soil. The success of the method highly varies by texture classes, which correspond with the findings of others (Rousseva, 1997). Hwang et al. (2002) also found the performance of PSD models to be texture dependent when they tested seven different models. They found the model of Fredlund et al. (2000) to perform best for 1387 soils from Korea. Shirazi et al. (2001) concluded that unifying the particle-size description into geometric mean and the geometric standard deviation of the particle-size offers a common language of soil texture research that is independent of classification systems. In practice, the log-linear interpolation has often been used to estimate missing particle-size classes for the FAO/USDA texture classification, but that method was shown to be unreliable (Nemes et al., 1999).

Till now, there seems to be no universally good technique to obtain equally accurate estimated/interpolated particle-size data for all soils. All techniques seem to have bottlenecks or seem to be valid only for a more or less well-defined range of soil textures. Often, the authors specify for which USDA texture classes a particular interpolation technique works well. In practice, however, in many cases texture class is exactly the information that we do not know, as we are seeking to interpolate the value, which can help determine the texture class of the soil. The combinations of the number and distribution of points on the measured PSD seem to be endless, which makes giving a general recommendation for the use of particular interpolation technique(s) difficult. It may be useful to first decide which approaches not to use rather than which approach to use.

5. EVALUATION OF DIFFERENT REPRESENTATIONS OF PARTICLE-SIZE DISTRIBUTION

In the following study, we intend to compare the usefulness of some of the most common representations of PSD in estimating water retention at -10, -33 and -1500 kPa and the AWC of the soil, which is defined as the difference between water contents at -33 and -1500 kPa matric potentials. To relate the above soil hydraulic properties to soil texture we developed a number of PTFs using data of three databases and group method of data handling (GMDH). We simulated the cutoff limit between silt and sand content being at 20, 50 and at 63 μm – as used in various countries. We also used the GMD and its standard deviation as predictors.

5.1. Soil data

The HYPRES (Wösten et al., 1999), UNSODA (Nemes et al., 2001) and HUNSODA (Nemes, 2002) databases were searched for soils for which PSD was characterized by at least six determined fractions and water retention by at least four $\theta(h)$ data pairs determined in the laboratory. From the selected soils only those were retained that had either the sequence of 2, 20 and 63 μm (used primarily in Germany) or of 2, 20 and 50 μm determined (used in many FAO/USDA conform systems). Data were filtered for obvious inconsistency in physical and hydraulic information.

The spline interpolation technique of Nemes et al. (1999) was used to interpolate the individual PSD curves. This technique uses the fit of a sixth degree non-parametric smoothing spline to estimate the cumulative distributions at a certain point. For those soils, for which the 2, 20 and 63 μm sequence was determined, we estimated the cumulative distribution at 50 μm. For the other group, we estimated the distribution at 63 μm. GMD and its standard deviation were calculated according to Shirazi and Boersma (1984). These measures were first calculated using all measured particle-size data available for each soil. Then, we calculated three alternatives to that, using only 2–20–2000, 2–50–2000 and 2–63–2000 μm data, thus only three fractions. Such fractions were used as these represent three of the internationally known systems to define the silt/sand boundary between fractions. Measured or estimated sand/silt data were used as applicable for each of the data sets.

Data in the different data sets were obtained from different sources and had different numbers and positions of points at the WRC. To obtain uniform description of all the WRCs, the volumetric soil water content, θ, as a function of matric potential, h, was described with the van Genuchten equation (van Genuchten, 1980):

$$\theta(h) = \theta_r + \frac{\theta_s - \theta_r}{[1 + (\alpha h)^n]^m} \tag{1}$$

where subscripts "r" and "s" refer to residual and saturated values, and α, n and m are curve shape parameters, where $m = 1 - 1/n$. Parameters of the above equation were fitted to the individual WRCs using the Simplex method (Nelder and Mead, 1965). Water contents at three different matric potentials (-10, -33, -1500 kPa) were then estimated for each soil using the fitted parameters. Two final data sets were formulated from the two separate groups of soils – i.e., for which PSD at 50 or 63 μm was estimated. There were 588 soils for which the 63 μm value was estimated (later referred to as Data Set 1). A randomized subset selection was performed a number of times on the other group to obtain a 588 size subset of that group (Data Set 2), which did not significantly differ from

Table 1
Summary statistics of selected soil properties of the two training and two test data sets

	Soil particles (%)				Water content (m³ m⁻³)		
	$p_{<2\mu m}$	$p_{<20\mu m}$	$p_{<50\mu m}$	$p_{<63\mu m}$	-10 kPa	-33 kPa	-1500 kPa
			[a]Data Set 1: 63 μm estimated				
MIN	0	0	1.96	2.68	0.042	0.023	0.008
MAX	69.60	96.00	100.00	100.00	0.725	0.605	0.420
MEAN	22.75	45.89	63.37	67.66	0.362	0.307	0.173
STD. DEV.	15.95	26.58	31.60	31.14	0.106	0.120	0.091
MEDIAN	19.90	48.05	75.50	80.93	0.380	0.329	0.180
			[b]Data Set 2: 50 μm estimated				
MIN	0	0	0	0	0.020	0.010	0.010
MAX	79.30	98.40	99.51	99.90	0.803	0.744	0.518
MEAN	23.85	42.50	61.07	64.99	0.345	0.298	0.186
STD. DEV.	18.96	28.03	32.73	33.38	0.139	0.145	0.118
MEDIAN	18.70	38.95	72.54	78.75	0.344	0.298	0.167

[a]RMSR of the fitted van Genuchten model: 0.0159 m³m⁻³; average number of points per WRC: 12.75.
[b]RMSR of the fitted van Genuchten model: 0.0121 m³ m⁻³; average number of points per WRC: 7.47.

the first set of 588 soils in any of the following properties: cumulative PSD at 2, 20, 50, 63 μm and water retention at matric potentials of − 10, − 33, − 1500 kPa as obtained from the van Genuchten parameters. Table 1 shows the summary statistics of selected soil properties of the two data sets used in this study.

5.2. Methods

We used GMDH (Farrow, 1984) to describe the relationships between input and output variables (see page 24). The method performs an automated selection of essential input variables and builds hierarchical polynomial regressions of necessary/desired complexity to estimate the output variable. Polynomials are built from some of the input variables and from sub-polynomials that are built from smaller subsets of input variables that may be better predictors of the output variable than some of the input variables alone. The best of such polynomials are included in the set of input variables that may again serve as input to obtain even better estimates. Examples for the application of this technique to estimate soil hydraulic properties can be found in, e.g., Pachepsky et al. (1998), Pachepsky and Rawls (1999) and Tomasella et al. (2003). For this application we used the commercial GMDH software ModelQuest (AbTech Corp., 1996). Values of non-problem specific variables were set to the default value in the software including a "cost-complexity factor" that will prevent from building models with excessive complexity. The jackknife cross-validation method (Good, 1999) was applied in this work to evaluate the performance of the developed PTFs. The data sets were randomly split into development/training and validation/testing subsets within the GMDH software using the default (i.e., 3:1) ratio in ModelQuest. The random splitting and subsequent model development and validation was performed four times, and the average accuracy of estimation was expressed by root-mean squared residuals (RMSR), which is defined as:

$$\text{RMSR} = \sqrt{(1/N) \sum_{i=1}^{N} (\theta_i - \hat{\theta}_i)^2} \qquad (2)$$

where N is the number of estimated and measured values, θ and $\hat{\theta}$ are measured and estimated water contents, respectively. RMSRs are presented in each case for the subsets used for model validation. As experimental water retention data at − 10, − 33 and − 1500 kPa matric potentials were not available for some of the soil samples, we used − for all soils − $\theta(h)$ values that were derived from the fitted van Genuchten model. This, however, means, that we introduced a certain amount of error to our subsequent calculations and model evaluations, which is a result of the non-perfect fit of the fitted model to the experimental data. Direct fitting of the van Genuchten model to the individual WRCs yielded averaged RMSR (calculated through the entire WRCs) of 0.0159 m^{3}m^{-3} for Data Set 1 and 0.0121 m^{3}m^{-3} for Data Set 2. As these errors are relatively small compared to the errors one usually obtains using PTFs, we neglect the effect of using such approximated data in the present analysis.

5.3. Results

We estimated water retention at − 10, − 33 and − 1500 kPa and the AWC of the soil − which is defined as the difference between water contents at − 33 and − 1500 kPa matric potentials − using solely soil PSD data as input. Some of the common representations of

Table 2
Results of the analysis by group method of data handling in terms of root-mean-squared residuals (RMSRs) ($m^3 \, m^{-3}$)

Input	Test data: 25% of Data Set 1				Test data: 25% of Data Set 2		
	−10 kPa, MEAN (STD)	−33 kPa MEAN (STD)	−1500 kPa MEAN (STD)	AWC MEAN (STD)	−33 kPa MEAN (STD)	−1500 kPa MEAN (STD)	AWC MEAN (STD)
	Block 1				Block 2		
	Training data: 75% of Data Set 1						
Group 1							
$p_{<2}$	0.074 (0.004)[a]	0.075 (0.004)[a]	0.050 (0.004)[a]	0.065 (0.005)[a]	0.098 (0.002)[a,c,d]	0.087 (0.003)[a]	0.061 (0.002)[a]
p_{2-20}	0.062 (0.003)[b]	0.063 (0.004)[b]	0.058 (0.004)[a]	0.062 (0.004)[a]	0.108 (0.002)[b,d]	0.102 (0.002)[a,b]	0.067 (0.005)[a]
p_{2-50}	0.061 (0.002)[b]	0.064 (0.003)[b]	0.062 (0.003)[b]	0.059 (0.004)[a]	0.117 (0.004)[b,e]	0.117 (0.003)[b,c,d]	0.065 (0.004)[a]
p_{2-63}	0.062 (0.002)[b]	0.065 (0.004)[b]	0.061 (0.002)[b]	0.059 (0.005)[a]	0.118 (0.002)[b]	0.119 (0.002)[b,c,d]	0.063 (0.003)[a]
Group 2							
$p_{<2}; p_{2-20}$	0.058 (0.002)[b]	0.057 (0.003)[a,c]	0.043 (0.003)[a,c]	0.059 (0.006)[a]	0.095 (0.003)[a,c]	0.084 (0.005)[a]	0.063 (0.004)[a]
$p_{<2}; p_{2-50}$	0.057 (0.003)[b]	0.056 (0.002)[b,c]	0.046 (0.003)[a,c]	0.057 (0.006)[a]	0.101 (0.002)[a,c,d]	0.091 (0.002)[a]	0.069 (0.005)[a]
$p_{<2}; p_{2-63}$	0.058 (0.002)[b]	0.058 (0.003)[b,c]	0.049 (0.003)[a,c]	0.056 (0.003)[a]	0.099 (0.001)[a,c,d]	0.090 (0.002)[a]	0.064 (0.002)[a]
Group 3							
GMD(all)	0.060 (0.006)[b]	0.062 (0.007)[b,d]	0.044 (0.003)[a,c]	0.057 (0.005)[a]	0.115 (0.013)[b,e,f]	0.102 (0.011)[a,c]	0.069 (0.008)[a]
GMD(20)	0.059 (0.003)[b]	0.059 (0.004)[b,c]	0.043 (0.003)[a,c]	0.057 (0.004)[a]	0.097 (0.002)[a,c,d]	0.084 (0.002)[a]	0.063 (0.004)[a]
GMD(50)	0.061 (0.003)[b]	0.057 (0.003)[b,c]	0.046 (0.003)[a,c]	0.062 (0.009)[a]	0.105 (0.006)[a,d,e]	0.092 (0.002)[a]	0.069 (0.009)[a]
GMD(63)	0.060 (0.002)[b]	0.057 (0.003)[b,c]	0.048 (0.003)[a,c]	0.057 (0.004)[a]	0.104 (0.004)[a,c,d,f]	0.094 (0.002)[a]	0.066 (0.003)[a]
Group 4							
GMD(all)$p_{<2}$	0.055 (0.005)[b]	0.056 (0.003)[b,c]	0.044 (0.003)[a,c]	0.054 (0.003)[a]	0.093 (0.002)[c]	0.085 (0.002)[a]	0.063 (0.003)[a]
GMD(all); p_{2-20}	0.058 (0.004)[b]	0.058 (0.004)[b,c]	0.044 (0.003)[a,c]	0.061 (0.012)[a]	0.098 (0.002)[a,c,d]	0.085 (0.003)[a]	0.064 (0.003)[a]
GMD(all); p_{2-50}	0.054 (0.006)[b]	0.053 (0.003)[c]	0.041 (0.002)[c]	0.052 (0.004)[a]	0.098 (0.003)[a,c,d]	0.087 (0.004)[a]	0.061 (0.003)[a]
GMD(all); p_{2-63}	0.052 (0.006)[b]	0.054 (0.002)[c,d]	0.042 (0.003)[c]	0.051 (0.004)[a]	0.099 (0.005)[a,c,d]	0.101 (0.025)[a,d]	0.060 (0.004)[a]
	Block 3				Block 4		
	Training data: 75% of Data Set 2						
Group 1							
$p_{<2}$	0.083 (0.004)[a,b]	0.083 (0.004)[a,b]	0.059 (0.004)[a]	0.070 (0.008)[a]	0.089 (0.002)[a]	0.080 (0.002)[a]	0.059 (0.004)[a]
p_{2-20}	0.077 (0.009)[a,c]	0.077 (0.007)[b]	0.074 (0.011)[a,c]	0.065 (0.006)[a]	0.101 (0.001)[b]	0.097 (0.001)[b]	0.056 (0.003)[a]
p_{2-50}	0.071 (0.004)[a]	0.073 (0.002)[b]	0.071 (0.004)[a,c]	0.062 (0.006)[a]	0.110 (0.003)[b]	0.109 (0.003)[c]	0.056 (0.002)[a]
p_{2-63}	0.068 (0.004)[a]	0.070 (0.004)[b]	0.068 (0.004)[a,c]	0.062 (0.005)[a]	0.111 (0.002)[b]	0.111 (0.002)[c]	0.055 (0.002)[a]

Group 2							
$p_{<2}; p_{2-20}$	0.073 (0.006)^a	0.070 (0.004)^b	0.063 (0.009)^{a,c}	0.076 (0.016)^a	0.087 (0.004)^a	0.077 (0.003)^a	0.055 (0.002)^a	0.063 (0.005)^a
$p_{<2}; p_{2-50}$	0.077 (0.009)^{a,c}	0.072 (0.003)^b	0.059 (0.004)^a	0.071 (0.002)^a	0.087 (0.003)^a	0.078 (0.002)^a	0.056 (0.002)^a	0.065 (0.010)^a
$p_{<2}; p_{2-63}$	0.075 (0.006)^a	0.072 (0.004)^b	0.059 (0.004)^a	0.067 (0.006)^a	0.087 (0.003)^a	0.078 (0.002)^a	0.056 (0.002)^a	0.059 (0.007)^a
Group 3								
GMD(all)	0.115 (0.016)^{b,d}	0.096 (0.019)^{a,b}	0.084 (0.013)^{b,c}	0.096 (0.012)^{a,b}	0.086 (0.004)^a	0.077 (0.002)^a	0.055 (0.003)^a	0.063 (0.011)^a
GMD(20)	0.074 (0.003)^a	0.072 (0.007)^b	0.054 (0.003)^a	0.074 (0.006)^a	0.087 (0.005)^a	0.078 (0.004)^a	0.054 (0.002)^a	0.061 (0.005)^a
GMD(50)	0.080 (0.004)^{a,c,d}	0.079 (0.002)^b	0.059 (0.005)^a	0.075 (0.012)^a	0.088 (0.003)^a	0.082 (0.005)^a	0.056 (0.002)^a	0.069 (0.015)^a
GMD(63)	0.080 (0.003)^{a,c,d}	0.078 (0.005)^b	0.061 (0.005)^a	0.064 (0.004)^a	0.087 (0.002)^a	0.080 (0.001)^a	0.056 (0.001)^a	0.059 (0.006)^a
Group 4								
GMD(all); $p_{<2}$	0.115 (0.016)^{b,c,d}	0.098 (0.010)^{a,b}	0.062 (0.005)^a	0.117 (0.013)^{a,b}	0.084 (0.004)^a	0.077 (0.002)^a	0.055 (0.002)^a	0.063 (0.009)^a
GMD(all); p_{2-20}	0.125 (0.010)^b	0.134 (0.036)^a	0.096 (0.019)^b	0.116 (0.011)^{a,b}	0.088 (0.006)^a	0.084 (0.006)^a	0.057 (0.004)^a	0.062 (0.008)^a
GMD(all); p_{2-50}	0.197 (0.030)^e	0.168 (0.048)^c	0.0159 (0.009)^d	0.149 (0.048)^{b,c}	0.090 (0.005)^a	0.077 (0.003)^a	0.054 (0.003)^a	0.065 (0.012)^a
GMD(all); p_{2-63}	0.151 (0.040)^b	0.158 (0.044)^c	0.0163 (0.009)^d	0.177 (0.062)^c	0.091 (0.006)^a	0.079 (0.002)^a	0.054 (0.003)^a	0.065 (0.009)^a

GMD: geometric mean diameter (and its standard deviation). Use of the same index letter within the same block and for the same estimated property means that differences are not significant at a 95% confidence level.

soil PSD were used along with additional alternatives to test their relevance. In total, 15 models were developed for each data set and each output variable. Results of the tests are shown in Table 2. The same models were developed using the development part (75%) of both data sets. All models were tested using data allocated for model validation (25%) from each data set separately. In the upper two blocks of Table 2, results are shown for the models developed from Data Set 1, and in the lower two blocks for the models developed from Data Set 2. Results using the testing subset of Data Set 1 are shown in the two blocks to the left, and those using the subset of Data Set 2 are to the right. Significance tests to evaluate differences between averaged RMSR values were performed within each block and for each estimated property separately. Significance was examined at 95% confidence level.

We grouped the 15 models into four groups. The first four models constitute Group 1. These models use only one particle fraction to estimate the soil hydraulic properties. Clay content ($p_{<2}$) is one of the input variables. We used three different definitions of the silt content as input, as defined in different systems (i.e., p_{2-20}; p_{2-50}; p_{2-63}). The second group consists of three models, which use clay content plus silt content, using one of the three definitions of silt/sand, as described above. The following four models – Group 3 – use GMD and its standard deviation as their inputs as calculated from all available particle-size data (GMD(all)), and using only 2–20–2000 μm (GMD(20)), 2–50–2000 μm (GMD(50)) and 2–63–2000 μm (GMD(63)) data. The last four models – Group 4 – use GMD and its standard deviation as calculated from all available PSD data, plus one additional particle fraction, clay or one of the definitions of silt used in this study. Differences between the accuracy of different models are often very small. Some trends can, however, be identified from the data.

Group 1. When tested on Data Set 1, regardless of which training data set was used, clay content was the worst predictor as a sole particle fraction in all cases except for water retention at −1500 kPa. Using Data Set 2 for testing, particle fractions between 2–50 and 2–63 μm seem to be the worst sole predictors for all three water retention points, although the difference from the accuracy obtained using 2–20 μm data is not significant in most cases. For this data set, clay content is the best predictor of all examined output variables, except for AWC in Block 4. Clay content was the best predictor of water retention at −1500 kPa in all cases, which corresponds to the common knowledge. It is interesting to note that the order of the goodness of predictors and the magnitude of RMSR clearly shows dependence on the testing data set but not on the training data set. While using the 2–50 and 2–63 μm fractions, either or both of the training and testing data sets included data that were not measured but interpolated. Usage of interpolated data was not a clear disadvantage. An example for the interpolated data being the best predictor can be seen, e.g., in Block 1, for AWC. Other examples exist in all four blocks of results that show interpolated data not being the worst predictor.

Group 2. Adding silt content to the clay content of soils resulted, in general, in a slight improvement in the estimation of all of the output variables. There are some cases, however, when the test results became worse than that of the model that uses only one of the inputs separately (as in Group 1). In case of Block 4, it may be the result of using interpolated data (AWC, $p_2 + p_{2-50}$). In case of Blocks 2 and 3, it may be the separate or combined result of cross-validation – using one of the data sets for training and the remainder of the other one for testing – and the use of interpolated data. There is no case, however, when test results in Group 2 would be significantly worse than in Group 1 for the same estimated property in the same block. RMSRs were in most cases larger for test Data Set 2, regardless of which training data set was used. Similarly to Group 1, results that involved the use of interpolated data either in the training or in the test data sets could not be clearly distinguished as the worst results in any of the blocks. In all cases, estimation errors of models in Group 2 were not significantly different from each other, for the same estimated property and within the same block.

Group 3. GMD can be calculated from different numbers of points, distributed differently on the PSD curve. Four of such possibilities are represented in Group 3. Test results are somewhat better in Blocks 1 and 4, although it is not true for all cases. It suggests that this representation of PSD works better for soils that originate from the same distribution as the training data set. This approach seems to be more sensitive to the training data set than the approach of Group 2 models. Similarly to the previous two groups, there is no indication that using interpolated data would hinder the application of these types of models. GMD calculated using all measured PSD data did not result in better estimates than when it was calculated from only three PSD points. This corresponds to the findings of Scheinost et al. (1997), i.e., that using more than three fractions to calculate GMD did not result in better water retention estimates than using only three PSD points. Using GMD from all data, however, resulted in unexpectedly high RMSRs in all four cases in Block 3 and for -10 and -33 kPa in Block 2. Such differences were statistically significant in some cases. This may be due to the different influence of those PSD points in Data Sets 1 and 2, which were not used in the other three GMD calculations and neither in the data pre-selections. No such phenomenon is seen in the other two blocks that used soils from the same distribution (Data Set) to train and test the models. Accuracy of Group 3 models that used three data points to calculate GMD did not significantly differ from the accuracy of Group 2 models, that also use three-point representation of PSD.

Group 4. The effect of using the two kinds of information (as in Groups 2 and 3) together is mixed. Results in Block 1 show improvements from the models that use the inputs separately in almost all cases. The same applies to Block 2 – that also used Data Set 1 for training. In Block 4 improvements are only seen for half of the cases, whereas in Block 3 almost all models gave results that are considerably worse than those of the models that use the inputs separately. The trend seen in Block 3 – that models that use interpolated data give the worst results – can not be generalized for the other three blocks. The generally weak performance of models in Block 3 can most logically be explained by the differences in PSD characteristics between Data Sets 1 and 2, which is emphasized by the combined use of two different kinds of indicators/measures. Using multiple indicators for PSD does not seem to deliver unambiguously better results. Even when estimation accuracy is improved in any of the blocks, the improvement is statistically not significant.

It may happen that the wrong silt/sand boundary is used in the PSD data when using a PTF. This could typically happen when one does not have enough information about the classification system that was used or in attempt to avoid the interpolation of the PSDs. We tested what results we may obtain if the wrong input data are given to a pedotransfer model. As an example, we show models that estimate the four selected soil hydraulic properties, using clay content along with silt content according to the FAO/USDA system (2–50 μm). The models were then tested using both test data sets and using all three definitions of the silt content (separately), as defined earlier, i.e., silt = 2–20 μm, silt = 2–50 μm and silt = 2–63 μm. Results are summarized in Table 3. In all four blocks, the second line ($p_{2-50-2000}$ as input) represents when the correct particle size data are used.

Table 3
Results, in terms of root-mean-squared residuals ($m^3 m^{-3}$), of the simulation of submitting errant particle-size data to the pedotransfer model

Input used to test model	Test data: 25% of Data Set 1				Test data: 25% of Data Set 1		
	−10 kPa MEAN (STD)	−33 kPa MEAN (STD)	−1500 kPa MEAN (STD)	AWC MEAN (STD)	−33 kPa MEAN (STD)	−1500 kPa MEAN (STD)	AWC MEAN (STD)
	Block 1				Block 2		
	Training data: 75% of Data Set 1						
$P_{2-20-2000}$	0.064 (0.005)[a]	0.066 (0.004)[a]	0.048 (0.004)[a]	0.061 (0.004)[a]	0.092 (0.002)[a]	0.084 (0.004)[a]	0.064 (0.002)[a]	0.071 (0.006)[a]
$P_{2-50-2000}$	0.057 (0.003)[a]	0.056 (0.002)[b]	0.046 (0.003)[a]	0.057 (0.006)[a]	0.101 (0.002)[a,b]	0.091 (0.002)[b]	0.062 (0.003)[a]	0.069 (0.005)[a]
$P_{2-63-2000}$	0.061 (0.004)[a]	0.061 (0.005)[a,b]	0.049 (0.004)[a]	0.059 (0.004)[a]	0.112 (0.010)[b]	0.096 (0.001)[c]	0.066 (0.002)[a]	0.074 (0.002)[a]
	Block 3				Block 4		
	Training data: 75% of Data Set 2						
$P_{2-20-2000}$	0.091 (0.008)[a]	0.085 (0.004)[a]	0.059 (0.004)[a]	0.099 (0.023)[a]	0.096 (0.004)[a]	0.086 (0.003)[a]	0.056 (0.002)[a]	0.093 (0.037)[a]
$P_{2-50-2000}$	0.077 (0.009)[a,b]	0.072 (0.003)[b]	0.059 (0.004)[a]	0.071 (0.002)[b]	0.087 (0.003)[b]	0.078 (0.002)[b]	0.056 (0.002)[a]	0.066 (0.010)[a]
$P_{2-63-2000}$	0.074 (0.007)[b]	0.071 (0.003)[b]	0.059 (0.004)[a]	0.066 (0.004)[b]	0.087 (0.003)[b]	0.077 (0.002)[b]	0.056 (0.002)[a]	0.062 (0.003)[a]

The trained model used clay content and silt content according to FAO/USDA (2–50 μm) as input to estimate the four selected soil hydraulic properties. Models are tested using three different representations of the silt/sand boundary along with clay content. Use of the same index letter within the same block and for the same estimated property means that differences are not significant at a 95% confidence level.

Using a test data set that is from the same distribution as the development data set (i.e., Blocks 1 and 4) seems to yield worse results when either of the incorrect particle fraction data is used. An exception from this is the estimation of AWC in Block 4, where the $p_{2-63-2000}$ data representation provided slightly smaller average errors. For soils that come from another distribution than the training data (i.e., Blocks 2 and 3) one of the other representations of silt provided slightly but consistently better estimations for most properties than the $p_{2-50-2000}$ data. In Block 2 it is always the $p_{2-20-2000}$ data, and in Block 3 it is always the $p_{2-63-2000}$ data. Note, that in Block 2, models were trained on data from Data Set 1, and tested on data from Data Set 2. It is the opposite way for Block 3. The improved estimation obtained using errant particle-size data clearly seems to be the result of the differences between the characteristics of the data of the two data sets. Particle-size information of Data Set 2, by the numbers, resembles the particle-size information $p_{2-50-2000}$ of Data Set 1 in its $p_{2-20-2000}$ rather than its $p_{2-50-2000}$ representation. The other way around it happens similarly with $p_{2-63-2000}$, the bias pointing the opposite way. This, however, was not apparent from the raw data of the two data sets (c.f. Table 1). Seemingly, one could obtain more reliable estimations using different particle-size data than how it is required by the pedotransfer model. However, it is very difficult to pre-determine and quantify the effect of using different data. In our example, we did not find a clear message in the original data that would have suggested whether the bias is towards one way or the other. All four blocks of Table 3, however, show examples where using the wrong data yielded a significant loss in estimation accuracy, meaning that the user risks losing accuracy by using the wrong data. Altogether, from the 16 comparisons (4 estimated properties by 4 blocks), 7 cases were found, where at least one of the errant representations of the silt/sand boundary in the PSD data yielded statistically significant loss in estimation accuracy. In two cases (Blocks 3 and 4, − 1500 kPa), such comparison could not be made, because the resulting optimized model used only clay content as input, thus any silt/sand representations remained ineffective.

6. SUMMARY

Overall, we could not point at one particular representation of the PSD that would clearly provide better results in estimating the selected soil hydraulic properties. Results are extremely heterogeneous. Although results in general were not considerably worse, instances for largest RMSRs were found while water content at − 10 and − 33 kPa were estimated. This can partly be explained by the wider range of water contents at these matric potentials. Another reason is that soil hydraulic properties at these matric potentials are largely affected by soil structure, which is not accounted for in any of our models. The accuracy of models using the representation of PSD by distinct particle-size classes (as in Group 2) seem to be more dependent on the test/application data set, rather than on the data set used to train the models. The use of models using GMD and its standard deviation seems to be safer for soils that originate from the same distribution as those used to train the models. The amount of soil particles in the 2–50 and 2–63 μm ranges do not seem to be as good predictors in Data Set 2 as the amount of particles in the 2–20 μm range. However, differences suggesting that are not statistically significant, such finding is not supported by Data Set 1, and thus cannot be generalized. Using GMD and its standard deviation to represent the entire PSD curve with two parameters (Group 3 models) did not result in significantly better estimates of the examined soil hydraulic properties, than models of Group 2. One possible reason for not finding significant differences between estimates of many of the models is that parameters of the different representations of the PSD are closely correlated. In practical terms, it means, that when PSD is represented differently in two models, the resulting models will be different, but the predictability of the hydraulic properties will not be significantly changed. Some differences in the estimations were apparently caused by the slightly different characteristics of the two data sets, despite of the efforts to establish two similar sets. We found no evidence suggesting that using interpolated data would reduce the accuracy of the estimation of these soil hydraulic properties. Misuse of the representation of the silt/sand boundary in a PTF was tested. It is not straightforward to pre-determine if using the wrong definition of the silt/sand boundary in the PSD data will result in significant loss of estimation accuracy; therefore, using such data should be avoided. Our study suggests that using interpolated data – which of course carries a certain magnitude of interpolation error – poses fewer risks in a PTF than using measured data with the wrong silt/sand boundary.

REFERENCES

AbTech Corp., 1996. ModelQuest. Users Manual. Version 4.0. Charlottesville, VA.

Ambroise, B., Reutenauer, D., Viville, D., 1992. Estimating soil water retention properties from mineral and organic fractions of coarse-textured soils in the Vosges Mountains of France. *In*: van Genuchten, M.Th., Leij, F.J., Lund, L. (Eds.), Proceedings of International Workshop on Indirect Methods for Estimating the Hydraulic Properties of Unsaturated Soils. University of California, Riverside, CA, pp. 453-462.

ArbeitsGemeinschaft Bodenkunde, 1982. Kartieranleitung Anleitung und Richtlinien zur Herstellung der Bodenkarte 1:25000. Hannover, Germany.

Aronovici, V.C., 1946. The mechanical analysis as an index of subsoil permeability. Soil Sci. Soc. Am. Proc. 11, 137-141.

Arya, L.M., Paris, J.F., 1981. A physicoempirical model to predict soil moisture characteristics from particle-size distribution and bulk density data. Soil Sci. Soc. Am. J. 45, 1023-1030.

Arya, L.M., Leij, F.J., van Genuchten, M.Th., Shouse, P.J., 1999a. Scaling parameter to predict the soil water characteristic from particle-size distribution data. Soil Sci. Soc. Am. J. 63, 510-519.

Arya, L.M., Leij, F.J., Shouse, P.J., van Genuchten, M.Th., 1999b. Relationship between the hydraulic conductivity function and the particle-size distribution. Soil Sci. Soc. Am. J. 63, 1063-1070.

Atterberg, A., 1905. Die rationale klassifikation der sande und kiese. Chem. Zeitung 29, 195-198.

Baize, D., 1993. Soil Science Analyses. Wiley, New York (in French).

Baumer, O., Kenyon, P., Bettis, J., 1994. MUUF v. 2.14 User's Manual. USDA-ARS, Grassland Soil & Water Research Lab, Tempe, TX.

Bedinger, M.S., 1961. Relationship between median grain size and permeability in the Arkansas River Valley, Arkansas. Professional Paper 424-C, US Department of the Interior, Geological Survey, Washington, DC.

Bittelli, M., Campbell, G.S., Flury, M., 1999. Characterization of particle-size distribution in soils with a fragmentation model. Soil Sci. Soc. Am. J. 63, 782-788.

Bloemen, G.W., 1980. Calculation of hydraulic conductivities of soils from texture and organic matter content. Z. Pflanzenernaehr. Bodenkd. 143, 581-615.

Briggs, L.J., McLane, J.W., 1907. The moisture equivalents of soils. USDA. Bur. Soils Bull., 45.

Briggs, L.J., Shantz, H.L., 1912. The wilting coefficient for different plants and its indirect determination. USDA. Bur. Plant Ind. Bull., 230.

Buchan, G.D., 1989. Applicability of the simple lognormal model to particle-size distribution in soils. Soil Sci. 147 (3), 155-161.

Buchan, G.D., Grewal, K.S., Robson, A.B., 1993. Improved models of particle-size distribution: an illustration of model comparison techniques. Soil Sci. Soc. Am. J. 57, 901-908.

Campbell, G.S., 1985. Soil Physics with BASIC: Transport Models for Soil-Plant System. Elsevier, New York, 150pp.

Campbell, G.S., Shiozawa, S., 1992. Prediction of hydraulic properties of soils using particle-size distribution and bulk density data. *In*: van Genuchten, M.Th., Leij, F.J., Lund, L. (Eds.), Proceedings of International Workshop on Indirect Methods for Estimating the Hydraulic Properties of Unsaturated Soils. University of California, Riverside, CA, pp. 317-328.

Carsel, R.F., Parrish, R.S., 1988. Developing joint probability distributions of soil water retention characteristics. Water Resour. Res. 24, 755-769.

Chang, H., Uehara, G., 1992. Application of fractal geometry to estimate soil hydraulic properties from the particle-size distribution. *In*: van Genuchten, M.Th., Leij, F.J., Lund, L. (Eds.), Proceedings of International Workshop on Indirect Methods for Estimating the Hydraulic Properties of Unsaturated Soils. University of California, Riverside, CA, pp. 125-138.

Clapp, R.B., Hornberger, G.M., 1978. Empirical equations for some soil hydraulic properties. Water Resour. Res. 14, 601-604.

Cosby, B.J., Hornberger, G.M., Clapp, R.B., Ginn, T.R., 1984. A statistical exploration of the relationships of soil moisture characteristics to the physical properties of soils. Water Resour. Res. 20, 682-690.

Dane, J.H., Puckett, W., 1992. Field soil hydraulic properties based on physical and mineralogical information. *In*: van Genuchten, M.Th., Leij, F.J., Lund, L. (Eds.), Proceedings of International Workshop on Indirect Methods for Estimating the Hydraulic Properties of Unsaturated Soils. University of California, Riverside, CA, pp. 389-403.

Dapples, E.C., 1975. Laws of distribution applied to sand sizes. Geol. Soc. Am. Mem. 142, 37-61.

De Jong, R., 1982. Assessment of empirical parameters that describe soil water characteristics. Can. Agric. Engng 24, 65-70.

El-Kadi, A.I., 1985. On estimating the hydraulic properties of soil. 2: a new empirical equation for estimating hydraulic conductivity for sands. Adv. Water Resour. 8, 148-153.

European Soil Bureau, 1998. Georeferenced Soil Database for Europe. European Soil Bureau and Joint Research Centre, EC. Ispra, Italy.

FAO, 1990. Guidelines for Soil Descriptions, Third edition. FAO/ISRIC, Rome, Italy.

Farrow, S.J., 1984. The GMDH algorithm. *In*: Farrow, S.J. (Ed.), Self-Organizing Methods in Modelling: GMDH Type Algorithms. Marcel Dekker, New York, pp. 1-24.

Filep, Gy., Ferencz, G., 1999. Javaslat a magyarországi talajok szemcseösszetétel szerinti osztályozásának pontosítására. (Recommendation for improving the accuracy of soil classification on the basis of particle composition). Agrokémia és Talajtan 48, 305-320, in Hungarian with English Abstract.

Fredlund, M.D., Fredlund, D.G., Wilson, G.W., 2000. An equation to represent grain size distribution. Can. Geotechnol. J. 37, 817-827.

Good, P.I., 1999. Resampling Methods: A Practical Guide to Data Analysis. Birkhäuser, Boston.

Grout, H., Tarquis, A.M., Wiesner, M.R., 1998. Multifractal analysis of particle size distributions in soil. Environ. Sci. Technol. 32, 1176-1182.

Gupta, S.C., Dowdy, R.H., Larson, W.E., 1977. Hydraulic and thermal properties of a sandy soil as influenced by incorporation of sewage sludge. Soil Sci. Soc. Am. J. 41, 601-605.

Hall, D.G., Reeve, M.J., Tomasson, A.J., Wright, V.F., 1977. Water retention, porosity and density of field soils. Technical Monograph No. 9. Soil Survey of England and Wales, Harpenden.

Handreck, K.A., 1983. Particle size and the physical properties of growing media for containers. Commun. Soil Sci. Plant Anal. 14, 209-222.

Hartge, K.H., 1969. Die ermittlung der wasserspannungskurve aus der könungssummenkurve und dem gesamtporenvolmen. Z. Kulturtech. Flurbereinig. 10 (1), 20-29.

Haverkamp, R., Parlange, J.-Y., 1986. Predicting the water-retention curve from particle-size distribution: I. Sandy soils without organic matter. Soil Sci. 142, 325-339.

Hodgson, J.M. (Ed.), 1974. Soil Survey Field Handbook. Soil Survey of England and Wales. Technical Monograph No. 5. Harpenden, UK.

Husz, G., 1967. The determination of pF-curves from texture using multiple regressions. Z. Pflanzenernaehr. Bodenkd. 116, 23-29.

Hwang, S.I., Powers, S.E., 2003a. Lognormal distribution model for estimating soil water retention curves for sandy soils. Soil Sci. 168, 156-166.

Hwang, S.I., Powers, S.E., 2003b. Using particle-size distribution models to estimate soil hydraulic properties. Soil Sci. Soc. Am. J. 67, 1103-1112.

Hwang, S.I., Kwang, P.L., Dong, S.L., Powers, S.E., 2002. Models for estimating soil particle-size distributions. Soil Sci. Soc. Am. J. 66, 1143-1150.

ISSS (International Society of Soil Science), 1929. Minutes of the first Commission Meetings, International Congress of Soil Science, Washington, 1927. Proc. Cong. Int. Soc. Soil Sci. 4, 215-220.

Jáky, J., 1944. Soil Mechanics. Egyetemi Nyomda, Budapest (in Hungarian).

Jamison, V.C., Kroth, E.M., 1958. Available moisture storage capacity in relation to texture composition and organic matter content of several Missouri soils. Soil Sci. Soc. Am. Proc. 22, 189-192.

Jarvis, N.J., Zavattaro, L., Rajkai, K., Reynolds, W.D., Olsen, P.-A., McGechan, M., Mecke, M., Mohanty, B.-P., Leeds-Harrison, P.B., Jacques, D., 2002. Indirect estimation of near-saturated hydraulic conductivity from readily available soil information. Geoderma 108, 1-17.

Jaynes, D.B., Tyler, E.J., 1984. Using soil physical properties to estimate hydraulic conductivity. Soil Sci. 138, 298-305.

Jonasson, S.A., 1992. Estimation of van Genuchten parameters from grain-size distribution. *In*: van Genuchten, M.Th., Leij, F.J., Lund, L. (Eds.), Proceedings of International Workshop on Indirect Methods for Estimating the Hydraulic Properties of Unsaturated Soils. University of California, Riverside, CA, pp. 443-451.

Katschinski, N.A., 1956. Die mechanische bodenanalyse und die klassifikation der böden nach ihrer mechanischen zusammensetzung. Rapports au Sixiéme Congrés de la Science du Sol. Paris, B, pp. 321–327.

Kern, J.S., 1995. Evaluation of soil water retention models based on basic soil physical properties. Soil Sci. Soc. Am. J. 59, 1134-1141.

Kozak, E., Pachepsky, Ya.A., Sokolowski, S., Sokolowska, Z., Stepniewski, W., 1996. A modified number-based method for estimating fragmentation dimensions of soils. Soil Sci. Soc. Am. J. 60, 1291-1297.

Kravchenko, A., Zhang, R., 1998. Estimating the soil water retention from particle-size distributions: a fractal approach. Soil Sci. 163, 171-179.

MacLean, A.H., Yager, T.U., 1972. Available water capacities of Zambian soils in relation to pressure plate measurements and particle size analysis. Soil Sci. 113, 23-29.

Matsushita, M., 1985. Fractal viewpoint of structure and accretion. J. Phys. Soc. Jpn 54, 857-860.

McDonald, R.C., Isbell, R.F., Speight, J.G., Walker, J., Hopkins, M.S., 1990. Australian soil and land survey field handbook, Second edition. Inkata Press, Melbourne.

McKenzie, N.J., Jacquier, D.W., 1997. Improving the field estimation of saturated hydraulic conductivity in soil survey. Aust. J. Soil Res. 35, 803-825.

McKenzie, N.J., Smettem, K.R.J., Ringrose-Voase, A.J., 1991. Evaluation of methods for inferring air and water properties of soils from field morphology. Aust. J. Soil Res. 29, 587-602.

Medina, H., Tarawally, M., del Valle, A., Ruiz, M.E., 2002. Estimating soil water retention curve in rhodic ferralsols from basic soil data. Geoderma 108, 277-285.

Minasny, B., McBratney, A.B., 2001. The Australian soil texture boomerang: a comparison of the Australian and USDA/FAO soil particle-size classification systems. Aust. J. Soil Res. 39, 1443-1451.

Minasny, B., McBratney, A.B., 2002. The Neuro-m method for fitting neural network parametric pedotransfer functions. Soil Sci. Soc. Am. J. 66, 352-361.

Minasny, B., McBratney, A.B., Bristow, K.L., 1999. Comparison of different approaches to the development of pedotransfer functions for water-retention curves. Geoderma 93, 225-253.

Mishra, S., Parker, J.C., Singhal, N., 1989. Estimation of soil hydraulic properties and their uncertainty from particle size distribution data. J. Hydrol. 108, 1-18.

Nelder, J., Mead, R., 1965. A simplex method for function minimization. Comput. J. 7, 308-313.

Nemes, A., 2002. Unsaturated soil hydraulic database of Hungary: HUNSODA. Agrokémia Talajtan 51, 17-26.

Nemes, A., Wösten, J.H.M., Lilly, A., Oude Voshaar, J.H., 1999. Evaluation of different procedures to interpolate particle-size distributions to achieve compatibility within soil databases. Geoderma 90, 187-202.

Nemes, A., Schaap, M.G., Leij, F.J., Wösten, J.H.M., 2001. Description of the unsaturated soil hydraulic database UNSODA version 2.0. J. Hydrol. 251, 151-162.

Nemes, A., Schaap, M.G., Wösten, J.H.M., 2003. Functional evaluation of pedotransfer functions derived from different scales of data collection. Soil Sci. Soc. Am. J. 67, 1093-1102.

Nielsen, D.R., Shaw, R.H., 1958. Estimation of the 15-atmosphere moisture percentage from hydrometer data. Soil Sci. 86, 103-105.

Pachepsky, Ya.A., Rawls, W.J., 1999. Accuracy and reliability of pedotransfer functions as affected by grouping soils. Soil Sci. Soc. Am. J. 63, 1748-1757.

Pachepsky, Ya.A., Shcherbakov, R.A., Várallyay, Gy., Rajkai, K., 1982. Soil water retention as related to other soil physical properties. Pochvovedenie 2, 42-52 (in Russian).

Pachepsky, Ya.A., Timlin, D., Várallyay, Gy., 1996. Artificial neural networks to estimate soil water retention from easily measurable data. Soil Sci. Soc. Am. J. 60, 727-773.

Pachepsky, Ya.A., Rawls, W.J., Giménez, D., Watt, J.P.C., 1998. Use of soil penetration resistance and group method of data handling to improve soil water retention estimates. Soil Tillage Res. 49, 117-126.

Paydar, Z., Cresswell, H.P., 1996. Water retention in Australian soils. II. Prediction using particle size, bulk density and other properties. Aust. J. Soil Res. 34, 679-693.

Posadas, A.N.D., Giménez, D., Bittelli, M., Vaz, C.M.P., Flury, M., 2001. Multifractal characterization of soil particle-size distributions. Soil Sci. Soc. Am. J. 65, 1361-1367.

Puckett, W.E., Dane, J.H., Hajek, B.F., 1985. Physical and mineralogical data to determine soil hydraulic properties. Soil Sci. Soc. Am. J. 49, 831-836.

Rajkai, K., Várallyay, G., 1992. Estimating soil water retention from simpler properties by regression techniques. *In*: van Genuchten, M.Th., Leij, F.J., Lund, L. (Eds.), Proceedings of International Workshop on Indirect Methods for Estimating the Hydraulic Properties of Unsaturated Soils. University of California, Riverside, CA, pp. 417-426.

Rajkai, K., Kabos, S., van Genuchten, M.Th., Jansson, P.-E., 1996. Estimation of water-retention characteristics from the bulk density and particle-size distribution of Swedish soils. Soil Sci. 161, 832-845.

Rawls, W.J., Pachepsky, Ya.A., 2002. Soil consistence and structure as predictors of water retention. Soil Sci. Soc. Am. J. 66, 1115-1126.

Rieu, M., Sposito, G., 1991. Fractal fragmentation, soil porosity, and water properties: I. Theory. Soil Sci. Soc. Am. J. 55, 1231-1238.

Rivers, E.D., Shipp, R.F., 1978. Soil water retention as related to particle-size in selected sandy and loamy soils. Soil Sci. 126, 94-100.

Rousseva, S.S., 1987. Studies on the soil erodibility of calcaric chernozems and cinnamonic forest soils. Ph.D. Thesis, N. Poushkarov Institute of Soil Science and Agrogeology, Academy of Agricultural Sciences, Sofia, Bulgaria (in Bulgarian).

Rousseva, S.S., 1997. Data transformations between soil texture schemes. Eur. J. Soil Sci. 48, 749-758.

Salter, P.J., Williams, J.B., 1969. The influence of texture on the moisture characteristics of soil. V. Relationships between particle-size composition and moisture contents at the upper and lower limits of available-water. J. Soil Sci. 20, 126-131.

Saxton, K.E., Rawls, W.J., Romberger, J.S., Papendick, R.I., 1986. Estimating generalized soil-water characteristics from texture. Soil Sci. Soc. Am. J. 50, 1031-1036.

Schaap, M.G., Bouten, W., 1996. Modeling water retention curves of sandy soils using neural networks. Water Resour. Res. 32, 3033-3040.

Schaap, M.G., Leij, F.J., van Genuchten, M.Th., 1998. Neural network analysis for hierarchical prediction of soil hydraulic properties. Soil Sci. Soc. Am. J. 62, 847-855.

Schaap, M.G., Leij, F.J., van Genuchten, M.Th., 2001. ROSETTA: a computer program for estimating soil hydraulic parameters with hierarchical pedotransfer functions. J. Hydrol. 251, 163-176.

Scheinost, A.C., Sinowski, W., Auerswald, K., 1997. Reginalization of soil water retention curves in highly variable soilscape: I. Developing a new pedotransfer function. Geoderma 78, 129-143.

Schuh, W.M., Cline, R.L., Sweeney, M.D., 1988. Comparison of a laboratory procedure and a textural model for predicting *in situ* soil water retention. Soil Sci. Soc. Am. J. 52, 1218-1227.

Serra-Whitting, C., Houot, S., Barriuso, E., 1996. Modification of soil water retention and biological properties by municipal solid waste disposal. Compost Sci. Util. 4, 44-52.

Shein, E.V., Dembovetsky, A.V., Guber, A.K., 1999. Pedotransfer functions: determination, substantiation, and use. Eurasian Soil Sci. 11, 1323-1331.

Shepard, J.C., 1993. Using a fractal model to compute the hydraulic conductivity function. Soil Sci. Soc. Am. J. 57, 300-306.

Shirazi, M.A., Boersma, L., 1984. A unifying quantitative analysis of soil texture. Soil. Sci. Soc. Am. J. 48, 142-147.

Shirazi, M.A., Boersma, L., Hart, J.W., 1988. A unifying quantitative analysis of soil texture: improvement of precision and extension of scale. Soil Sci. Soc Am. J. 52, 181-190.

Shirazi, M.A., Boersma, L., Johnson, C.B., 2001. Particle-size distributions: comparing texture systems, adding rock, and predicting soil properties. Soil Sci. Soc. Am. J. 65, 300-310.

Skaggs, T.H., Arya, L.M., Shouse, P.J., Mohanty, B.P., 2001. Estimating particle-size distribution from limited soil texture data. Soil Sci. Soc. Am. J. 65, 1038-1044.

Smettem, K.R.J., Bristow, K.L., 1999. Obtaining soil hydraulic properties for water balance and leaching models from survey data. 2. Hydraulic conductivity. Aust. J. Agric. Res. 50, 1259-1262.

Smettem, K.R.J., Gregory, P.J., 1996. The relation between soil water retention and particle-size distribution parameters for some predominantly sandy Western Australian Soils. Aust. J. Soil Res. 34, 695-708.

Soil Survey Staff, 1951. Soil Survey Manual, USDA Handbook No. 18. US Government Printing Office, Washington, DC.

Soil Survey Staff, 1975. Soil Taxonomy: A Basic System of Soil Classification for Making and Interpreting Soil Surveys, USDA/SCS Agricultural Handbook No. 436. U.S. Government Printing Office, Washington, DC.

Soil Survey Staff, Soil Conservation Service, 1991. State Soil Geographic Data Base (STATSGO), Data User Guide, Misc. Publ. 1492. U.S. Government Printing Office, Washington, DC.

Terzagi, K., Peck, R.B., 1967. Soil Mechanics in Engineering Practice. Wiley, New York.

Tietje, O., Tapkenhinrichs, M., 1993. Evaluation of pedo-transfer functions. Soil Sci. Soc. Am. J. 57, 1088-1095.

Tomasella, J., Hodnett, M.G., 1998. Estimating soil water retention characteristics from limited data in Brazilian Amazonia. Soil Sci. 163, 190-202.

Tomasella, J., Pachepsky, Ya.A., Crestana, S., Rawls, W.J., 2003. Comparison of two techniques to develop pedotransfer functions for water retention. Soil Sci. Soc. Am. J. 67, 1085-1092.

Turcotte, D.L., 1986. Fractals and fragmentation. J. Geophys. Res. 91, 1921-1926.

Tyler, S.W., Wheatcraft, S.W., 1989. Application of fractal mathematics to soil water retention estimation. Soil Sci. Soc. Am. J. 53, 987-996.

Tyler, S.W., Wheatcraft, S.W., 1992. Fractal aspects of soil porosity. *In*: van Genuchten, M.Th., Leij, F.J., Lund, L. (Eds.), Proceedings of International Workshop on Indirect Methods for Estimating the Hydraulic Properties of Unsaturated Soils. University of California, Riverside, CA, pp. 53-63.

van Genuchten, M.Th., 1980. A closed-form equation for predicting the hydraulic conductivity of unsaturated soils. Soil Sci. Soc. Am. J. 44, 892-898.

Vanoni, V.A., 1980. Sedimentation Engineering. Sedimentation Committee, Hydraulic Division, American Society of Civil Engineers, New York.

Vereecken, H., Maes, J., Feyen, J., 1990. Estimating unsaturated hydraulic conductivity from easily measured soil properties. Soil Sci. 149, 1-12.

Walker, P.H., Chittleborough, D.J., 1986. Development of particle-size distributions in some Alfisols of southeastern Australia. Soil. Sci. Soc. Am. J. 50, 394-400.

Williams, J.R., Prebble, E., Williams, W.T., Hignett, C.T., 1983. The influence of texture, structure and clay mineralogy on the soil moisture characteristic. Aust. J. Soil Res. 21, 15-31.

Williams, J., Ross, P., Bristow, K., 1992. Prediction of the Campbell water retention function from texture, structure, and organic matter. *In*: van Genuchten, M.Th., Leij, F.J., Lund, L. (Eds.), Proceedings of International Workshop on Indirect Methods for Estimating the Hydraulic Properties of Unsaturated Soils. University of California, Riverside, CA, pp. 427-442.

Wösten, J.H.M., Lilly, A., Nemes, A., Le Bas, C., 1999. Development and use of a database of hydraulic properties of European soils. Geoderma 90, 169-185.

Wösten, J.H.M., Veerman, G.J., de Groot, W.J.M., Stolte, J., 2001. Waterretentie – en doorlatendheidskarakteristieken van boven – en ondergronden in Nederland: de Staringreeks. (Water retention and hydraulic conductivity characteristics of top- and subsoils of the Netherlands: the Staring-series). ALTERRA Report Nr. 153. ALTERRA. Wageningen, The Netherlands (in Dutch).

Wu, L., Vomocil, J.A., Childs, S.W., 1990. Pore size, particle size, aggregate size, and water retention. Soil Sci. Soc. Am. J. 54, 952-956.

Wu, Q., Borkovec, M., Sticher, H., 1993. On particle-size distribution in soils. Soil Sci. Soc. Am. J. 57, 883-890.

Zeiliguer, A.M., Pachepsky, Ya.A., Rawls, W.J., 2000. Estimating water retention of sandy soils using the additivity hypothesis. Soil Sci. 165, 373-383.

Zhuang, J., Jin, Y., Miyazaki, T., 2001. Estimating water retention characteristic from soil particle-size distribution using a non-similar media concept. Soil Sci. 166, 308-321.

Chapter 5

SIMPLE PARAMETRIC METHODS TO ESTIMATE SOIL WATER RETENTION AND HYDRAULIC CONDUCTIVITY

D.J. Timlin[1,*], R.D. Williams[1], L.R. Ahuja[2] and G.C. Heathman[3]

[1]USDA-ARS Alternate Crops and Systems Laboratory, Bldg 001, Rm 342 BARC-W, 10300 Baltimore Ave., Beltsville, MD 20705, USA

[2]USDA-ARSL, Great Plains Systems Research Laboratory, 2150 Centre Ave., Bldg. D, Suite 200, Fort Collins, CO 80256, USA

[3]National Soil Erosion Research Laboratory, 275 S. Russell Street, West Lafayette, IN 47907, USA

*Corresponding author: Tel.: +301-504-6255; fax: +301-504-5823

1. INTRODUCTION

Information on the hydraulic properties of soils is required by a wide range of disciplines, from agricultural sciences to ecology. These hydraulic properties include soil water retention, i.e., the soil water content–soil matric potential relationship, and the soil water conductivity–soil matric potential relationship that is often determined from the former (van Genuchten and Nielsen, 1985). These dependencies have application in predicting infiltration, drainage and available water for plant growth, estimating other soil properties such as unsaturated hydraulic diffusivity, quantifying management effects on soil properties, and watershed modeling.

Laboratory and field procedures used to determine the soil water retention curve are tedious and time consuming. In addition, the highly variable nature of soil hydraulic properties in spatial domains means that large numbers of samples have to be collected to properly characterize field or watershed level systems. Furthermore, published information for soils around the world may have data on organic matter content, bulk density and texture, but the hydraulic properties data may be incomplete or missing. For these reasons, a great deal of research has been devoted to development of approaches to estimate the soil water characteristic either from simpler soil properties, and/or limited data. A summary of pedotransfer functions and their current status is given in Pachepsky et al. (1999). Early approaches used by Rawls et al. (1982) and Rawls and Brakensiek (1982) among others were based on regression equations relating soil texture, organic matter, and bulk density to water retention at different matric potentials ranging from -4 to -1500 kPa. These equations were developed from a national database of 2543 horizons of 500 soils from 18 states in the U.S. The form of the equations came from work by Gupta and Larson (1979). Investigators at that time had also shown that the parameters of soil water characteristic curves may be estimated from similar regressions using limited experimental data by assuming certain functional forms for these curves (McQueen and Miller, 1974; de Jong, 1983; Saxton et al., 1986; Wösten and van Genuchten, 1988).

The pedotransfer functions developed from large, national level databases and methods used to calculate water content as a function of specific matric potentials (Rawls et al., 1982) have been shown to give reasonable results (Timlin et al., 1996a,b; Ahuja et al., 1989) for a number of different applications. Ahuja et al. (1989) concluded that the regression-based method had too much error to characterize a spatially variable soil at a small watershed scale. When measured soil water content at two matric potentials was used with the regression method, however, the errors were greatly reduced. This led to the development of a scaling method that used a measured value of soil water content at one matric potential and a complete reference soil water release curve as an alternative to regression-based functions to generate an estimated soil water release curve.

2. ESTIMATING SOIL WATER CONTENTS AND SOIL WATER RETENTION

2.1. A scaling method to estimate soil water retention curves

The dependence of volumetric water content, θ, on pressure head, h, for any specified site and depth within a watershed can be constructed with reasonable accuracy from a knowledge of soil bulk density (hence porosity) and either -33 or -10 kPa soil water content (Ahuja et al., 1985; Williams et al., 1992; Williams and Ahuja, 1992), using the extended similar-media scaling theory (Miller and Miller, 1956; Warrick et al., 1977; Simmons et al., 1979). Matric potential for a fixed degree of saturation (S) at site i, $h_i(S)$ is related to a scaled mean matric potential value for a soil or a small watershed, $h_m(S)$ as:

$$\alpha_i h_i(S) = h_m(S) \tag{1}$$

where α_i is a scaling factor for site i that applies to any arbitrary value of S.

Assume that below the air-entry matric potential, the $h_m(S)$ can be expressed by a power-form equation (Campbell, 1974):

$$h_m(S) = AS^{-M} \tag{2}$$

with A and M constants. Substituting Equation (2) in (1) and rearranging yields:

$$S(h_i) = \left(\frac{\alpha_i h_i}{A}\right)^{-\frac{1}{M}} \tag{3}$$

Equation (3) can describe the variation in S at different sites at a fixed value of h when the scaling factor for that site, α_i, is known:

$$S_i(h) = \left(\frac{\alpha_i h}{A}\right)^{-\frac{1}{M}} \tag{4}$$

Note that Equation (4) describes $\theta(h)$ pairs from different locations, each described by a unique value of α_i and a specific value of h.

Equation (4) has two unknown constants and a set of α_i values that need to be determined. Suppose that for one given value of h (e.g., -33 kPa) the values of S_i at all different sites are known, and the mean of these S_i values is $\bar{S}$. At the site where S_i is closest to $\bar{S}$, the complete $S(h)$ function provides an adequate approximation of the exponent M of the scaled mean water retention function [$S(h_m)$] in the form of Equation (4). Then solving Equation (4) gives

$$\alpha_i = (A/h\cdot) S_i^M \tag{5}$$

where $h\cdot$ is the known fixed value (e.g., -33 kPa) for all sites. Now impose the condition that the mean of α_i values for N sites is equal to 1.0, or that the sum of all α_i values is equal to N

$$\sum_{i=1}^{N} \alpha_i = N = (A/h\cdot) \sum_{i=1}^{N} S_i^{-M} \tag{6}$$

which yields

$$A = \frac{Nh\cdot}{\sum_{i=1}^{N} S_i^{-M}} \tag{7}$$

With A determined, all different α_i values for the site can then be found. Thus, with knowledge of a reference curve for a soil type or a group of soils, $h_m(S)$, and one measured value of $h_i(S)$, one can obtain α_i, and then the entire $h_i(S)$ curve. The procedure is illustrated in Figure 1, where h is replaced by suction τ ($\tau = -h$). The $h_i(S)$ curve can then be converted back to $h_i(\theta)$, from the knowledge of θ_s or bulk density. The details of this method are given in Ahuja et al. (1985).

For the same soils as used for testing regression models, the mean relative error of this scaling method ranged from -0.96 to 9.78%, and the standard deviation of errors was also reduced compared to regression-based methods where the mean relative error ranged from 8 to 29% (Ahuja et al., 1985). Evaluation of this method for several other soils is given in Williams et al. (1992) and Williams and Ahuja (1992). An additional advantage of using the scaling method is that the variability of $\theta(h)$ among sites is described by the distribution of scaling factors. This information is very useful for spatial modeling purposes. The reference curve for the soil type or the watershed required in the method may be replaced by a representative $\theta(h)$ curve for the textural class of the site (Heathman et al., 2003). Table 1 provides the parameters for a representative curve for each of 11 textural classes from Rawls et al., 1982.

2.2. The One-Parameter Gregson–Hector–McGovan (GHM) Model

The GHM One-Parameter Model is based on a strong linear correlation ($r^2 > 0.95$) observed between the slope b and intercept a of the log–log linear form of the soil water

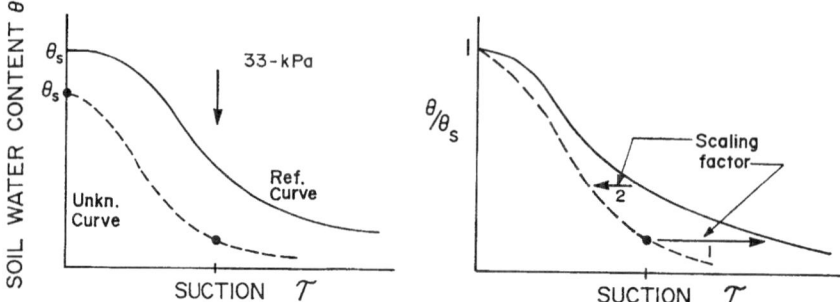

Figure 1. The similar media type scaling method to estimate complete soil water retention curve from bulk density and 33-kPa water content using a known reference curve for a related soil textural class. Only one value of water content on the unknown curve is known (here 33 kPa). Knowledge of the scaling factor, α_i, allows one to match the two curves having a common intercept of 1. This provides values of $S_i(h)$ (θ_i/θ_s) for the unknown curve for any arbitrary value of h. With knowledge of θ_s, possibly from bulk density, if not measured directly, S_i can be converted to values of water content, θ.

retention curve (Brooks and Corey, 1964) in the matric potential range below the air-entry value ψ_b (Gregson et al., 1987; Ahuja and Williams, 1991):

$$\ln[-\psi(\theta)] = a_i + b_i \ln(\theta - \theta_r) \tag{8}$$

where θ_r is the residual soil water content and ψ is the absolute value of matric potential.

Linear regression with log-transformed data was used to determine the parameters a_i and b_i for the soil water release curve (Equation (8)). The parameters a_i vs. b_i of Equation (8) from a wide variety of soils are themselves linearly related to each other:

$$a_i = p + qb_i \tag{9}$$

The secondary parameters, p and q, define the relationship between the slopes and intercepts (a_i and b_i) of linearized moisture release curves given in Equation (8).

Ahuja and Williams (1991) showed that $\psi(\theta)$ data of several different soils were described quite well by Equations (8) and (9). In agreement with Gregson et al. (1987), Williams et al. (1992) found a highly significant ($r = 0.82$) linear relationship between the slope (a) and intercept (b) in Equation (8) for 366 $\psi(\theta)$ relationships from Lane, OK, U.S.A. (Figure 2). These data were aggregated by soil texture and depth. The p and q values calculated for these soils was -1.287 and 0.949, respectively. Gregson et al. (1987) reported these values to be -0.982 for p and 0.585 for q. These values of the parameters p and q appeared to be general enough to adequately describe soil moisture release curves in general when either a or b is known (Gregson et al., 1987).

Although Ahuja and Williams (1991) found the a vs. b relationships (Equation (9)) for all soils were fairly close to each other, it was still better to divide the soils into four textural groups having slightly different a vs. b relationships. The p and q values for the textural groups are provided in Table 2.

Substituting Equation (9) into (8) yields a One-Parameter Model (Equation (10)), provided that an approximate value of θ_r for the soil type under consideration is known or can be estimated from information in the literature. In practice, however, the model is not

Table 1
Representative Campbell water retention model parameters for USDA soil textural classes

Texture	b			h_b(cm)			ϕ(cm^3 cm^{-3})		
	Rawls et al. (1982)	Clapp and Hornberger (1978)	de Jong (1983)	Rawls et al. (1982)	Clapp and Hornberger (1978)	de Jong (1983)	Rawls et al. (1982)	Clapp and Hornberger (1978)	de Jong (1983)
Sand	1.44	4.05	3.21	15.98	12.1	8.39	0.44	0.40	0.40
Loamy sand	1.81	4.38	4.09	20.58	9.0	12.68	0.44	0.41	0.39
Sandy loam	2.64	4.90	4.78	30.20	21.8	26.03	0.45	0.44	0.39
Loam	4.51	5.39	6.60	40.12	47.8	26.63	0.46	0.45	0.49
Silt loam	4.74	5.30	7.34	50.87	78.6	19.41	0.50	0.49	0.48
Silty clay loam	3.13	7.75	7.95	59.41	35.6	13.26	0.40	0.48	0.57
Sand clay loam	4.13	7.12	–	56.43	29.9	–	0.46	0.42	–
Clay loam	5.65	8.52	9.84	70.33	63.0	10.21	0.46	0.48	0.50
Sandy clay	4.48	10.40	–	79.48	15.3	–	0.43	0.43	–
Silty clay	7.87	10.40	10.26	76.54	49.0	16.98	0.48	0.49	0.53
Clay	6.06	11.40	12.71	85.60	40.5	12.24	0.48	0.48	0.53

From Rawls et al. (1991). $(\theta/\phi) = (h_b/h)^{1/b}$.

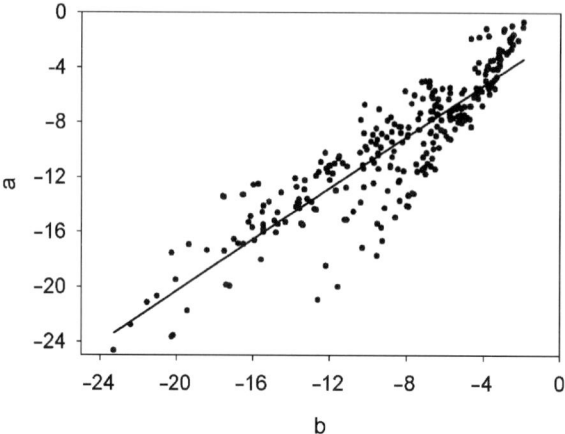

Figure 2. Relationship of slope (*a*) to intercept (*b*) of Equation (9) for soils from Lane, OK (Williams et al., 1992). The solid line is the least squares fit to the data. The equation is $a = -1.287 + 0.949 \times b$.

sensitive to the residual water content (Ahuja and Williams, 1991).

$$\ln[-\psi(\theta)] = p + b_i[\ln(\theta - \theta_r) + q] \tag{10}$$

Equation (10) can therefore, be used to estimate the entire $\psi(\theta)$ relationship below the air-entry value of ψ, simply from one measured value on the $\psi(\theta)$ curve for which b has been determined. Note here that the parameter b_i is specific for a particular soil water retention curve while the parameters p and q are defined for a grouping of water retention curves based on soil texture. Given values of p and q for a soil type or textural group and a known (ψ, θ) pair at -33 kPa, b is determined as:

$$b = \frac{[\ln(\psi_{-33\ kPa}) - p]}{[\ln(\theta_{-33\ kPa} - \theta_r)]} + q \tag{11}$$

Table 2
Average *p* and *q* values for texture groups

Group	Textural ranges	$p \ln(kPa)$	$q \ln(cm^3\ cm^{-3})$
1	Loam, silty clay loam, clay loam, silt loam	1.415	0.839
2	Sandy loam, sandy clay loam, loamy sand	0.343	1.072
3	Clay, sandy clay	0.897	0.955
4	Sand	0.541	1.469
Australian and British soils	A mixture of textures	-0.982	0.585

Average *p* and *q* values for several Australian and British soils pooled together, reported by Gregson et al. (1987), are given in the same units for comparison.

The parameter, b, is obtained from knowledge of the $\psi(\theta)$ pair and known values of p and q. Hence, the name of the technique: One-Parameter Model (for the parameter b). The known or estimated, from bulk density, saturated water content (θ_{sat}) value bounds the $\psi(\theta)$ curve and enables determination of the air-entry (ψ_b) value. For matric potentials greater than ψ_b up to the saturated water content (ψ_{sat}) water contents are estimated using a linear interpolation.

To estimate water content at different matric potentials using the One-Parameter Model (Equation (8)), the average soil water content at -33 kPa for each texture class was used in Equation (11) as the one $\psi(\theta)$ value required by the model (Williams et al., 1992). Using this $\psi(\theta)$ value, b was estimated for each textural grouping using the p and q for the texture grouping (Ahuja and Williams, 1991) discussed earlier. Soil water content was then calculated for each matric potential. Results are shown in Figure 3.

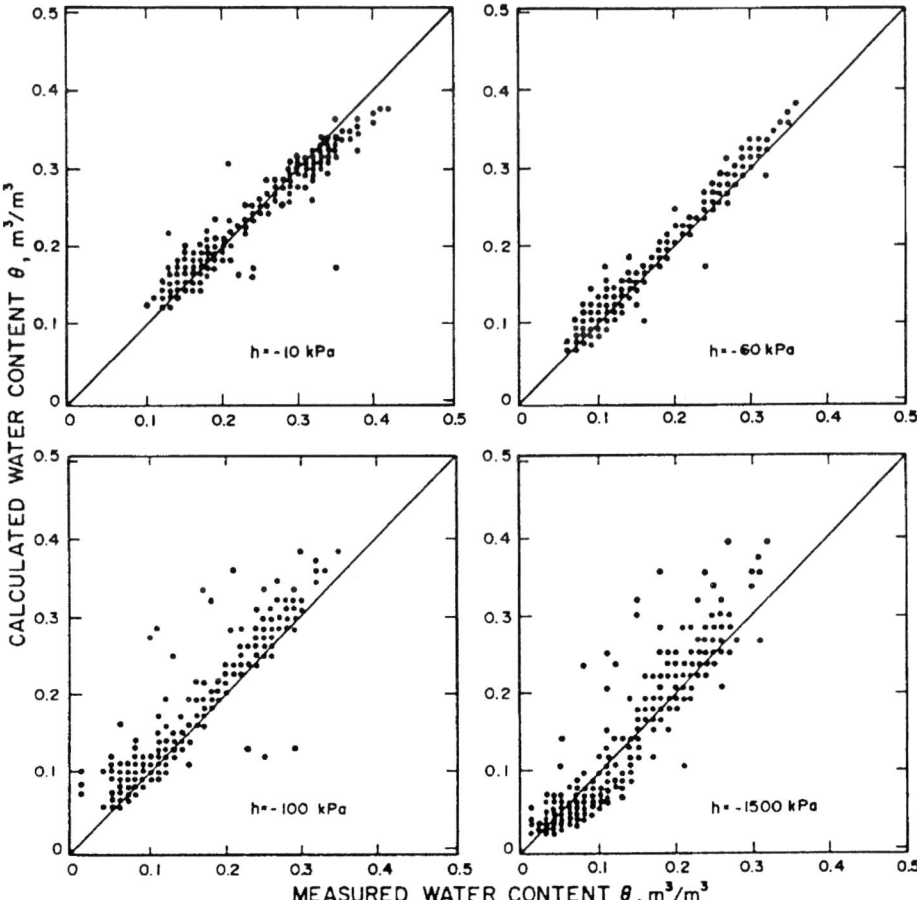

Figure 3. Soil water content and matric potential (ψ) calculated by the GHM model, against the measured values for 366 soil cores from Lane, OK, USA (Williams et al., 1992).

The similar media scaling approach was compared to the GHM One-Parameter Model with a and b values determined from textural class given in Table 2 (Williams and Ahuja, 1992). Overall, for the eight soils the root mean square error (RMSE) for the similar media scaling technique ranged from 0.020 to 0.058 m^3 m^{-3} whereas that for the GHM method ranged from 0.0 15 to 0.044 m^3 m^{-3} indicating an improvement of the GHM method over similar media scaling. Both methods are an improvement over regression-based methods but need one matching [$\psi(\theta)$] pair to be used. The value of water content at saturation as derived from bulk density can be used as a matching value, however, if only bulk density is available. What is surprising is that through numerous applications of the GHM method, the p and q values derived from many diverse data sets are quite similar (Williams et al., 1992; Williams and Ahuja, 2000) and similar to those first published by Gregson et al. (1987). This relationship holds much better, however, when matric potential is given as a function of water content (Williams et al., 1992) rather than the inverse (i.e., water content dependency on matric potential) which is the general convention in the soils literature.

2.3. Air-entry potential and saturated water content and the GHM Model

Because the One-Parameter approach for the $\psi(\theta)$ is based on the existence of a strong linear relationship between the intercept and slope of the log–log linear form of the relation between ψ and θ (Equation (8)), it can be shown geometrically that the linear relationship between the intercept and slope (a and b in Equation (9)) will be perfect ($R^2 = 1.0$) and unique if all the log–log straight-line functions pass through a single point (Williams and Ahuja, 1993). Based on Equation (9), the coordinates of this point will be $(-q, p)$. Thus, for near perfect conditions we hypothesized the presence of a convergence point in the ψ–θ plane. However, in soils the linear relationship between intercept and slope is not perfect and unique, and the convergence occurs within a narrow region around the point $(-q, p)$.

We can rearrange Equation (8) and use the air-entry potential and the water content at that potential (usually taken to be the saturated water content in the Brooks–Corey equation):

$$a_i = \ln(-\psi_b) - b_i[\ln(\theta_{sat})] \tag{12}$$

This equation applies to one given $\psi(\theta)$ curve where the matching pair is (ψ_b, θ). Note that according to Equation (9), a_i and b_i are linearly related, i.e., $a_i = p + qb_i$. For an ensemble of n $\psi(\theta)$ curves ($i = 1 \ldots n$), Equation (9) defines $a_i = p + qb_i$ where p and q can be determined from least square regression. The similarity between Equations (12) and (9) implies that the value of p should be equivalent to an effective average value of the $\ln(-\psi_b)$, the logarithm of the air-entry potential, and the value of q to an effective average value of $[\ln(\theta_{sat})]$, the negative logarithms of saturated water content. The results in Figure 4 indicate that most of the data points in both p and q plots do approximately fall along a 1:1 line for our soils (Ahuja and Williams, 1991). This model was shown to estimate water contents well for soils with a wide range in textures (Williams and Ahuja, 1993, 2000, and 2003). These estimates, using p and q values based on air-entry potential and saturated water contents only were at least as good or better than the estimates based on group p and q values. This One-Parameter Model that requires only one known $\psi(\theta)$ value, was also shown to be consistently more precise that texture-based regression models (Williams et al, 1992).

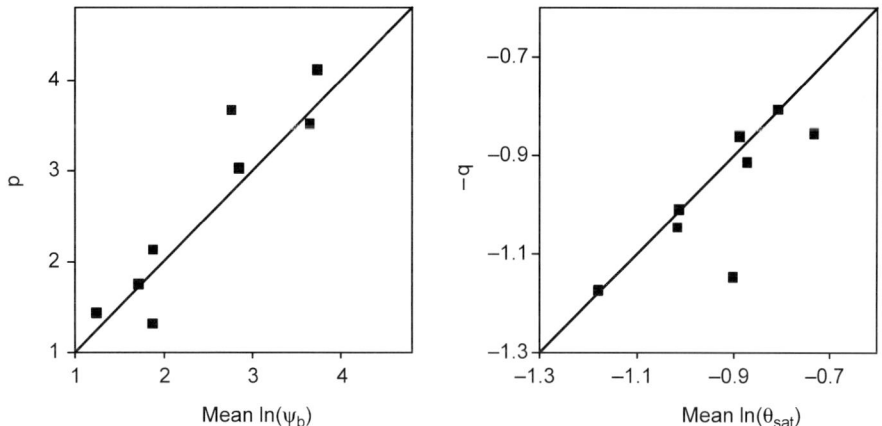

Figure 4. Relationship between p and mean $\ln(-\psi_b)$ and between $-q$ and mean $\ln(\theta_{sat})$ for eight U.S. Soils.

2.4. The GHM One-Parameter Model with generalized parameters

The relationship between air-entry potential and saturated water content with the p and q values given above was shown to be valid for the eight U.S. soils from which the data (air-entry potential and saturated water content) were obtained (Ahuja and Williams, 1991; Williams and Ahuja, 1993). This led us to develop a more generalized approach to determine p and q (Equation (9)) by using the mean values of air-entry potential and saturated water contents from textural class data as given by Rawls et al. (1982). This generalized approach uses the textural class means from a wide range of soils to determine the values of p and q rather than from a more limited range of eight soils.

The parameters for the Brooks–Corey equation (e.g., pore-size distribution, λ, and air-entry potential, ψ_b) were used for this generalized approach. Rawls et al. (1982) have published data on parameters for the Brooks–Corey equation as a function of soil texture class (Table 1). These parameters can be related to the parameters in the GHM model (Equation (8)). We begin by rearranging Equation (8):

$$-b \ln(\theta - \theta_r) = a - \ln(-\psi)$$

$$\ln(\theta - \theta_r) = -\frac{a}{b} + \frac{1}{b} \ln(-\psi) \tag{13}$$

$$(\theta - \theta_r) = \exp -\frac{a}{b}(-\psi)^{\frac{1}{b}}$$

Equation (8) also implies:

$$\ln(-\psi_b) = a + b \ln(\theta_s - \theta_r) \tag{14}$$

where ψ_b is the air-entry potential ($-$kPa) and θ_s is the total porosity (cm^3 cm^{-3}). Rearranging this equation, as was done in Equation (13), we can express Equation (14) as:

$$(\theta_s - \theta_r) = \exp\left(\frac{a}{b}\right)(-\psi_b)^{\frac{1}{b}} \tag{15}$$

Combining Equations (13) and (15) we have

$$\frac{\theta - \theta_r}{\theta_s - \theta_r} = \left(\frac{\psi_b}{\psi}\right)^{-\frac{1}{b}} \tag{16}$$

Equation (16) is equivalent to the following form of the Brooks–Corey equation for $\psi < \psi_b$:

$$\frac{\theta - \theta_r}{\theta_s - \theta_r} = \left(\frac{\psi_b}{\psi}\right)^{\lambda} \tag{17}$$

where λ is the pore-size distribution index.
Comparison of Equations (16) and (17) yields

$$b = \frac{-1}{\lambda} \tag{18}$$

Knowing this relationship, the intercept, a, can be estimated with Equation (14) by substituting $-1/\lambda$ for the slope, b, and rearranging:

$$a = \ln(-\psi_b) + \left(\frac{1}{\lambda}\right)\ln(\theta_s - \theta_r) \tag{19}$$

Equations (18) and (19) can be used to determine the slope and intercept of a log–log expression of the $\psi(\theta)$ retention curve. The necessary information on the average ψ_b, λ, θ_s and θ_r for a soil or group of soils is available from Rawls et al. (1982) who have provided the mean values of these properties for eleven textural classes (Table 1). For each textural class, the intercept a was obtained using Equation (19) and known mean values of ψ_b, λ, θ_s and θ_r for the class. The results are given in Table 3. The values of a and $(-1/\lambda)$ from 11 textural classes were used in Equation (9) to obtain the generalized p and q values for all classes. The relationships are shown in Figure 5.

There was a strong linear relationship between the textural class mean intercept (a) and slope (b) that could be described by Equation (9) (Figure 5). Based on the hydrologic properties for the 11 textural classes, p was equal to -0.5236 in kPa and q equal to 0.6691 cm^3 cm^{-3} with r^2 of 0.86. Using Gregson et al. (1987) p and q values, -0.982 ln kPa and 0.5852 ln cm^3 cm^{-3}, respectively, and the calculated slope for each texture class based on Equation (12), a line was superimposed on the results for the textural classes (Figure 5, dashed line). Both lines are similar; showing the universal nature of the relationship, a vs. b. This supports our earlier conjecture that a common set of p and q values may be used for scaling across soil types and textural classes.

Table 3
Textural values of p and q based on mean air-entry pressure (ψ_b) and total porosity (θ_{sat})

Texture class	Air-entry pressure[a] (ψ_b)	p[b]	Total porosity[c] (θ_{sat})	q[d]
Sand	0.726	−0.320	0.437	0.828
Loamy sand	0.869	−0.140	0.437	0.828
Sandy loam	1.466	0.382	0.453	0.792
Loam	1.115	0.110	0.463	0.770
Silt loam	2.076	0.730	0.501	0.691
Sandy clay loam	2.808	1.320	0.398	0.921
Clay loam	2.589	0.951	0.464	0.768
Silty clay loam	3.256	1.180	0.471	0.753
Sandy clay	2.917	1.071	0.430	0.844
Silty clay	3.419	1.229	0.479	0.736
Clay	3.730	1.316	0.475	0.744

[a](ψ_b), geometric mean of the air-entry pressure for each textural class, as reported by Rawls et al. (1982) (cm).
[b]p value calculated as $\ln(\psi_b)$.
[c](θ_{sat}), mean total porosity or saturated water content for each textural class, as reported by Rawls et al. (1982) (cm^3 cm^{-3}).
[d]q value calculated as $\ln(\theta_{sat})$ and taken as positive.

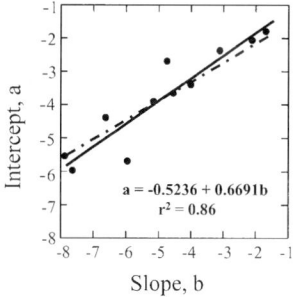

Figure 5. Relationship of the intercept, a, and slope b, calculated with Equations (18) and (19) and the mean hydrologic parameters for each texture class. Using the p and q values of Gregson, D., Hector, D.J., and McGowan, M. (J. Soil Sci. 38, 483, 1987) the slope (dash line) was calculated for each textural class based on Equation (8).

2.4.1. Implementation of the GHM One-Parameter Model with generalized parameters

Using Equation (17) and the mean hydrologic parameters in Table 3, soil water contents were calculated for each of the 11 textural classes at matric potentials −5, −10, −20, −50, −500, −1000, and −1500 kPa. This provided a fairly complete soil water release curve. The data were scaled using Equation (8) and inverted to give θ as the dependent variable as:

$$\ln(\theta - \theta_r) = \frac{(\ln[-\psi_i(\theta)] - p)}{b} - q \qquad (20)$$

where the textural class $b = 1/\lambda$ and the generalized p and q values calculated from data in Table 3.

Using the Brooks–Corey equation (Equation (17)) and the average hydrologic parameters of the texture classes in Tables 1 and 3, we calculated the soil water characteristic curve for each texture class. These soil water characteristic curves were scaled using Equation (20); the slope, b, was estimated using the average θ (Equation (18)); and the generalized p and q values were based on the slope–intercept relationship determined in Figure 4. The results of scaling the $\psi(\theta)$ across texture classes are presented in Figure 6. The scaling technique with the One-Parameter Model coalesced the data into a tight group surrounding the 1:1 line.

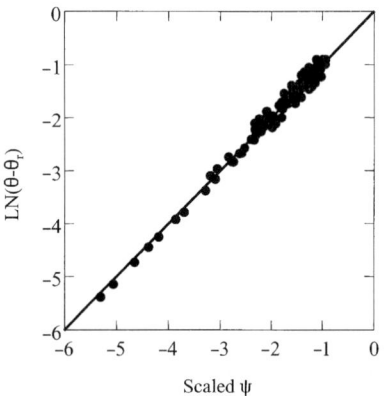

Figure 6. Measured values of $\ln(-r)$ vs. values calculated by Equation (20) using textural class values of p and q for the mean water release curves for each soil texture class given in Rawls et al. (1982).

We scaled the soil water retention curve for four soils using the One-Parameter Model and generalized p and q values. These soils range in texture from sandy to clay loam and encompass a range of scaling efficiencies (in terms of how well the original data could be reproduced) from best (Renfrow sandy loam) to worst (Pima clay). We used the water content at -33 kPa as the known ψ–θ pair required by the model, and the generalized p and q values in Figure 2. The model estimated the soil water content quite well (Figure 7). Using the generalized p and q values the mean errors in estimating θ_s for Lakeland, Pima, Renfrow and Teller soils were 0.026, 0.012, -0.047, and 0.004 $cm^3\ cm^{-3}$, respectively, while the RMSEs were 0.032, 0.052, 0.066, and 0.025 $cm^3\ cm^{-3}$, respectively. For Lakeland and Pima soils these errors are very similar to those obtained with the group p and q values (Table 2), while the mean error and the RMSE were larger for Renfrow (compared to $-0.006\ cm^3\ cm^{-3}$) and Teller (compared to 0.012 $cm^3\ cm^{-3}$), respectively. Overall, results for these soils using the generalized p and q are quite similar to those reported earlier for individual and group values (Ahuja and Williams, 1991; Williams and Ahuja, 1992).

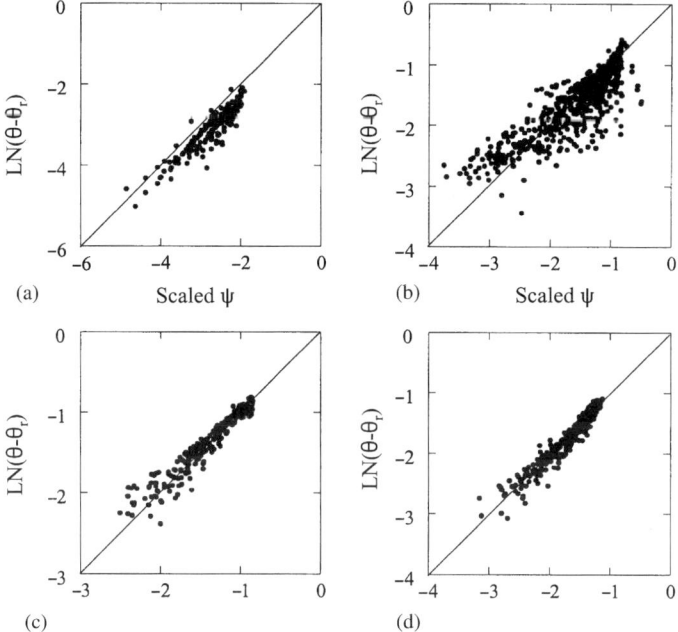

Figure 7. Results of scaling the soil water characteristic for Lakeland (A), Pima (B), Renfrow (C) and Teller (D) soils using the One-Parameter Model and generalized p and q values.

2.5. Use of available water capacity with the GHM One-Parameter Model

Application of the GHM One-Parameter Model requires a matching ψ–θ pair to generate a moisture release curve. In some cases these data are not available, as for example, SOILS-5 database or the USDA-NRCS soil survey (Soil Survey Staff, 2003) where only available water capacity (AWC) has been recorded. For this reason Williams and Ahuja (1993) investigated the possibility of calculating a known ψ–θ pair from AWC to calculate a value of b. An empirical polynomial regression equation was developed to calculate the b value for the GHM model given only AWC. The use of this relationship to determine b and hence p and q values resulted in errors that were only slightly larger than those if θ at the -33 kPa matric potential had been used to calculate b (Williams and Ahuja, 1993). In order to make the GHM more useful, however, broader relationships between AWC and the slope of the log–log model need to be determined on a textural or group basis.

3. HYDRAULIC CONDUCTIVITY

3.1. Determining saturated hydraulic conductivity, K_{sat}

Saturated hydraulic conductivity, K_{sat} [$K(h)$ at $h \geq 0$)], is probably the most important soil hydraulic parameter to describe water flow processes in soils. This parameter is difficult to obtain as it is highly sensitive to soil conditions, such as compaction, macropores, sample size, temperature, and entrapped air, and thus is highly variable. In a field, the K_{sat} may vary between one and two orders of magnitude. Scale effects, i.e., differences between measurement scales of K_{sat} and the soil data used as predictors are much more important for estimation of K_{sat} than for water retention. Because of this, it is much more meaningful to try to determine a reasonably accurate distribution of K_{sat} (e.g., mean and standard deviation) in a field, rather than determine highly accurate point values.

3.1.1. Predicting saturated conductivity from effective porosity

Studies have shown that K_{sat} is strongly related to effective porosity (Ahuja et al., 1984, 1989). Effective porosity, ϕ_e, was defined as the total porosity minus the volumetric soil water content at 33 kPa tension, and was related to K_{sat} by a generalized Kozeny–Carman equation:

$$K_{sat} = B\phi_e^n \tag{21}$$

where B and n are constants. Figure 8 shows the relationship for nine soils combined (473 data points). These soils came from diverse locations in Hawaii, Arizona, Oklahoma, and several states of the Southeast U.S. The correlation of this relationship in Figure 8 for all the soils was as good as for any one soil individually, which indicates that Equation (21) is applicable across soil types. Realizing that the measurement of K_{sat} be subject to an error as large as one order of magnitude, due to unknown effects of entrapped air and

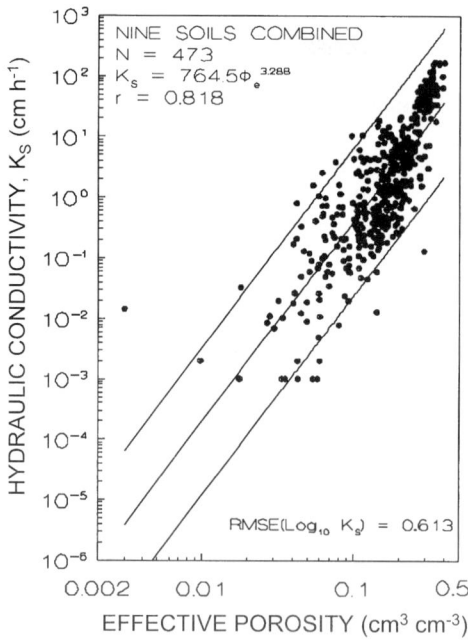

Figure 8. Effective porosity vs. saturated hydraulic conductivity for a range of soils (Ahuja, 1984).

macropores, the empirical equation given in Figure 8 is useful in estimating K_{sat} from the simpler measurements of effective porosity for a soil. Tests of this equation against field data for soils from Hawaii and Oklahoma and core data for several Korean soils as well as some soils from Indiana have shown good results (Ahuja, unpublished data; Franzmeier, 1991).

Later, work by Timlin et al. (1999) slightly improved Ahuja's K_{sat} equation by inclusion of the Brooks–Corey pore-size distribution index (λ) into the Kozeny–Carman equation. The best improvements for K_{sat} were obtained when λ was included in the term for the coefficient (B) for the modified Kozeny–Carman equation (Equation (21)) and a constant exponent of 2.54 was used.

$$K_{sat} = 0.000259(10^{0.6\lambda})\phi_e^{2.54} \tag{22}$$

The next best form was when λ was included in the exponent for ϕ_e. The two best models appeared to better preserve the mean, standard deviation and range of the original data than did the original equation. The coefficient of determination (r^2) for log(K_s) increased from 0.70 to 0.73 and the RMSE of log(K_{sat}) decreased from 0.60 to 0.57. The use of λ improved the fit for larger values of K_s ($>2.5 \times 10^{-5}$ m s^{-1}) (Figure 9).

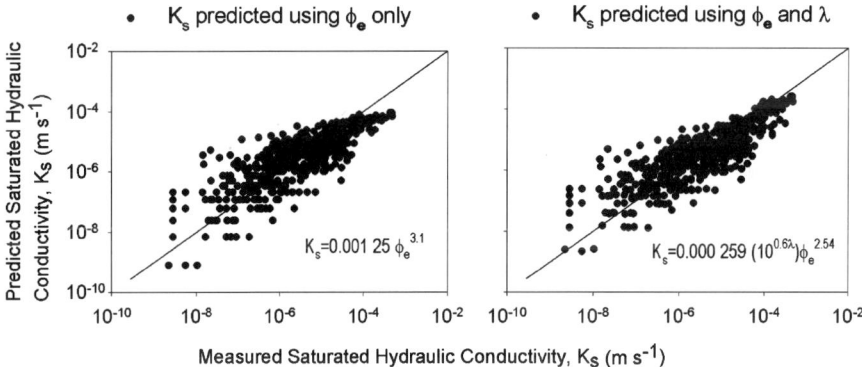

Figure 9. Predicted and measured saturated conductivities by the original equation (Equation (21)) of Ahuja et al. (1984) and the equation modified by inclusion of Brooks–Corey parameters (Equation (22)) (Timlin et al., 1999).

Work of Ahuja et al. (1993) has shown that an average K_{sat} of a soil profile is related to drainage of the surface soil in two days after wetting, i.e., the change in soil water content of the surface soil in two days, through Equation (21) with parameter B and n changed. Thus, measurements of soil bulk density to determine θ_s and water content at several locations in a field two days after a soaking rain can provide an estimate of spatial distribution of average profile K_{sat} in the field. Ahuja et al. (1993) give details and analysis of this technique, and several examples of K_{sat} estimations.

3.2. Relationships for unsaturated hydraulic conductivity

3.2.1. Extending the One-Parameter Model to unsaturated hydraulic conductivity–matric potential relationships

Just as for the $\psi(\theta)$ relationships, a linear relationship has been observed to exist between log K and log$(-\psi)$ below the air-entry value for a variety of soils (Brooks and Corey, 1964). An interesting recent finding was that the slope and intercept of this log–log linear relationship are themselves linearly related, and that this latter linear relationship is approximately unique across several soil types evaluated (Ahuja and Williams, 1991). For ψ less than the air-entry value we took the basic equation in terms of ln(K) as the dependent variable

$$\ln(K_i) = A_i + B_i \ln(-\psi) \tag{23}$$

where i refers to individual data points, A_i and B_i are coefficients of the equation. The equation described the experimental data well. The coefficients A_i and B_i were found to be highly correlated with each other:

$$A_i = P + QB_i \tag{24}$$

and the constants P and Q were also approximately the same for all soil types. Figure 10 shows the relationship between A_i and B_i for a wide range of soils from the Southeastern U.S., California and Hawaii which appears to be quite good. Substituting Equation (24) in Equation (23) gives:

$$\ln(K_i) = P + B_i[\ln(-\psi) + Q] \tag{25}$$

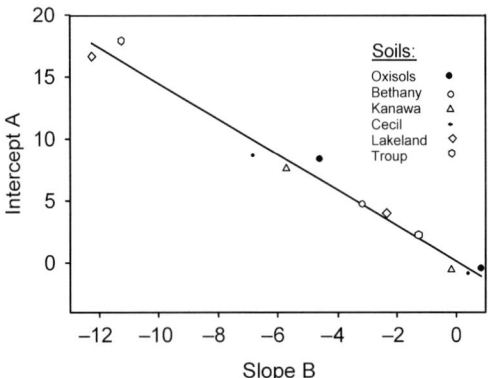

Figure 10. Fitted (Equation (24)) intercept vs. slope relationships for the unsaturated hydraulic conductivity function $K(\psi)$ of different soils and for all soils combined. For individual soils, only the end points of relations are shown. The units of K and ψ are cm h^{-1} and kPa $\times$ 10^{-1}, respectively (Ahuja and Williams, 1991).

Table 4
Results for regression of Equation (24), $A = P + QB$ where A and B are the intercept and slope of the ln K vs. ln$(-\psi)$ relationship (Equation (23))

Soil	R^2 for Equation (24)	Log(cm h^{-1})		Log(kPa × 10)	
		P	SE[a] for P	Q	SE for Q
Oxisols	0.952	−1.094	0.388	−1.787	0.058
Bethany	0.933	−2.791	0.309	−1.863	0.092
Konawa	0.951	−2.564	0.512	−1.680	0.089
Cecil	0.915	−2.283	0.784	−1.531	0.074
Lakeland	0.841	−0.767	1.015	−1.452	0.088
Troup	0.993	−1.818	0.413	−1.785	0.037
All soils	0.945	−1.819	1.016	−1.635	0.027

[a]Standard error.

Rearranging gives:

$$\text{scaled } \ln[K_i(\psi)] = \frac{\ln(K_i) - P}{B_i} - Q = \ln(-\psi) \qquad (26)$$

As with the $\psi(\theta)$ relationships, this relationship is a result of convergence of log K vs. log(ψ) lines for all soils to within a narrow region around a point (Ahuja and Williams, 1991). If the correlation between A_i and B_i in Equation (24) is perfect, scaled ln(K_i) and ln$(-\psi)$ should fall on a 1:1 line for all $K(\psi)$ relationships. Since Equation (26) contains only one unknown location dependent parameter (B_i), the entire $K(\psi)$ curve, below the air-entry value can be estimated from one measured or known value of the K, ψ pair on the curve. The value of K_{sat} and the air-entry value, ψ_b, would be one such known pair that may be commonly measured. An air-entry value can be obtained using regression equations derived by Rawls et al. (1982) (Table 4). The ψ_b obtained from an estimated $\psi(\theta)$ curve could also be used. In fact, if the wetting-front capillary potential, ψ_w, a parameter in the Green–Ampt type infiltration equations, defined as:

$$\int_{-\infty}^{0} \frac{K(\psi)}{K_{sat}} d\psi \qquad (27)$$

is known, by an iterative process one can determine ψ_b. This suggests that the complete $K(\psi)$ function may be estimated from just the field-measured infiltration-time data, since these data can be used to determine both K_{sat} and ψ_w or ψ_b of a soil horizon. Finally, the knowledge of K_{sat} provides a closure of the complete $K(\psi)$ relationship.

Utilizing this apparently unique relationship between slope and intercept of log K vs. log$(-\psi)$, one can easily define the $K(\psi)$ curve below the air-entry value if the air-entry value is known. As in Equation (7) a single pair of measured values (K, ψ) are needed to determine all the parameters. Estimations of unsaturated conductivity for two soils are given in Figure 11. The method works well for these soils.

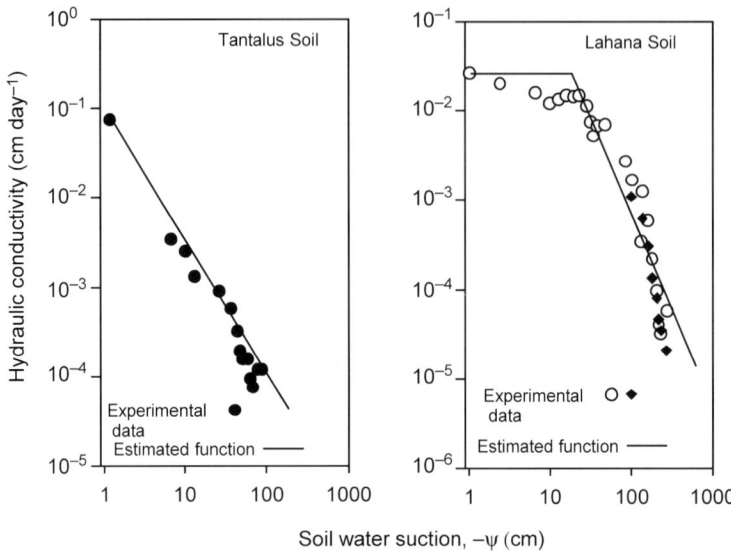

Figure 11. Estimates of the hydraulic conductivity $K(\psi)$ relations of two Hawaii soils based on the Gregson et al. (1987) approach using known saturated hydraulic conductivity, K, and wetting-front suction, ψ_w (Equation (27)).

4. APPLICATIONS OF PEDOTRANSFER FUNCTIONS FOR SIMULATION MODELS

The GHM One-Parameter Model (using texture group p and q, Table 2) was used to estimate the soil moisture release curve and available water holding capacity for use in the soybean model, GLYCIM (Timlin et al., 1996b). The predicted water contents for a database of soils from Mississippi (Timlin et al., 1996a) along with those predicted using the method of Rawls et al., 1982 are shown in Figure 12. The standard error for the One-Parameter Model was 0.058 vs. 0.054 cm^3 cm^{-3} for the method of Rawls et al. (1982). This was a completely independent database from the one the p and q values were derived. The Rawls method had a lower standard error but the GHM One-Parameter Model predictions were more evenly distributed about a 1:1 line (Figure 12). The soybean yields simulated with hydraulic properties estimated using the One-Parameter Model were closer to yields simulated using measured soil hydraulic properties and had a slightly lower mean square error (777 vs. 799 kg ha^{-1}). This result was attributed to better estimation of available water holding capacity by the One-Parameter Model when compared to the method of Rawls et al. (1982).

Starks et al. (2003) investigated the use of a hierarchy of limited soil input data, ranging from soil textural class of soil horizons alone, to measured soil texture and bulk densities of horizons, additional lab or field measurement of -33 kPa soil water content, to additional field measurement of average saturated hydraulic conductivity. These five modeling scenarios, along with meteorological and plant information, were input to the

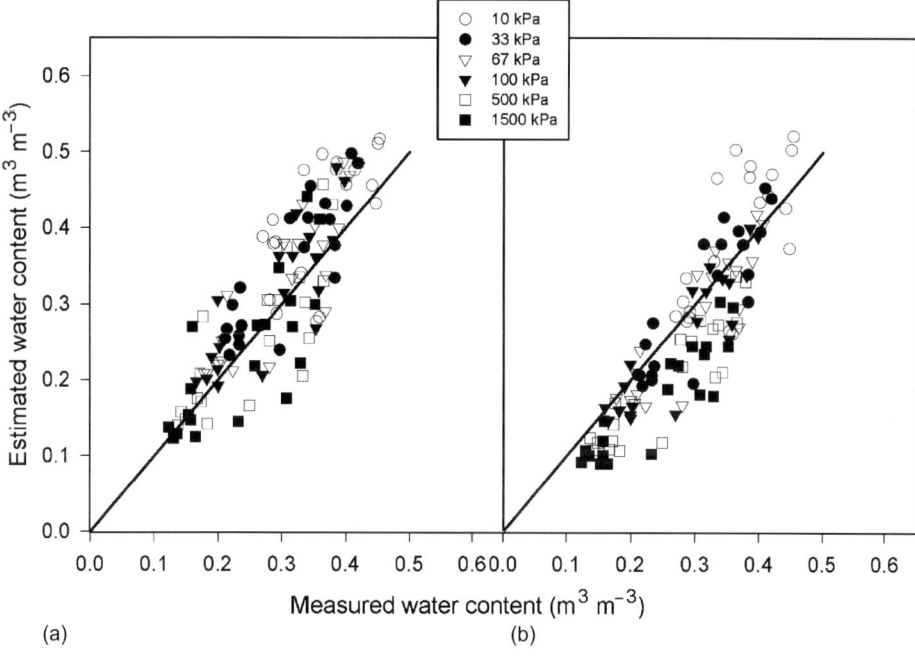

Figure 12. Water contents for seven Mississippi Soils estimated by (a) the One-Parameter Model and (b) using soil texture (Rawls et al., 1982).

Root Zone Water Quality Model (RZWQM) (Ahuja et al., 2000) to estimate water content in the surface 0–60 over a 30-day period in 1997 at the Little Washita River Experimental Watershed in Oklahoma. The estimated water contents were compared with time-domain reflectometry (TDR) profile measurements and gravimetric sampling of soil surface moisture. In addition to the five scenarios using limited input data, a more detailed set of data based on laboratory measured soil water retention curves and field measured saturated conductivity was supplied to the model for all Brooks–Corey function parameters (full description mode). Estimates of root zone soil water content using detailed input of soil hydraulic properties were compared to estimates obtained using minimum input data. A hierarchy of input data were evaluated. The minimum input data consisted of aggregations of soil hydraulic properties based on soil texture class, the next level included textural class data calculated from detailed soil textural component hydraulic properties data. These methods used textural class based parameters compiled by Rawls et al. (1982) for the Brooks–Corey equation. The next level incorporated a measured $\theta(\psi)$ pair at $\psi = -33$ or -10 kPa. For this level of data the scaling method was used to estimate $\theta(\psi)$ curves. Overall, Heathman et al. (2003) reported reasonable agreement between TDR-measured and RZWQM-predicted average water contents for 0–60 cm depths. Surprisingly, the smallest errors in the predicted water contents were achieved using either the textural class only or the hydraulic properties determined *in situ*, with RMSEs ranging from 0.012 to 0.018 m^3 m^{-3}. In some scenarios, the water content simulations using estimated soil hydraulic properties performed better than those with laboratory determined soil hydraulic properties. Hence, the use of textural class-name data only to describe soil hydraulic properties in the model provided adequate estimates of average profile soil water contents. In further research Heathman et al. (2003) used data assimilation to incorporate measured values of surface water content into simulations to update predicted water contents. When used with estimated soil properties data assimilation further improved estimates of water content.

5. SUMMARY

We have outlined several methods for estimating soil hydraulic properties from easily available data including soil texture and bulk density. The scaling method was introduced to improve upon the regression models of Rawls et al. (1982) to predict water contents at specific matric potentials using soil texture. The improvement of the scaling method comes from incorporation of a matching water content–matric potential $\psi(\theta)$ pair to scale the water release curve initially calculated using Rawls's method. A One-Parameter Model (GHM One-Parameter Model) was next introduced to parameterize soil water release curves using only one parameter. The method is based on a log linear relationship between water content and matric potential where the slopes and intercepts of the relationships are themselves linearly related. The parameters for this linear relationship (i.e., slope and intercept) were initially shown to be consistent for a wide range of soils so that a single slope and intercept could be used to generate any moisture release curve given a matching water content–matric potential pair. Later, we found that the estimations of water contents could be improved if the slope and intercept values were aggregated by soil texture groupings. We further generalized the parameter slope and intercept values (p and q) for textural class by casting the GHM One-Parameter Model into a Brooks–Corey form. The p and q values could be determined from the air-entry pressure and saturated water contents of the textural class groupings of Brooks–Corey parameters given by Rawls et al. (1982).

A simple equation to predict saturated hydraulic conductivity from soil porosity near saturation was shown to give good results over a wide range of soils. This relationship was improved slightly by including the pore-size distribution parameter from the Brooks–Corey equation. We also presented an application of the GHM model to estimate unsaturated hydraulic conductivity. When the unsaturated conductivity was expressed as a log-linear relationship with water content, a relationship between the slopes and intercepts of this function was found, similar to that for the GHM model for water retention data.

The GHM One-Parameter Model was used to estimate water contents for a database of soils from Mississippi using previously determined group p and q (slope and intercept) values based on textural groupings. The estimates were slightly better than those from Rawls et al. (1982) regression methods. The use of water contents estimated from limited data were tested in the RZWQM. The model was used to predict water contents in a watershed in Oklahoma. The predictions of water content by the simulation model using data from soil textural class to estimate soil hydraulic properties compared surprisingly well with predictions using lab-measured soil hydraulic properties and in some cases were better. Predictions were further improved when infrequent measurements of surface soil water content were used to update predictions by the simulation model.

There are still few to no dependable methods to estimate soil hydraulic properties easily over large areas and in large numbers. There have been considerable advances in methods to measure water content over large areas through remote sensing (Jackson et al., 1999) and improved instrumentation. Combined with methods to easily measure soil texture there is potential to greatly expand the usefulness of pedotransfer functions. Further improvement in pedotransfer functions is also possible by including landscape parameters such as slope and curvature (Pachepsky et al., 2001; Timlin et al., 2003). An advantage of using landscape parameters is that scale issues in pedotransfer functions can be more readily addressed. As a result pedotransfer functions will be expected to become even more useful in the future.

REFERENCES

Ahuja, L.R., Williams, R.D., 1991. Scaling water characteristic and hydraulic conductivity based on Gregson–Hector–McGowan approach. Soil Sci. Soc. Am. J. 55, 308-319.

Ahuja, L.R., Naney, J.W., Green, R.E., Nielsen, D.R., 1984. Macroporosity to characterize spatial variability of hydraulic conductivity and effects of land management. Soil Sci. Soc. Am. J. 48, 699-702.

Ahuja, L.R., Naney, J.W., Williams, R.D., 1985. Estimating soil water characteristics from simpler soil properties or limited data. Soil Sci. Soc. Am. J. 49, 1100-1105.

Ahuja, L.R., Cassel, D.K., Bruce, R.R., Barnes, B.B., 1989. Evaluation of spatial distribution of hydraulic conductivity using effective porosity data. Soil Sci. 148, 404-411.

Ahuja, L.R., Wendroth, O., Nielsen, D.R., 1993. Relationship between initial drainage of surface soil and average profile saturated conductivity. Soil Sci. Soc. Am. J. 57, 19-25.

Ahuja, L.R., Rojas, K.W., Hanson, J.D., Shaffer, M.J., Ma, L. (Eds.), 2000. Root Zone Water Quality Model: Modeling Management Effects on Water Quality and Crop Production. Water Resources Publications, Highlands Ranch, CO, 360 pp.

Brooks, R.H., Corey, A.T., 1964. Hydraulic Properties of Porous Media, Hydrol. Pap. No. 3. Colorado State Univ., Ft. Collins.

Campbell, G.S., 1974. A simple method for determining unsaturated conductivity from moisture retention data. Soil Sci. 117, 311-314.

Clapp, R.B., Hornberger, G.M., 1978. Empirical equations for some soil hydraulic properties. Water Resour. Res. 14 (4), 601-604.

de Jong, R., 1983. Soil water desorption curves estimated from limited data. Can. J. Soil Sci. 63, 697-703.

Franzmeier, D.P., 1991. Estimation of hydraulic conductivity from effective porosity data for some Indiana soils. Soil Sci. Soc. Am. J. 55, 1801-1803.

Gregson, D., Hector, D.J., McGowan, M., 1987. A one-parameter model for the soil water characteristic. J. Soil Sci. 38, 483.

Gupta, S.C., Larson, W.E., 1979. Estimating soil water retention characteristics from particle size distribution, organic matter percent, and bulk density. Water Resour. Res. 15, 1633-1635.

Heathman, G.C., Starks, P.J., Ahuja, L.R., Jackson, T.J., 2003. Assimilation of surface soil moisture to estimate profile soil water content. J. Hydrol. 279, 1-17.

Jackson, T.J., Le Vine, D.M., Hsu, A.Y., Oldak, A., Starks, P.J., Swift, C.T., Isham, J.D., Haken, M., 1999. Soil moisture mapping at regional scales using microwave radiometry: the Southern Great Plains hydrology experiment. IEEE Trans. Geosci. Remote Sensing 37, 2136-2151.

McQueen, I.S., Miller, R.F., 1974. Approximating soil moisture characteristics from limited data: empirical evidence and tentative model. Water Resour. Res. 10, 521-527.

Miller, E.E., Miller, R.D., 1956. Physical theory for capillary flow phenomena. J. Appl. Phys. 27, 324.

Pachepsky, Y.A., Rawls, W.J., Timlin, D.J., 1999. The current status of pedotransfer functions: their accuracy, reliability, and utility in field- and regional-scale modeling, Assessment of Non-point Source Pollution in the Vadose Zone, Vol. 108, pp. 223–234.

Pachepsky, Ya., Tomlin, D.J., Rawls, W.J., 2001. Soil water retention as related to topographic variables. Soil Sci. Soc. Am. J. 65, 1787-1795.

Rawls, W.J., Brakensiek, D.L., 1982. Estimating soil water retention from soil properties. J. Am. Soc. Civ. Eng. Irrig. Drain Div. 108, 166-171.

Rawls, W.J., Brakensiek, D.L., Saxton, K.E., 1982. Estimation of soil water properties. Trans. ASAE 25, 1316.

Rawls, W.J., Gish, T.J., Brakersiek, D.L., 1991. Estimating soil water retention from soil physical properties and characteristics. Adv. Agron. 16, 213-234.

Saxton, K.E., Rawls, W.J., Romberger, J.S., Papendick, R.I., 1986. Estimating generalized soil-water characteristics from texture. Soil Sci. Soc. Am. J. 50, 1031-1036.

Simmons, C.S., Nielsen, D.R., Biggar, J.W., 1979. Scaling of field-measured soil water properties. Hilgardia 47, 77.

Soil Survey Staff, 2003. National Soil Survey Characterization Data. Soil Survey Laboratory, National Soil Survey Center. USDA-NRCS, Lincoln, NE, USA.

Starks, P.J., Heathman, G.C., Ahuja, L.R., Ma, L., 2003. Use of limited soil property data and modeling to estimate root zone soil water content. J. Hydrol. 272, 131-147.

Timlin, D.J., Pachepsky, Ya., Alcock, B., Whisler, F., 1996a. Indirect estimation of soil hydraulic properties to predict soybean yield using GLYCIM. Agric. Syst. 52, 331-353.

Timlin, D.J., Williams, R.D., Ahuja, L.R., 1996b. Methods to estimate soil hydraulic parameters for regional-scale applications of mechanistic models. *In*: Corwin, D., Loage, K. (Eds.), Applications of GIS to the Modeling of Non-point Source Pollutants in the Vadose Zone. ASA Special publication ASA, CSSA, and SSSA, Madison, WI, pp. 185-203.

Timlin, D.J., Ahuja, L.R., Pachepsky, Ya., Williams, R.D., Gimenez, D., Rawls, W., 1999. Use of Brooks–Corey parameters to improve estimates of saturated hydraulic conductivity from effective porosity. Soil Sci. Soc. Am. J. 63, 1086-1092.

Timlin, D.J., Pachepsky, Ya., Walthall, C., 2003. A mix of scales: topographic information, point samples and yield maps. Scaling Methods in Soil Physics. Bridging Scales in Soil Physics. CRC Press, Boca Raton, FL, pp. 227-241.

van Genuchten, M.T., Nielsen, D.R., 1985. On describing and predicting the hydraulic properties of unsaturated soils. Ann. Geophys. 3, 615-628.

Warrick, A.W., Mullen, G.J., Nielsen, D.R., 1977. Scaling field-measured soil hydraulic properties using a similar-media concept. Water Resour. Res. 13, 355.

Williams, R.D., Ahuja, L.R., 1992. Evaluation of similar-media scaling and a one-parameter model for estimating the soil water characteristic. J. Soil Sci. 43, 237.

Williams, R.D., Ahuja, L.R., 1993. Using a one-parameter model to estimate the soil water characteristic. *In*: Wang, Sam S.Y. (Ed.), Advances in Hydro-Science and Engineering, Vol. I, Part A. Center for Computational Hydroscience and Engineering, School of Engineering, University of Mississippi, MI, pp. 485.

Williams, R.D., Ahuja, L.R., 2000. Using the Gregson one-parameter model to estimate the soil-water retention. Conference Proceedings of the Fourth International Conference on Soil Dynamics, Adelaide, South Australia. March 26–30 (CD-ROM format, no pagination).

Williams, R.D., Ahuja, L.R., 2003. Scaling and estimating the soil water characteristic using a one-parameter model. *In*: Pachepsky, Ya., Radcliffe, D.E., Selim, H.M. (Eds.), Scaling Methods in Soil Physics. CRC Press, Boca Raton, FL, pp. 35-48.

Williams, R.D., Ahuja, L.R., Naney, J.W., 1992. Comparison of methods to estimate soil water characteristics from soil texture, bulk density, and limited data. Soil Sci. 153, 172-184.

Wösten, J.M.H., van Genuchten, M.Th., 1988. Using texture and other soil properties to predict the unsaturated soil hydraulic functions. Soil Sci. Soc. Am. J. 52, 1762-1770.

Chapter 6

EFFECT OF SOIL ORGANIC CARBON ON SOIL HYDRAULIC PROPERTIES

W.J. Rawls[1,*], A. Nemes[1] and Ya. Pachepsky[2]

[1]USDA-ARS Hydrology & Remote Sensing Lab, Bldg. 007, Rm. 104, BARC-W, Beltsville, MD 20705-2350, USA

[2]Environmental Microbial Safety Laboratory, USDA-ARS-BA-ANRI-EMSL, Bldg. 173, Rm. 203, BARC-EAST, Powder Mill Road, Beltsville, MD 20705, USA

*Corresponding author: Tel.: +1-301-504-8745; fax: +1-301-504-8931

1. INTRODUCTION

Soil hydraulic properties govern soil functioning in ecosystems and greatly affect soil management. Data on these properties are used in research and applications in hydrology, agronomy, meteorology, ecology, environmental protection, and many other soil-related fields. These properties are measured in some soil survey programs; however, it is impractical because of cost and time to measure these properties for all applications especially in large-scale applications. Hydraulic soil properties need to be estimated from other readily available soil properties. Regression equations for such estimation are often called pedotransfer functions (PTFs). The primary soil hydraulic properties that soil organic carbon affects are porosity, soil water retention and hydraulic conductivity. The following will be a summary of the effect of organic carbon or organic matter on these properties and how these effects can be incorporated into PTFs. Organic matter content is approximately equal to 1.724 times organic carbon content (Nelson and Sommers, 1982).

2. BULK DENSITY/POROSITY

Adams (1973) showed that the amount of organic carbon had a significant effect on the bulk density of soils and presented the following equation to describe the effect.

$$bd = 100/((om/bd_{om}) + ((100 - om)/bd_m)) \tag{1}$$

where bd is the soil bulk density (g/cm^3), om, the percent by weight of organic matter, bd_{om}, the average bulk density of the organic matter (0.224 g/cm^3), and bd_m, the bulk density of the mineral matter (g/cm^3).

Since the only unknown in Equation (1) is the bulk density of the mineral matter, Rawls (1983) developed from over 2700 soils a contour map of mineral bulk densities

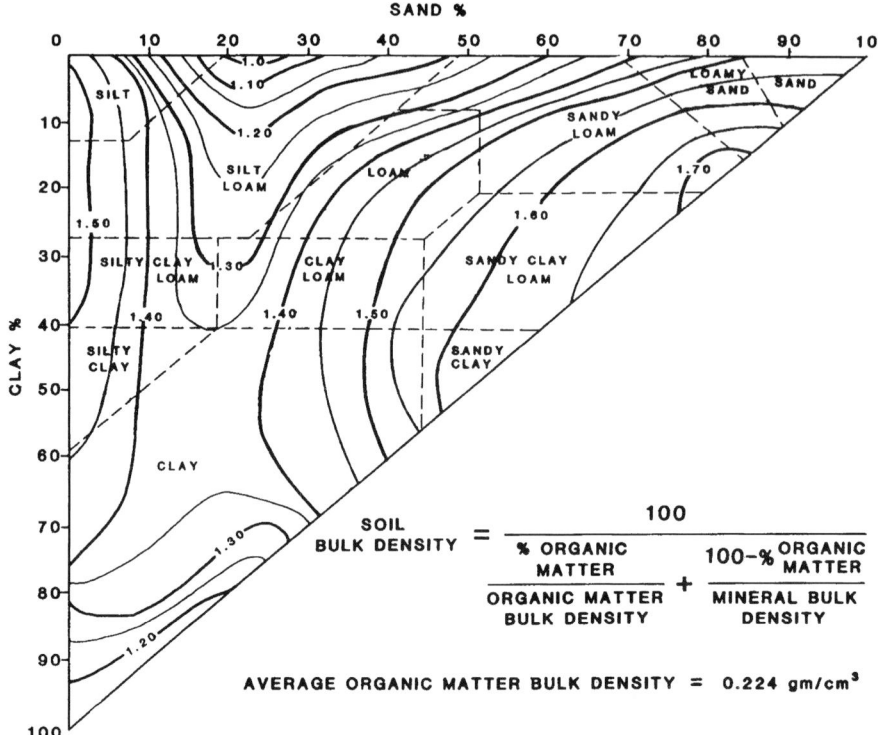

Figure 1. Mineral bulk density (g/cm^3).

overlaid on the USDA texture triangle (Figure 1). Use of Figure 1 in conjunction with the % sand and % clay the effect of organic carbon on bulk density can be determined.

The following model was developed to predict soil bulk density from % sand, % clay and % organic carbon using over 2100 A soil horizons taken from the USDA-NRCS Soil Characterization Database.

$$\begin{aligned}\rho_b = {} & 1.36411 + 0.185628(0.0845397 + 0.701658w - 0.614038w^2 - 1.18871w^3 \\ & + 0.0991862y - 0.301816wy - 0.153337w^2y - 0.0722421y^2 + 0.392736wy^2 \\ & + 0.0886315y^3 - 0.601301z + 0.651673wz - 1.37484w^2z + 0.298823yz \\ & - 0.192686wyz + 0.0815752y^2z - 0.0450214z^2 - 0.179529wz^2 \\ & - 0.0797412yz^2 + 0.00942183z^3) \end{aligned} \quad (2)$$

where: $x = -1.2141 + 4.23123\,\text{sand}$ $(0.004 < \text{sand}\,(P_{50-2000\,\mu m}) < 0.952(g/g))$; $y = -1.70126 + 7.55319\,\text{clay}$ $(0.002 < \text{clay}(P_{<2\mu m}) < 0.807(g/g))$; $z = -1.55601 + 0.507094\,\text{OM}$ $(1 < \text{Organic matter content} < 14.70(\%))$; $w = -0.0771892 + 0.256629x + 0.256704x^2 - 0.140911x^3 - 0.0237361y - 0.098737x^2y - 0.140381y^2 + 0.0140902xy^2 + 0.0287001y^3$.

The RMSR of the above model is 0.1322 g/cm³. Soil porosity can be derived from the soil bulk density using the following equation:

$$\varphi = 1 - (\rho_h/\rho_p) \qquad (3)$$

where φ is the ratio of voids to the total volume of sample (cm³/cm³), ρ_p, the particle density (typically = 2.65 g/cm³), ρ_b, the soil bulk density (g/cm³).

3. SOIL WATER RETENTION

Extensive research has shown that water retention is a complex function of soil structure and composition (Rawls et al., 1991; Wösten et al., 2001). Soil organic carbon content and composition affect both soil structure and adsorption properties and, therefore, water retention may be affected by soil organic carbon. Reports on the effect of soil organic carbon on soil water retention are contradictory as seen in Table 1. Hudson (1994)

Table 1
Observed effect of organic carbon content on soil water retention at two water potentials

Authors	−33 kPa	−1500 kPa
Rawls et al. (2003)	Yes	Yes
Bell and van Keulen (1995)	No	Yes
Hudson (1994)	Yes	Yes
Danalatos et al. (1994)	No	No
Bauer and Black (1981)	Yes	Yes
Beke and McCormick (1985)	No	Yes
De Jong (1983)	Yes	Yes
McBride and MacIntosh (1984)	No	Yes
Riley (1979)	Yes	Yes
Lal (1979)	No	No
Calhoun et al. (1973)	Yes	No
Petersen et al. (1968)	No	Yes
Salter and Haworth (1961)	No	No
Jamison and Kroth (1958)	Yes	Yes

found for three texture classes (sand, silt loam, silty clay loam) that as the soil organic carbon increased the volume of water held at field capacity increased at a much greater rate than that held at wilting point. Rawls and Brakensiek (1982) and Rawls et al. (1983) found it useful to include the organic carbon content in the list of inputs to PTFs to estimate water contents at both −33 and −1500 kPa. Bell and van Keulen (1995) saw the need to use both organic carbon content and pH in estimating water content at wilting point. Beke and McCormick (1985) and Petersen et al. (1968) found it useful to employ data on organic carbon content to estimate water content at −1500 kPa, but not at −33 kPa. In contrast, the use of organic carbon content improved PTFs at −33 kPa but not at −1500 kPa in the work of Calhoun et al. (1973). Viville et al. (1986) indicated that the differences in water retention within soil profiles correlated with profiles of the organic carbon content. Hollis et al. (1977) found the organic carbon content to be the most influential soil variable to estimate water content at −5 kPa. Lal (1979) and Danalatos et al. (1994) did not find any effect of organic carbon content on water retention; the latter authors attributed that to the generally low organic carbon content in their samples. Similarly, Puckett et al. (1985) did not use organic carbon in PTFs because of its low level in samples. Bauer and Black (1981) found that the effect of organic carbon on water retention in disturbed samples was substantial in sandy soil and marginal in medium- and fine-texture soils. De Jong (1983) experimented with disturbed soil samples and found that the increase in organic carbon content meant higher water content at all suctions. Similar observations were made by several other authors (Jamison and Kroth, 1958; Petersen et al., 1968; Riley, 1979; Ambroise et al., 1992; Kern, 1995). Salter and Haworth (1961) argued that organic carbon might not be an important predictor to estimate water content at specific suctions, but it is an important factor if water contents at field capacity and wilting point are measured directly. McBride and MacIntosh (1984) found that organic carbon content affected water retention at −1500 kPa only when this content was larger than 5%. Most of the cited authors worked with small number of soils from a specific region. Kay et al. (1997) compared relative effects of organic carbon on water retention using PTFs developed in different regions and found large regional differences. Rawls et al. (2003) found that the relationship of soil water retention to organic carbon content is affected by proportions of textural components and that soil water retention at −33 kPa is affected by the organic carbon more strongly than water retention at −1500 kPa. Also, water retention of soils with coarse texture is substantially more sensitive to the amount of organic carbon as compared with fine-textured soils and the effect of changes in organic carbon content on soil water retention depends on the proportion of textural components and the amount of organic carbon present in the soil. At low carbon contents, an increase in carbon content leads to an increase in water retention in coarse soils and to decrease in water retention in fine-textured soils. At high carbon contents, increase in carbon contents results in an increase in water retention of all textures.

Rawls et al. (2003) with additional studies have produced the most detailed analysis of the effect of organic carbon on the water retention at −33 and −1500 kPa and these analyses will be presented in detail.

3.1. Data

Two subsets of data were extracted from the USDA-NRCS National Soil Characterization database (Soil Survey Staff, 1997). The methods of soil analysis are given in the Soil Survey Staff (1996).

The first subset of data was reported by Rawls et al. (2003) and consisted of about 12,000 samples. This subset is referred below as "all horizons." The samples had data on soil texture, organic carbon content, water retention at −33 and −1500 kPa, bulk density at −33 kPa, and taxonomic characterization. Sandy loams, loams and silt loams were represented best and together constituted more than 60% of all samples. Silts, sands, and sandy clay loams were represented each with less than 300 samples. Mollisols and Alfisols were better represented than other taxonomic orders, whereas Spodosols, Oxysols, and Histosols were represented relatively poorly. The organic carbon content in the samples encompassed a range from 0.1 to 20%. The second subset consisted of 2149 samples. This subset of data called below "A horizons" was extracted according to the following criteria: mineral soil horizons were selected from the contiguous United States having horizon notation "A", "A1" and "Ap" (and their derivatives), with the condition that the top of the horizon was at the soil surface. Organic carbon content of the selected soils had to be between 0.6 and 8.7%, and bulk density was not allowed to be below 0.5 g/cm^3 or above 2.0 g/cm^3. Soils were filtered for obvious inconsistency in physical and hydraulic data. The samples had data on soil texture, organic carbon content, water retention at -33 and -1500 kPa, bulk density at -33 kPa, and taxonomic characterization. Almost 25% of the samples belonged to pedons that had no information on soil taxonomy. Mollisols and Alfisols were particularly well represented. Aridisols, Entisols, Ultisols, Vertisols and Inceptisols were all well represented, whereas Andosols, Spodosols and Oxisols were poorly represented in the selected data set. In terms of texture groups, silt loams contributed 34.2% to the data set, and soils with loam, silty clay loam and sandy loam texture were represented with 14.8, 13.4 and 12.9%, respectively. Other textures were represented with 1.5–6.5%, with no sandy clays in the selected data. Samples in the data set were well distributed over the United States, representing 46 states.

3.2. Methods to quantify the effect of organic carbon content on water retention

To develop PTFs a combination of regression tree analysis and group method of data handling (GMDH) were used. The following is a brief description of the techniques.

Regression tree modeling is an exploratory technique that can be used to uncover structure in data (Clark and Pregibon, 1992). Regression trees can use both categorical and numerical variables as predictors (Breiman et al., 1993). The resulting model partitions data first into two groups, which are then further split into subgroups, and so on. This can be done to a required level of partitioning, providing as homogeneous subgroups as possible at each level. The final fit of the model to data looks like a tree with two branches originating in each node. The technique had previously been used for the estimation of soil hydraulic properties by, e.g., Rawls and Pachepsky (2002), van Lanen et al. (1992), McKenzie and Jacquier (1997) and McKenzie and Ryan (1999). Regression trees in this study were developed using the SPLUS software (Mathsoft, 1999).

GMDH (Farrow, 1984) is a powerful tool to describe complex input–output relationships. The method performs an automated selection of essential input variables and builds hierarchical polynomial regressions of necessary/desired complexity to estimate the output variable. Polynomials are built from some of the input variables and from sub-polynomials that are built from smaller subsets of input variables that may be better predictors of the output variable than some of the input variables alone. The best of such polynomials are included in the set of input variables that may again serve as input to obtain even better estimates. The GMDH was used to develop the equations to relate water retention to contents of textural components and organic carbon. GMDH combines advantages of regression analysis and artificial neural networks (Hecht-Nielsen, 1990). The GMDH constructs a flexible equation of neural-network type to relate the inputs to outputs, and at the same time has a built-in algorithm to retain only essential input variables (Farrow, 1984). The GMDH has been recently used to develop PTFs (Pachepsky and Rawls, 1999; Gimènez et al., 2001). For this application we used the commercial

GMDH software ModelQuest (AbTech Corp, 1996). Values of non-problem specific variables were set to the default value in the software.

The advantage of regression trees is the transparency of results, and that the relative importance of inputs can be easily assessed. The GMDH algorithm provides equations that can be used both for predictions and for sensitivity analysis. Regression trees were used to explore data and to help in grouping of soils in order to obtain more accurate model using GMDH. Both the regression trees and the GMDH are iteratively built models with progressively increasing complexity. The processes have to be stopped to prevent overfitting, otherwise the predictive capability of the resulting models with respect to new data will be deplorable. The jack-knife cross-validation method (Good, 1999) was applied with both algorithms in this work. The database was randomly divided, 10 times, into development and testing subsets in 9:1 proportion, and the average accuracy of estimating water retention was expressed by root-mean squared residuals (RMSR). The number of terminal nodes in regression tree analysis and the number of iterations in the GMDH algorithm were varied to provide the minimum average RMSR in the validation datasets.

To quantify the PTF accuracy RMSR was used. The data sets were randomly split to a development and a validation subset using the default ratio in ModelQuest (i.e., 3:1) to split the data in each case. RMSR was calculated in each case for the validation subset. The influence of OM content on the estimates – i.e., the sensitivity of the developed models to changes in OM content – was assessed globally, as provided by the ModelQuest software.

3.3. Regression trees

3.3.1. Predictors: soil texture class and organic carbon content

The regression tree analysis on the "all horizons" data set for the soil water retention at -33 kPa is shown in Figure 2. The first split divides soils by their textural class. Sands, loamy sands, and sandy loams form one large group, and soils with finer texture form another one. The OM content is the most important splitting variable in coarse-textured soils. Finer soils are further split in the group with fine texture (loam, silt loam, sandy clay loam, silt, clay loam, and sandy clay) and another group with very fine texture (silty clay loam, silty clay, and clay). The organic carbon content is the next important split variable in soils with fine texture, whereas soils with very fine texture continue to be partitioned by textural class. Only silty clay and clay soils do not show a need to use organic carbon content to make the soil groups more homogeneous by their water retention at -33 kPa.

The regression tree analysis on the "all horizons" data set for the soil water content at -1500 kPa is shown in Figure 3. Soil textural class and organic carbon content are predictor variables. The first split divides soils by their textural class. Sands, loamy sands, sandy loams, loam, silt loam, sandy clay loam and silt form one large group, and clay loam, silty clay loam, sandy clay, silty clay and clay form another one. The second and third splits are also based on texture class except for the third split of loam, silt loam and silt group when OM content is the most important splitting variable. Essentially the eleven texture classes were grouped into seven groups. Organic carbon is primary split for the fourth split with limited grouping of textures.

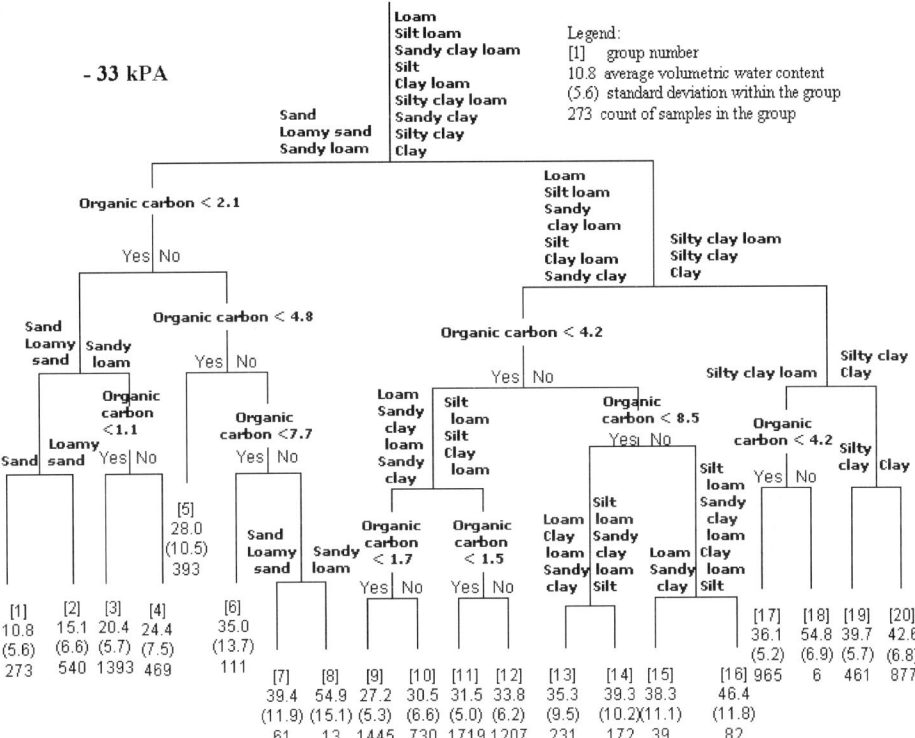

Figure 2. Effect of soil texture classes and organic carbon content on water retention at −33 kPa.

3.3.2. *Predictors: soil texture class, soil taxonomic order and organic carbon content*

Overall, from the used input information, soil texture class is by far the most important factor in determining soil water retention at − 33 kPa. Soils are split by texture class in two or three branching levels. Coarse-textured soils (sand, loamy sand, sandy loam) are first split from other textures. Sands and loamy sands are further split by soil taxonomic order, regardless of their organic carbon content. For loamy sands, organic carbon content is the next most important splitting factor in estimating water retention. Finer textured soils are further split by texture at two levels. Clearly, these splits result in estimating larger average water contents for finer textured soils. For all texture groups except clays, organic carbon content is shown to be the next most important factor. For the coarsest (sand, loamy sand) and finest (clay) textured soils organic carbon content does not seem to play a big role in determining water content. For all other texture classes organic carbon content is the next factor to further split soils into more homogeneous groups.

Overall, from the used input information, soil texture class is by far the most important factor in determining soil water retention at − 1500 kPa. Soils are split by texture class in the first two split in the regression tree for all soils. Texture class is the third splitting factor to separate sand and loamy sand soils from soils with sandy loam and silt textures. In all but one cases taxonomic order is the most important factor after soil texture class.

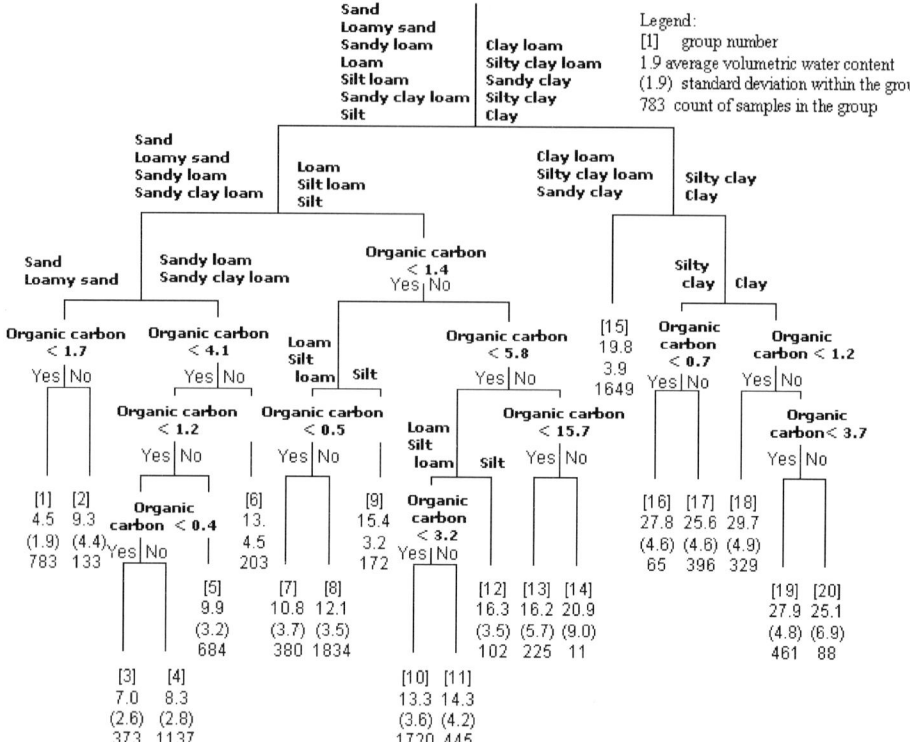

Figure 3. Effect of soil texture classes and organic carbon content on water retention at −1500 kPa.

Organic carbon content is less important as a splitting factor for all soils except those with clay loam and silty clay loam texture.

3.3.3. Predictors: soil taxonomic order and organic carbon content

Overall, from the used input information, taxonomic order is by far the most important factor in determining soil water retention at −33 kPa. Vertisols, Andisols and Spodosols belong to one node, which displays larger average water content. We have to note, that the number of Andisols and Spodosols was very low to allow general conclusions. This group is further split by organic carbon content. Interestingly, organic carbon content was negatively correlated with the amount of retained water. Further analysis unveiled, however, that it was simply because most Vertisols had low organic carbon content whereas almost all Spodosols and Andisols belonged to the group with larger organic carbon contents (data not shown). Vertisols have the largest clay content, which is known for its positive effect on the amount of retained water. For the other soils orders (i.e., Alfisols, Aridisols, Entisols, Inceptisols, Mollisols, Oxisols and Ultisols) organic carbon content was the next important split variable at two levels of split. Among soils with the smallest organic carbon content Entisols and Ultisols have lower average water retention than the other soil orders. Aridisols and Ultisols have a smaller amount of water retained at

−33 kPa than the other soil orders when organic carbon content is between 1.28 and 9.7%. We note, that similarly to Andisols and Spodosols, information obtained for Oxisols may not be enough to draw general conclusions due to the low number of available samples.

Overall, from the used input information, taxonomic order is by far the most important factor in determining soil water retention at −1500 kPa. Vertisols were first separated from all other soil orders. Vertisols have on average the largest clay content. All other soil orders were split by soil order information again at the next branching level. Andisols, Inceptisols, Oxisols and Mollisols have, on average, larger water content at −1500 kPa than Alfisols, Aridisols, Entisols, Spodosols and Ultisols. In each of these groups, organic carbon content is the most important factor to determine water content for the next two levels, except for one case. In all cases (except for Vertisols) increasing OM content had a positive effect on the amount of retained water.

3.3.4. Predictors: sand, silt, clay and organic carbon content

Overall, from the used input information in determining soil water retention at −33 kPa the main dividing criterion was sand content. The second level of splitting was on sand and clay content. Organic carbon content was the next splitting criterion for all soils except for those with the sand content greater than about 67%. At this level of branching, influence of organic carbon content seems to be the smallest for the coarsest and the finest textured soils.

Overall, from the used input information in determining soil water retention at −1500 kPa the main dividing criterion was clay content in the first three branching levels. Organic carbon content appears in the fourth level for soils with clay content less than about 15%, thus coarse-textured soils. The increase in organic carbon content indicates an increase in the average water content. In soils with clay content larger than 15%, the average amount of retained water increases as clay content increases.

3.3.5. Summary

Regression tree results provided a useful visualization of relationships in the database in this work and allowed a preliminary ranking of input soil properties to estimate water retention. The differences in relative importance of organic carbon content in coarse and fine-textured soils could be clearly seen. Organic carbon content appeared to be an important soil property to improve estimation of soil water retention from soil texture. Including OM content improved the accuracy of regression tree predictions (Table 2) as noted by the 15 and 10% decrease in root mean squared residuals that was achieved at −33 and −1500 kPa, respectively. The structure-forming effect of organic carbon is affecting the water retention at water content close to field capacity to larger extent than water retention close to the wilting point. The water retention of organic carbon itself is a probable reason of its effect on water retention at −1500 kPa, although the organic carbon is also known to modify the availability of adsorption sites of clay minerals to water (Cristensen, 1996). Using sand, silt, and clay contents resulted in slight improvement over using just textural class information (Table 2). Such improvement was more pronounced for water retention at −1500 kPa than at −33 kPa. The textural composition is a significant factor affecting the importance of organic carbon content in estimating water retention. Organic carbon content is a leading variable to group coarse soils by their water retention. Using the taxonomic order as a predictor improves predictions as compared with using only textural class or textural composition for the −33 and −1500 kPa matric potential. However, using

Table 2
Root-mean squared residuals of the water retention predictions using regression trees

Predictors	Volumetric water content at -33 kPa, θ_{33}, %	Volumetric water content at -1500 kPa, θ_{1500}, %	$\theta_{33} - \theta_{1500}$	Slope
No textural information				
Organic carbon	9.0	6.9	6.6	0.092
Textural class				
Textural class	7.4	3.9	6.6	0.077
Textural class + taxonomic order	6.9	3.8	6.3	0.076
Textural class + organic carbon	6.4	3.7	6.1	0.076
Textural class + organic carbon + taxonomic order	6.3	3.6	5.9	0.074
Textural composition				
Clay + silt + sand	7.0	3.4	6.2	0.069
Clay + silt + sand + organic carbon	6.2	3.1	5.8	0.069
Clay + silt + sand + organic carbon + taxonomic order	5.9	3.1	5.8	0.067
Texture + bulk density				
Clay + silt + sand + bulk density	5.6	3.1	5.6	0.067

organic carbon with textural class or textural composition makes predictions substantially better and a further addition of the taxonomic order does not bring much improvement.

As noted in Table 2, bulk density was a better complimentary predictor of water retention at -0.33 kPa as compared with organic carbon content, therefore, if available it may be a preferable predictor of water retention. Bulk density values can represent the effect of organic carbon on water retention if the primary effect of organic carbon is changing bulk density. Because the correlation between organic carbon content and bulk density is relatively low, improvements by adding bulk density indicates that bulk density incorporated other structural information that has an influence on water retention.

3.4. Group method of data handling

GMDH was used on the "A horizon" data set to estimate water retention at -33 and -1500 kPa from (1) sand and clay content and (2) sand, clay and OM content to examine the influence of organic carbon content as an input parameter. Soils of the data set were first used without further grouping. Next, we grouped soils by soil taxonomy order and separately by soil texture class. Results are shown in Table 3.

Table 3
Root-mean squared residuals (m³/m³) of the water retention estimations using group method of data handling using estimators: (1) sand and clay content and (2) sand, clay and organic carbon content

Soil group	Sample size	Volumetric water content at −33 kPa		Influence of organic carbon (%)	Volumetric water content at −1500 kPa		Influence of organic carbon (%)
		Sand and clay	Sand and clay organic carbon		Sand and clay	Sand and clay organic carbon	
All	2149	0.0490	0.0469	18	0.0265	0.0245	8
Mollisols	669	0.0447	0.0447	18	0.0224	0.0200	18
Alfisols	394	0.0447	0.0490	27	0.0173	0.0173	22
Aridisols	125	0.0678	0.0686	6	0.0224	0.0224	9
Entisols	122	0.0510	0.0500	6	0.0316	0.0316	15
Ultisols	102	0.0458	0.0400	22	0.0283	0.0245	18
Vertisols	98	0.0490	0.0490	0	0.0557	0.0557	0
Inceptisols	89	0.0520	0.0510	18	0.0265	0.0283	11
SiL	736	0.0490	0.0500	16	0.0200	0.0200	13
L	318	0.0480	0.0490	30	0.0300	0.0265	27
SicL	288	0.0387	0.0332	15	0.0245	0.0300	57
SL	277	0.0632	0.0678	71	0.0224	0.0224	67
C	140	0.0529	0.0600	40	0.0387	0.0608	31
cL	130	0.0592	0.0557	39	0.0458	0.0656	31
SiC	83	0.0447	0.0557	14	0.0374	0.0447	46
S	66	0.0557	0.0959	18	0.0265	0.0173	66
Si	40	0.1400	0.0877	63	0.0374	0.0316	43
SCL	39	0.0529	0.0872	18	0.0173	0.0387	32
lS	32	0.0469	0.0469	75	0.0412	0.0141	86

3.4.1. No split of the data

The average value of RMSR was 0.049 m^3/m^3 for estimating water retention at −33 kPa using soil textural information for all soils. Adding organic carbon content as an input to the model resulted in a small but not significant improvement of the model (RMSR = 0.0469). Organic carbon content had an overall weight of 18% in the model. For water contents at −1500 kPa we obtained a similar trend (RMSR = 0.0265 and 0.0245, respectively) for the two models, but organic carbon content had only 8% of influence on the model outcome.

3.4.2. Split by taxonomic order

The number of samples in each group can be consulted in Table 3. New models were developed after soil grouping. Seven orders had enough data to allow the development of predictive regression networks. The average RMSR was between 0.0447 and 0.0678 m^3/m^3 for models that did not use organic carbon content as input, and between 0.0447 and 0.0686 m^3/m^3 for models that used organic carbon content. Improvements in the estimation of water contents at −33 kPa by adding organic carbon content as input were only visible for Entisols, Ultisols and Inceptisols, however, the extent of improvement was not significant. For the other soil orders estimations did not improve. Despite this, organic carbon content appears in the model for all orders except for Vertisols.

We used the same grouping also to estimate water content at −1500 kPa. The range of average RMSR was between 0.0173 and 0.0557 m^3/m^{-3} with and without organic carbon content being an input to the models. Slight improvements are seen for Mollisols and Ultisols. Organic carbon content did not appear as a selected input in the model for Vertisols. For both water retention points, organic carbon content appeared in the model most significantly for Mollisols, Alfisols and Ultisols.

3.4.3. Split by texture classes

Grouping the 2149 soils by USDA texture classes (USDA, 1951) prior to model development resulted a very uneven distribution of soils in the groups. Sandy clays were not represented, and sands, silts, sandy clay loams and loamy sands were poorly represented (Table 3). As the data sets are further split into development and validation data sets, results derived from small groups of soils may be very unreliable.

The average RMSR was between 0.0387 and 0.140 m^3/m^{-3} for models that did not use OM contents as input, and between 0.0332 and 0.0959 m^3/m^{-3} for models that used OM contents. Improvements in the estimation of water contents at −33 kPa by OM content were only visible for silty clay loams, clay loams and silts, however, the extent of improvement was only significant for silts, a texture group with only 40 soils. For other textures estimations did not improve. Despite this, organic carbon content appears in the model for all textures, most importantly for loamy sands (75%), sandy loams (71%) and silts (63%).

The same grouping was used to estimate water content at −1500 kPa. The range of average RMSR was between 0.0173 and 0.0458 m^3/m^3 without OM content and between 0.0173 and 0.0656 m^3/m^3 with organic carbon content being an input to the models. Improvements are seen for soils with loam, sand, silt and loamy sand textures. Organic carbon content does appear as a selected input in the model for all textures. It is most significant in coarse-textured soils (sands, loamy sands, sandy loams).

3.5. Pedotransfer models

Application of the GMDH with the "all horizons" data set resulted in the following equations to estimate water retention at -33 kPa (θ_{33}) and -1500 kPa (θ_{1500}), in % volume.

$$\theta_{33} = 29.7528 + 10.3544(0.0461615 + 0.290955x - 0.0496845x^2 + 0.00704802x^3$$
$$+ 0.269101y - 0.176528xy + 0.0543138x^2y + 0.1982y^2 - 0.060699y^3$$
$$- 0.320249z - 0.0111693x^2z + 0.14104yz + 0.0657345xyz - 0.102026y^2z$$
$$- 0.04012z^2 + 0.160838xz^2 - 0.121392yz^2 - 0.0616676z^3) \qquad (4)$$

$$\theta_{1500} = 14.2568 + 7.36318(0.06865 + 0.108713x - 0.0157225x^2 + 0.00102805x^3$$
$$+ 0.886569y - 0.223581xy + 0.0126379x^2y - 0.017059y^2 + 0.0135266xy^2$$
$$- 0.0334434y^3 - 0.0535182z - 0.0354271xz - 0.00261313x^2z - 0.154563yz$$
$$- 0.0160219xyz - 0.0400606y^2z - 0.104875z^2 + 0.0159857xz^2 - 0.0671656yz^2$$
$$- 0.0260699z^3)$$

where $x = -0.837531 + 0.430183 \times$ Organic carbon; $y = -1.40744 + 0.0661969 \times$ Clay; $z = -1.51866 + 0.0393284 \times$ Sand; $0.02 <$ Organic Carbon < 28.44, $0.0 <$ Clay < 90, $0.7 <$ Sand < 95.

The RMSR of those equations is 0.063 and 0.031 m^3/m^3 for -33 and -1500 kPa, respectively. The RMSR were higher, 0.068 and 0.034 m^3/m^3, respectively, when only clay and sand content were used.

The GMDH equations were used to generate isolines of water content at -33 kPa for various values of organic carbon contents as shown in Figure 4. It can be seen that, for the same proportion of clay and sand, the water retention mostly increases as the organic carbon content increases. However, a decrease in water retention with the increase in organic carbon content value can be seen for fine-textured soils with high clay content. Where water retention increases in parallel with organic carbon content, the largest increment in water contents occurs in coarse-textured soils. Isolines of water contents at -1500 kPa for various organic carbon contents are shown in Figure 5. A substantial increase in water retention with the increase of organic carbon content can be observed in samples with low clay contents. The opposite trend can be seen in soils with very large clay contents.

The GMDH equations allow one to estimate the sensitivity of water retention to changes in organic carbon content for different levels of the initial organic carbon content in soil. An example of such estimation is shown in Figure 6 where changes in water content at -33 kPa per 1% increase in organic carbon content are shown for three levels of the organic carbon content before the change. At low organic carbon content of 1%, the

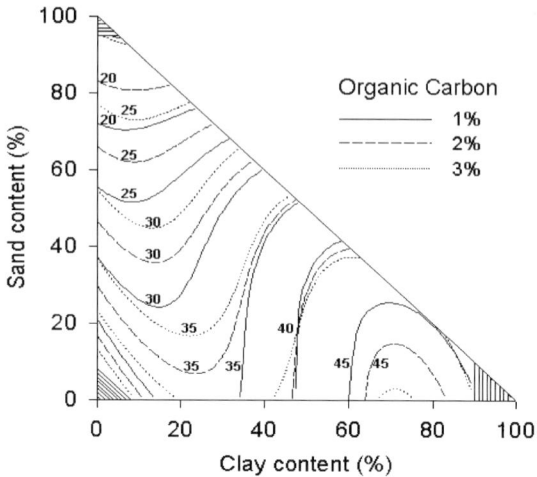

Figure 4. Effect of organic carbon on water retention at −33 kPa.

sensitivity is the highest. Water retention dramatically increases (decreases) in soils with low (high) clay content. An intermediate value of the initial organic carbon content of 3% makes the changes less dramatic but clay contents of about 50% continue to separate regions of the textural triangle in which changes in water retention occur in the same or in opposite direction with changes in organic carbon content. The high initial value of the

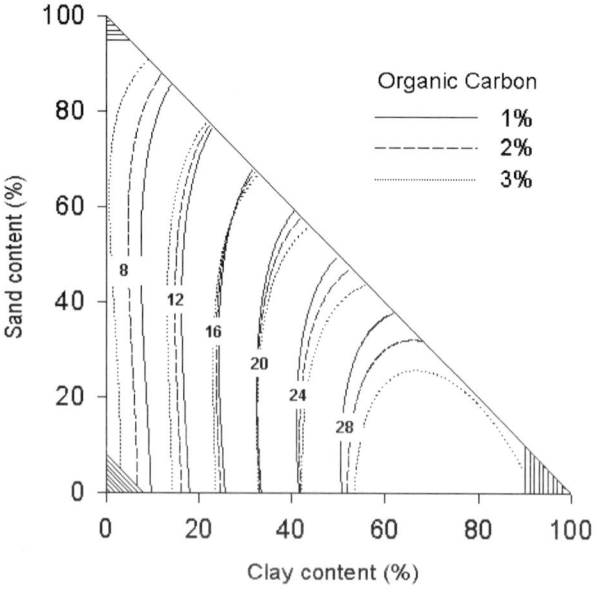

Figure 5. Effect of organic carbon on water retention at −1500 kPa.

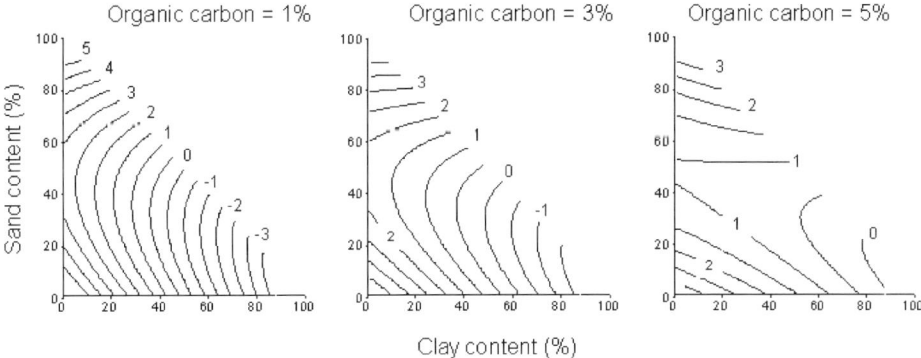

Figure 6. Changes in water retention at −33 kPa (vol%) per 1% change in organic carbon content with various initial organic carbon contents.

organic carbon content of 5% results in a different sensitivity pattern. An increase in organic carbon content leads to the increase in water retention practically for all textures, although high clay content leads to comparatively smaller increases. The tendency of silty soils to respond to the changes in organic carbon content with changes in their water retention similarly to sandy soils having the same clay contents can be clearly seen at 5% initial organic carbon content value. The same tendency can be traced at lower initial organic carbon content values.

The sensitivity analysis also shows that water retention of coarse-textured soils is much more sensitive to changes in organic carbon content as compared with fine-textured soils. These results concur with data of Khaleel et al. (1981) who reviewed experiments on application of organic waste and found that the increase in gravimetric water content at −33 kPa is larger for coarse-textured soils than those for fine-textured soils. Figure 4 shows that clay content alone or sand content alone are not satisfactory predictors of the effect of the organic carbon content on water retention at −33 kPa. For example, the 35% isoline with 2% organic carbon content in Figure 4 is applicable to samples with 20% of sand and 5% of clay, 10% of sand and 20% of clay, and 20% of sand and 35% of clay. Ten percent of sand provides an increase in water retention at low clay contents and a decrease in water retention at high clay contents of more than 50%. The decrease in water retention in heavy clay soils with increasing organic carbon content seen in Figures 4–6 was not previously reported in literature. This result may be database-specific since our data for high clay content soils are relatively sparse. It may also be related to the fact that many of the soils with high clay content in the database are Vertisols in which increase in organic carbon content decreases bulk density and decreases the volumetric water content, although gravimetric water content may actually increase.

3.6. Summary

The sensitivity of water retention to changes in organic carbon content decreases as the initial organic carbon content increases (Figure 5). A similar conclusion can be drawn from equations presented by Khaleel et al. (1981) for water retention of soils amended with organic waste in soils in the USA, England, India, and Germany. Their equations also show that the relative increase becomes smaller as the increase in OM content grows. A similar pattern can be observed in Figure 6. The reduction of the effect of increasing organic carbon content on water content at -5 kPa with the increase in the original value of the organic carbon content was reported by Hollis et al. (1977). Water retention of peat soils will probably present a limit case for the increase of organic carbon content in samples. Soil survey databases contain both data on soils in natural ecosystems and on agricultural soils showing similar response of soil water retention to changes in organic carbon content. Organic amendments caused changes in soil water retention quantitatively and qualitatively similar to those reported in this work (Gupta et al., 1977; Unger and Stewart, 1974; Kladivko and Nelson, 1979).

4. SATURATED HYDRAULIC CONDUCTIVITY

Examples of internationally published PTFs that use OM content as input are those of Vereecken et al. (1990), Tamari et al. (1996), Wösten (1997) and Wösten et al. (1999).

Wagner et al. (2001) reviewed eight well-known and accepted PTFs used for predicting saturated hydraulic conductivity from routinely available soils data. One equation was based solely on particle-size information, three equations used organic carbon as an input variable, three equations used bulk density or porosity as an input variable which, as shown in Equation (1), incorporates organic carbon effects and one equation used both porosity and the slope of the water retention curve which both incorporate the effects of organic carbon. The equation that uses both porosity and the slope of the water retention curve was proposed by Ahuja et al. (1984). It is a generalized Kozeny–Carman (Carman, 1956) equation relating the matrix saturated hydraulic conductivity to effective porosity in the following form

$$K_s = C\phi_e^m \tag{5}$$

where K_s is the saturated hydraulic conductivity (mm/h); ϕ_e, the effective porosity (m^3/m^3) (total porosity, φ, minus water content at -33 kPa pressure head, θ_{33}) and C and m are empirically derived constants. Rawls et al. (1998) redefined the exponent in Equation (5) as three minus the Brooks–Corey pore-size distribution index (λ). Ahuja et al. (1985) showed that the slope of the water retention curve can be adequately obtained from a simple log–log plot through two points. Since water content at -33 and -1500 kPa are usually measured values or can be easily predicted from other soil properties, the Brooks–Corey pore-size distribution index (λ) can be obtained by fitting a log–log plot of water content vs. pressure head using only the -33 and -1500 kPa water contents. Fitting data for 26 soil texture/porosity classes resulted in C equal to 1930, resulting in the following equation:

$$K_s = 1930\phi_e^{3-\lambda} \tag{6}$$

where K_s is the saturated hydraulic conductivity (mm/h); ϕ_e, the effective porosity (m^3/m^3) (total porosity, φ, minus water content at -33 kPa pressure head, θ_{33}). Using Equation (6), we developed Figure 7 illustrating the effect that organic carbon (OC)

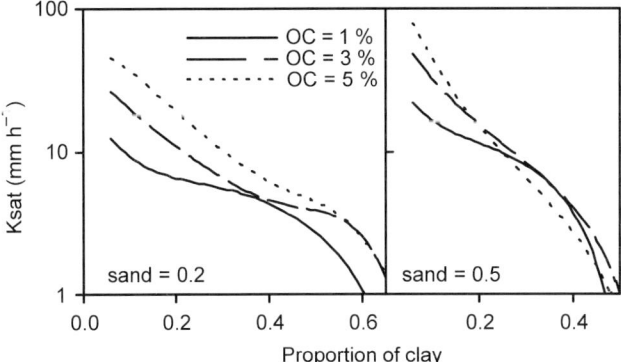

Figure 7. Effect of organic matter and texture on saturated hydraulic conductivity.

has on the matrix saturated hydraulic conductivity with texture. As can be seen in Figure 7 for the proportion of clay less than about 0.3, as organic carbon increases the saturated hydraulic conductivity increases. With the proportion greater than 0.3 the effect of organic carbon is varied indicating a very complex relationship.

5. CONCLUSIONS

Modeling of changes of organic carbon content in soils and related changes in ecosystem productivity attracts significant attention with regard to climate changes and management changes. Existing models lack the feedback effect of organic carbon content accumulation on water retention and saturated hydraulic conductivity. Results presented in this chapter can be used in those models to improve their predictive ability.

REFERENCES

AbTech Corp., 1992–1996. ModelQuest. Users Manual. Version 4.0. Charlottesville, VA.
Adams, W.A., 1973. The effect of organic matter on the bulk and true densities of some uncultivated podzolic soils. J. Soil Sci. 24, 10-17.
Ahuja, L.R., Naney, J.W., Green, R.E., Nielsen, D.R., 1984. Macroporosity to characterize spatial variability of hydraulic conductivity and effects of land management. Soil Sci. Soc. Am. J. 48, 699-702.
Ahuja, L.R., Naney, J.W., Williams, R.D., 1985. Estimating soil water characteristics from simpler properties or limited data. Soil Sci. Soc. Am. J. 49, 1100-1105.
Ambroise, B., Reutenauer, D., Viville, D., 1992. Estimating soil water retention properties from mineral and organic fractions of coarse-textured soils in the Vosges mountains of France. *In*: van Genuchten, M.Th., Leij, F.J., Lund, L.J. (Eds.), Indirect Methods for Estimating the Hydraulic Properties of Unsaturated Soils. University of California, Riverside, CA, pp. 453-462.

Bauer, A., Black, A.L., 1981. Soil carbon, nitrogen, and bulk density comparison in two cropland tillage systems after 25 years and in virgin grassland. Soil Sci. Soc. Am. J. 45, 1166-1170.

Beke, G.L., McCormick, M.J., 1985. Predicting volumetric water retention for subsoil materials from Colchester County, Nova Scotia. Can. J. Soil Sci. 65, 233-236.

Bell, A.M., van Keulen, H., 1995. Soil pedotransfer functions for four Mexican soils. Soil Sci. Soc. Am. J. 59, 865-871.

Breiman, L., Friedman, J.H., Olshen, R.A., Stone, C.J., 1993. Regression Trees. Chapman and Hall/CRC, Boca Raton, FL.

Calhoun, F.G., Hammond, L.C., Caldwell, R.E., 1973. Influence of particle size and organic matter on water retention in selected Florida soils. Soil Crop Sci. Fla. Proc. 32, 111-113.

Carman, P.C., 1956. Flow of Gases Through Porous Media. Academic Press, Inc., New York.

Clark, L.A., Pregibon, D., 1992. Tree-based models. *In*: Hastie, T.J. (Ed.), Statistical Models in S. Wadsworth. Pacific Grove, California, pp. 377-419.

Cristensen, B.T., 1996. Carbon in primary and secondary organomineral complexes. *In*: Carter, M.R., Stewart, B.A. (Eds.), Structure and Organic Matter Storage in Agricultural Soils. Lewis Publishers, Boca Raton, New York, pp. 97-165.

Danalatos, N.G., Kosmas, C.S., Driessen, P.M., Yassoglou, N., 1994. Estimation of the draining soil moisture characteristics from standard data as recorded in soil surveys. Geoderma 64, 155-165.

De Jong, R., 1983. Soil water desorption curves estimated from limited data. Can. J. Soil Sci. 63, 697-703.

Farrow, S.J., 1984. The GMDH algorithm. *In*: Farrow, S.J. (Ed.), Self-Organizing Methods in Modeling: GMDH-type Algorithms. Marcel Dekker, New York, pp. 1-24.

Gimènez, D., Rawls, W.J., Pachepsky, Y., Watt, J.P.C., 2001. Prediction of a pore distribution factor from soil textural and mechanical parameters. Soil Sci. 166, 79-88.

Good, P.I., 1999. Resampling Methods: A Practical Guide to Data Analysis. Birkhäuser, Boston.

Gupta, S.C., Dowdy, R.H., Larson, W.E., 1977. Hydraulic and thermal properties of a sandy soil as influenced by incorporation of sewage sludge. Soil Sci. Soc. Am. J. 41, 601-605.

Hecht-Nielsen, R., 1990. Neurocomputing. Addison-Wesley, Reading, MA.

Hollis, J.M., Jones, R.J.A., Palmer, R.C., 1977. The effect of organic matter and particle size on the water retention properties of some soils in the West Midlands of England. Geoderma 17, 225-238.

Hudson, B.D., 1994. Soil organic matter and available water capacity. J. Soil Water Conserv. 49, 189-194.

Jamison, V.C., Kroth, E.M., 1958. Available moisture storage capacity in relation to texture composition and organic matter content of several Missouri soils. Soil Sci. Soc. Am. Proc. 22, 189-192.

Kay, B.D., da Silva, A.P., Baldock, A.P., 1997. Sensitivity of soil structure to changes in organic carbon content: predictions using pedotransfer functions. Can. J. Soil Sci. 77, 655-667.

Kern, J.S., 1995. Evaluation of soil water retention models based on basic soil physical properties. Soil Sci. Soc. Am. J. 59, 1134-1141.

Khaleel, R., Reddy, K.R., Overcash, M.R., 1981. Changes in soil physical properties due to organic waste applications: a review. J. Environ. Qual. 10, 133-141.

Kladivko, E.J., Nelson, D.W., 1979. Changes in soil properties from application of anaerobic sludge. J. Water Pollut. Control Fed. 51, 325-332.

Lal, R., 1979. Physical properties and moisture retention characteristics of some Nigerian soils. Geoderma 21, 209-223.

Mathsoft, 1999. SPLUS 2000 Professional. User's Manual. Mathsoft, Cambridge, MA.

McBride, R.A., MacIntosh, E.E., 1984. Soil survey interpretations from water retention data: 1. Development and validation of a water retention model. Soil Sci. Soc. Am. J. 48, 1338-1343.

McKenzie, N.J., Jacquier, D.W., 1997. Improving the field estimation of saturated hydraulic conductivity in soil survey. Aust. J. Soil Res. 35, 803-825.

McKenzie, N.J., Ryan, P.J., 1999. Spatial prediction of soil properties using environmental correlation. Geoderma 89, 67-94.

Nelson, D.W., Sommers, L.E., 1982. Total carbon, organic carbon and organic matter. In: Page, A.L., Miller, R.H., Keeny, D.R. (Eds.), Methods of Soil Analysis, Part 2, Chemical and Microbiological Properties. Agronomy Society of America, Soil Science Society of America, Madison, WI.

Pachepsky, Ya.A., Rawls, W.J., 1999. Accuracy and reliability of pedotransfer functions as affected by grouping soils. Soil Sci. Soc. Am. J. 63, 1748-1757.

Petersen, G.W., Cunningham, R.L., Matelski, R.P., 1968. Moisture characteristics of Pennsylvania soils. II. Soil factors affecting moisture retention within a textural class- silt loam. Soil Sci. Soc. Am. Proc. 32, 866-870.

Puckett, W.E., Dane, J.H., Hajek, B.F., 1985. Physical and mineralogical data to determine soil hydraulic properties. Soil Sci. Soc. Am. J. 49, 831-836.

Rawls, W.J., 1983. Estimating soil bulk density from particle size analysis and organic matter content. Soil Sci. 135, 123-125.

Rawls, W.J., Brakensiek, D.L., 1982. Estimating soil water retention from soil properties. J. Irrig. Drain. Div., Proc. ASCE 198 (IR2), 166-171.

Rawls, W.J., Pachepsky, Ya.A., 2002. Soil consistence and structure as predictors of water retention. Soil Sci. Soc. Am. J. 66, 1115-1126.

Rawls, W.J., Brakensiek, D.L., Soni, B., 1983. Agricultural management effects on soil water processes. Part I. Soil water retention and Green–Ampt parameters. Trans. ASAE 26, 1747-1752.

Rawls, W.J., Gish, T.J., Brakensiek, D.L., 1991. Estimating soil water retention from soil physical properties and characteristics. Adv. Soil Sci. 16, 213-234.

Rawls, W.J., Gimenez, D., Grossman, R., 1998. Use of soil texture, bulk density and the slope of the water retention curve to predict saturated hydraulic conductivity. Trans. ASAE 4, 983-988.

Rawls, W.J., Pachepsky, Y.A., Ritchie, J.C., Sobecki, T.M., Bloodworth, H., 2003. Effect of soil organic carbon on soil water retention. Geoderma 116, 61-76.

Riley, H.C.F., 1979. Relationship between soil moisture holding properties and soil texture, organic matter content, and bulk density. Agric. Res. Exp. 30, 379-398.

Salter, P.J., Haworth, F., 1961. The available-water capacity of a sandy loam soil. I. The effects of farmyard manure and different primary cultivations. J. Soil Sci. 12, 335-342.

Soil Survey Staff, 1996. Soil Survey Laboratory Methods Manual, Soil Survey Investigations Report No. 42. Version 3.0. USDA-NRCS, Lincoln, NE.

Soil Survey Staff, 1997. National Characterization Data. Soil Survey Laboratory, National Soil Survey Center, and Natural Resources Conservation Service, Lincoln, NE.

Tamari, S., Wösten, J.H.M., Ruiz-Suárez, J.C., 1996. Testing an artificial neural network for predicting soil hydraulic conductivity. Soil Sci. Soc. Am. J. 60, 1732-1741.

Unger, P.W., Stewart, B.A., 1974. Feedlot waste effects on soil conditions and water evaporation. Soil Sci. Soc. Am. Proc. 38, 954-957.

USDA (United States Department of Agriculture), 1951. Soil Survey Manual, U.S. Dept. Agriculture Handbook No. 18. USDA, Washington, DC.

van Lanen, H.A.J., van Diepen, C.A.J., Reinds, G.J., de Koning, G.H.J., 1992. A comparison of qualitative and quantitative physical land evaluations, using an assessment of the potential for sugar-beet growth in the European Community. Soil Use Manag. 8, 80-89.

Vereecken, H., Maes, J., Feyen, J., 1990. Estimating unsaturated hydraulic conductivity from easily measured soil properties. Soil Sci. 149, 1-12.

Viville, D., Ambroise, B., Korosec, B., 1986. Variabilité spatiale des propriétés texturales et hydrodynamiques des 5015 dans le bassin versant du Ringelbach (Vosges, France). *In*: Vogt, H., Slaymaker, O. (Eds.), Erosion Budgets and their Hydrologic Basis, Z. Geomoroh. N.F., Suppl.-Bd 60, p. 2140.

Wagner, B., Tarnawski, V.R., Hennings, V., Müller, U., Wessolek, G., Plagge, R., 2001. Evaluation of pedo-transfer functions for unsaturated soil hydraulic conductivity using an independent data set. Geoderma 102, 275-297.

Wösten, J.H.M., 1997. Pedotransfer functions to evaluate soil quality. *In*: Gregorich, E.G., Carter, M.R. (Eds.), Soil Quality for Crop Production and Ecosystem Health. Developments in Soil Science, Vol. 25. Elsevier, Amsterdam, pp. 221-245.

Wösten, J.H.M., Lilly, A., Nemes, A., Le Bas, C., 1999. Development and use of a database of hydraulic properties of European soils. Geoderma 90, 169-185.

Wösten, J.H.M., Pachepsky, Ya.A., Rawls, W.J., 2001. Pedotransfer functions: bridging the gap between available basic soil data and missing soil hydraulic characteristics. J. Hydrol. 251, 123-150.

Chapter 7

USING SOIL MORPHOLOGICAL ATTRIBUTES AND SOIL STRUCTURE IN PEDOTRANSFER FUNCTIONS

A. Lilly[1,*] and H. Lin[2]

[1]Macaulay Land Use Research Institute, Craigiebuckler, Aberdeen AB15 8QH, Scotland, United Kingdom

[2]Department of Crop and Soil Sciences, 116 A.S.I. Building, The Pennsylvania State University, University Park, PA 16802, USA

*Corresponding author: Tel.: +44(0)1224-498200

1. INTRODUCTION

Soil morphological attributes such as pedogenic horizon type, color, texture, structure, consistence, mottles, concretions and clay films are routinely described in the field during the inspection of soil profiles. These descriptions are generally qualitative or semi-quantitative, categorical assessments of the visual and tactile qualities of the soil. Soil structure, along with the abundance and size of plant roots and visible pores, are attributes that are often recorded in these field descriptions, but the quantity, size, shape, and continuity of visible pores are generally less well documented because of the complexity and dynamic nature of soil pores.

As soil morphology is often a key component of soil classification (and of characterization), there is a considerable volume of morphological data available within the databases of many national and regional soil survey organizations. In most cases, the volume of these data is vastly greater than measured soil hydraulic properties despite the necessity of these data for environmental modelling. It has been shown that it is possible to develop rule-based relationships between soil hydraulic properties and soil morphology for a variety of purposes (e.g., McKeague et al., 1982; Boorman et al., 1995) or to improve equation-based PTFs (e.g., Lin et al., 1999b; Pachepsky and Rawls, 1999). Although soil morphological data are mainly suited to the development of class PTFs, continuous PTFs could also be developed using quantified soil morphological attributes as inputs.

Pedotransfer functions that are based on the statistical relationship between soil hydraulic properties and soil texture are not readily transferable to other bioclimatic zones (McKeague et al., 1991; Wösten et al., 2001; O'Connell and Ryan, 2002). This may be due, in a large part, to the fact that soils with similar soil textures, but influenced by different moisture or thermal regimes, will not necessarily develop the same structure and pore architectures. Wagner et al. (1998) reported that predictions of soil hydraulic properties from a number of models based only on soil texture gave poor results in structured soils. However, one of the models that incorporated a measurement of retained moisture content (an additional measure of soil porosity) gave better predictions. In general, PTFs that relate soil hydraulic properties to texture alone cannot correctly predict saturated hydraulic conductivity (*Ksat*) or other hydraulic values for soils that contain large cracks, worm holes, or root channels. There is evidence that many clayey soils, especially those having strong, fine blocky or granular structure, and with a large number of biopores or cracks, had *Ksat* values as great as or even greater than coarse-textured soils (e.g., O'Neal, 1949, 1952; McKeague et al., 1982; Coen and Wang, 1989; Bouma, 1991; Lin et al., 1997). Furthermore, shrinking or swelling upon drying or wetting creates dynamic hydraulic behavior in active clayey soils that cannot be accounted for by texture alone (e.g., Lin et al., 1998). As soil structure is a function of the interaction between soil texture and bioclimate, PTFs that predict soil hydraulic properties based on structure should have a greater degree of transferability than regional, texture-based models.

The term soil structure has been used in US soil surveys, and elsewhere, to refer to the natural organization of soil particles into individual units (called peds) separated by planes of weakness. The peds are generally described according to their shape (platy, prismatic, columnar, angular/subangular blocky, or granular), their size range (very fine, fine, medium, coarse, or very coarse) and grade or distinctness (weak, moderate, or strong). Additionally, the internal surface features of the peds are also described, consisting of (1) coats of a variety of substances unlike the adjacent soil material and covering part or all of the surfaces, (2) concentration of material on surfaces caused by the removal of other materials, and (3) stress formations in which thin layers at the surfaces have undergone reorientation or packing by stress or shear (Soil Survey Division Staff, 1993). The kinds of structural surface features include clay films, clay bridges, sand or silt coatings, other coatings, stress surfaces, and slickensides. In general, pores are considered separately from soil structure, hence, in the US, the concept of soil structure is sometimes referred to as "*pedality*". Although soil pedality and macroporosity are related to one another, many soils have interpedal, intrapedal, and/or transpedal pores, which are not necessarily well represented by pedality. These pores, along with biopores and structural cracks, are critical in determining the hydraulic properties of field soils.

Soil morphological attributes, including soil structure, have long been used to infer soil hydraulic properties. Soil scientists have been successful in using descriptive morphological information to make qualitative judgments about a number of soil hydraulic properties, notably *Ksat* (e.g., O'Neal, 1949; King and Franzmeier, 1981; McKeague et al., 1982; Coen and Wang, 1989; Soil Survey Division Staff, 1993). For example, soil structure, as described by soil surveyors, provides clues about the macroporosity of the soil and the dominant pathways of water movement through the soil in saturated and near-saturated conditions. Thus, soil hydraulic properties such as *Ksat* often lend themselves to prediction from field characterizations of soil structure. National and regional soil surveys have often made use of soil morphological data for semi-quantitative estimates of soil hydraulic properties that are of relevance to land resource evaluation. In addition, soil morphological data are perhaps ideally suited to grouping soils by their hydrological functioning. As demonstrated by Wösten et al. (1990), class PTFs that use well-defined soil horizons as "carriers" of physical information allow efficient use of soil morphological data (Bouma, 1992).

While qualitative soil morphological attributes have been widely applied, quantification of such data is generally lacking. Rawls et al. (1993) noted that a quantitative description of the effects of soil morphological properties on soil water movement is yet to be established. So far, limited studies have demonstrated the potential for quantifying soil macromorphology through field observations or soil micromorphology through thin sections. As illustrated by Bouma and Anderson (1973) and Lin et al. (1999a,b), micro- and macro-morphometric data could be used to quantitatively derive soil hydraulic parameters. The utilization of soil structure descriptors in a quantitative fashion would be a step forward towards incorporating soil structure into PTFs and the modeling of flow and transport in soils. Furthermore, quantification of soil morphology would enhance the understanding of the relationships among different morphological features and permit a better assessment of soil profile descriptions in relation to water movement in soils and over landscapes. Such information could vastly improve the value of existing soil surveys.

In this chapter, we will review the contribution of soil morphology including soil structure to the derivation and improvement of PTFs that predict soil hydraulic properties in various soils around the world. We categorize the approaches into (1) qualitative or semi-quantitative and (2) quantitative methods. We conclude this chapter with some suggestions on how soil morphological data can be further used to improve PTFs.

2. USING SOIL MORPHOLOGY AND STRUCTURE IN ESTIMATING SOIL HYDRAULIC PROPERTIES

2.1. Qualitative or semi-quantitative approaches

2.1.1. Predictions of hydraulic conductivity

Some early papers on the use of soil morphology to predict hydraulic properties were by O'Neal (1949, 1952) who developed a system to estimate soil permeability (equivalent to *Ksat*) in US soils on the basis of primary and secondary structures as described in the field. Soil permeabilities were grouped into seven classes and ranged from approximately 3 cm day^{-1} to 610 cm day^{-1}. O'Neal (1949, 1952) produced a set of guidelines, presented as a table. Initially soils were grouped by soil structure type followed by the relative size of the horizontal and vertical cracks and fissures between peds, the degree of overlap in these fissures, the presence of visible pores such as wormholes and, finally, texture. These groups were further modified by secondary attributes such as silt content, aggregate stability, and mottling in order to derive estimates of soil permeability. O'Neal (1949, 1952) recognized that structure development depended largely on climate and while soil texture may be the same, soils would develop different structures under different climatic conditions. However, the measurements of hydraulic conductivity used to calibrate the field estimations were made on soil cores of only approximately 7.5 cm in diameter but ped sizes ranged from 0.1 to over 10 cm, therefore, there is some doubt if the measurements of conductivity were representative of flow in the voids between peds for all soils. In some early work in New Zealand, McDonald and Julian (1965) attempted to quantify soil porosity in the field in relation to texture, structure and roots by measuring pore space from photographs taken of a smoothed, horizontal soil surface. However, their main concern was estimating if soils had sufficient aeration for plant growth rather than estimating hydraulic conductivity.

King and Franzmeier (1981) measured the *Ksat* of 25 US soil series with a range of parent materials. The conductivity measurements were made using piezometers installed to below the water table. They initially attempted to group the soil horizons by soil structure type, size, and grade to derive class PTFs for the prediction of *Ksat* but this approach produced too many classes. They also found difficulties in dealing with soil horizons that had multiple ped sizes and that the *Ksat* values in these groupings were too variable. They then grouped the soil horizons by texture, parent material (either lacustrine, glacial till, water-modified glacial till, or loess) and by horizon pedogenesis. It would be expected that the different depositional environments would impart inherent characteristics such as density and porosity to the parent drifts and that these characteristics would subsequently be modified by pedogenesis. King and Franzmeier (1981) found that the measured *Ksat* values for C horizons of both glacial till and water-modified till were within a narrow range, perhaps because there was little structure development at depth within these soil parent materials. Although the conductivities of the loess and lacustrine deposits were more variable, there was no overlap in *Ksat* values of these parent material groups. Differences in structural development and ped sizes allowed these authors to distinguish two groups of fragic B horizons that could be assigned different conductivity values. However, the piezometer method may underestimate the hydraulic conductivity in soils with predominantly coarse vertical fractures (prismatic or columnar structures) such as those found in fragic horizons.

Building on the earlier work of O'Neal (1949, 1952) and Nowland (1981), McKeague et al. (1982) developed a system to estimate *Ksat* of Canadian soils. They measured the hydraulic conductivity of 78 horizons with an air-entry permeameter. The conductivity was measured within 1 m of a soil profile pit where detailed morphological descriptions of the soil horizons had been made. These measured *Ksat* values were then grouped into eight classes ranging from <0.4 cm day^{-1} to >1200 cm day^{-1}. Starting with previously developed guidelines that were continuously modified during the work, independent estimates of the hydraulic conductivity class of each horizon from the soil morphology were made by two researchers. The guidelines allowed the allocation of a soil horizon to a conductivity class based on soil texture, structure, presence of fracture cracks, and biopores (pores attributed to burrowing animals such as worms). They paid particular attention to whether the fracturing or biopores extended vertically through the whole horizon. Although these two independent assessors correctly estimated the *Ksat* of a horizon to within one class over 80% of the time, there was a marked bias between them with one assessor overestimating by one class and the other underestimating. An important conclusion from this work was that better predictions of *Ksat* were obtained from estimates of macroporosity (pores >60 μm) and soil structure than from texture, especially in horizons where the texture was finer than sandy loam. Some clayey soils with well developed structures and with many biopores had greater *Ksat* than sands. Apart from inconsistencies between assessors, the estimation of the effectiveness of biopores and fracturing between peds in conducting water also proved to be difficult, resulting in misclassification of some horizons. The presence of stones and cemented horizons also caused difficulties in the estimation of *Ksat*.

Wang et al. (1985) applied a similar approach to that of McKeague et al. (1982) and developed separate guidelines for the estimation of both vertical and horizontal *Ksat* although they used different class intervals (*Ksat* ranging from <0.12 cm day^{-1} to >1200 cm day^{-1}). The soil morphological properties they investigated were soil texture, consistence, structure, and the presence of biopores as well as stratification within a horizon (reflecting the mode of deposition of the parent material). The accuracy of the predictions was similar to those of McKeague et al. (1982) with *Ksat* assessed correctly within one class for 83% of the 18 horizons. They also found that it was difficult to estimate accurately *Ksat* for compact or cemented horizons and to assess the contribution of the voids between peds to flow rates (primarily due to the problem of estimating the width of the fractures). In 1989, Coen and Wang made further modifications to the McKeague et al. (1982) methodology and applied the revised system to soils in Alberta, Canada. They tested the role of soil texture and structure as the main determinants of vertical *Ksat*. While they also found that 83% of the estimated values were within one class of the mean measured *Ksat* values, more than half the soil horizons sampled were in only one of two *Ksat* classes and there was a wide variability in the measured *Ksat* values.

Overall, the work of O'Neal (1949, 1952), McKeague et al. (1982), Wang et al. (1985), and Coen and Wang (1989) have highlighted the role of soil structure (in particular, interpedal voids) and biopores in determining the *Ksat* of North American soils. There were also indications that it was difficult to predict *Ksat* from morphological attributes alone for soils that shrink and swell and for fragic (indurated) horizons. Hollis and Woods (1989) also found that biopores, particularly their orientation and size, as well as soil texture were important criteria in determining the *Ksat* of UK soils and reported that near-vertically orientated stones increased the *Ksat* due to preferential flow. O'Connell and Ryan (2002), working in Australia, found that the presence of angular stones also resulted in greater hydraulic conductivities than expected, which they attributed to preferential flow. Like others, Hollis and Woods (1989) encountered problems in determining useful methods for predicting the *Ksat* of fragic horizons.

In Scotland, Lilly (2000) related over 600 measurements of field-saturated hydraulic conductivity (*Kfs*), as measured by the Guelph permeameter (Reynolds and Elrick, 1986), to the size, shape, and grade of structural aggregates. He identified 49 unique combinations of primary and secondary soil structures. Like King and Franzmeier (1981), Lilly (2000) found that grouping hydraulic conductivities by soil structure alone resulted in too many classes to be useful so he ranked the soil structures according to the measured *Kfs*. This ranking revealed natural groupings in the conductivity data that could be related to ped size and orientation of fracturing. He then calculated the geometric mean and range of *Kfs* for each of the four groups identified to derive a set of class PTFs. However, as there were limited data for soil horizons with small ped sizes, further work needs to be undertaken in order to refine this method and test its wider applicability.

In New Zealand, Griffiths et al. (1999) attempted to develop methods to predict minimum, maximum and mean *Ksat* from soil morphological data. Unusually, they also developed methods to estimate near-saturated hydraulic conductivity so that preferential flow could be estimated separately from matrix flow. In addition to the more usual morphological attributes of soil structure (grade, size, and shape), they also assessed the degree of aggregate packing (using a shear vane), surface roughness, and ped face coatings. In order to improve the quantitative estimate of soil structure, up to a maximum of seven discrete ped sizes were identified and the relative proportions of each were determined in each horizon. Platy structures (like those often associated with induration or fragipans) were treated separately. Tracer dyes were used to identify and quantify the proportion of macropores within the horizons. Soil structure, the area and continuity of macropores, and packing density as measured by a shear vane were the main attributes used to categorize pedal soils while apedal soils were classed according to the area and continuity of macropores, the degree of packing, and soil texture. Unsaturated hydraulic conductivity for the soil matrix was related to the degree of packing and soil texture.

The ability of researchers to correctly assign horizons to conductivity classes based on these morphological criteria was very variable and ranged from zero to 87%. However, a large proportion of the predictions were again within one class of the measured value. The authors acknowledge that morphological descriptors in existing soil survey databases cannot be translated into the new descriptors they used, which will limit the wider applicability of their results.

In summary, the prediction of *Ksat* from soil morphological attributes is not as robust or well developed as those PTFs that rely primarily on statistical correlations with soil texture. One reason may be that there is a lack of consistent methods employed to measure *Ksat* and to describe soil morphology. For example, McKeague et al. (1982) used an air entry permeameter, King and Franzmeier (1981) used a piezometer, Lilly (2000), Hollis and Woods (1989) used the Guelph permeameter while Griffiths et al. (1999) measured *Ksat* with a double ring infiltrometer. Each of these measurement methods differs in the volume of soil measured and, therefore, in their potential to measure a representative pore volume. Standardization of measurement techniques may help improve morphology-based PTFs. In order to fully utilise the vast store of soil morphological data in soil survey databases, work is needed to relate the standard, routinely determined, soil morphological attributes to hydraulic conductivity values measured using consistent and comparable methodologies.

2.1.2. Predictions of moisture retention

Many of the PTFs that make use of soil morphology have concentrated on the prediction of available water capacity (AWC), for example, Hall et al. (1977), McKeague (1987) and McKenzie and MacLeod (1989). Only a few papers have addressed the prediction of the soil moisture retention curve ($\theta|h|$), e.g., Williams et al. (1983), Köhne et al. (2002) and O'Connell and Ryan (2002). Although AWC is widely used in agronomic models, there is no consistency in the method of calculation between different countries. This limits the development of robust, widely applicable methods to estimate AWC from either soil texture or soil morphology.

As with *Ksat*, it would seem logical that soil moisture retention should be related to soil structure. In Canada, a comparison by De Jong and McKeague (1987) between texture-based statistical models for predicting field capacity (moisture retained at -5 or -33 kPa), permanent wilting point (moisture retained at -1500 kPa), and AWC showed differences in predictive ability, particularly for field capacity. The PTFs that were compared were derived from moisture retention data for soils from the Province of Ontario (McBride and MacIntosh, 1984) and from data held within a national Canadian dataset (De Jong and Loebel, 1982). Those PTFs derived from the national dataset produced less accurate predictions than those PTFs derived from the regional dataset when compared to a new set of measured data collected from soils in Ontario. Although, this again shows that PTFs are not always readily transferable, the data in the national dataset comprised moisture retention characteristics derived from disturbed soil samples, making an evaluation of the accuracy of the PTFs difficult. De Jong and McKeague (1987) concluded that the addition of data on soil structure would improve the PTFs. Similarly, McKeague (1987) reported work to refine and validate guidelines to estimate AWC, air capacity (defined as volume of air filled pores at -5 kPa) and permanent wilting point (moisture retained at -1500 kPa) based on data from 24 horizons from Canadian soils. It was found that AWC could be estimated from soil morphology to within 5% of the measured value;

however, when these guidelines were applied to Tanzanian soils derived from volcanic materials (McKeague et al., 1991), AWC was poorly estimated. Even though McKeague et al. (1991) changed the upper limit of AWC from -5 kPa to -10 kPa (which would have the effect of reducing AWC), the PTF over estimated the AWC and the field capacity of the Tanzanian soils while under estimating the moisture retained at -1500 kPa. This illustrates the difficulties in extrapolation from humid temperate to tropical soils and to soils with different parent materials. In general, PTFs should be seen as an interpolation technique rather than extrapolation tool.

Abbaspour and Moon (1992) developed regression equations to predict soil moisture retention of soils in British Columbia based on the guidelines of McKeague (1987) with the addition of other routinely recorded soil morphological attributes such as soil color, consistence and horizon pedogenesis. They found that soil color (chroma), type of horizon (a function of pedogenic processes) and structure shape were good predictors of saturated water content (θ_s). In British Columbia, like most recently glaciated areas, soil color, particularly hue, is primarily inherited from parent rocks and not an indication of long-term weathering. However, soils with high values of chroma (the strength of the hue) in these landscapes is often an indication of freely draining soils, which are generally quite porous. Soil color, structure shape, and horizon type were important predictors for field capacity (-33 kPa) while texture, color, and structure were the main predictors of permanent wilting point (-1500 kPa).

In Australia, McKenzie and MacLeod (1989) used tactile soil morphological properties (consistence, hand texture) and visual (color, structure) to estimate the moisture constants at field capacity (-10 kPa or θ_{10}), permanent wilting point (-1500 kPa or θ_{1500}) and AWC. They found that percentage clay content (taken as the proportion of clay at the mid-point of the range that defines the soil texture class) explained most of the variation in θ_{10} with soil color (hue) also contributing. Unlike soils in glaciated areas, McKenzie and MacLeod (1989) suggested that the color of Australian soils is indicative of soil age, weathering and clay mineralogy. This relationship between clay mineralogy and soil color may explain the importance of soil color in accounting for part of the variation in retained moisture capacities. They also found that the best estimator of AWC was texture class. Although soil structural properties (such as ped size) correlated well with AWC in univariate analyses, they were not significant contributors in multivariate analyses. This may be partly explained by autocorrelation between soil structural development, soil age and clay content. Later, McKenzie et al. (2000) doubted if conventional soil morphological attributes could be used to predict soil hydrological properties but, where they were, they suggested that a high degree of correlation should be established rather than assumed.

Also in Australia, O'Connell and Ryan (2002) developed PTFs to predict θ_s, θ_{1500} and $Ksat$ in order to derive the $\theta|h|$ and $K|\theta|$ relationships for modeling work within a forested catchment. They used the standard morphological descriptions associated with most soil surveys to predict moisture retention and hydraulic conductivity. Stone content proved to be important in determining $Ksat$ in some specific pedological horizons and geomorphological conditions. They also gave a useful illustration of why PTFs are not always transferable. Organic carbon (or organic matter) content is often used as a predictor in PTFs with the assumption that the organic matter both promotes and maintains soil structure. In Australian soils, the widespread occurrence of charcoal as a result of periodic bush fires increased the proportion of organic carbon in soils but this would have little effect on the development or maintenance of soil structure.

Baumer (1992) developed a computerized expert system to predict $\theta|h|$ and $K|h|$. Initially, θ_s, residual water content (θ_r), air entry value (termed h_b), θ_{33}, and θ_{1500} were estimated from a wide range of data including soil morphology, stone content, texture, clay activity and macroporosity. These estimated values were then used to generate values for $Ksat$ and subsequently to derive a set of Mualem-van Genuchten parameters to describe $\theta|h|$ and $K|h|$.

One of the problems with fitting Mualem-van Genuchten parameters to soil hydraulic data is that this procedure often under predicts $Ksat$ particularly where the $K|h|$ curve declines steeply at matric potentials just less than zero. This decline is not always matched by the $\theta|h|$ curve. Köhne et al. (2002) developed a technique whereby the soil is treated as a dual permeability medium. The volume of macropores, including those pores associated with fracturing and estimated from soil structure descriptions, was used to apportion flow in this model. Although not a direct use of soil morphology in predictive PTFs, this technique may be useful for improving those texture-based PTFs that predict the Mualem-van Genuchten parameters (e.g., Wösten et al., 1999).

In summary, the use of PTFs to predict soil moisture retention characteristics for agronomy and land resource evaluations is well established but these are generally based on soil texture. This can limit their wider applicability as interactions between climate and soil can result in quite different soil structures despite similarities in soil texture. In some instances the moisture retention characteristics were derived from disturbed soil samples and therefore cannot be related to morphological properties such as soil structure or macroporosity. Few studies have attempted to relate the moisture retained within a soil to morphological characteristics. There is also a lack of consistence in the definition of AWC, which inhibits the use of any PTFs beyond their original geographic or bioclimatic region.

2.1.3. Grouping and classification of soil hydrological functions and pedotransfer rules (PTRs)

Grouping soils by soil morphological attributes has potential to improve PTFs. Williams et al. (1983) used a numerical classification to group soil retention curves of 78 soil horizons from 17 profiles (representing 12 Australian Great Soil Groups) into eight classes based on their similarity. Such grouping provided a useful means of bringing together soils that retain water in a similar manner. Franzmeier (1991) grouped $Ksat$ of some Indiana soils by soil classes, called lithomorphic classes, based on origin of parent material, type of soil horizon and soil texture. Danalatos et al. (1994) estimated moisture retention characteristics from routine soil survey data collected from 105 horizons in 34 representative soils from Greece. They found that grouping soil horizons by soil types and structures improved the coefficient of determination for their PTFs to estimate moisture retention from θ_s and clay content. They also found that two major groups of soil horizons could be distinguished among the horizons they studied, namely those that were well-structured and those that were structureless or weakly structured. Batjes (1996) used hierarchical pedotransfer rules (PTRs) and functional grouping to predict AWC for the main soil types of the FAO–UNESCO world soil map using soil unit type, horizon textural class and organic matter class. He showed that the PTR-derived AWC values had a better correlation with measured AWC values than was the case for the PTF-derived AWC values. Pachepsky and Rawls (1999) examined the accuracy and reliability of PTFs as affected by grouping soils based on US soil Great Groups, textural class, soil moisture and temperature regimes. They found that grouping improved the accuracy of PTFs in most cases, with textural grouping and grouping by soil moisture regimes yielding better results for moisture retained at −33 kPa than grouping by soil Great Groups or by temperature regimes in the dataset they studied (447 pedons from Oklahoma). However, none of the grouping criteria proved to be clearly superior in Pachepsky and Rawls' study. They also found that although PTFs developed from the groups were more accurate than the PTFs developed from the whole dataset, the former were not more reliable than the latter as assessed by cross-validation (dividing the available dataset into development and validation subsets).

Wösten et al. (1990) have shown that use of soil horizons to predict hydraulic properties for simulation modeling can be quite successful, as compared with results obtained with direct measurements. Class PTFs that use well-defined soil horizons as "carriers" of physical information allow efficient use of soil morphological data because soil horizons can be determined easily and reproducibly by pedologists (Bouma, 1992). Wösten et al. (1985) and Wösten et al. (1987) used this concept of building blocks to provide information on the water retention and hydraulic conductivity of Dutch soil horizons and pedons, the *Staringreeks* or *Staring Series*. However, as noted by Wösten et al. (1985), Breeuwsma et al. (1986) and Bouma (1992), pedogenic differences, as expressed by horizon designations, do not necessarily correspond with functional differences, such as differences in physical and hydraulic properties. Bouma (1992) suggested a more promising three-stage protocol that attempts to calculate hydraulic parameters directly from physical or morphological data. First, good measurements of soil hydraulic properties should be made that take soil morphological attributes (such as soil horizon and soil structure described in the field) into account. Secondly, measured hydraulic properties should be expressed in terms of coefficients as defined, for example, by van Genutchen equations and, thirdly, relate those coefficients by regression analysis to readily available soil properties (such as texture, bulk density, and organic matter content) or to more qualitative groupings of soil horizons based on pedological expertise.

Quisenberry et al. (1993) devised a system of classifying soils that is somewhat analogous to the development of PTRs but the property predicted was mainly water flow pathways and patterns (uniform flow or different types of preferential flow). Much of this work makes use of subsoil structure assessments as well as surface soil texture and clay mineralogy. Quisenberry et al. (1993) suggested that flow is rarely uniform within structured soils and that a better understanding of soil hydrology and solute transport can be gained from a study of preferential pathways of water movement in structured soils. The work of Quisenberry et al. (1993), however, has been limited to soils in South Carolina in the US and is descriptive in nature. A more comprehensive and quantitative approach may be worth pursuing.

In the UK a soil hydrological classification based on soil morphological attributes has also been developed to predict water movement through soils and substrates (Boorman et al., 1995; Lilly et al., 1998). Termed HOST (Hydrology of Soil Types), this classification drew on the combined morphological datasets of the then Soil Survey of England and Wales and the Soil Survey of Scotland. River hydrological indices were used to validate classification. Much of the initial work that lead to the development of the HOST classification came from the derivation of regression-based PTFs to predict water retention (e.g., Reeve et al., 1973; Hall et al., 1977; Hollis et al., 1977) and from the use of morphological attributes to estimate packing density, air capacity (pores >60 μm), and

AWC (e.g., Hodgson, 1974, 1997). This work, and that by others such as Thomasson (1975, 1978) and Avery (1980), also led to the development of the PTFs and PTRs used in UK land resource evaluation systems for Scotland (Bibby et al., 1982) and for England and Wales (MAFF, 1988). The latter publication in particular sets out the PTRs that underpin much of the HOST classification.

In one of the UK's Agricultural Land Classification system (MAFF, 1988), texture-based class PTFs were developed to predict AWC (θ_5 or $\theta_{10}-\theta_{1500}$) for each soil texture class and modified according to soil structure (type, size, and grade) and consistence such that soil horizons with moderate and poor structures were deemed to have less available water. There are also complex PTRs to determine if a soil horizon is slowly permeable ($Ksat < 10$ cm day^{-1}), which influences a range of hydrological properties of the whole soil. First, the physical properties of the soil such as texture, ped type, ped size, consistence, and the presence of biopores are assessed. A slowly permeable horizon is recognized from the various combinations of these properties (Figure 1) and its presence within the profile (pedon) is confirmed by evidence of gleying in either that horizon or the one immediately above. Gleying is strictly defined in terms of soil color, and the presence of mottling or gleying on ped faces (Avery, 1980). Soils with high value but low chroma (grey and pale colors) are also classed as gleyed.

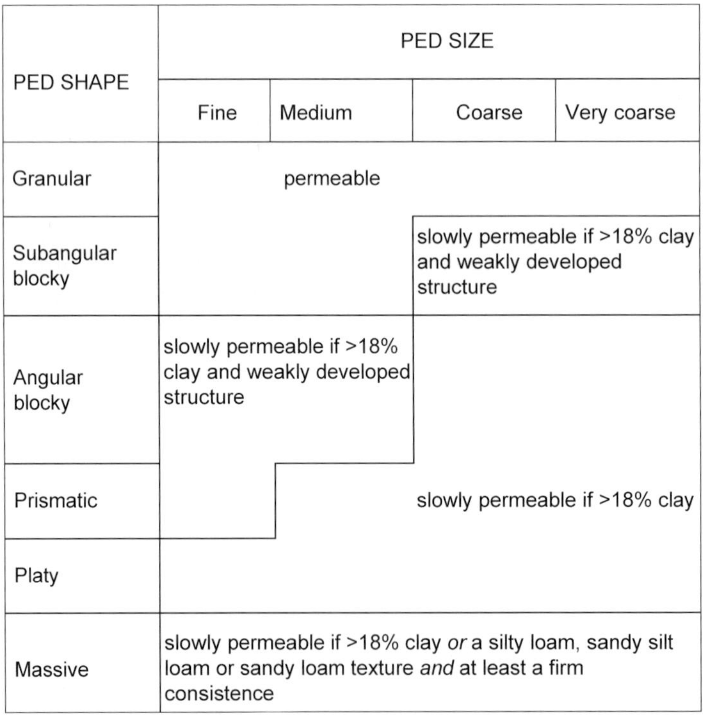

PED SHAPE	PED SIZE			
	Fine	Medium	Coarse	Very coarse
Granular	permeable			
Subangular blocky			slowly permeable if >18% clay and weakly developed structure	
Angular blocky	slowly permeable if >18% clay and weakly developed structure			
Prismatic			slowly permeable if >18% clay	
Platy				
Massive	slowly permeable if >18% clay *or* a silty loam, sandy silt loam or sandy loam texture *and* at least a firm consistence			

Figure 1. Diagrammatic representation of the combinations of structure, texture and consistence that are characteristic of slowly permeable layers ($Ksat < 10$ cm day^{-1}) in British soils (After MAFF, 1988).

These PTRs developed for land resource evaluation were used in the development of HOST, a national scale soil hydrological classification (Figure 2). The rules were applied to over 24,000 soil profiles in order to develop an attribute database of semi-quantitative soil attributes from which the hydrological responses and flow pathways within UK soils could be predicted. These attributes were: the presence or absence of an organic surface layer, substrate hydrogeology, the depth to a slowly permeable layer, the depth to gleying and air capacity values (volume of pores that drain under the influence of gravity). Information on the spatial distribution of soils within catchments was obtained from existing 1:250,000 scale soil maps of Scotland, Wales and England. Multiple regression analyses of these properties against two hydrological indices (Base Flow Index and Standard Percentage Runoff) were used to develop the soil hydrological classification that comprises 29 classes and encompasses all UK soils. This classification is used by applied hydrologists for

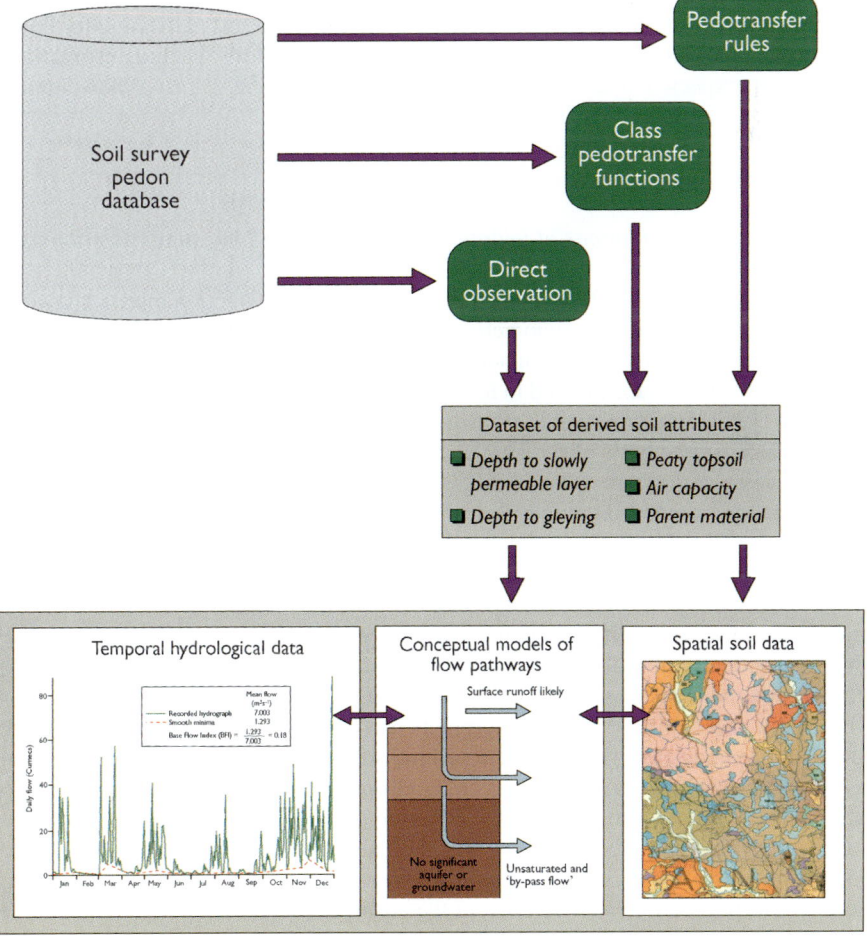

Figure 2. Utilization of soil morphology and pedotransfer rules to determine soil hydrological functioning in UK soils.

predicting river levels in ungauged catchments, designing spillways and by soil scientists to apportion flow in transport models. The same basic techniques of PTRs were applied to the 1: 1,000,000 digital soil map of Europe to generate similar soil attributes at the continental scale (Jones and Hollis, 1996).

Whilst many of the more process-based models simulate only uniform water flow within soils, the grouping or classification of soil hydrological functioning from soil morphological data allows the incorporation of flow pathways and preferential movement into hydrological models. Clearly there would be an improvement in the prediction of catchment (or watershed) scale hydrological models if both approaches could be combined.

Being able to group and even classify soils prior to, during and after the development of PTFs would enhance the accuracy and reliability of the application of PTFs. A description of the groups in terms of both soil morphology and hydraulic properties is a valuable means of developing simple predictive PTFs for field soils. Furthermore, a grouping or classification of soils (particularly when linked to soil map units) for flow and transport characteristics would be useful for many applications, such as estimating the magnitude of expected hydraulic properties and determining *a priori* how important preferential flow is in a given soil.

2.2. Quantitative approaches

2.2.1. Quantitative calculations of hydraulic conductivity and moisture retention using micromorphometric data

Bouma and Anderson (1973), using a planar-void pore-interaction model, showed that *Ksat* calculated with micromorphometric data (including the number, length, width of planar voids based on point counts made in thin sections) agreed well with *Ksat* measured *in situ* with the double-tube method in four pedal subsoil horizons in Wisconsin. Anderson and Bouma (1973) further showed that estimating *Ksat* from seven soil peels from a pedal argillic horizon on the basis of the similar micromorphometric data yielded reproducible results that were in the same order of magnitude as values measured *in situ*. These results indicated that micromorphometric soil structure characteristics can be used to predict *Ksat* of structured soils. Bouma and Denning (1974) also tested the validity of the pore interaction model in sands and found that hydraulic conductivity of coarse porous media with packing voids could be calculated not only with accepted physical, but also with morphometric techniques if matching factors were used. Bouma et al. (1979) predicted hydraulic conductivities of six Dutch clay soils using the similar micromorphologic techniques on the basis of staining patterns in thin sections.

Based on the capillary model, morphometric point-count data were also used by Bouma and Denning (1974) to calculate soil moisture retention curves. They found good agreement between calculated and measured adsorption moisture retention curves in soil matric potential range from 0 to about -30 cm in sands with few or no fine particles. As the content of fine particles increases, the capillary model became unrealistic, and the moisture retention curve calculated using morphometric data deviated considerably from the desorption curve as well as from the adsorption curve for matric potentials less than -30 cm.

Though successful, the above micromorphological methods are more laborious than direct measurements of soil hydraulic properties. As Bouma (1992) pointed out, these micromophological techniques serve not as an alternative to direct physical measurement, but rather to increase the understanding of the underlying processes.

2.2.2. Quantification of macromorphological attributes in developing PTFs

Lin et al. (1999a,b) developed a system to quantify soil macromorphological features commonly used in soil surveys, and demonstrated that the use of such quantified morphological data yielded comparable predictive power as soil physical properties in estimating soil hydraulic parameters. In their study, a point scale system was developed as a means of quantifying the class variables of soil morphology, including textural class, initial moisture state, pedality, macroporosity, and root density (Table 1). A hypothetical structureless and nearly impermeable clay soil was taken as the reference in their point scale system. This reference clay, assigned one point, was assumed to be massive, contain no macropores and no roots, and at a fully swollen, saturated state. For each morphological feature quantified, descriptive classes were scaled (rated) relative to the reference clay in terms of increased capacity to transmit water vertically. Based on functional or empirical relationships between relevant soil morphological attributes and hydraulic properties, points were assigned to each morphological class for the amount of possible increase in permeability relative to the reference clay. Optimal points that yielded the best correlation with the measured steady-state infiltration rates were selected using optimization techniques (Lin et al., 1999a). The total optimal points of a morphological feature were then divided by its maximum value to obtain an index ranging from 0 to 1 (called morphometric indices). Such indexing provided equal weight for each morphological feature in their final comprehensive evaluations.

After quantifying soil morphological descriptors, Lin et al. (1999b) used principal components analysis to investigate the structure of the relationships among the morphological variables. A subset of variables were then selected through statistical procedures to construct optimal PTFs for estimating soil hydraulic properties, including hydraulic conductivity at zero potential (K_0) (which is similar to Kfs), steady-state water flow rates in macropores (v_{macro}), mesopores (v_{meso}), and micropores (v_{micro}), and a soil structure/texture parameter α_{macro} (Table 2). Their results showed that macropore quantity and size were crucial in characterizing hydraulic parameters in macropore flow region (i.e., K_0 and v_{macro}); initial moisture state, root density, and ped grade had impact on mesopore flow rate (v_{meso}); whereas textural class placed major impact on micropore flow rate (v_{micro}). The classical approach of using particle-size distribution, dry bulk density, and organic carbon content was found to be insufficient for predicting the hydraulic parameters controlled by macropores and/or mesopores in their study (Table 2), primarily because organic carbon content and bulk density did not adequately reflect macroporosity in their soil samples. Also noteworthy is that no single morphometric index appeared to provide adequate estimation of all the hydraulic parameters; rather, the estimation of soil hydraulic parameters required the use of different combinations of morphometric indices, depending on the flow domain to be included (see Table 2).

Another significance of the morphological quantification system developed by Lin et al. (1999a) was that it permitted the examinations of the interrelationships among different soil morphological attributes that would otherwise be difficult to do with the qualitative descriptors. As shown in Table 3, the morphometric index of macroporosity (MI_p) had significant positive correlation with both the morphometric indices of initial moisture state (MI_m) and pedality (MI_s), but not with texture (MI_t), indicating that the drier the soil or the

Table 1
Optimal points of soil morphological classes with respect to increased capacity to transmit water vertically relative to the reference clay. These points were obtained based on the 96 soil horizons of varying textures and structures investigated by Lin et al. (1999a)

Texture and initial moisture		Pedality				Macroporosity				Root density		
Textural class	Points	1. Ped grade class	Points			1. Quantity class	Points			1. Quantity class	Points	
clay	1	(massive)	0			very few	1			few or very few	1	
silty clay	2	weak	1			few	3			common	16	
sandy clay	3	moderate	5			common	10			many	25	
silty clay loam	4	strong	25			many	28					
clay loam	5	(single grain)	50			very many	60					
sandy clay loam	6											
		2. Ped size class[a]	Points			2. Macropore size class	Points			2. Root size class	Points	
loam	10	very coarse	1			very fine	1			very coarse	1	
silt loam	13	coarse or medium	3			fine	9			coarse or medium	13	
sandy loam	15	fine or very fine	18			medium	49			fine or very fine	43	
silt	19					coarse	60					
loamy sand	24					very coarse	70					
sand	27					extremely coarse	75					
Moisture class	Points	3. Ped shape class	Points			3. Macropore type class	Points					
saturated	1	(massive)	0			vugh	1					
wet	3	platy	1			channel	8					
moist	7	prismatic	10			fracture	10					
dry	30	blocky	10			packing-void	25					
very dry	65	granular or (single grain)	30									

[a] In describing platy peds, *thin* is used instead of *fine* and *thick* instead of *coarse*.

Table 2
Comparison of goodness-of-fit (R^2_{adj}) for three types of PTFs developed from different predictor variables using multivariate regression. The response hydraulic parameters include hydraulic conductivity at zero potential (K_0), steady-state water flow rates in macropores (v_{macro}), mesopores (v_{meso}), and micropores (v_{micro}), and a soil structure/texture parameter αv_{macro}. Parameter coefficients of the PTFs developed from the selected morphological attributes and their relative importance are shown in the last two columns

Hydraulic parameter	R^2_{adj}			PTFs developed from the selected key morphological attributes								Relative importance of the attributes (Based on the magnitudes of the type II sum of squares)
	(1) PTFs using key morphological attributes[a]	(2) PTFs using textural class and pedality[b]	(3) PTFs using physical properties[c]	Parameter coefficients								
				Intercept	MI_t	MI_m	MI_{sg}	MI_{st}	MI_{pq}	MI_{ps}	MI_{rq}	MI_{rs}	
K_0	0.860	0.340	0.194	−22.43	−38.26		33.82	21.07	47.54	102.36	45.45	33.27	$MI_{pq} > MI_{ps} > MI_{sg} > MI_t > MI_{rq} > MI_m > MI_{st}$
v_{macro}	0.856	0.347	0.202	−24.71	−45.68		32.52	23.42	52.38	110.13	50.92	35.84	$MI_{pq} > MI_{ps} > MI_t > MI_{sg} > MI_{rq} > MI_s > MI_m$
v_{meso}	0.464	0.155	0.181	1.89		10.93	1.49						$MI_m > MI_{rq} > MI_{rs} > MI_{sg}$
v_{micro}	0.685	0.585	0.466	0.14	2.03	0.73		−0.68			7.77	−5.04	$MI_t > MI_m > MI_{rq} > MI_{st}$
α_{macro}	0.532	0.330	0.344	27.83	−60.25		15.10	31.99		45.62	0.41	27.76	$MI_{ps} > MI_t > MI_{rs} > MI_{sg} > MI_{st}$

Source: Lin et al. (1999b).

[a]The predictor variables include selected key morphometric indices as shown in the last two columns. Variable selection was based on the maximum R^2_{adj} that gave the smallest Mallow's C_p (Lin et al., 1999b). Initial inputs include the morphometric indices of textural class (MI_t), initial moisture state (MI_m), ped grade (MI_{sg}), ped size (MI_{ss}), ped shape (MI_{st}), macropore quantity (MI_{pq}), macropore size (MI_{ps}), macropore type (MI_{pt}), root abundance (MI_{rq}), and root size (MI_{rs}). Morphological attributes without parameter coefficients indicated were not selected. Morphometric indices of ped size (MI_{ss}) and macropore type (MI_{pt}) were not selected in all PTFs, and thus not shown in the table.

[b]The predictor variables include only the morphometric indices of textural class (MI_t), ped grade (MI_{sg}), ped size (MI_{ss}), and ped shape (MI_{st}).

[c]The predictor variables include the mass fractions of clay, silt and sand separates, organic carbon content, and dry bulk density.

Table 3
The Pearson correlation coefficients among the five quantified soil macromorphological features. The overall morphometric indices of pedality, macroporosity and root density were the products of their individual attributes

Morphometric indices	Initial moisture (MI_m)	Pedality (MI_s)	Macroporosity (MI_p)	Root density (MI_r)
Texture (MI_t)	0.057	$-0.173^†$	-0.151	0.348***
Initial moisture (MI_m)		0.0131	0.635***	0.111
Pedality (MI_s)			0.222*	0.249*
Macroporosity (MI_p)				0.001

Source: Lin et al., 1999a.
Significant levels are marked at $p < 0.001$ (***), <0.01 (**), <0.05 (*), and <0.1 (†).

stronger the pedality, the greater the macroporosity. This was not surprising as the majority of the 96 soil horizons that Lin et al. (1999a) studied were active clay soils with shrink–swell phenomena. Although not statistically significant, the negative sign of the correlation coefficients in Table 3 between the MI_p and MI_t and between the MI_s and MI_t appeared to suggest that the fine-textured soils tended to have greater macroporosity and stronger pedality than the coarse-textured soils. The MI_s also showed positive correlation with the morphometric index of root density (MI_r), implying positive relationships between root development and pedality.

While the studies of Lin et al. (1999a,b) were a first step towards quantitative use of soil macromorphological attributes for developing PTFs, we note that some morphological classes not well represented by their sample population would require further data for application in a wider range of soils. In addition, many morphological features examined by Lin et al. (1999a,b) are related to land use and dynamic nature of soil properties (e.g., root density, pedality, macroporosity, and initial moisture state). Hence, there is a potential significance to expand and enhance this to address land use-dependent and dynamic soil properties that are being recognized by the US National Cooperative Soil Survey program (e.g., Grossman et al., 2001).

2.2.3. Other quantitative uses of qualitative morphological attributes in PTFs

New ways of quantitatively utilizing qualitative morphological descriptors in developing PTFs have been recently explored. These include regression tree (RT) (e.g., McKenzie and Jacquier, 1997; Rawls and Pachepsky, 2002a) and group method of data handling (GMDH) (e.g., Pachepsky et al., 1998; Pachepsky and Rawls, 1999). Although categorical soil morphological variables could potentially be used in logistical regressions, or be converted to numerical numbers in standard regressions through "dummy coding", or used as inputs in artificial neural networks (ANN), these efforts have not yet occurred in the literature. This may be in part due to the difficulties in using consistent morphological descriptors as well as the complex link between these attributes and soil hydraulic properties.

Regression tree portrays a nonlinear and conditional relationship between soil hydraulic properties and basic soil parameters, which is performed by successive binary partitioning of a dataset. Optimum partitioning of soil survey databases with regression trees could be used to find the best predictors and the best grouping of soil samples (Rawls and Pachepsky, 2002a). However, it would be desirable if sufficient physical meaning and logic sequence of the regression tree branches could be developed and justified.

Williams et al. (1983) used a numerical classification similar to regression tree analysis to determine the main soil morphological attributes that influence the water retention curves in Australian soils. The soil moisture retention curves were broadly associated with specific texture classes but other properties such as soil structure were also important. The presence or absence of pedality and structure grade were found to be strongly associated with the moisture retention groups. But the shape and size of peds had only weak associations with the differences in soil moisture retention curves in their study.

Working with 99 horizons from tropical soils in Australia, McKenzie and Jacquier (1997) tested a wide range of soil morphological and measured properties for their ability to estimate $Ksat$. Useful morphological descriptors included hand texture, structure grade, areal porosity, bulk density, dispersion index, and horizon type. Although they found univariate relationships between $Ksat$ and individual soil morphological properties, there was wide scatter in the data. The data were then analyzed using regression trees, which gave more plausible models than multiple regressions. The first split in the data was made on the basis of hand texture rather than soil structure with the clay horizons being separated from the remainder. While regression tree analysis is useful in illustrating the soil properties that effect $Ksat$, it may be difficult to draft a set of guidelines that can be applied in the field. For example, bulk density greater than or less than 1.26 g cm^{-3} was a discriminating factor in one branch of the tree while in another it was density greater than or less than 1.55 g cm^{-3}. This shows that there would be a degree of sensitivity in the order that properties in the field were assessed and that guidelines would have to be very carefully drafted. Regression tree analyses can also result in rather sharp delineations, for example, McBratney et al. (2002) describes a regression tree where the predicted $Ksat$ increased 35 fold when the bulk density increased above a threshold.

Also using regression tree analysis, Rawls and Pachepsky (2002a) examined soil structure and consistence as predictors for moisture retained at -33 and -1500 kPa from over 2000 samples in the US. They found that plasticity class, structure grade, and dry consistency class were the leading separators and that using soil structural and consistence parameters along with textural classes provided a small, yet significant improvement in accuracy of water retention estimates as compared with estimation from texture alone. Where texture was not included in the analysis, the regression tree first split the data on the basis of plasticity; however, where soil texture was included, the data were split on the basis of texture. This may simply indicate the close relationship between soil plasticity and texture. Structure grade was also a useful predictor for moisture retention at -33 kPa with soils of weak and moderate grades having greater retention than those with a strong grade. Structure shape tended to be important only in the later stages of the analysis but unsurprisingly, it was a more important discriminator for moisture retention at -33 kPa than at -1500 kPa.

The role of soil aggregates in determining the van Genuchten parameters associated with water retention (i.e., α, n) as well as water retained at -10, -33 and -1500 kPa was examined by Guber et al. (2003). Using regression tree analysis on over 100 samples from Russia, Ukraine, and Uzbekistan, they found that the proportion and size of aggregates helped partition data at some stage for all but water contents at −10 kPa. The authors stated that while they were not trying to develop PTFs from this method, as few datasets comprised data on aggregate distribution, it does illustrate the influence of aggregation on soil water retention. Soil particle size class was also a partitioning factor in many cases; however, there was an overlap between aggregate size (<0.25 mm) and particle size and as soil aggregates are composed of individual soil particles there is a degree of autocorrelation between soil texture and soil structure.

When the number of predictor variables is very large, or the relationship between inputs and outputs is very complex, GMDH could successfully compete with statistical regression in finding an appropriate relationship between a set of inputs and an output (Hecht-Nielsen, 1990). Through iterations, GMDH produces a hierarchical network of polynomial equations with good accuracy and with only significant predictor variables. The advantage of GMDH is that it automates the finding of essential input variables to be included in PTFs, and unlike ANN, presents an explicit adductive network of equations (Pachepsky et al., 1998). Because only numerical variables can be used in a polynomial equation, categorical variables such as soil taxonomic units and textural classes may be used to group soils prior to GMDH execution. Pachepsky and Rawls (1999) demonstrated this for 447 pedons from the USDA-National Resource Conservation Service soils database from Oklahoma. They found that grouping by soil Great Group, textural class, and soil moisture and temperature regimes improved the accuracy of PTFs developed by GMDH in most cases as compared to the PTFs developed from the whole database. Pachepsky et al. (1998) also used the GMDH to improve soil water retention estimates using penetration resistance (related to soil structure) as a useful additional input to the PTFs that were based on soil texture and bulk density for 180 soil samples from New Zealand.

ANN have been found to provide more accurate nonlinear PTFs than polynomial regressions (e.g., Schaap and Bouten, 1996; Pachepsky et al., 1996; Tamari et al., 1996; Minasny et al., 1999). It has been suggested that ANN can combine quantitative, descriptive, and ranked data in making predictions or classifications (Levine and Kimes, 1997). This would be especially useful when soil description, classification, and quantitative data are mixed together; however, no such effort has been published in the literature. On the other hand, ANNs have some major drawbacks that include the lack of an explicit procedure for selecting essential input variables and the lack of physical interpretation of the "black-box" type of nonlinear input-output relationship. Levine and Kimes (1997) suggested that, when used with a genetic algorithm, ANNs are able to identify, from a large number of possible inputs, only those variables that behave synergistically to produce the best network performance.

A final note on the new methods to develop PTFs – with or without categorical soil morphological attributes as inputs – is that the success of any mathematical/statistical techniques will be heavily dependent on the quality, quantity, and appropriateness of the original data stored in the databases. An interesting example was provided by Wösten et al. (2001) where the accuracy of predicting water content at −33 kPa using ANN, GMDH and RT was shown to be quite comparable. The fact that these methods yield comparable results indicate that major progress in PTFs is not to be expected from new statistical methods but rather from better data.

3. FUTURE IMPROVEMENTS

Soil morphological attributes, especially those related to soil structure and macroporosity, provide clues and/or estimates of pore volume that water both flows through and is held in. It would therefore seem logical that these assessments could be used to predict soil hydraulic properties. However, it is clear that the methods for predicting soil hydraulic properties from soil morphology (both tactile and visual) are not as well developed as those that utilize soil texture data in deriving PTFs. The volume of soil morphological data residing in regional, national, and international soil survey databases is potentially as great as soil particle size data. Therefore, a large resource of soils data remains under used. Based on the above review, we would like to suggest some possible enhancements to the use of morphological attributes (including soil structure) in future PTFs developments and applications. These suggestions are centred along the need to provide better data rather than better statistical methods.

3.1. Standardization of soil morphology descriptions and hydraulic measurements

One possible reason that soil morphology is not fully utilized in predicting soil hydraulic properties is the lack of a consistent approach to developing robust relationships. For example, different researchers have used a variety of methods to determine *Ksat*. These values have been related to a wide variety of morphological attributes but, in some cases, these attributes were not those commonly recorded. This means that the derived PTFs are not readily transferable and, in many cases, cannot even be tested against data in other regions or countries. The use of a reasonably standardized set of morphological attributes that are described or quantified in a more objective manner would help improve the PTFs based on soil morphology. In the meantime, there is also a clear need to use consistent methods of measuring soil hydraulic properties so that these values themselves are comparable among different soils and across databases.

While there is a degree of consistency in the definitions of size ranges of the various types of soil structure (FAO, 1990; Soil Survey Division Staff, 1993), this is not the case for the field assessment of soil porosity. For example, pore size classes in the US system are different from both the Canadian and British classes (Soil Survey Division Staff, 1993; McKeague et al., 1986; Hodgson, 1997). In the US soil surveys, pores visible to human eyes (macropores) are described in terms of size, quantity, shape, and vertical continuity but the Canadian system describes the percentage total porosity and air porosity of mineral soils in three classes. A provision is made for describing voids in terms of abundance and size in five size classes that range from micro (<0.1 mm) to coarse ($5-10$ mm) as well as orientation, distribution, morphology, continuity and type. In the British system, pores and fissures are recorded from observation or estimated from soil texture. Fissures are classed into five classes and are largely related to the cracks found between structural peds (Hodgson, 1997). Soil pores are separated into macropores (>0.06 mm) and micropores (<0.06 mm). Four classes of macropore are described and the volume of pores is also estimated. The porosity class for pores <0.06 mm but >0.0002 are estimated from soil texture class, packing density and horizon type. From the above, it is apparent that a standardized set of descriptions of soil porosity would be required to enhance the incorporation of soil morphology into future PTFs.

3.2. Quantification of soil morphology including soil structure

A greater use of quantitative and reproducible approaches should be pursued if we are to make better use of soil morphological data in PTFs. Bouma (1990) pointed out that the use of relatively simple macromorphological methods should be encouraged to allow widespread morphological measurements under field conditions.

Besides the work of Bouma and Anderson (1973), Lin et al. (1999a,b), and a few others as described above, other quantifications of soil macromorphological attributes in association with soil hydraulic properties are noted here. These studies were not intended to develop PTFs, but have relevance to the enhancement of the future use of quantitative soil morphology in developing or applying PTFs. Simpson and Cunningham (1978) quantified soil morphological data using a relative rating scale to show differences in morphological properties that could be related to hydraulic properties and that had been significantly altered due to wastewater irrigation. They assigned low scores to those soil morphological features usually associated with wetter soil moisture regimes (i.e., lower permeability). They found that the occurrence of a shallow seasonal water table due to irrigation has significantly altered all soil mottling properties (i.e., increased the abundance, size, and contrast of mottles and reduced the depth to mottles). In addition, they found that the size of structural units increased and the moist consistence became firmer in the subsoils after 13 years of irrigation (reflecting apparent degradation of soil structure). However, Simpson and Cunningham (1978) did not provide clear justifications for the way they assigned scores to various soil morphological features.

In an attempt to develop quantitative soil hydromorphological indicators of water table behavior, Galusky et al. (1998) examined the depth to gleying (d-gley), depth to soil chroma of 3 to 4 (d-34), and the depths to redox concretions (d-conc) and depletions (d-depl) in 29 sites in the coastal plain of Maryland. They found that d-34 correlated the most highly with average monthly water table levels. This correlation was greatest for the month of March (when seasonal water table levels in Maryland are generally at their highest) and decreased during the summer months. The d-gley also was highly correlated with late winter/early spring water table levels. However, they did not find good correlations between d-conc or d-depl with average monthly water table levels. Soil morphology is sensitive to long-term, average monthly water table depths, and thus could be used to estimate statistical (e.g., monthly average) and stochastic (e.g., probabilistic) properties of the monthly water table regime. The depth to gleying has long been used as a crude indicator of the mean position of the wet-season water table (Franzmeier et al., 1983). Soil chroma of three to four also has been found to be associated with prolonged periods of water table saturation (Franzmeier et al., 1983; Evans and Franzmeier, 1986) in the US. Similarly, the depths to redox concretions and depletions have been associated with water table fluctuations (e.g., Vepraskas, 1992). A similar system exists in the UK where, in the absence of direct measurement, soils can be assigned to one of six soil wetness classes that describe the height and duration of water logging by applying a set of PTRs (Jarvis et al., 1984; Lilly and Matthews, 1994; Lilly et al., 2003). Many of the properties used in the HOST classification (Boorman et al., 1995) such as the depth to a slowly permeable layer, depth to gleying are also used in determining the soil water regime and these properties are derived from PTRs applied to soil morphological properties.

3.3. Derivation of PTFs for soils with unusual characteristics

Through soil morphology, there is also a potential to extrapolate PTF outputs from well characterized to less well characterized soil types. Pedotransfer functions based on many existing datasets of soil morphology may be a cheap and rapid method to generate much needed hydrologic data for some special soils where few data are available. For example, little work has been done to develop PTFs for predicting water contents or hydraulic conductivities in organic soils (peats or histosols). This may be because much of the work in relating soil morphology to soil hydraulic properties was developed for land resource evaluations that have, in the past, been aimed primarily at the classification and grading of agricultural lands. However, there is a growing need to develop and run models that simulate processes in organic soils related to issues such as carbon sequestration and the release of dissolved organic carbon in light of global climate change. It is recognized that changes to the water regime in these soils could have a major impact on global carbon mitigation strategies but there are few data and limited PTFs to predict the hydraulic properties needed to test theories through modeling. Schwärzel et al. (2002) reported differences in AWC and $K|h|$ of peat soils attributed to pedogenic changes in peat layers and the associated changes in both the volume and size of macropores. Holden et al. (2001) also found that macroporosity in UK peat soils varied with depth and type of surface vegetation. In addition, vertic, fragic, and stony horizons seem to present difficulties when attempting to estimate hydraulic properties from their morphology. Many researchers merely report the problems but there seems to be little effort to try to resolve them. Apart from some work in Australia, there also seems to be a lack of progress in developing morphology-based PTFs for tropical soils.

3.4. Grouping soils based on terrain and geomorphology

Recent work by Romano and Palladino (2002) in Italy and by Rawls and Pachepsky (2002b) in the US have shown that topographic attributes can be used to modify and improve existing PTFs that predict moisture retention. Although different approaches (multiple regression and regression trees) were used to analyze the data, similar results were apparent. In Italy, slope curvature was considered as an important modifier of water retention perhaps due to increased porosity resulting from soil creep on convex slopes. In the US, Rawls and Pachepsky (2002b) suggested that the topography-related redistribution of finer soil particles due to erosion and subsequent deposition might affect water retention. They also suggested that increased root development associated with a more favorable water regime in topographic depressions may increase the volume of biopores. However, in wetter regions of the world such depressions often contain anaerobic soils associated with waterlogging. In these soils root development can be restricted along with the formation of biopores. When applying PTFs to soil map units there is often a tendency to assume that the unit has a uniform texture. Grouping soils based on terrain and geomorphology allow the spatial disaggregation of map units and the ability to refine PTFs at a local scale, e.g., for hydrological studies in small catchments.

4. SUMMARY

This chapter has shown that it is possible to develop both PTFs and PTRs to predict soil hydraulic properties and soil hydrological functioning from soil morphology including soil structure. By doing so, a considerable volume of soil morphological data residing in national and regional databases could be of great use in providing information on soil hydrology. However, while there is scope for further developments, this may require an acceptance of both standardized methodologies to describe and quantify soil morphological attributes and to measure soil hydraulic properties in the field.

ACKNOWLEDGEMENTS

H. Lin's contribution to this work was partially supported by the USDA-NRI grant #2002-35102-12547 while the Scottish Executive Environment and Rural Affairs Department funded A. Lilly.

REFERENCES

Abbaspour, K.C., Moon, D.E., 1992. Relationships between conventional field information and some soil properties measured in the laboratory. Geoderma 55, 119-140.

Anderson, J.L., Bouma, J., 1973. Relationship between saturated hydraulic conductivity and morphometric data of an Argillic horizon. Soil Sci. Soc. Am. Proc. 37, 408-413.

Avery, B.W., 1980. Soil classification for England and Wales [Higher categories]. Soil Survey of England and Wales Technical monograph No. 14. Lawes Agricultural Trust. Rothamsted Experimental Station, Harpenden.

Batjes, N.H., 1996. Development of a world data set of soil water retention properties using pedotransfer rules. Geoderma 71, 31-52.

Baumer, O.W., 1992. Predicting unsaturated hydraulic parameters. *In*: Van Genuchten, M.Th., Leij, F.J., Lund, L.J. (Eds.), Indirect Methods for Estimating the Hydraulic Properties of Unsaturated soils, Proceedings of an International Conference on Indirect Methods for Estimating the Hydraulic Properties of Unsaturated soils, Riverside, California, pp. 341-354.

Bibby, J.S., Douglas, H.A., Thomasson, A.J., Robertson, J.S., 1982. Land Capability Classification for Agriculture, Soil Survey of Scotland Monograph. The Macaulay Institute for Soil Research Institute for Soil Research, Aberdeen.

Boorman, D.B., Hollis, J.M., Lilly, A., 1995. Hydrology of soil types: a hydrologically-based classification of the soils of the United Kingdom, Institute of Hydrology Report No. 126. Institute of Hydrology, Wallingford.

Bouma, J., 1990. Using morphometric expressions for macropores to improve soil physical analyses of field soils. Geoderma 46, 3-13.

Bouma, J., 1991. Influence of soil macroporosity on environmental quality. *In*: Sparks, D.L. (Ed.), Advances in Agronomy, Vol. 46. Academic Press, Inc, New York, pp. 1-37.

Bouma, J., 1992. Effects of soil structure, tillage, and aggregation upon soil hydraulic properties. *In*: Wagenet, R.J., Bavege, P., Stewart, B.A. (Eds.), Interacting Processes in Soil Science. Lewis, London, pp. 1-36.

Bouma, J., Anderson, J.L., 1973. Relationships between soil structure characteristics and hydraulic conductivity. *In*: Bruce, R.R. (Ed.), Field Soil Water Regime. Soil Science Society of America Special Publication, 5, pp. 77-105.

Bouma, J., Denning, J.L., 1974. A comparison of hydraulic conductivities calculated with morphometric and physical methods. Soil Science Society of America Proceedings 38, 124-127.

Bouma, J., Jonggerius, A., Schoonderbeek, D., 1979. Calculation of saturated hydraulic conductivity of some pedal soils using micromorphometric data. Soil Sci. Soc. Am. J. 43, 261-264.

Breeuwsma, A., Wösten, J.H.M., Vleeshouwer, J.J., van Slobbe, A.M., Bouma, J., 1986. Derivation of land qualities to access environmental problems from soil surveys. Soil Sci. Soc. Am. J. 50, 186-190.

Coen, G.M., Wang, C., 1989. Estimating vertical saturated hydraulic conductivity from soil morphology in Alberta. Can. J. Soil Sci. 69, 1-16.

Danalatos, N.G., Kosmas, C.S., Driessen, P.M., Yassoglou, N., 1994. Estimation of the draining soil moisture characteristic from standard data as recorded in routine soil surveys. Geoderma 64, 155-165.

De Jong, R., Loebel, K., 1982. Empirical relations between soil components and water retention at 0.33 and 15 atmospheres. Can. J. Soil Sci. 62, 343-350.

De Jong, R., McKeague, J.A., 1987. A comparison of measured and modelled water retention data. Can. J. Soil Sci. 67, 697-703.

Evans, C.V., Franzmeier, D.P., 1986. Saturation, aeration, and color patterns in a toposequence of soils in north-central Indiana. Soil Sci. Soc. Am. J. 50, 975-980.

FAO (Food and Agriculture Organisation), 1990. Guidelines for soil description, Third edition. FAO/ISRIC, Rome.

Franzmeier, D.P., 1991. Estimation of hydraulic conductivity from effective porosity data for some Indiana soils. Soil Sci. Soc. Am. J. 55, 1801-1803.

Franzmeier, D.P., Yahner, J.E., Steinhardt, G.C., Sinclair, H.R. Jr., 1983. Color patterns and water-table levels in some Indiana soils. Soil Sci. Soc. Am. J. 47, 1196-1202.

Galusky, L.P., Rabenhorst, M.C., Hill, R.L., 1998. Toward the development of quantitative soil morphological indicators of water table behavior. *In*: Rabenhorst, M.C., Bell, J.C., McDaniel, P.A. (Eds.), Quantifying Soil Hydromorphology, Soil Science Society of America Special Publication No. 54. Soil Science Society of America Inc, Madison, WI, pp. 77-93.

Griffiths, E., Webb, T.H., Watt, J.P.C., Singleton, P.L., 1999. Development of soil morphological descriptors to improve field estimation of hydraulic conductivity. Aust. J. Soil Res. 37, 971-982.

Grossman, R.B., Harms, D.S., Seybold, C.A., Herrick, J.E., 2001. Coupling use-dependent and use-invariant data for soil quality evaluation in the United States. J. Soil Water Conserv. 56, 63-68.

Guber, A.K., Rawls, W.J., Shein, E.V., Pachepsky, Ya.A., 2003. Effect of soil aggregate size distribution on water retention. Soil Sci. 168, 223-233.

Hall, D.G.M., Reeve, M.J., Thomasson, A.J., Wright, V.F., 1977. Water retention, porosity and density of field soils, Soil Survey Technical Monograph No. 9. Soil Survey of England and Wales, Lawes Agricultural Trust (Soil Survey of England and Wales). Rothamsted Experimental Station. Harpenden.

Hecht-Nielsen, R., 1990. Neurocomputing. Alison-Wesley, Reading, MA.

Hodgson, J.M., 1974. Soil Survey Field Handbook: Describing and Sampling Soil Profiles, Soil Survey Technical Monograph No.5, Soil Survey of England and Wales. Lawes Agricultural Trust (Soil Survey of England and Wales). Rothamsted Experimental Station. Harpenden.

Hodgson, J.M., 1997. Soil Survey Field Handbook: Describing and Sampling Soil Profiles, Soil survey Technical Monograph No. 5, Third edition. Soil Survey and Land Research Centre, Silsoe, England.

Holden, J., Burt, T.P., Cox, N.J., 2001. Macroporosity and infiltration in blanket peat: the implications of tension disc infiltrometer measurements. Hydrol. Process. 15, 289-303.

Hollis, J.M., Woods, S.M., 1989. The measurement and estimation of saturated soil hydraulic conductivity, Research Report for the Ministry of Agriculture, Fisheries and Food. Soil Survey and Land Research Centre, Silsoe, England.

Hollis, J.M., Jones, R.J.A., Palmer, R.C., 1977. The effects of organic matter and particle size on the water retention properties of some soils in the west Midlands of England. Geoderma 17, 225-238.

Jarvis, R.A., Bendelow, V.C., Bradley, R.I., Carroll, D.M., Furness, R.R., Kilgour, I.N.L., King, S.J., 1984. Soils and their use in northern England, Soil Survey of England and Wales Bulletin No. 10. Lawes Agricultural Trust (Soil Survey of England and Wales). Rothamsted Experimental Station, Harpenden.

Jones, R.J.A., Hollis, J.M., 1996. Pedotransfer rules for environmental interpretations of the European Union Soil Database. *In*: Le Bas, C., Jamagne, M. (Eds.), Soil databases to support sustainable development, EUR 16371 EN. INRA/Joint Research Centre, Orleans, France.

King, J.J., Franzmeier, D.P., 1981. Estimation of saturated hydraulic conductivity from soil morphological and genetic information. Soil Sci. Soc. Am. J. 45, 1153-1156.

Köhne, J.M., Köhne, S., Gerke, H.H., 2002. Estimating the hydraulic functions of dual-permeability models from bulk soil data. Water Resour. Res. 38, 26.

Levine, E., Kimes, D.S., 1997. Evaluating water holding capacity across spatial scales with neural network, Pedometrics'97. Proceedings of an international workshop. University of Wisconsin, Madison, WI.

Lilly, A., 2000. The relationship between field-saturated hydraulic conductivity and soil structure: development of class pedotransfer functions. Soil Use Manag. 16, 56-60.

Lilly, A., Matthews, K.B., 1994. A soil wetness class map for Scotland: new assessments of soil and climate data for land evaluation. Geoforum 25, 371-379.

Lilly, A., Boorman, D.B., Hollis, J.M., 1998. The development of a hydrological classification of UK soils and the inherent scale changes. Nutrient Cycling Agroecosyst. 50, 299-302.

Lilly, A., Ball, B.C., McTaggart, I.P., Horne, P.L., 2003. Spatial and temporal scaling of nitrous oxide emissions from the field to the regional scale in Scotland. Nutrient Cycling Agroecosyst. 66, 241-257.

Lin, H.S., McInnes, K.J., Wilding, L.P., Hallmark, C.T., 1997. Low tension water flow in structured soils. Can. J. Soil Sci. 77, 649-654.

Lin, H.S., McInnes, K.J., Wilding, L.P., Hallmark, C.T., 1998. Macroporosity and initial moisture effects on infiltration rates in Vertisols and vertic intergrades. Soil Sci. 163, 2-8.

Lin, H.S., McInnes, K.J., Wilding, L.P., Hallmark, C.T., 1999a. Effects of soil morphology on hydraulic properties: I. Quantification of soil morphology. Soil Sci. Soc. Am. J. 63, 948-954.

Lin, H.S., McInnes, K.J., Wilding, L.P., Hallmark, C.T., 1999b. Effects of soil morphology on hydraulic properties: II. Hydraulic pedotransfer functions. Soil Sci. Soc. Am. J. 63, 955-961.

MAFF, 1988. Agricultural land classification of England and Wales: Revised guidelines and criteria for grading the quality of agricultural land. Ministry of Agriculture, Fisheries and Food Publications, Alnwick, England.

McBratney, A.B., Minasny, B., Cattle, S.R., Vervoort, R.W., 2002. From pedotransfer functions to soil inference systems. Geoderma 109, 41-73.

McBride, R.A., MacIntosh, E.E., 1984. Soil survey interpretations from water retention data: I Development and validation of a water retention model. Soil Sci. Soc. Am. J. 48, 1338-1343.

McDonald, D.C., Julian, R., 1965. Quantitative estimation of soil total porosity and macroporosity as part of the pedological description. N. Z. J. Agric. Res. 8, 927-946.

McKeague, J.A., 1987. Estimating air porosity and available water capacity from soil morphology. Soil Sci. Soc. Am. J. 51, 148-152.

McKeague, J.A., Wang, C., Topp, G.C., 1982. Estimating saturated hydraulic conductivity from soil morphology. Soil Sci. Soc. Am. J. 46, 1239-1244.

McKeague, J.A., Wang, C., Coen, G.M., 1986. Describing and interpreting the macrostructure of Mineral soils – A preliminary report. Land Resource Research Institute, Ottawa, Ontario, Research Branch Agricultural Canada.

McKeague, J.A., Wang, C., Ross, G.J., Modestus, K., 1991. Test of guidelines for field estimates of water retention properties of soils derived from volcanic materials in northern Tanzania. Can. J. Soil Sci. 71, 1-10.

McKenzie, N.J., Jacquier, D., 1997. Improving the field estimation of saturated hydraulic conductivity in soil survey. Aust. J. Soil Res. 35, 803-825.

McKenzie, N.J., MacLeod, D.A., 1989. Relationships between soil morphology and soil properties relevant to irrigated and dryland agriculture. Aust. J. Soil Res. 27, 235-258.

McKenzie, N.J., Cresswell, H.P., Ryan, P.J., Grundy, M., 2000. Contemporary land resource survey requires improvements in direct soil measurement. Commun. Soil Sci. Plant Anal. 31, 1553-1569.

Minasny, B., McBratney, A.B., Bristow, K.L., 1999. Comparison of different approaches to the development of pedotransfer functions for water-retention curves. Geoderma 93, 225-253.

Nowland, J.L., 1981. Soil water regime classification. *In*: Day, J.H. (Ed.), Proceedings of the 3rd meeting of the Expert Committee on Soil Survey, Ottawa, March 1981. Agriculture Canada, Ottawa, pp. 64-74.

O'Connell, D.A., Ryan, P.J., 2002. Prediction of three key hydraulic properties in a soil survey of a small forested catchment. Aust. J. Soil Res., 191-206.

O'Neal, A.M., 1949. Soil characteristics significant in evaluating permeability. Soil Sci. 67, 403-409.

O'Neal, A.M., 1952. A key for evaluating soil permeability by means of certain field clues. Proc. Sci. Soc. Am. 16, 312-315.

Pachepsky, Ya.A., Rawls, W.J., 1999. Accuracy and reliability of pedotransfer functions as affected by grouping soils. Soil Sci. Soc. Am. J. 63, 1748-1757.

Pachepsky, Ya.A., Timlin, D.J., Várallyay, G., 1996. Artificial neural networks to estimate soil water retention from easily measurable data. Soil Sci. Soc. Am. J. 60, 727-773.

Pachepsky, Ya.A., Rawls, W.J., Gimenez, D., Watt, J.P.C., 1998. Use of soil penetration resistance and group methods of data handling to improve soil water retention estimates. Soil Tillage Res. 49, 117-126.

Quisenberry, V.L., Smith, B.R., Philips, R.E., Scott, H.D., Nortcliff, S., 1993. A soil classification system for describing water and chemical transport. Soil Sci. 156, 306-315.

Rawls, W.J., Pachepsky, Ya.A., 2002a. Soil consistence and structure as predictors of water retention. Soil Sci. Soc. Am. J. 66, 1115-1126.

Rawls, W.J., Pachepsky, Ya.A., 2002b. Using field topographic descriptors to estimate soil water retention. Soil Sci. 167, 423-435.

Rawls, W.J., Ahuja, L.R., Brakensiek, D.L., Shirmohammadi, A., 1993. Infiltration and soil water movement. In: Maidment, D.R. (Ed.), Handbook of Hydrology. McGraw-Hill Inc, New York, p. 51.

Reeve, M.J., Smith, P.D., Thomasson, A.J., 1973. The effect of density on water retention properties of field soils. J. Soil Sci. 24, 355-367.

Reynolds, W.D., Elrick, D.E., 1986. A method for simultaneous in situ measurement in the vadose zone of field-saturated hydraulic conductivity, sorptivity and the conductivity-pressure head relationship. Ground Water Monit. Rev. 6, 84-95.

Romano, N., Palladino, M., 2002. Prediction of soil water retention using soil physical data and terrain attributes. J. Hydrol. 265, 56-75.

Schaap, M.G., Bouten, W., 1996. Modelling water retention curves of sandy soils using neural networks. Water Resour. Res. 32, 3033-3040.

Schwärzel, K., Renger, M., Sauerbrey, R., Wessolek, G., 2002. Soil physical characteristics of peat soils. J. Plant Nutr. Soil Sci. 165, 471-486.

Simpson, T.W., Cunningham, R.L., 1978. Soil morphologic and hydraulic changes associated with wastewater irrigation, Research project technical completion report. Institute for Research on Land and Water Resources, The Pennsylvania State University, University Park, PA.

Soil Survey Division Staff, 1993. Soil survey manual, US Department of Agriculture Handbook No. 18. Government Printing Office, Washington, DC.

Tamari, S., Wösten, J.H.M., Ruiz-Suarez, J.C., 1996. Testing an artificial neural network for predicting soil hydraulic conductivity. Soil Sci. Soc. Am. J. 60, 1732-1741.

Thomasson, A.J. (Ed.), 1975. Soils and field drainage, Soil Survey Technical Monograph No.7. Soil Survey of England and Wales. Rothamsted Experimental Station, Lawes Agricultural Trust, Harpenden.

Thomasson, A.J., 1978. Towards an objective classification of soil structure. J. Soil Sci. 29, 38-46.

Vepraskas, M.J., 1992. Redoximorphic features for identifying aquic conditions, North Carolina Agricultural Research Service Technical Bulletin No. 301, NC, USA.

Wagner, B., Tarnawski, V.R., Wessolek, G., Plagge, R., 1998. Suitability of models for the estimation of soil hydraulic parameters. Geoderma 86, 229-239.

Wang, C., McKeague, J.A., Topp, G.C., 1985. Comparison of estimated and measured horizontal Ksat values. Can. J. Soil Sci. 65, 707-715.

Williams, J., Prebble, R.E., Williams, W.T., Hignett, C.T., 1983. The influence of texture, structure and clay mineralogy on the soil moisture characteristic. Aust. J. Soil Res. 21, 15-32.

Wösten, J.H.M., Bouma, J., Stoffelsen, G.H., 1985. The use of soil survey data for regional soil water simulation models. Soil Sci. Soc. Am. J. 49, 1238-1245.

Wösten, J.H.M., Bannink, M.H., Beuving, J., 1987. Water Retention and Hydraulic Conductivity Characteristics of Top- and Sub-soils in the Netherlands: the Staring Series, Report 1932. Soil Survey Institute, Wageningen, The Netherlands (In Dutch).

Wösten, J.H.M., Schuren, C.H.J.E., Bouma, J., Stein, A., 1990. Comparing four methods to generate soil hydraulic functions in terms of their effect on simulated soil water budgets. Soil Sci. Soc. Am. J. 54, 827-832.

Wösten, J.H.M., Lilly, A., Nemes, A., Le Bas, C., 1999. Development and use of a database of hydraulic properties of European soils. Geoderma 90, 169-185.

Wösten, J.H.M., Pachepsky, Ya.A., Rawls, W.J., 2001. Pedotransfer functions: bridging the gap between available basic soil data and missing hydraulic characteristics. J. Hydrol. 251, 123-150.

Chapter 8

SOIL AGGREGATES AND WATER RETENTION

A. Guber[1,*], Ya. Pachepsky[1], E. Shein[2] and W. J. Rawls[3]

[1]Environmental Microbial Safety Laboratory, USDA-ARS-BA-ANRI-EMSL, Bldg. 173, Rm. 203, BARC-EAST, Powder Mill Road, Beltsville, MD 20705, USA

[2]119992 Moscow, Leninskie Gory, Moscow State University, Soil Science Faculty, Russia

[3]USDA-ARS Hydrology & Remote Sensing Lab, Bldg. 007, Rm. 104, BARC-W, Beltsville, MD 20705-2350, USA

*Corresponding author: Tel.: +1-301-504-5656; fax: +1-301-504-6608

1. INTRODUCTION

Soil aggregate composition is an important characteristic of soil structure and as such has been expected to affect soil water retention. Wittmuss and Mazurak (1958) studied water retention of samples made of individual aggregate fractions of silty clay soil. Water retention of samples from aggregates 0.07–0.15, 0.15–0.3, 0.30–0.60, 0.60–1.20, 1.20–2.4, and 2.4–4.8 was similar whereas aggregates less than 0.07 mm had distinctly different retention for matric potential from 0 to −1.5 MPa. Those authors noted that the texture of soil in aggregates became finer as the size of aggregates decreased. Tamboli et al. (1964) worked with aggregates from silt loam soil and observed similarity in water retention of aggregates 2–3, 3–5, 5–9.5 and 9.5–12 mm. Water retention of aggregates 1–2 and 1–0.5 mm was lower than for the larger fractions and decreased with the aggregate size. No variation in textural composition with aggregate size was found. Amemiya (1965) studied model samples composed of individual fractions of soil aggregates from silt loam, silty clay loam and clay loam soils in the range of matric potentials from −0.02 to −1.2 Mpa. He found water retention to be similar for samples from aggregate fractions 1–2, 2–3, and 3–5 mm, and substantial difference between water retention of samples from aggregates 0.5–1 mm and from larger fractions. Combination of various aggregate fractions in one sample led to both increase and decrease in soil water retention as compared to the water retention of samples made of individual fractions. Abrol and Palta (1970) observed monotonous decrease in water retention of aggregate fractions with aggregate size in the range of matric potential from −0.045 to −1 MPa. Chang (1968) compared water retention of <0.25 cm and 1–4.8 cm fractions from clay soil. Water retention of the smaller fraction was smaller at matric potentials from 0 to −1 kPa. As the matric potential decreased further, smaller fraction had larger water retention. Those differences were viewed as a result of differences in intra-aggregate and inter-aggregate water retention.

Data on model systems composed of aggregates indicated a potential effect of aggregate size distributions of soils on their water retention. Wu et al. (1990); Wu and Vomocil (1998) had found similarity of cumulative pore, aggregate, and particle size distributions, and used this similarity to transform one of those distributions to another. Our objective is to show how aggregate size distributions affect soil water retention from saturation to the wilting point. We use the van Genuchten approximation of gravimetric water retention data and report effects of both aggregate size and particle size distributions on parameters of the van Genuchten equation. A similar study has been reported for water contents at specific soil water potentials (Guber et al., 2003)

2. SOIL DATABASE

Soil properties were studied on samples of Podzoluvisols, Planosols, Chernozems, Fluvisols, Calcisols, and Gleysols (FAO-UNESCO, 1974; Stolbovoi and Sheremet, 2000) represented with 9, 7, 102, 3, 11, and 10 samples, respectively. Samples were taken in Moscow and Voronezh provinces of Russia, Krasnohvardiis'ka province of Ukraine, and Tashkent and Fergana provinces of Uzbekistan. Soils were sampled at arable lands in A, E, B and transitional horizons. Texture of the samples was measured with the pipette method (Gee and Bauder, 1986) after dispersion with sodium pyrophosphate $Na_4P_2O_7$. Particles diameter groups were <0.001 mm, 0.001–0.005, 0.005– 0.01, 0.01–0.05, 0.05–0.25, >0.25 mm. Those fractions are referred to as clay, fine silt, medium silt, coarse silt, fine sand, and coarse sand, respectively, in this work. The dataset included samples of loam, silt loam, silty clay loam and silty clay soils, represented with 11, 28, 47, and 56 samples respectively. Dry aggregate size distribution was determined by sieving into following aggregate diameter groups: <0.25, 0.25–0.5, 0.5–1, 1–2, 2–3, 3–5, 5–7, 7–10, and >10 mm.

Cumulative distributions were used in data analysis. Particle size variables were $d_{<0.001}$ = the percentage of particles smaller than 0.001 mm, $d_{<0.005}$ = the percentage of particles smaller than 0.005 mm, etc. Aggregate size variables were $a_{<0.25}$ = the percentage of aggregates smaller than 0.25 mm, $a_{<0.5}$ = the percentage of particles smaller than <0.5 mm, etc.

Water retention was measured in sand–kaolin boxes (Stakman et al., 1969; Varallyay and Mironenko, 1979) for matric potentials within range from 0 to -50 kPa and by vapor equilibration above K_2SO_4, KCl and $Ca(NO_3)_2$ solutes at -2.76, -20.5 and -81.6 MPa. Gravimetric water contents W were used in the analysis in this work. Parameters of water retention were found by fitting van Genuchten equation (van Genuchten, 1980) to measured data:

$$W = Wr + (Ws - Wr)/[1 + (\alpha \psi)^n]^m; \qquad m = 1 - 1/n \tag{1}$$

Here Ws and Wr are saturated and residual water contents, respectively, α, n and m are shape-defining parameters. Equation (1) gave a satisfactory approximation of the gravimetric water retention data in our dataset. Root-mean square errors were between 1.5 and 3.5% in 50% of cases. The Kolmogorov-Smirnov test showed that statistical distributions of RMSE did not differ between textural classes.

Results of grouping samples by textural classes are shown in Table 1. Distributions of all parameters of water retention are similar for silty clays and silty clay loams that have

Table 1
Values of van Genuchten parameters at different probability levels grouped by textural classes

Probability	Loam			Silt loam			Silty clay			Silty clay loam		
	$\alpha * 100$ (cm^{-1})	n	Ws	$\alpha * 100$ (cm^{-1})	n	Ws	$\alpha * 100$ (cm^{-1})	n	Ws	$\alpha * 100$ (cm^{-1})	n	Ws
0.1	nd	nd	nd	1.30	1.14	0.344	0.27	1.17	0.343	0.27	1.17	0.344
0.25	0.63	1.15	0.343	1.90	1.19	0.417	0.44	1.18	0.376	0.40	1.18	0.370
0.5	1.30	1.24	0.401	2.90	1.26	0.514	0.91	1.21	0.400	0.90	1.21	0.396
0.75	2.20	1.30	0.420	5.70	1.26	0.587	1.70	1.23	0.440	1.40	1.23	0.419
0.9	nd	nd	nd	10.10	1.31	0.633	2.70	1.25	0.478	2.40	1.26	0.457

relatively low mean values of saturated water content Ws and parameter α. Those soils have also the largest average exponent n as compare with loam and silt loam. The largest mean values of the parameters Ws, α and n were found in silt loam soils.

The aggregate size distributions in the database are characterized in Table 2. Aggregate fractions can be divided into three groups by the average content of the fraction. Aggregates smaller than 1 mm form the first group with average content of 3.4–3.8%, and aggregates in the range 1–10 mm form the second group with average contents of 9.9–13.0%. Large soil aggregates (>10 mm) represent a separate group, which has great variation in contents and average content of 32.8%.

Table 2
Percentage of aggregate fractions at different probability levels

Probability	Aggregate size (mm)								
	<0.25	0.25–0.5	0.5–1.0	1.0–2.0	2.0–3.0	3.0–5.0	5.0–7.0	7.0–10.0	>10
0.1	1	1	1	5	5	8	6	6	6
0.25	1	1	1	7	7	9	7	8	21
0.5	2	2	2	11	10	12	9	9	33
0.75	6	4	4	14	13	16	10	11	45
0.9	8	9	7	17	18	19	15	19	53

3. REGRESSION TREE MODELING

Relationships between soil water retention and distributions of aggregate sizes and particle sizes were established using regression trees (see Section 2 of Chapter 2). The algorithm first finds a partition of the database into two of the most homogeneous subsets. For that, all possible splits of ranges of all input variables are compared in terms of the non-homogeneity of resulting subsets. The input variable that provides the maximum improvement of the non-homogeneity, is thought to have the greatest effect on the output variable and is used for partitioning. The procedure of partitioning is repeated for each of the two subsets, and the next two splitting variables are defined. The result of the algorithm work is a tree-like structure in which two new branches appear after each split. Regression trees were built with the software package SPLUS (MathSoft, 1999). This software uses the sum of squared differences between average in a subset and individual values in this subset as a measure of non-homogeneity within the group. The minimum size of five samples in a subset was set to prevent forming groups that are too small. We terminated the tree-building when the non-homogeneity decreased two times from the value of the whole dataset.

Regression tree for $\lg(\alpha)$ is presented in Figure 1. The total database has first been partitioned by the content of clay. The first group of samples contains the loam and silt loam samples that have the largest values $\lg(\alpha)$ and the lowest clay content. The second group of samples has a finer texture. The splitting variable here is the content of aggregates larger than 10 mm. Samples with high contents of aggregates > 10 mm are then grouped according contents of fine textural fractions. Soils with the negligible sand content form a separate group (node [5]). Soils with high "clay + fine silt" content are grouped at node [4]. Low "clay + fine silt" contents allow further separation into groups with low and high clay content (nodes [2] and [3], respectively). The aggregate size distribution remains the governing parameter for values α when the percentage of aggregates > 10 mm is relatively low (see right branch of the tree in Figure 1). Contents of the largest aggregates continue to be grouping variables as splitting progresses in this part of the database. The largest values of α are found in samples where fraction 7–10 mm is represented well, i.e., constitutes more than 9% (node [7]).

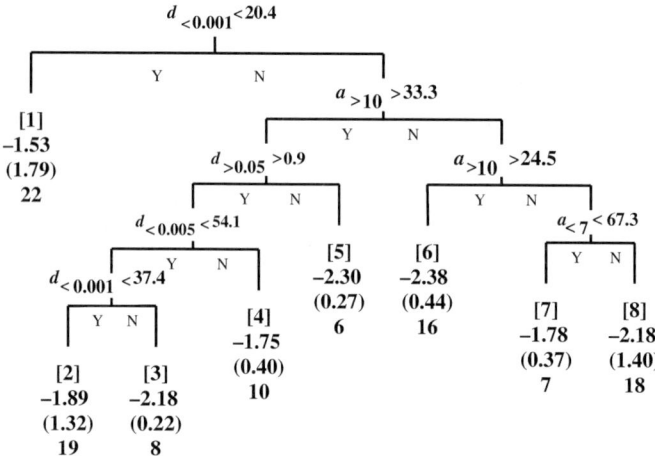

Figure 1. Regression tree to group samples according their $\lg(\alpha)$; "y" and "n" mean "yes" and "no" answers to the condition above the branching point; columns below the terminal nodes contain the node number in brackets, the average $\lg(\alpha)$ for the group, the standard deviation of $\lg(\alpha)$ within groups in parentheses, and the number of samples in the group.

The best partitioning of samples by their van Genuchten's parameter n could be done using the content of aggregates smaller than 1 mm (Figure 2). Contents of small aggregates continue to be the most important partitioning parameter for the samples where the percentage of smallest aggregates relatively is small (<10%) and fine silt and particles sand fraction constitute a small part of soil textural particles (nodes [1]–[4]). Particle size distribution becomes a source of partitioning parameters for the samples in the right

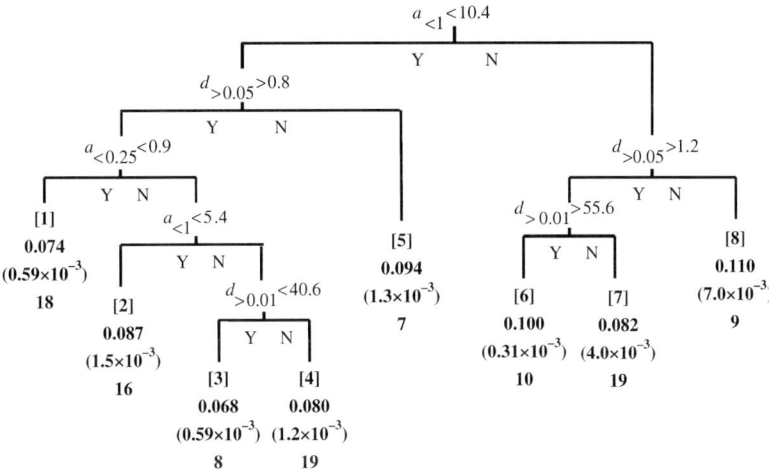

Figure 2. The same as in Figure 1 but for the logarithm of the van Genuchten parameter n.

branch of the tree where the content of aggregates < 1 mm is relatively high, i.e., larger than 10.4%. A singular node [8] denotes the group of samples with a negligible sand percentage. The content of coarse textural fractions $d_{>0.01}$ is important for the final partitioning in this group.

The first grouping of samples by their saturated water contents Ws has been made according to their clay fraction content $d_{<0.001}$ (Figure 3). The left big branch of the tree

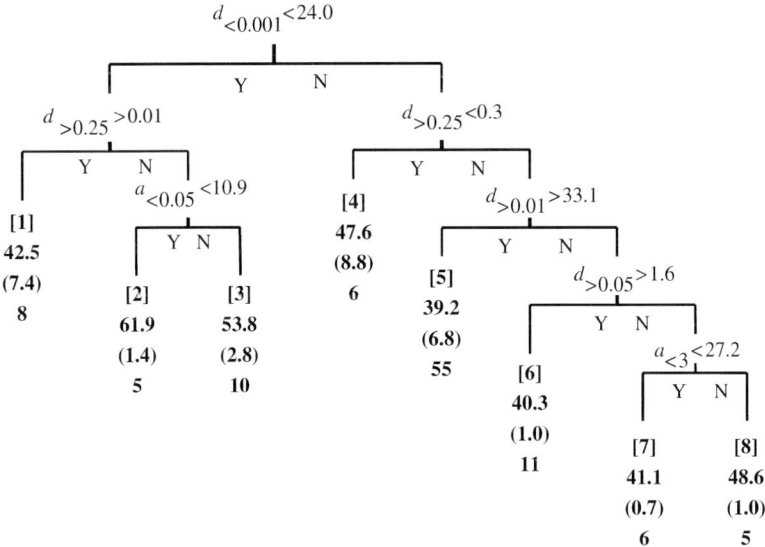

Figure 3. The same as in Figure 1 but for the saturated water contents Ws in the van Genuchten equation.

encompasses loam and silt loam samples whereas the right big branch includes "silty clay and silty clay loam" samples. The second partition in both groups occurs according the content of the coarse sand $d_{>0.25}$. Low coarse sand contents are associated with low Ws in the "loam + silt loam" group (node [1]) and with relatively high values Ws in the "silty clay + silty clay loam" group (node [4]). Low percentage of small aggregates ($a_{<0.05}$) is related the highest Ws in loam and silt loam samples (node [2]). An increase in the small aggregate content leads to smaller Ws values (node [3]). Partitioning by texture dominates further grouping in the "silty clay + silty clay loam" group (nodes [5]–[7]). A small group of silt clay samples with large content of aggregates <3 cm is separated into node [8] with the highest Ws.

The overall accuracy of estimating van Genuchten parameters with the developed regression trees is shown in Table 3 and Figure 4. In spite of giving the same, averaged over a group value to all measured values in the group, regression trees correctly reflect trends.

Table 3
Statistics of water retention parameters obtained with regression trees

Statistics	α (cm^{-1})	n	Ws
Mean	0.0157	1.217	43.7
Standard deviation	0.0154	0.045	8.3
Minimum	0.0015	1.148	28.4
Maximum	0.0940	1.462	68.1
RMSE	0.0102	0.030	4.6

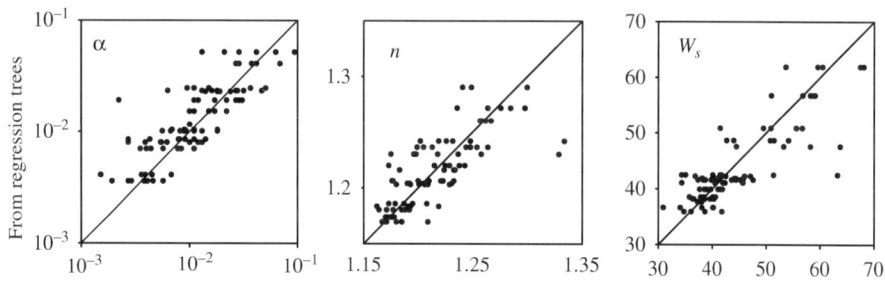

Figure 4. Comparison between van Genuchten parameter values from measured water retention and from regression trees.

4. DISCUSSION AND CONCLUSION

Soil aggregate composition provided important grouping parameters for water retention parameters. Contents of either small aggregates (<0.25 mm, <0.5 mm, <1 mm) or large aggregates (>7 mm, >10 mm) were the splitting variables in most cases for parameters of the van Genuchten equation. Results demonstrate that water retention reflects a complex interaction of texture and structure jointly affecting soil water retention. In silty clay and silty clay loam samples, texture was defining values of α where the amount of large aggregates was relatively low, less than 33%. No effect of texture on the values of α could be seen in samples with high contents of coarse aggregates (Figure 1). Parameter n was affected only by texture if the content of aggregates <1 was sufficiently large. However, small aggregate contents continued to be splitting variables where the amount of small aggregates was low. One may conclude that if van Genuchten parameter is affected by a parameter from the end of the aggregate size distribution, then further partitioning involves other parameters from the same end of the aggregate size distribution and adds more details about this end of the distribution in the regression tree.

Ranges of aggregate sizes affecting the van Genuchten parameters α and n may be related to the role of those parameters in defining shapes of water retention curves.

The effect of van Genuchten parameters on the shape of water retention curves computed from Equation (1) is shown in Figure 5. Parameter n influences the steepness of the curves in coordinates "water content vs. lg(suction)." Value of n is affected mostly by the small aggregate contents (Figure 2), and larger values on correspond to the larger percentages of small aggregates. A probable reason for that is that large number of small aggregates creates an hierarchical structure that empties gradually as the suction increases; such aggregates would fill space between large aggregates and large particles. Another possible reason for that is a similarity between external water retention of small aggregates and large sand particles. Such similarity was shown by Wittmuss and Mazurak (1958) in studies of columns made from aggregate fractions and from textural particles. They indicated that small aggregates less than 0.3 mm behave in terms of water retention in the same way as soil particles whereas larger aggregates have water retention different from water retention of particles of the same size.

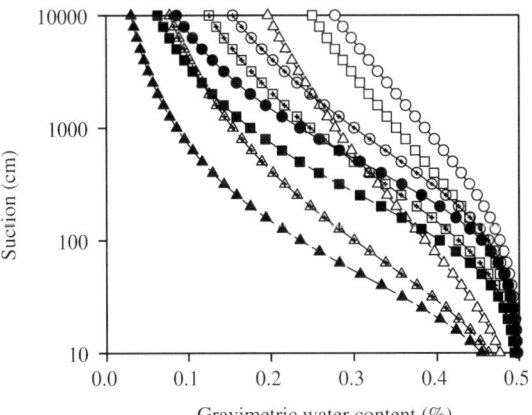

Figure 5. Examples of the effect of van Genuchten parameters α and n on the shape of water retention curves. Values of Ws and Wr were set to 0.5 and 0.0, respectively. Circles squares and triangles denote $\alpha = 0.005$, $\alpha = 0.01$, and $\alpha = 0.015$ cm^{-1}, respectively; hollow, dotted, and filled symbols denote $n = 1.1$, $n = 1.3$, and $n = 1.5$, respectively.

Increase in parameter α causes the faster loss of water with an increase in suction values. We have observed an increase in values of α with the increase of the percentage of large aggregates (Figure 1). The larger the amount of large aggregates the more likely is the presence of large intra-aggregate pores that empty at low suctions. Regression tree analysis gave results similar to the results of Gupta and Ewing (1998) who reported a modeling study on effect of aggregate-size distribution of water retention. They found that soils with dominant aggregate sizes in the range 12–50 mm had an inflection point distinctly different from that in soils with dominant aggregate sizes 2–3.3 mm, whereas much less difference was found between soils with dominant aggregate sizes 0.053–0.25 mm and 2–3.3 mm. The decrease in saturated water content Ws value in silt loam and loam samples with the increase of small aggregate contents also concurs with modeling results of the Gupta and Ewing (1998) who predicted flatter water retention curves in the low matric potential range for smaller dominant aggregate fractions.

We realize that some results of the regression tree analysis may be specific to our dataset. Effect of the database content on estimation of water retention from other soil properties has been documented (De Jong, 1983; Schaap and Leij, 1998). Comparison of retention parameter distributions shown in Figure 2 with distributions in other databases is problematic because of the definition of clay particles as particles <0.001 mm; the majority of texture classifications use the threshold of 0.002 mm. It remains to be seen whether and how the grouping results and the effect of aggregate distribution parameter on water retention will hold for other databases. This question presents an interesting avenue to explore because of the relatively strong effect of aggregate distribution on water retention parameters observed in this work.

All textural fractions, but not all aggregate fractions were used to partition samples by their water retention with regression trees. Separation of aggregates 1–3 mm into groups 1–2 and 2–3 mm, as well as separation of aggregates 3–7 mm into groups 3–5 and 5–7 mm, was made in measurements of aggregate size distributions but was not reflected in regression trees. One reason for that could be a similarity in water retention of the fractions 1–2 and 2–3 mm or 3–5 and 5–7 mm. Earlier similarity in retention of fractions 1–2 and 2–3 mm was observed by Amemiya (1965) and similarity in water retention of fractions 3–5 and 5–9.5 was observed by Tamboli et al. (1964). The intermediate fractions 1–2, 2–3, 2–5, 5–7 mm probably could be used if the dataset was larger.

Parameters of the aggregate size composition appeared to be important to split the dataset into homogeneous subsets using regression trees. This indicates that aggregate size distributions, if available in soil databases or feasible to measure, can be useful in estimating parameters of soil water retention from other soil properties.

REFERENCES

Abrol, I.P., Palta, J.P., 1970. A study of the effect of aggregate size and bulk density on moisture retention characteristics of selected soils. Agrochimica XIV 2–3, 157-165.

Amemiya, M., 1965. The influence of aggregate size on soil moisture content capillary conductivity relations. Soil Sci. Soc. Am. Proc. 29, 744-748.

Chang, R.K., 1968. Component potentials and hysteresis in water retention by compacted clay soil aggregates. Soil Sci. 105, 172-176.

De Jong, R., 1983. Soil water desorption curves estimated from limited data. Can. J. Soil Sci. 63, 697-703.
FAO-UNESCO. 1974. Soil Map of the World 1:5,000,000. Volume 1. Legend. UNESCO, Paris
Gee, G.W., Bauder, J.W., 1986. Particle-size analysis. In: Klute, A. (Ed.), Methods of Soil Analysis. Part 1. Physical and Mineralogical Methods, pp. 399-404.
Guber, A.K., Rawls, W.J., Shein, E.V., Pachepsky, Y.A., 2003. Effect of soil aggregate size distribution on water retention. Soil Sci. 168, 223-233.
Gupta, S.C., Ewing, R.P., 1998. Modeling water retention characteristics and surface roughness of tilled soils. In: van Genuchten, M. Th., Leij, F. J., Lund, L. J. (Eds.), Indirect methods for estimating the hydraulic properties of unsaturated soils. Proceedings of the International Workshop on Indirect Methods for Estimating the Hydraulic Properties of Unsaturated Soils. Riverside, California, October 11–13, 1989, pp. 379-388.
Mathsoft, 1999. SPLUS 2000 Professional. User's Manual.
Schaap, M.G., Leij, F.J., 1998. Database-related accuracy and uncertainty of pedotransfer functions. Soil Science 163, 765-779.
Stakman, W.P., Valk, G.A., van der Harst, G.G., 1969. Determination of Soil Moisture Retention Curves. I Sand-Box Apparatus. II Pressure Membrane Apparatus. ICW, Wageningen, The Netherlands.
Stolbovoi, V.S., Sheremet, B.N., 2000. Correlation between the legends of the 1:2.5 M soil map of the Soviet Union and the FAO soil map of the world. Eurasian Soil Sci. 3, 239-248.
Tamboli, P.M., Larson, W.E., Amemiya, M., 1964. Influence of aggregate size on moisture retention. Iowa Acad. Sci. 71, 103-108.
van Genuchten, M.Th., 1980. A closed-form equation for predicting the hydraulic conductivity of unsaturated soils. Soil Sci. Soc. Am. J. 44 (5), 892-898.
Varallyay, Gy., Mironenko, E.V., 1979. Soil–water relationships in saline and alkali conditions. In: Kovda, V.A., Szabolcs, I. (Eds.), Modelling of soil salinization and alkalinization. Agrokemia es Talajtan, 28, 33-82.
Wittmuss, H.D., Mazurak, A.P., 1958. Physical and chemical properties of aggregates in a Brunizem soil. Soil Sci. Soc. Am. Proc. 22, 1-5.
Wu, L., Vomocil, J.A., 1998. Predicting the soil water characteristic from the aggregate-size distribution. In: van Genuchten, M. Th., Leij, F. J., Wu, L. (Eds.), Characterization and measurement of the hydraulic properties of unsaturated porous media: proceedings of the International Workshop on Characterization and Measurement of the Hydraulic Properties of Unsaturated Porous Media. Riverside, California, October 22–24, 1997, pp.139-145.
Wu, L., Vomocil, J.A., Childs, S.W., 1990. Pore size, particle size, aggregate size, and water retention. Soil Sci. Soc. Am. J. 54, 952-956.

Chapter 9

UTILIZING MINERALOGICAL AND CHEMICAL INFORMATION IN PTFs

A. Bruand

Institut des Sciences de la Terre d'Orléans (ISTO), Bâtiment Géosciences, Université d'Orléans, BP 6759 45067, Orléans Cedex 2, France
Tel.: +33-2-38-41-70-24; fax: +33-2-38-41-73-08

Soil structure is known to reflect mineralogical composition of clay fraction and soil chemical composition. Because soil hydraulic properties are likely to depend on soil structure, chemical and mineralogical compositions have long been expected to be important predictors of soil ability to retain and transmit water (Rawls et al., 1991). Nevertheless, the quantitative, pedotransfer-type information about the effect of clay mineralogy and chemistry of soil solution on soil hydraulic properties remains relatively scarce. This chapter summarizes available data and ideas that may show future research directions in this field.

1. MINERALOGICAL COMPOSITION OF THE CLAY FRACTION

When the pressure head h decreases, the proportion of water retained within the porosity that is controlled by the arrangement of the clay-sized particles, increases. As a result, the percentage of water retention variations accounted for by the clay content increases when h decreases. This proportion varies according not only to the clay content, but also according to soil clay characteristics such as the predominant clay mineral, the size of elementary particles and the nature of cations at the surfaces of the particles (Tessier and Pédro, 1987; Quirk, 1994).

Tessier (1984) and Tessier et al. (1992) studied water retention properties of pure clays and demonstrated that the clay fabric at a given water potential was closely related to the characteristics of the elementary clay particles. They also showed that the differences between the water retention of kaolinite, illite and montmorillonite decreased with h. Ali and Biswas (1968) demonstrated that the difference between water retention of montmorillonite and illite is not large (about 9%) at -1500 kPa but amounts to 240% at -10 kPa suction. Bruand and Zimmer (1992) studied water retention properties of clayey soils and discussed the role of both the clay mineralogy and stress history.

Clay mineralogy was recognized by Williams et al. (1983) as a source of grouping criteria before developing PTFs for Australian soils. The mineralogy of the clay-size fraction was accurately described by X ray diffraction thus enabling quantification of illite, montmorillonite, kaolinite, interstratified material, quartz, iron oxides, vermiculite and halloysite content. Results showed that the presence of montmorillonite, often in quite small amounts, could serve as a discriminating property. On the contrary, Puckett et al. (1985) studied the water retention properties and K_s of soils in the Lower Coastal Plain of Alabama and showed no significant correlation between these hydraulic properties and mineralogical properties of the clay fraction. As emphasized by Puckett et al. (1985), this lack of correlation might be related to the small variation of mineralogical composition of the clay-size fraction of these soils that consisted of hydroxy interlayered vermiculite, kaolinite, and gibbsite. Danalatos et al. (1994) discussed the question of using mineralogy as grouping criteria and did not find any significant difference with no grouping of the 105 horizons from Greek soils.

Gaiser et al. (2000) investigated the effect of contrasting clay mineral composition on water retention and PTFs characteristics for soils from semiarid tropical regions. They analyzed water retention at -33 and -1500 kPa of 663 horizons from NE Brazil and SE Niger. They showed that PTFs for soils containing predominantly low-activity clay (CEC $<$ 24 cmol/kg clay, LAC soils) differed considerably from those established for non LAC soils. Hodnett and Tomasella (2002) used a set of 771 horizons from the IGBP-DIS soils database from ISRIC in Wageningen and compared the van Genuchten soil retention parameters for temperate and tropical soils. They showed differences between montmorillonitic and kaolinitic soils and suggested to include predictors related to mineralogy in PTFs.

Bruand and Tessier (2000) studied the water retention properties of clayey subsoils' horizons according to the variation of clay characteristics. The horizons developed on a large range of age and facies of calcareous or calcium-saturated clayey sediments. Results showed that the water-retention properties of the clay varies greatly from one soil to another with respect to the clay fabric that depends on the CEC, the size of elementary particles and hydric stress history of the clay.

Content of free iron oxides was a dominating property that affected soil water retention in the work of El Ashkar et al. (1956). Positive correlations were found between the content of free iron oxide and the water retained at -33 and -150 kPa and the water retained at -150 kPa increased more rapidly with increase in free iron oxides than it did at -33 kPa.

2. CATION EXCHANGE CAPACITY

In soils with a small organic carbon content, the cation exchange capacity can be used as an estimator of the clay mineralogy. The cation exchange capacity usually increases as the swelling of clays content does.

Bruand and Zimmer (1992) showed that the pore volume resulting from clay-particles packing in clay soils (clay content $>30\%$) was closely related to the CEC of the clay phase. The pore volume developed by the clay phase (v_v^{clay} in cm^3 per g of clay) was shown to increase with the cation exchange capacity of the clay phase (cec^{clay} in cmol$_+$ per g of clay) as indicated by the following relationship:

$$v_v = 0.275 + 0.067 cec^{clay}$$

As a consequence, when cec^{clay} does not vary for a group of soils, Bruand (1990) showed that PTFs can be established with clay content as single predictor.

Arrouays and Jamagne (1993) investigated the water-retention properties of loamy soils and recorded a positive correlation between the water content at field capacity, or permanent wilting point, and CEC. Tessier et al. (1999) showed that the CEC can be considered as a parameter characterizing the amount and nature of the <2 μm fraction in a soil. Relationships between the water retained at a given water potential and CEC were established (Figure 1). The closeness of these relationships increased when the water potential decreased. Pachepsky and Rawls (1999) found the CEC of clay fraction to be essential predictor of water retention at -33 and -1500 kPa. In addition, Karathanasis and Hajek (1982) and Lenhard (1984) showed that the mineralogical compositions become a dominant factor when the water content at very small h (i.e., $<-10^4$ kPa) is studied.

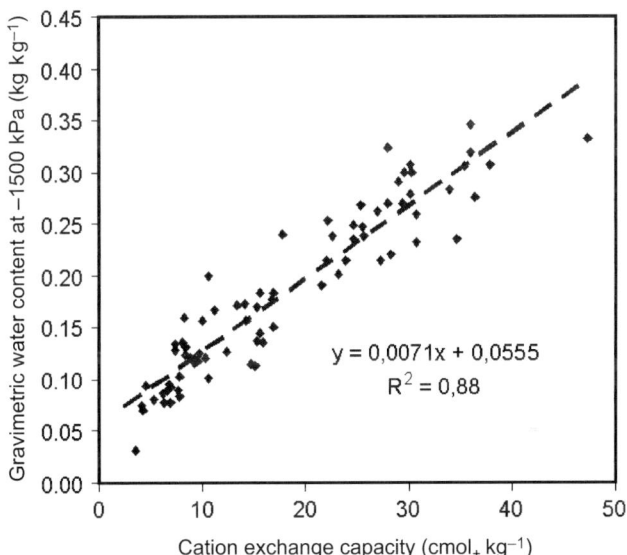

Figure 1. Gravimetric water content at -1500 kPa as a function of the cation exchange capacity.

3. SOIL CHEMICAL PROPERTIES

Consequences of chemical and mineralogical characteristics for soil hydraulic properties are often linked, the effect of chemical characteristics on hydraulic properties being closely related to the mineralogical composition of the clay fraction.

Rajkai and Várallyay (1992) showed that for salt-affected soils, water retention was primarily affected by soil chemical properties, while soil physical variables were found to play only a secondary role. PTFs for water retention were developed with the exchangeable sodium Na_{exc} and bulk density as predictors for $-1 > h > -50$ kPa, with Na_{exc} and the clay content for $h = -250$ kPa and the total salt content and the clay content for $h = -1500$ kPa. Thus, they found chemical properties and bulk density, but not the textural components to be the necessary inputs in their PTF to estimate water retention in salt-affected soils.

To estimate the effect of salinity and solution composition on soil hydraulic properties, values of the SAR and total electrolyte concentration C or solute ionic strength I were used. Lenhard (1984) was able to estimate 95% of variation in changes in water retention of clay samples could be estimated from SAR and C values. Changes in soil-water retention were attributed to soil mineralogical composition in the work of Jayawardane and Beattie (1978). Lima et al. (1990) demonstrated that values of C and SAR affect parameters of van Genuchten's equation (Van Genuchten, 1980) in a regular manner and can be in principle used to modify these parameters. Baumer et al. (1994) suggested to include the product of SAR and clay content in PTFs to estimate residual water contents and water content at −1500 kPa, and to include SAR in the PTF to estimate the bubbling pressure in the Brooks–Corey equation.

Nashshineh-Pour et al. (1970) studied the effect of electrolyte composition on K_{sat} of several soils in Texas. They concluded that the most significant single soil characteristic is soil mineralogy. Changes in K_{sat} as affected by sodium adsorption and electrolyte concentration were related to clay mineralogy, clay content and bulk density in the work of Frenkel et al. (1978). Results showed plugging of the pores by dispersed clay particles was the major cause of reduced K_{sat}. The sensitivity to excessive exchangeable sodium and small electrolyte concentration increased with clay content and bulk density. The kaolinitic soil was less sensitive than montmorillonitic and vermiculitic soils, the difference between the latter remaining small.

Saturated hydraulic conductivity of soil decreased with decreasing C and increasing SAR values in studies of McNeal and Coleman (1966), Lagerwerff et al. (1969), Pupisky and Shainberg (1984) and Russo and Bresler (1977). Pachepsky (1989) assembled a data base of published experimental data and found that the combined effect of the exchangeable sodium content and ionic strengths of solution on saturated hydraulic conductivity could be expressed using a single variable, the excessive thickness of hydrated layer, Δ, that he defined in terms of the proportion of exchangeable sodium in the sum of exchangeable cations Y_{Na} and ionic strengths I, mol L^{-1} as:

$$\Delta = Y_{Na}(12/I^{0.37} - 5)$$

The saturated hydraulic conductivity K_{sat} changed as a function of Δ:

$$K_{sat} = \frac{K_{sat,0}}{1 + \left(\dfrac{\Delta}{\Delta_{0.5}}\right)^q}$$

where parameters Δ, $\Delta_{0.5}$ and q were soil-specific, $K_{sat,0}$ was the hydraulic conductivity in absence of exchangeable sodium. Dane and Klute (1977) and Lima et al. (1990) observed the decrease in unsaturated hydraulic conductivity as the total concentration decreased.

4. CONCLUDING REMARKS

Soil-water management remains the most important issue in the regions of the World where soil salinity and alkalinity has developed or is developing. Much needs to be done to provide water management models with reliable estimates of inter-relations between water and salt movement in soils of these regions.

REFERENCES

Ali, M.H., Biswas, D., 1968. Soil water retention and release as related to mineralogy of soil clays. Proc. 55th Indian Sci. Congr. 3, 633.

Arrouays, D., Jamagne, M., 1993. Sur la possibilité d'estimer les propriétés de rétention en eau de sols limoneux lessivés hydromorphes du Sud-Ouest de la France à partir de leurs caractéristiques de constitution. Comptes Rendus de l'Académie d'Agriculture de France 79, 111-121.

Baumer, O., Kenyon, P., Bettis, J., 1994. MUUF v2.14 User's Manual. USDA Natural Resource Conservation Service, Lincoln, Nebraska.

Bruand, A., 1990. Improved prediction of water-retention properties of clayey soils by pedological stratification. J. Soil Sci. 41, 491-497.

Bruand, A., Tessier, D., 2000. Water retention properties of the clay in soils developed on clayey sediments: significance of parent material and soil history. Eur. J. Soil Sci. 51, 679-688.

Bruand, A., Zimmer, D., 1992. Relation entre la capacité d'échange cationique et la volume poral dans les sols argileux: incidences sur la morphologie des assemblages élémentaires. C.R. Acad. Sci. 315, 223-229.

Danalatos, N.G., Kosmas, C.S., Driessen, P.M., Yassoglou, N., 1994. Estimation of the draining soil moisture characteristics from standard data as recorded in soil surveys. Geoderma 64, 155-165.

Dane, J.H., Klute, A., 1977. Salt effects on the hydraulic properties of a swelling soil. Soil Sci. Soc. Am. J. 41, 1043-1049.

El Ashkar, M.A., Bodman, G.B., Peters, D.B., 1956. Sodium hyposulfite-soluble iron oxide and water retention by soils. Soil Sci. Soc. Am. Proc. 20, 352-365.

Frenkel, H., Goertzen, J.O., Rhoades, J.D., 1978. Effects of clay type and content, exchangeable sodium percentage, and electrolyte concentration on clay dispersion and soil hydraulic conductivity. Soil Sci. Soc. Am. J. 42, 32-39.

Gaiser, Th., Graef, F., Carvalho Cordiero, J., 2000. Water retention characteristics of soils with contrasting clay mineral composition in semiarid tropical regions. Aust. J. Soil Res. 38, 523-536.

Hodnett, M.G., Tomasella, J., 2002. Marked differences between van Genuchten soil water-retention parameters for temperate and tropical soils: a new water retention pedotransfer functions developed for tropical soils. Geoderma 108, 155-180.

Jayawardane, N.S., Beattie, J.A., 1978. Effect of salt solution composition on moisture release curves of soils. Aust. J. Soil Res. 17, 89-99.

Karathanasis, A.D., Hajek, B.F., 1982. Quantitative evaluation of water absorption on soil clays. Soil Sci. Soc. Am. J. 46, 1321-1325.

Lagerwerff, J.V., Nakayama, F.S., Frere, M.H., 1969. Hydraulic conductivity related of porosity and swelling of soil. Soil Sci. Soc. Am. Proc. 33, 3-11.

Lenhard, R.J., 1984. Effects of clay–water interactions on water retention in porous media. Ph.D. Thesis, Oregon State University, Corvallis, OR, 145 pp.

Lima, L.A., Grimser, M.E., Nielsen, D.R., 1990. Salinity effects on Yolo loam Hydraulic properties. Soil Sci. 150, 451-458.

Nashshineh-Pour, B., Kunze, G.W., Carson, C.D., 1970. The effect of electrolyte composition on hydraulic conductivity of certain Texas soils. Soil Sci. 110, 124-127.

McNeal, B.L., Coleman, N.T., 1966. Effect of solution composition on soil hydraulic conductivity. Soil Sci. Soc. Am. Proc. 30, 308-312.

Pachepsky, Y.A., 1989. Effect of the composition of soil solutions and exchange cations on water retention and hydraulic conductivity. Soviet Soil Sci. 21, 90-103.

Pachepsky, Y.A., Rawls, W.J., 1999. Accuracy and reliability of pedotransfer function as affected by grouping soils. Soil Sci. Soc. Am. J. 63, 1748-1759.

Puckett, W.E., Dane, J.H., Hajek, B.F., 1985. Physical and mineralogical data to determine soil hydraulic properties. Soil Sci. Soc. Am. J. 49, 831-836.

Pupisky, H., Shainberg, I., 1984. Effect of salt content on hydraulic conductivity of a sandy soil. Soil Sci. Soc. Am. J 48, 429-433.

Quirk, J.P., 1994. Interparticle forces: a basis for the interpretation of soil physical behavior. Adv. Agron. 53, 121-183.

Rajkai, K., Várallyay, G., 1992. Estimating soil water retention from simpler properties by regression techniques. *In*: van Genuchten, M.Th., Leij, F.J., Lunds, L.J. (Eds.), Methods for Estimating the Hydraulic Properties of Unsaturated Soils, Proceedings of the International Workshop on Indirect Methods for Estimating the Hydraulic Properties of Unsaturated Soils, Riverside, California, 11–13 October 1989, pp. 417-426.

Rawls, W.J., Gish, T.J., Brakensiek, D.L., 1991. Estimating soil water retention from soil physical properties and characteristics. Adv. Soil Sci. 16, 213-234.

Russo, D., Bresler, E., 1977. Analysis of the saturated–unsaturated hydraulic conductivity in a mixed Na-Ca soil system. Soil Sci. Soc. Am. J. 41, 706-710.

Tessier, D. 1984. Etude expérimentale de l'organisation des matériaux argileux, Thèse Doctorat d'Etat, Université de Paris 6, France, 247 pp.

Tessier, D., Bigorre, F., Bruand, A., 1999. La capacité d'échange: outil de prévision des propriétés physiques des sols. C.R. Acad. Agric. Fr. 85, 37-46.

Tessier, D., Lajudie, A., Petit, J.C., 1992. Relation between the macroscopic behavior of clays and their microstructural properties. Appl. Geochem. (Suppl. 1), 151-161.

Tessier, D., Pédro, G., 1987. Mineralogical characterization of 2:1 clays in soils: importance of the clay texture. *In*: Schultz, L.G., van Olphen, H., Mumpton, F.A. (Eds.), Proceedings of the International Clay Conference, Denver, 1985. The Clay Minerals Society, Bloomington, IN, pp. 78-84.

Van Genuchten, M.Th., 1980. A closed-form equation for predicting the hydraulic conductivity of unsaturated soils. Soil Sci. Soc. Am. J. 44, 892-898.

Williams, J.R., Prebble, E., Williams, W.T., Hignett, C.T., 1983. The influence of texture, structure and clay mineralogy on the soil moisture characteristic. Aust. J. Soil Res. 21, 15-31.

Chapter 10

PRELIMINARY GROUPING OF SOILS

A. Bruand

Institut des Sciences de la Terre d'Orléans (ISTO), Bâtiment Géosciences, Université d'Orléans, BP 6759 45067, Orléans Cedex 2, France
Tel.: +33-2-38-41-70-24; fax.: +33-2-38-41-73-08

1. ORIGIN OF THE VARIABILITY AND GROUPING STRATEGY

Grouping soils appeared very early as a way to increase the reliability and applicability of PTFs. Indeed, as noticed by Hodnett and Tomasella (2002), it might be never possible to develop a very reliable "universal" PTF for all soils because the worldwide range of soil properties is so great. Grouping emerged as a strategy to stratify the resulting variability, thus enabling the development of closer PTFs between hydraulic properties and easily available basic soil properties.

Most studies used an *a priori* grouping, without any analysis of the variability of hydraulic properties prior grouping. Few works discussed the variability observed prior to grouping and PTFs development. Williams et al. (1983) examined the variability of water retention curves for 78 horizons from Australian soils and then developed eight groups of water retention curves (Figure 1). They identified the soil characteristics that provided the grouping. In particular, they showed that grade of structure and particle size distribution were the soil properties most consistently associated with the groups of soil having similar water retention curves. On the other hand, the size and shape of the peds had only weak association with the differences in the water retention. Williams et al. (1983) developed PTFs separately for each of the groups. Data of water retention for 1448 soils in the United States were analyzed by Cosby et al. (1984) to construct a grouping based on the soil hydraulic behavior analogous to the texture classification. Cosby et al. (1984) also discussed the variability of water retention and showed that besides soil texture, the size and the shape of the structure accounted for the highest percentage of variance. The moist consistency alone accounted for similar variance to structure while the land use, the drainage class, the slope, and the root abundance accounted for very little proportion of variance. The position in the soil profile (horizons A, B and C) accounted for intermediate proportion of variance. Wösten et al. (1986) analyzed the variability of the hydraulic properties according to soil functioning. They measured the water retention properties, the saturated and the unsaturated hydraulic conductivity in 25 C horizons with sand texture and 23 C horizons with a clay loam and silty clay loam texture, as distinguished in the Dutch soil survey. Two groups of horizons were distinguished based on three functional properties: (i) the travel time from soil surface to water table, (ii) the water table allowing a defined upward-flux density, and (iii) the downward-flux density at a defined air content; they found two groups of horizons.

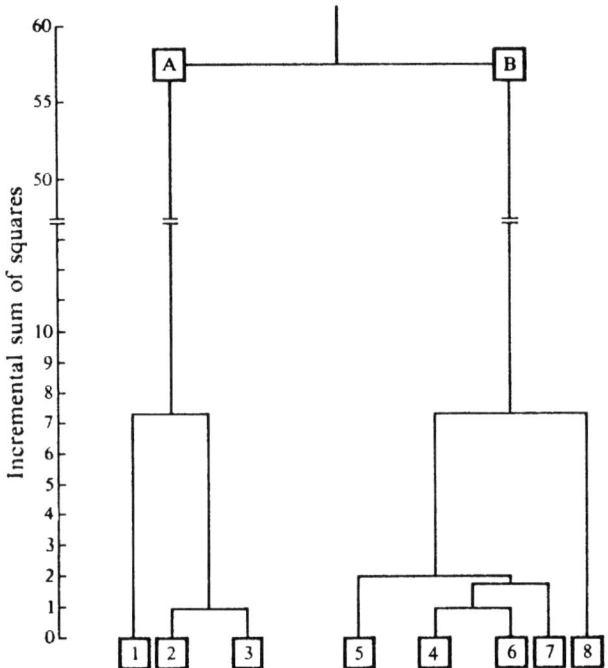

Figure 1. Dendrograms showing the relationship between the eight groups of water retention curves using the incremental sum of squares. Sets A and B are groups of soils that differed in their pedality and secondarily in their texture (after Williams et al., 1983).

The small number of studies examining the origin of the variability in hydraulic properties and its consequences for PTF development can be easily explained. Water retention and conductivity are closely related to the geometry of the pore network, the latter depending on the nature and assemblage of the elementary soil particles. Based on that premise, most researchers used preliminary grouping by particle-size distribution characteristics (texture as a global expression of the particle size distribution, clay, silt and sand content) and then by structure characteristics (structure type and grade, bulk density, consistency) without any discussion of the pertinence of the criteria used.

2. GROUPING CRITERIA

2.1. Genetic grouping

Developing PTFs for soils in a climatic region is the grouping that is implicitly based on genetic criteria. It can be considered at world scale. Thus, analyzing data of soils from West Africa and Brazil, Gaiser et al. (2000) established PTFs for water retention at -33 and -1500 kPa of soils in semiarid tropical regions and showed the significance of clay mineralogy. Tomasella and Hodnett (1998) studied the water retention of 613 soils from the Brazilian Amazonia and developed PTFs that predict the Brooks and Corey parameters. The authors suggested that these PTFs would be more adapted to soils under the tropics than most PTFs already published and developed for soils from temperate regions. Using water retention data from the IGBP-DIS database, Hodnett and Tomasella (2002) selected 771 horizons from 249 soil profiles in 22 countries under the tropics. They showed that averaged parameters of the Van Genuchten model (1980) were significantly different for most textural classes when compared to those recorded for soils from the temperate regions. Hodnett and Tomasella (2002) used these averaged values and multiple linear regression to establish class and continuous PTFs, respectively. Their results imply that PTFs might be developed for other climatic regions.

The PTFs developed for soils in a country located within a single climatic region can be considered as PTFs that are established for soils developed under similar climatic conditions and showing pedological similarities. Several studies were carried out for particular groups of soils within a single country. Thus, Pidgeon (1972) and Jamagne et al. (1977) developed PTFs enabling prediction of water at field capacity and permanent wilting point for ferrallitic soils in Uganda and Luvisols and Cambisols in France, respectively. On the other hand, Bruand (1990) showed less variability of the water retention properties of French clayey soils when grouping the soils by soil family, i.e., having the same pedogenetic origin and developed from a specific parent material.

Van den Berg et al. (1997) reviewed literature on PTFs for Ferralsols and discussed the necessity to have PTFs for soil groupings at world level. To exemplify such an approach, they investigated water retention at -10 and -1500 kPa of Ferralsols and related soils from South America, Africa and South East Asia, and developed PTFs using the multiple linear regression. On the other hand, Tomasella and Hodnett (1997) showed that for K_{unsat} of Brazilian soils, the parameters of the generalized Kozeny–Carman equation and Brooks–Corey equation derived in temperate soils could be applied to most soil studied. Tomasella and Hodnett (1997) suggested the possibility of generalizing the hydraulic conductivity PTFs for a greater variety of soils, and even across great soil groups. That appears to be less probable for the water retention PTFs. Indeed, as indicated above, Hodnett and Tomasella (2002) showed that averaging Van Genuchten parameters across textural class led to significant differences between tropical and temperate soils for most textures. They also showed that continuous PTFs developed for tropical soils without any grouping performed better than class PTFs based on soil types. Hodnett and Tomasella (2002) suggested that such difference in performance was observed because the continuous PTFs took into account a minimum of six soil variables while the soil type PTFs used only the averaged parameters for the Van Genuchten model (1980).

2.2. Horizon-based grouping

Because elementary constituents and structure vary with depth, grouping by horizon type has been used in several studies. Petersen et al. (1968) studied Pennsylvania soils and showed that water contents retained at -33 and -1500 kPa were generally the greatest in the A, less in the B and the smallest in the C horizon. No significant difference between the water retention of cultivated and uncultivated horizons was found. Differences in structure grade, gleying and clay accumulation intensity also did not seem to case differences in water retention. The authors concluded that their results were more a reflection of coarse fragment content than of the other horizon characteristics studied. Reeve et al. (1973) examined the available water capacity of 158 horizons from soils of England and Wales grouped in 5 textural classes. They observed a decrease in available water capacity in B and C horizons whereas in A horizons, the available water capacity tended to increase with bulk density, silty soils being an exception. Hall et al. (1977) grouped topsoils (A horizons) and subsoils (E, B and C horizons) and developed PTFs for water retention at 5 values of pressure head for $-5 \leq h \leq -1500$ kPa. These PTFs were regression equations with clay, silt, sand and organic carbon content, and bulk density as input variables (Table 1). The regression intercept was greater in topsoils and the regression coefficients for the clay and silt content were smaller as compared with subsoil. Other coefficients did not demonstrate systematic differences between the two groups. Grouping by separating topsoils and subsoils was justified by Hall et al. (1977) by invoking differences of structure that give different parameters in the regression equations. Working at the scale of the 1:50,000 mapping unit, Wösten et al. (1995) measured hydraulic properties of the soils classified as sandy, siliceous, mesic Typic Haplaquods. They grouped topsoils (A horizons) and subsoils (B and C horizons) and developed PTFs for $\theta(h)$ and $K(h)$.

Table 1
Regression equations developed for topsoils and subsoils, and corresponding to PTFs for the water content at -50 hPa (θ_{50}), -100 hPa (θ_{100}), -400 hPa (θ_{400}), -2000 hPa (θ_{2000}), and $-15,000$ hPa ($\theta_{15,000}$) (modified after Hall et al., 1977)

Regression equations

Topsoils
$\theta_{50} = 47.00 + 0.25\,(\%cl) + 0.10\,(\%si) + 1.12\,(OC) - 16.52 D_b$
$\theta_{100} = 37.47 + 0.32\,(\%cl) + 0.12\,(\%si) + 1.15\,(OC) - 1.25 D_b$
$\theta_{400} = 26.66 + 0.36\,(\%cl) + 0.12\,(\%si) + 1.00\,(OC) - 7.64 D_b$
$\theta_{2000} = 8.70 + 0.45\,(\%cl) + 0.11\,(\%si) + 1.03\,(OC)$
$\theta_{15,000} = 2.94 + 0.83\,(\%cl) - 0.0054\,(\%cl)^2$

Subsoils
$\theta_{50} = 37.20 + 0.35\,(\%cl) + 0.12\,(\%si) - 11.73 D_b$
$\theta_{100} = 27.87 + 0.41\,(\%cl) + 0.15\,(\%si) - 8.32 D_b$
$\theta_{400} = 20.81 + 0.45\,(\%cl) + 0.13\,(\%si) - 5.96 D_b$
$\theta_{2000} = 7.57 + 0.48\,(\%cl) + 0.11\,(\%si)$
$\theta_{15,000} = 1.48 + 0.84\,(\%cl) - 0.0054\,(\%cl)^2$

%cl is the clay content as percentage, %si the silt content as percentage, OC the organic carbon content as percentage and D_b the bulk density.

Pachepsky et al. (1996) used data on the water contents at 8 pressure heads for 100 Aquic Ustoll soil samples. They showed that grouping of data according to three horizon classes (horizons A, B and C) increases the precision of water retention estimates. Using the 5521 hydraulic properties from 1777 soils of the European database HYPRES (http://www.macaulay.ac.uk/hypres/), Wösten et al. (1999) used separating topsoil from subsoil as primary grouping. Then the groups were further subdivided by texture to develop PTFs for the Mualem–Van Genuchten parameters (Van Genuchten, 1980) of the $\theta(h)$ and $K(h)$ relationships (Table 2). The optimized Mualem–Van Genuchten parameters

Table 2
Class PTFs developed for topsoils and subsoils according to the texture classes of the FAO guidelines (FAO, 1990) using the European database HYPRES (Mualem–Van Genuchten parameters for the fits on the geometric mean values of θ and K at 14 pressure heads, after Wösten et al., 1999)

	θ_r	θ_s	α	n	m	l	K_s
Topsoils							
Coarse	0.025	0.403	0.0383	1.3774	0.2740	1.2500	60.000
Medium	0.010	0.439	0.0314	1.1804	0.1528	−2.3421	12.061
Medium fine	0.010	0.430	0.0083	1.2539	0.2025	−0.5884	2.272
Fine	0.010	0.520	0.0367	1.1012	0.0919	−1.9772	24.800
Very fine	0.010	0.614	0.0265	1.1033	0.0936	2.5000	15.000
Subsoils							
Coarse	0.025	0.366	0.0430	1.5206	0.3424	1.2500	70.000
Medium	0.010	0.392	0.0249	1.1689	0.1445	−0.7437	10.755
Medium fine	0.010	0.412	0.0082	1.2179	0.1789	0.5000	4.000
Fine	0.010	0.481	0.0198	1.0861	0.0793	−3.7124	8.500
Very fine	0.010	0.538	0.0168	1.0730	0.0680	0.0001	8.235
Organic[a]	0.010	0.766	0.0130	1.2039	0.1694	0.4000	8.000

[a]No distinction is made between topsoils and subsoils for organic soils (Histic layers, FAO, 1990).

were determined to fit the geometric mean values of θ and K at 14 pressure heads within each of the 11 classes. No distinction of horizon type and texture was made for organic soils that correspond to the Histic layers as defined in the FAO guidelines (FAO, 1990). There was no difference of θ_r between topsoils and subsoils for any of the textural classes. Values of θ_s were greater in topsoils, α, n and m were smaller in topsoils for the coarse texture and greater for the other textures as compared with subsoils. Differences in values of l and K_{sat} could not associated with the horizon.

2.3. Texture grouping

Texture grouping is the most common grouping found in literature. The early PTFs were developed by grouping soils by texture and enabled prediction of permeability (Diebold, 1954) or available water capacity solely (Reeve et al., 1973). Jamagne et al. (1977) used measurements of water retention for soils from Northern France and proposed values of volumetric water content at field capacity and −1500 kPa for the 15 textural classes of the Soil Survey of France. The study by Petersen et al. (1968) on water retention at −33 and −1500 kPa for Pennnsylvania soils is also among the earliest works where PTFs have been generated for several pressure heads after grouping by texture. Hall et al. (1977) used topsoil and subsoil as primary grouping criteria and then texture to develop PTFs that predict single value of the volumetric water content at −5 and −1500 kPa for the 10 textural classes of the Soil Survey of England and Wales. Rawls et al. (1982) used data from 1323 soils with about 5350 horizons, from 32 states of USA, to develop PTFs for the water retention curve and the saturated hydraulic conductivity (K_{sat}) after grouping soil samples according to the 11 USDA texture classes. Those PTFs were the averaged values for the parameters of the Brooks and Corey equation (1964) and K_s. Saxton et al. (1986) divided the texture triangle into grids of 10% sand and 10% clay increments. They used the resulting 55 grid midpoints to generate PTFs that corresponded to averaged water contents for 10 pressure heads from -10 to -1500 kPa for 44 of the 55 sections of the texture triangle.

Researchers also used grouping by broad textural classes to develop PTFs. Working on Portuguese soils, Gonçalves et al. (1997) showed that grouping using the three main textural classes of the FAO triangle significantly increased the prediction accuracy for the water retention and unsaturated hydraulic conductivity. Williams et al. (1999) used the Gregson et al. (1987) one-parameter function and proposed average p and q values for four texture groups. They obtained estimates of the water retention that were better than those from the regression models using texture and bulk density. Bruand et al. (2002, 2003) developed class-PTFs for the water retention properties of French soils after grouping by texture and proposed fitted parameters for the Van Genuchten model (1980) (Figure 2). Texture grouping was also used for smaller areas. Salchow et al. (1996) developed PTFs for water content at -33 and -1500 kPa using 108 horizons of alluvial soils in southern Ohio. Horizons were first grouped in four USDA textural classes. Then PTFs that enabled

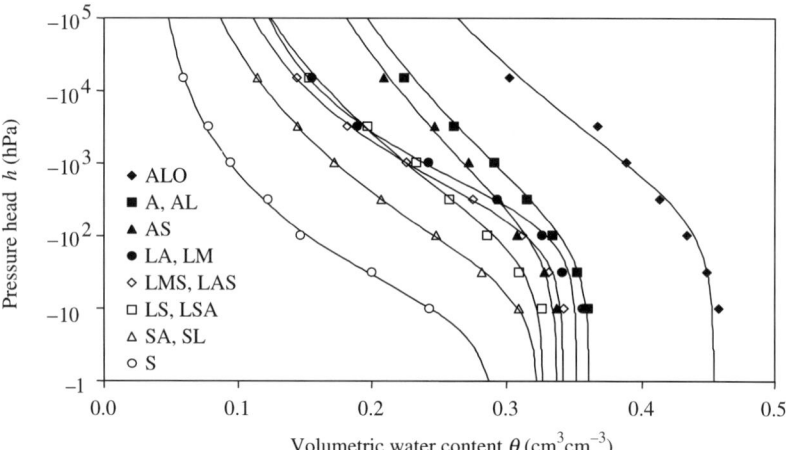

Figure 2. Water retention curves obtained with class PTFs developed for French soils after grouping by texture; ALO – heavy clay; A, AL – clay to loamy-clay; AS – sandy clay; LA, LM – clayey loam to loam; LMS, LAS – sandy loam to sandy clay loam; LS, LSA – sandy loam to clay sand loam; SA, SL – clayey sand to loamy sand; S – sand; modified after Bruand et al., 2002.

prediction of the field capacity, permanent wilting point, available water capacity and K_s were developed using sand, silt, clay, organic matter content and bulk density as predictor.

Databases of hydraulic properties that were developed with data from one or several countries were used to group the soil using texture prior to PTFs development. Leij et al. (1996) used 780 horizons of the International Unsaturated Soil Database (UNSODA) (http://www.ussl.ars.usda.gov) (Leij et al., 1996; Nemes et al., 2001) and proposed average parameters θ_s, θ_r, α, n and K_s for the 11 classes of the USDA soil textural triangle.

Those authors also showed that uncertainty of errors in hydraulic properties was exacerbated because data were collected, compiled and applied by different individuals. Large databases are particularly well adapted to the application of grouping techniques prior to PTFs development, but Leij et al. (1999) have pointed out the difficulty of gathering a large number of consistently measured hydraulic properties and avoiding large volume of incorrect data. Using the 5521 hydraulic properties from 1777 soils of the European database HYPRES (http://www.macaulay.ac.uk/hypres/), Wösten et al. (1999) developed class PTFs for the Mualem–Van Genuchten parameters of the $\theta(h)$ and $K(h)$ relationships after grouping by texture (Table 2) (see Section 2.2). The database of hydraulic properties of Hungarian soils (HUNSODA) was used by Nemes (2002) to propose class PTFs for the Van Genuchten parameters of the $\theta(h)$ relationship (Table 3). After preliminary grouping by separating topsoils and subsoils, values of θ_s, θ_r, α and n were proposed for the 5 textural classes of the FAO triangle and the 11 textural classes of the USDA triangle.

Table 3
Van Genuchten parameters for Hungarian soils after grouping according to the FAO and the USDA texture classes (after Nemes et al., 2001)

	θ_r	θ_s	n	α
FAO texture classes				
Coarse	0.00966	0.414814	0.027478	1.534133
Medium	0.00000	0.438973	0.009746	1.228564
Medium fine	0.00000	0.447729	0.002281	1.251066
Fine	0.00000	0.450373	0.000823	1.254555
Very fine	0.00000	0.525737	0.000883	1.226032
USDA texture classes				
Sand	0.01300	0.408743	0.023771	1.875734
Loamy sand	0.00000	0.413930	0.022367	1.412027
Sandy loam	0.00000	0.424590	0.016445	1.251622
Sandy clay loam	0.00000	0.430524	0.029298	1.192810
Clay	0.00000	0.498629	0.000670	1.252291
Clay loam	0.00000	0.430199	0.002402	1.246581
Loam	0.00000	0.423860	0.006519	1.245827
Silty loam	0.00000	0.458333	0.009931	1.230832
Silt	0.00000	0.463677	0.003128	1.282823
Silty clay loam	0.00000	0.435508	0.001765	1.239395
Silty clay	0.00000	0.453244	0.000854	1.246492

2.4. Grouping based on structure and bulk density

Williams et al. (1992) developed PTFs for the parameters of the Campbell (1974) water retention model for a wide range of soils from Australia and United States. They separated massive and structured soils before grouping by texture. Danalatos et al. (1994) showed that separating well-structured horizons from structureless to weakly structured horizons on the other led to close relationship between the clay content and the γ parameter of the Driessen and Konijn equation (1992) for the water retention curve.

Incorporating other soil properties in the regression analyses produced only a slightly greater R^2-value. Lin et al. (1997) measured the *in situ* steady-state infiltration for 96 horizons of Texas soils and showed that PTFs could be developed incorporating morphological features. They did not group the horizons using these morphological features, but their results showed clearly that a quantification of morphological features and their combination might result in promising grouping criteria. Lilly (2000) developed PTFs for field K_{sat} by grouping soils using soil structure. A total of 627 field K_{sat} measured for various soils of Scotland were distributed among 49 structure groups, each corresponding to a unique combination of primary and secondary structures according to the terminology and classes of the FAO Guidelines (FAO, 1990). The PTFs proposed by Lilly (2000) are geometric means of K_{sat} that vary from 0.06 to 1036.8 cm day^{-1}, with quartile ranges attached for each of the 49 classes of structure.

Bulk density was very early recognized as significant for water retention (Petersen et al., 1968) and hydraulic conductivity (Diebold, 1954). Considering the effect of bulk density on the water retention of French soils, Bruand et al. (1996) developed PTFs for clayey soils (clay content > 30%) using bulk density as single predictor variable. The clod bulk density was superior to the horizon bulk density because the latter included macropores that do not intervene in water retention and vary in tilled topsoils with time and management (Bruand et al., 2003). Thus, the clod bulk density was used as grouping criteria within every texture class (Figure 3). This enabled to propose class PTFs using information about texture and structure as grouping criteria.

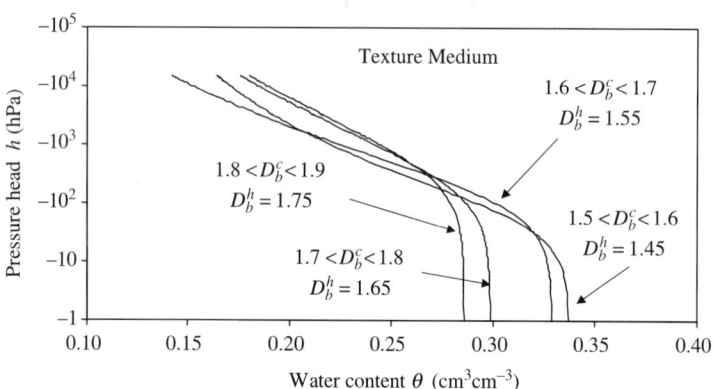

Figure 3. Water retention curves computed for the texture medium (FAO triangle) using the horizon bulk density (D_b^h) and class PTFs that enable prediction of the volumetric water content at seven pressure heads after preliminary grouping by texture and then by clod bulk density (D_b^c) (modified after Bruand et al., 2003).

McKenzie and Jacquier (1997) measured K_{sat} on 99 horizons from 36 soils in South-Eastern Australia. They showed that grouping soils by visual estimation of the areal porosity using pore charts enabled satisfactory prediction of K_{sat}. A more quantitative system of measurement provided only slightly better predictions. They also showed that regression trees gave more plausible predictive models than standard multiple regressions and suggested that it was because regression trees provided a realistic portrayal of the non-additive and conditional nature of the relationships between morphology and K_{sat}.

2.5. Parent material grouping

Parent material was rarely used as grouping criteria. Jamagne et al. (1977) separated soils developed in sedimentary clays from those developed in clays resulting from weathering within the heavy clay class of the French texture triangle. Puckett et al. (1985) established PTFs for Ultisols developed in unconsolidated sediments of the Lower Coastal Plain of Alabama in which the clay fraction consisted mainly in vermiculite, kaolinite and gibbsite. The authors suggested that these PTFs can be used for soils developed in this type of parent material with similar mineralogical composition. Bastet (1999) grouped 597 soils from France by type of parent material. Among parent materials, Bastet (1999) proposed PTFs for soils developed in recent quaternary alluviums, old quaternary alluviums, marly limestones, marls, aeolian loams, sandstones, clays resulting from decarbonatation, molasses and detritical sediments.

2.6. Consecutive grouping

As already mentioned above, numerous studies used combined grouping. Most used preliminary grouping by horizon type and then by texture or by texture and then by bulk density.

Thus, Williams et al. (1983) examined the relative importance of texture, structure, organic matter and clay mineralogy to group the water retention curves over a pressure head range from -4 to -1507 kPa for an extensive group of Australian soils. They studied 78 horizons from 17 profiles representing 12 Australian Great Soil Groups as defined in Prebble (1970). Structure development and texture had the greatest importance for grouping the water retention curves in 8 groups. For each group, Williams et al. (1983) developed PTFs that are parameters of a power law relationship between h and θ. Rawls et al. (1999) accumulated and analyzed the US national database of about 1000 data sets on K_s values. Results of grouping these data by texture and then into two porosities classes are shown in Figure 4.

Wösten et al. (1999) separated topsoils and subsoils and then proposed class PTFs for the parameters of the Mualem–Van Genuchten model for the five texture classes of the FAO triangle. Bruand et al. (2002) studied water retention of French soils and developed class PTFs by grouping soils by texture (8 classes based on the 13 classes of the French triangle) and then by bulk density within every texture class. Values of θ at seven pressure heads and Van Genuchten fitted parameters were proposed (Table 4). Using a similar approach, Bruand et al. (2003) developed class PTFs using preliminary grouping by texture according to five classes of the FAO triangle.

3. GROUPING DECREASES THE NUMBER OF PREDICTORS

Grouping leads to less variability within each resulting group of soils and, as a consequence, results in PTFs using a smaller number of soils characteristics as predictors. Danalatos et al. (1994) showed that their PTFs developed for representative Greek soils with 6 soil characteristics could be simplified into PTFs with clay content as single predictor if the applicability of the PTFs was restricted to group of soils with similar

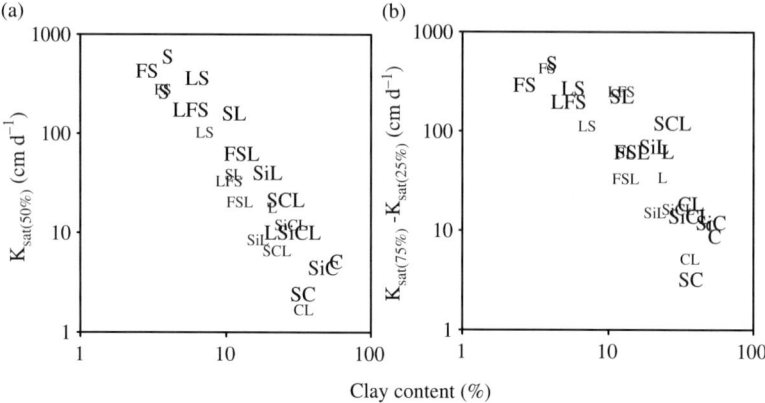

Figure 4. Saturated hydraulic conductivity K_{sat} in the national K_{sat} database as grouped by texture and porosity; a – median values, b – the difference between 75 and 25% quartiles. Textural classes: S – sand, FS – fine sand, LS – loamy sand, LFS – loamy fine sand, SL – sandy loam, FSL – fine sandy loam, L – loam, SiL – silty loam, SCL – sandy clay loam, CL – clay loam, SiCL – silty clay loam, SC – sandy clay, SiC – silty clay, C – clay; the high porosity data are shown using a larger font (after Pachepsky and Rawls, 2003).

mineralogy. Bruand (1990) determined the water retention at $h = -33$ and -1500 kPa of 40 French clayey B horizons. Among the latter, 18 horizons originated from various contrasting soil families, 13 were B_t horizons from one soil family and 9 were B_w horizons from another soil family. Bruand (1990) showed that for horizons originating from contrasting soil families, accurate PTFs were developed with the reciprocal of bulk density as single predictor because it enabled to take both clay content and clay fabric into account. For horizons originating from a single soil family, accurate PTFs were established with either the reciprocal of the bulk density or the clay content as single predictor because of the close relationship between the bulk density and the clay content in the absence of clay fabric variation. The applicability of those PTFs was shown by Bruand et al. (1994) for another group of B_t horizons originating from a single soil family with clay content and bulk density ranging from 50 to 73% and from 1.30 to 1.47, respectively. Arrouays and Jamagne (1993) also showed that θ at -1500 kPa could be accurately estimated using the clay content as single predictor for soils from the South-West of France.

4. COMPARISON OF GROUPINGS AND IMPROVEMENT OF PREDICTION AFTER GROUPING

King and Franzmeier (1981) determined K_{sat} *in situ* with the piezometer method for 25 soil series in Indiana. Grouping using both texture and soil structure was compared to grouping using texture, origin of the parent material and type of genetic horizon. The second grouping resulted in more homogeneous classes. Salchow et al. (1996) improved the closeness of PTFs after grouping into four textural classes (silty clay loam, silt loam, loam and sandy loam). Pachepsky et al. (1996) showed for Aquic Ustoll that grouping by

Table 4
Class PTFs based on combined grouping using the texture according to the FAO texture triangle and the clod bulk density (modified after Bruand et al., 2003)

Texture class	Class of D_b^c	D_b^h	$\theta_{1.0}$ (cm³cm⁻³)	$\theta_{1.5}$ (cm³cm⁻³)	$\theta_{2.0}$ (cm³cm⁻³)	$\theta_{2.5}$ (cm³cm⁻³)	$\theta_{3.0}$ (cm³cm⁻³)	$\theta_{3.5}$ (cm³cm⁻³)	$\theta_{4.2}$ (cm³cm⁻³)	θ_s (cm³cm⁻³)	θ_r (cm³cm⁻³)	n	α
Very Fine	[1.2–1.3]	1.25	0.531	0.514	0.490	0.465	0.428	0.418	0.329	0.527	0.0100	1.0849	0.0098
	[1.3–1.4]	1.15	0.484	0.473	0.451	0.428	0.393	0.384	0.303	0.481	0.0001	1.0868	0.0083
		1.35	0.493	0.486	0.467	0.447	0.416	0.401	0.321	0.488	0.0002	1.0930	0.0042
	[1.4–1.5]	1.25	0.456	0.450	0.433	0.414	0.385	0.371	0.298	0.452	0.0006	1.0923	0.0043
		1.45	0.489	0.477	0.464	0.445	0.422	0.386	0.318	0.481	0.0001	1.1055	0.0028
		1.35	0.455	0.444	0.432	0.415	0.393	0.359	0.296	0.448	0.0001	1.1066	0.0027
Fine	[1.3–1.4]	1.35	0.459	0.429	0.419	0.390	0.369	0.332	0.270	0.449	0.0007	1.0975	0.0088
		1.25	0.425	0.398	0.388	0.361	0.341	0.325	0.250	0.415	0.0010	1.0927	0.0086
	[1.4–1.5]	1.45	0.441	0.422	0.400	0.381	0.348	0.323	0.274	0.441	0.0002	1.0802	0.0194
		1.35	0.410	0.393	0.373	0.355	0.324	0.301	0.255	0.410	0.0007	1.0811	0.0180
	[1.5–1.6]	1.55	0.383	0.378	0.366	0.350	0.326	0.295	0.259	0.383	0.0006	1.0854	0.0062
		1.45	0.358	0.353	0.342	0.328	0.305	0.276	0.242	0.358	0.0001	1.0864	0.0059
	[1.6–1.7]	1.65	0.381	0.363	0.353	0.333	0.312	0.302	0.264	0.384	0.0003	1.0558	0.0377
		1.55	0.358	0.341	0.332	0.313	0.293	0.284	0.248	0.361	0.0002	1.0560	0.0367
	[1.7–1.8]	1.75	0.366	0.364	0.341	0.315	0.310	0.292	0.263	0.377	0.0005	1.0518	0.0560
		1.65	0.345	0.343	0.322	0.297	0.292	0.276	0.239	0.352	0.0001	1.0583	0.0333
Medium Fine	[1.4–1.5]	1.45	0.381	0.365	0.348	0.313	0.264	0.220	0.193	0.377	0.1402	1.3325	0.0068
		1.35	0.355	0.340	0.324	0.292	0.246	0.205	0.180	0.352	0.1309	1.3332	0.0068
	[1.5–1.6]	1.55	0.372	0.357	0.340	0.307	0.262	0.212	0.181	0.369	0.1002	1.2653	0.0068
		1.45	0.348	0.334	0.318	0.287	0.245	0.199	0.170	0.345	0.0943	1.2631	0.0070
	[1.6–1.7]	1.65	0.370	0.358	0.343	0.323	0.281	0.236	0.196	0.367	0.0435	1.1707	0.0056
		1.55	0.347	0.336	0.322	0.304	0.264	0.222	0.185	0.344	0.0583	1.1875	0.0053
Medium	[1.5–1.6]	1.55	0.356	0.340	0.312	0.274	0.231	0.206	0.175	0.360	0.1125	1.2472	0.0170
		1.45	0.334	0.318	0.292	0.257	0.216	0.193	0.164	0.338	0.1036	1.2423	0.0176
	[1.6–1.7]	1.65	0.350	0.338	0.319	0.286	0.241	0.193	0.152	0.350	0.0120	1.1862	0.0078
		1.55	0.329	0.318	0.299	0.268	0.226	0.181	0.143	0.329	0.0088	1.1820	0.0082

Table 4. Continued

[1.7–1.8]	1.75	0.322	0.310	0.299	0.282	0.261	0.226	0.184	0.317	0.0002	1.1231	0.0049
	1.65	0.304	0.292	0.282	0.266	0.246	0.212	0.173	0.299	0.0005	1.1245	0.0048
[1.8–1.9]	1.85	0.311	0.300	0.287	0.272	0.265	0.239	0.181	0.302	0.0003	1.1276	0.0026
	1.75	0.294	0.284	0.271	0.257	0.250	0.226	0.172	0.286	0.0009	1.1240	0.0028
Coarse [1.6–1.7]	1.65	0.315	0.277	0.210	0.182	0.142	0.114	0.089	0.352	0.0334	1.2429	0.0843
	1.55	0.296	0.260	0.197	0.171	0.133	0.121	0.084	0.339	0.0328	1.2286	0.1123
[1.7–1.8]	1.75	0.280	0.252	0.193	0.154	0.121	0.100	0.086	0.294	0.0695	1.4180	0.0339
	1.65	0.264	0.238	0.193	0.154	0.100	0.094	0.081	0.272	0.0711	1.5179	0.0257
[1.8–1.9]	1.85	0.303	0.281	0.257	0.226	0.183	0.165	0.128	0.310	0.0008	1.1434	0.0304
	1.75	0.287	0.266	0.243	0.214	0.173	0.156	0.121	0.294	0.0008	1.1435	0.0307

D_b^c is the bulk density of centimetric sized clods and D_b^h the bulk density of the horizon.

Table 5
Mean root square errors of the water content ($m^3\ m^{-3}$) estimates before and after grouping samples by horizon for Aquic Ustoll soils. Artificial neural networks with seven hidden units were used in all estimations (after Pachepsky et al., 1996)

Matric potential KJ m^{-3}	All samples ($N = 100$)	Samples grouped by horizons		
		Horizon A ($N = 43$)	Horizon B ($N = 32$)	Horizon C ($N = 25$)
-0.1	0.017	0.017	0.008	0.004
-1.0	0.018	0.018	0.006	0.005
-3.2	0.019	0.017	0.005	0.007
-10	0.022	0.018	0.005	0.006
-20	0.023	0.018	0.006	0.006
-50	0.024	0.016	0.006	0.009
-250	0.025	0.014	0.014	0.015
-1600	0.022	0.011	0.017	0.009
Combined	0.022	0.016	0.009	0.008

N = number of samples.

horizon type (A, B and C horizon) increases the precision of water retention estimates when compared to absence of grouping (Table 5). They suggested that improvement was related to differences in organic matter content among horizons that are known to affect soil water retention. Pachepsky and Rawls (1999) studied water retention at -33 and -1500 kPa and compared four criteria to group 447 soils from the Oklahoma National Resource Service Database: (i) soil great group, (ii) soil moisture regime, (iii) soil temperature regime, and (iv) soil textural class. Results showed that grouping improved the accuracy of PTFs in most cases, but none of the grouping criteria could be considered to be the best. However, there was no improvement of reliability for PTFs developed in groups when compared to PTFs developed from the whole database. Bruand et al. (2003) established PTFs after grouping by texture and after grouping by both texture and bulk density. They showed smaller mean error of prediction and standard deviation of prediction with the PTFs developed after grouping by both texture and bulk density.

5. CONCLUSION

Grouping enables a decrease in the variability of soil characteristics such as mineralogy, organic matter composition and type and development of structure, and thus leads to closer relationship between the hydraulic properties and the remaining variability of soil characteristics. Among soil characteristics used as grouping criteria, texture and bulk density appear to be the most efficient criteria to improve accuracy of PTFs; texture provides information on the size and reactivity of the elementary particles, and bulk density on the arrangement of the elementary particles. Thus, preliminary grouping by texture, even by using a limited number of texture classes, and then by bulk density can be recommended. Finally, several studies also show that parent material could be also used as grouping criteria in order to improve PTFs accuracy and reliability.

REFERENCES

Arrouays, D., Jamagne, M., 1993. Sur la possibilité d'estimer les propriétés de retention en eau de sols limoneux lessivés hydromorphes du Sud-Ouest de la France à partir de leurs caractéristiques de constitution. C. R. Acad. Agric. Fr. 79, 111-121.
Bastet, G., 1999. Estimation des propriétés de rétention en eau des sols à l'aide de fonctions de pédotransfert: développement de nouvelles approches. Thèse de Doctorat de l'Université d'Orléans, Orléans, France.
Brooks, R.H., Corey, A.T., 1964. Hydraulic properties of porous media. Colorado State University Hydrology, Paper No. 3, 27 pp.
Bruand, A., 1990. Improved prediction of water-retention properties of clayey soils by pedological stratification. J. Soil Sci. 41, 491-497.
Bruand, A., Baize, D., Hardy, M., 1994. Predicting of water retention properties of clayey soil using a single soil characteristic. Soil Use Manag. 10, 99-103.
Bruand, A., Duval, O., Gaillard, H., Darthout, R., Jamagne, M., 1996. Variabilité des propriétés de retention en eau des sols: importance de la densité apparente. Etud. Gestion Sols 3, 27-40.
Bruand, A., Pérez Fernàndez, P., Duval, O., Quétin, P., Nicoullaud, B., Gaillard, H., Raison, L., Pessaud, J.F., Prud'Homme, L., 2002. Estimation des propriétés de rétention en eau des sols: utilisation de classes de pédotransfert après stratification texturale et texturo-structurale. Etud. Gestion Sols 9, 105-125.
Bruand, A., Pérez Fernàndez, P., Duval, O., 2003. Use of class pedotransfer functions based on texture and bulk density of clods to generate water retention curves. Soil Use Manag. 19, 232-242.
Campbell, G.S., 1974. A simple model for determining unsaturated conductivity from moisture retention data. Soil Sci. 117, 311-314.
Cosby, B.J., Hornberger, G.M., Clapp, R.B., Ginn, T.R., 1984. A statistical exploration of the relationships of soil moisture characteristics to the physical properties of soils. Water Resour. Res. 20, 682-690.
Danalatos, N.G., Kosmas, C.S., Driessen, P.M., Yassoglou, N., 1994. Estimation of the draining soil moisture characteristics from standard data as recorded in soil surveys. Geoderma 64, 155-165.
Diebold, C.H., 1954. Permeability and intake rates of medium textures soils in relation to silt content and degree of compaction. Soil Sci. Soc. Am. Proc. 18, 339-343.
Driessen, P.M., Konijn, N.T., 1992. Land-Use Systems Analysis, INRES, Agric. Univ. of Wageningen, 230 pp.
Food and Agriculture Organisation (FAO), 1990. Guidelines for soil description, 3rd Ed. FAO/ISRIC, Rome.
Gaiser, Th., Graef, F., Carvalho Cordiero, J., 2000. Water retention characteristics of soils with contrasting clay mineral composition in semiarid tropical regions. Aust. J. Soil Res. 38, 523-536.

Gonçalves, M.C., Pereira, L.S., Leij, F.J., 1997. Pedo-transfer for estimating unsaturated hydraulic properties of Portuguese soils. Eur. J. Soil Sci. 48, 387-400.

Gregson, K., Hector, D.J., McGowan, M., 1987. A one-parameter model for the soil water charactcristic. J. Soil Sci. 38, 483-486.

Hall, D.G.M., Reeve, M.J., Thomasson, A.J., Wright, V.F., 1977. Water retention, porosity and density of field soils, Technical Monograph No. 9. Soil Survey of England & Wales, Harpenden, 75 pp.

Hodnett, M.G., Tomasella, J., 2002. Marked differences between van Genuchten soil water-retention parameters for temperate and tropical soils: a new water retention pedo-transfer functions developed for tropical soils. Geoderma 108, 155-180.

Jamagne, M., Bétrémieux, R., Bégon, J.C., Mori, A., 1977. Quelques données sur la variabilité dans le milieu naturel de la réserve en eau des sols. Bull. Information Tech. 324–325, 627-641.

King, J.J., Franzmeier, D.P., 1981. Estimation of saturated hydraulic conductivity from soil morphological and genetic information. Soil Sci. Soc. Am. J. 45, 1153-1156.

Lin, H.S., McInnes, K.J., Wilding, L.P., Hallmark, C.T., 1997. Low tension water flow in structured soils. Can. J. Soil Sci. 77, 649-654.

Lilly, A., 2000. The relationship between field-saturated hydraulic conductivity and soil structure: development of class pedotransfer functions. Soil Use Manag. 16, 56-60.

Leij, F., Alves, W.J., van Genuchten, M.Th., Williams, J.R., 1996. The UNSODA unsaturated soil hydraulic database. User's manual version 1.0. EPA/600/R-96/095, National Risk Management Laboratory, Office of Research and Development, Cincinnati, OH.

Leij, F., Alves, W.J., van Genuchten, M.Th., Williams, J.R., 1999. The UNSODA unsaturated soil hydraulic database. In: Van Genuchten, M.Th., Leij, F.J., Wu, L. (Eds.), Proceedings of the International Workshop on Characterization and Measurement of the Hydraulic Properties of Unsaturated Porous Media, Riverside, California, October 22–24, 1997, pp. 1269-1281.

McKenzie, N., Jacquier, D., 1997. Improving the field estimation of saturated hydraulic conductivity. Aust. J. Soil Res. 35, 803-825.

Nemes, A., 2002. Unsaturated soil hydraulic database of Hungary: HUNSODA. Agrokémia és Talajtan. 51, 17-26.

Nemes, A., Schaap, M.G., Leij, F.J., Wösten, J.H.M., 2001. Description of the unsaturated soil hydraulic database UNSODA Version 2.0. J. Hydrol. 251, 151-162.

Pachepsky, Y.A., Rawls, W.J., 1999. Acuracy and reliability of pedotransfer function as affected by grouping soils. Soil Sci. Soc. Am. J. 63, 1748-1759.

Pachepsky, Y.A., Rawls, W.J., 2003. Soil structure and pedotransfer functions. Eur. J. Soil Sci. 54, 443-452.

Pachepsky, Y.A., Timlin, D., Varallyay, G., 1996. Artificial neural networks to estimate soil water retention from easily measurable data. Soil Sci. Soc. Am. J. 60, 727-733.

Petersen, G.W., Cunningham, R.L., Matelski, R.P., 1968. Moisture characteristics of Pennsylvania soils. II. Soil factors affecting moisture retention within a textural class-silt loam. Soil Sci. Soc. Am. Proc. 32, 866-870.

Pidgeon, J.D., 1972. The measurement and prediction of available water capacity of Ferralitic soils in Uganda. J. Soil Sci. 23, 431-441.

Prebble, R.E., 1970. Physical properties from 17 soil groups in Queensland, CSIRO Australian Division of Soils, Technical Memorandum 10/70.

Puckett, W.E., Dane, J.H., Hajek, B.F., 1985. Physical and mineralogical data to determine soil hydraulic properties. Soil Sci. Soc. Am. J. 49, 831-836.
Rawls, W.J., Brakensiek, D.L., Saxton, K.E., 1982. Estimation of soil water properties. Trans. Am. Soc. of Agric. Engng 25, 1316-1328.
Rawls, W.J., Pachepsky, Y.A., Gimenez, D., Elliott, R., 1999. Development of STATGO pedotransfer functions using a group method data. In: Van Genuchten, M.Th., Leij, F.J., Wu, L. (Eds.), Proceedings of the International Workshop on Characterization and Measurement of the Hydraulic Properties of Unsaturated Porous Media, Riverside, California, October 22–24, 1997, pp. 1333-1342.
Reeve, M.J., Smith, P.D., Thomasson, A.J., 1973. The effect of density on water retention properties of field soils. J. Soil Sci. 24, 355-367.
Salchow, E., Lal, R., Fausey, N.R., Ward, A., 1996. Pedotransfer functions for variable alluvial soils in Southern Ohio. Geoderma 73, 165-181.
Saxton, K.E., Rawls, W.J., Romberger, J.S., Papendick, R.I., 1986. Estimating generalized soil-water characteristics from texture. Soil Sci. Soc. Am. J. 50, 1031-1036.
Tomasella, J., Hodnett, M.G., 1997. Estimating unsaturated hydraulic conductivity of Brazilian soils using soil-water retention data. Soil Sci. 162, 703-712.
Tomasella, J., Hodnett, M.G., 1998. Estimating soil water retention characteristics from limited data in Brazilian Amazonia. Soil Sci. 163, 190-202.
Van den Berg, M., Klant, E., van Reeuwijk, L.P., Sombroek, G., 1997. Pedotransfer functions for the estimation of moisture retention characteristics of Ferralsols and related soils. Geoderma 78, 161-180.
Van Genuchten, M.Th., 1980. A closed-form equation for predicting the hydraulic conductivity of unsaturated soils. Soil Sci. Soc. Am. J. 44, 892-898.
Williams, J.R., Prebble, E., Williams, W.T., Hignett, C.T., 1983. The influence of texture, structure and clay mineralogy on the soil moisture characteristic. Aust. J. Soil Res. 21, 15-31.
Williams, J., Ross, P., Bristow, K., 1992. Prediction of the Campbell water retention function from texture, structure, and organic matter. In: van Genuchten, M.Th., Leij, F.J., Lund, L.J. (Eds.), Proceedings of the International Workshop on Indirect Methods for Estimating the Hydraulic Properties of Unsaturated Soils. University of California, Riverside, CA, pp. 427-442.
Williams, R.D., Ahuja, L.R., Rawls, W.J., 1999. Estimating soil water retention using the Gregson one-parameter function. In: Van Genuchten, M.Th., Leij, F.J., Wu, L. (Eds.), Proceedings of the International Workshop on Characterization and Measurement of the Hydraulic Properties of Unsaturated Porous Media, Riverside, California, October 22–24, 1997, pp. 1011-1018.
Wösten, J.H.M., Bannink, M.H., De Gruijter, J.J., Bouma, J., 1986. A procedure to identify different groups of hydraulic-conductivity and moisture-retention curves for soil horizons. J. Hydrol. 86, 133-145.
Wösten, J.H.M., Finke, P.A., Jansen, M.J.W., 1995. Comparison of class and continuous pedotransfer functions to generate soil hydraulic characteristics. Geoderma 66, 227-237.
Wösten, J.H.M., Lilly, A., Nemes, A., Le Bas, C., 1999. Development and use of a database of hydraulic properties of European soils. Geoderma 90, 169-185.

PART III

HYDROLOGICAL AND PHYSICAL PARAMETERS

Chapter 11

PEDOTRANSFER FUNCTIONS FOR SOIL EROSION MODELS

D. Flanagan

USDA-Agricultural Research Service, National Soil Erosion Research Laboratory, 1196 Building SOIL, Purdue University, 275 S. Russell Street, West Lafayette, IN 47907, USA
Tel.: +1-765-494-7748; fax: +1-765-494-5948

1. INTRODUCTION

Some of the most practical applications of soil pedotransfer functions are in the realm of runoff and soil erosion prediction equations or models, for use by field conservationists and environmental planners in estimating sediment losses from farm fields, rangelands, forest harvest regions, and other land uses. To these end users, it is important to have sound technology that provides reasonable representation across the wide range of soils across the United States (or their country or region of interest) that soil erosion by water may occur on. This section will discuss some historical soil pedotransfer functions for erosion prediction, some pedotransfer functions for modern erosion models, and procedures to develop infiltration and erodibility parameters for these types of models.

2. HISTORY OF EARLY U.S. EROSION RESEARCH

Research on soil erosion by water has been conducted in the United States for about 90 years on field experimental plot and small watersheds. The earliest soil erosion research in the United States was conducted on overgrazed rangeland in central Utah beginning in 1912 (Sampson and Weyl, 1918). Field erosion plot research began in 1917 at the Missouri Agricultural Experiment Station in Columbia, Missouri (Miller, 1926).

In 1929, the United States Congress provided an appropriation of $160,000 for field research on soil erosion. This resulted in the establishment of 10 experiment stations at Guthrie, OK, Temple, TX, Hays, KS, Tyler, TX, Bethany, MO, Statesville, NC, Pullman, WA, Clarinda, IA, La Crosse, WI, and Zanesville, OH.

The federal erosion research stations conducted experimental studies that used plot design based on the work of Miller at the University of Missouri. The most common plots were 6 ft wide by 72.6 ft long, which comprised 1% of an acre. Research studies examined a variety of factors affecting erosion, including slope steepness, slope length, type of crop and crop rotations, conservation practices such as contouring, etc. The results from these studies as well as from additional research sites added in the 1940s and 1950s provided a large database of information on runoff and soil loss as affected by location (climate), slope, soil, and management conditions.

Early erosion researchers beginning in the 1940s also developed mathematical equations to estimate the amount of soil erosion and the impact of the use of alternative cropping management practices and/or conservation practices. Zingg (1940) conducted extensive experiments on the effect of slope length and steepness on erosion and developed a prediction equation:

$$A = CS^{1.4}L^{0.6} \tag{1}$$

where A was the average soil loss per unit of area, C was a constant, S was land slope (%), and L was slope length (ft).

This work was followed by Smith (1941), who added cropping and support practice factors to Zingg's function. Smith's equation was $A = CS^{1.4}L^{0.6}P$, where P was the ratio of soil loss with a mechanical conservation practice to soil loss without the practice. The C factor in this equation included the effects of soil, weather, and cropping system. Smith used this equation to create a graphic procedure to select conservation practices in the Midwest.

Browning et al. (1947) added soil erodibility and management factors to the Smith equation, and used it throughout Iowa beginning in the 1940s. The equation was rewritten as:

$$L = \frac{A_1^{5/3}}{PC} S^{-7/3} \tag{2}$$

where L is the slope length limit (ft), A_1 is the permissible soil loss (e.g., 5 t ac^{-1}), P is the conservation practice factor (1.0 if no practices), S is slope steepness in percent, and C is a constant expressing the effects of weather, soil, crop rotation, degree of erosion, or soil treatment on soil losses (for Browning's base Marshall soil at Clarinda, Iowa, the value of C was 0.035). Browning et al. (1947) also listed in their paper comparative erodibility factors for 12 other soils relative to the Marshall soil. Thus, this was one of the first instances of a limited pedotransfer function for soil erosion prediction.

Problems with application of the various equations arose when erosion predictions had to be made outside the climate region and soils used in their original database. There was a need for a widely applicable equation that could be applied over a broad geographic region and take into account climatic and soil variation.

In 1954, the USDA-ARS National Runoff and Soil Loss Data Center was created at Purdue University under the direction of Walt Wischmeier, and was to be the central location for the soil erosion data that had been collected across the U.S. since the 1930s. Part of the Center's task was to locate and assemble the wide assortment of experimental data from across the U.S, and then also to utilize this data in further development of erosion prediction equations. A substantial database of the measured runoff and soil loss data was eventually created from 47 research stations in 24 of the 37 states east of the Rocky Mountains as well as Pullman, Washington and Mayaguez, Puerto Rico, totaling over 10,000 plot years.

3. THE UNIVERSAL SOIL LOSS EQUATION

Wischmeier utilized the database to determine relationships between rainfall characteristics and soil loss, as well as effects of slope length, slope gradient and soil factors. The work ultimately resulted in creation of the Universal Soil Loss Equation (USLE), first published in USDA Agriculture Handbook 282 (Wischmeier and Smith, 1965). The USLE is:

$$A = R \times K \times L \times S \times C \times P \tag{3}$$

where A is the average annual soil loss (t ac^{-1}), R is the rainfall and runoff factor (100 ft t in ac^{-1} h^{-1}), K is the soil erodibility factor (0.01 t ac h ac^{-1} ft-t^{-1} in^{-1}), L is a slope-length factor, S is a slope-steepness factor, C is a cropping-management factor, and P is supporting erosion control practice factor.

The soil erodibility factor, K, is the soil loss per unit of R for a unit plot. For USLE, a unit plot was 72.6 ft long on a uniform 9% slope maintained in continuous tilled fallow. Table 1 in Agriculture Handbook 282 listed computed K values for 23 soils at erosion research stations. These values were based largely upon work of Olson and Wischmeier (1963), which evaluated data from fallow and cropped plots. For application of the USLE on a multitude of other soils, values of K were assigned at joint ARS–SCS regional workshops, based upon comparisons of soils to the original 23 soils' characteristics. However, a more scientific approach for obtaining K values for soils was needed.

In order to determine inherent soil erodibility values as a function of soil properties, a five-year field, laboratory and statistical study was conducted by Wischmeier and Mannering (1969). Rainfall simulation was used on 55 U.S. Corn Belt soils, and an empirical relationship was derived to compute K values based upon 24 soil parameter terms. The equation they developed for K was:

$$\begin{aligned}
K = 0.013[&18.82 + 0.62(\%\text{Silt}/\%\text{OM}) + 0.043(\%\text{Silt} \times \text{Reaction}) \\
& - 0.07(\%\text{Silt} \times \text{SS}) + 0.0082(\%\text{Silt} \times \%\text{Sand}) - 0.10(\%\text{Sand} \times \%\text{OM}) \\
& - 0.214(\%\text{Sand} \times \text{AI}) + 1.73(\text{Clay ratio}) - 0.0062(\text{Clay ratio} \times \%\text{Silt}) \\
& - 0.26(\text{Clay ratio} \times \%\text{OM}) - 2.42(\text{Clay ratio}/\%\text{OM}) \\
& + 0.30(\text{Clay ratio} \times \text{AI}) - 0.024(\text{Clay ratio}/\text{AI}) \\
& - 21.5(\text{AI}) - 0.18(\text{ASM}) + 1.0(\text{Increase in acidity below plow zone}) \\
& + 5.4(\text{Structure}) + 4.4(\text{SS}) + 0.65(\text{Structure change below plow layer}) \\
& - 0.39(\text{Thickness of "granular" material}) \\
& + 0.043(\text{Depth from "friable" to "firm"}) - 2.82(\text{Loess} = 1, \text{other} = 0) \\
& + 3.3(\text{Over calcareous base} = 1, \text{other} = 0) + 3.29(\%\text{OM} \times \text{AI}) \\
& - 1.38(\text{Reaction} \times \text{Structure})] \tag{4}
\end{aligned}$$

where OM is the organic matter, AI, aggregation index, SS, structure strength, and ASM, antecedent soil moisture. This complex pedotransfer function accounted for 98% of the total experimental variance on the 55 soils studied.

While the equation developed by Wischmeier and Mannering (1969) explained most of the experimental variance, the large number of terms made it difficult to apply by most field users. Further work by Wischmeier et al. (1971) resulted in a simplified equation and soil erodibility nomograph that can be used to calculate K values for soils with less than 70% silt and very fine sand:

$$100K = 2.1[\%\text{Silt} \times (100 - \%\text{Clay})]^{1.14}(10^{-4})(12 - \%\text{OM}) + 3.25(b-2) + 2.5(c-3) \tag{5}$$

where b is the soil structure code used in soil classification, and c is the profile permeability class. The equation and nomograph were also presented in Agriculture Handbook 537 (Wischmeier and Smith, 1978), allowing for a rapid graphical solution for K by field users of USLE.

In the Revised Universal Soil Loss Equation (RUSLE) handbook (Renard et al., 1997), all available published global data (225 soils) were used to derive a relationship for soil erodibility as:

$$K = 7.594 \left\{ 0.0034 + 0.0405 \exp\left[-\frac{1}{2}\left(\frac{\log(D_g) + 1.659}{0.7101}\right)^2 \right] \right\} \tag{6}$$

where

$$D_g = \exp\left(0.01 \sum f_i(\ln(m_i))\right) \tag{7}$$

with D_g being the geometric mean particle diameter (mm), f_i the primary particle size fraction in percent, and m_i is the arithmetic mean of the particle size limits of the size fraction. Coefficient of determination (r^2) for this relationship was reported to be 0.983.

4. PARAMETERIZATION OF EROSION PREDICTION MODELS

4.1. Erosion prediction models

Where the development of the USLE had almost solely been an activity within the United States, beginning in the 1960s and 1970s scientists and conservationists in many countries utilized or tried to utilize USLE and develop the necessary databases to apply it. Also, a variety of soil erosion prediction tools and models began to be developed both inside and outside the U.S. beginning in the 1970s and 1980s. The increasing power of computing systems allowed for development and application of more physical process-based soil erosion models or model components.

Outside the U.S., Elwell (1978) developed the Soil-Loss Estimation Model for South Africa (SLEMSA), as an alternative and simpler approach to USLE. Major considerations with SLEMSA were climate and cropping-management representations that were different from those in the U.S. In Australia, Rosewell and Edwards (1988) developed the SOILOSS computer program, to evaluate changes in land management on soil erosion by water. SOILOSS used a somewhat different approach to compute rainfall energy than that of the USLE, and contained the USLE soil erodibility nomograph and new experimentally obtained values for New South Wales in Australia. Many of the approaches used in SOILOSS are similar to those present in RUSLE.

Other erosion prediction tools developed outside the U.S. include EUROSEM (Morgan et al., 1998), GUEST (Misra and Rose, 1996), and the Hairsine–Rose (1992a,b) model. EUROSEM (European Soil Erosion Model) is a fully dynamic process-based water erosion model that utilizes the runoff and sediment routing components of KINEROS (Woolhiser et al., 1990). The Hairsine–Rose model computes erosion as an instantaneous process of sediment entrainment, deposition and re-entrainment. GUEST (Griffith University Erosion System Template) uses many of the same concepts of entrainment, deposition and re-entrainment, but was designed as a practical conservation tool for field application. All of these models require input parameterization to represent soil conditions.

Major erosion prediction models developed in the U.S. during the same time were CREAMS (Knisel, 1980), EPIC (Williams et al., 1984), RUSLE (Renard et al., 1997) and Water Erosion Prediction Project (WEPP) (Flanagan and Nearing, 1995). EPIC (Erosion Productivity Impact Calculator) was developed to quantify the economic costs of soil erosion and the effects of soil conservation practices, and it can use USLE or alternately modifications of USLE for sediment predictions (various rainfall and/or runoff erosivity factors). RUSLE is a revision of the USLE, with improved subfactor component calculations, and user-friendly interfaces for application at the field-office level. WEPP was designed to be a physical process-based computer simulation model with user-friendly interfaces and extensive databases for field applications on hillslope profiles and small watersheds. Erosion estimates in WEPP are a function of daily rainfall and detachment by raindrops and flowing water, and the model can also predict sediment deposition and delivery off-site.

The USDA-ARS CREAMS (Chemicals, Runoff and Erosion from Agricultural Management Systems) model (Knisel, 1980) was the first to separate erosion processes into those on rill and interrill areas (Foster et al., 1980). The CREAMS model computed sediment detachment by raindrop impact and by flowing water, sediment transport and sediment deposition. However, CREAMS still relied upon USLE erodibility values for parameterization and generally used SCS Curve Number procedures to estimate runoff. To adequately simulate the sediment transport processes, however, information on the sediment characteristics was needed, including size fractions, diameters and densities.

4.2. Sediment particle fractions and particle composition

In the initial CREAMS model (Knisel, 1980) released in 1980, sets of equations to estimate sediment particle fractions and characteristics were included. However, a more thorough set of equations was presented by Foster et al. (1985) which utilized a larger data set from 28 soils in creating the final equations. The approach assumes that sediment is composed of five size classes: primary clay, primary silt, small aggregates, large aggregates, and primary sand.

Fraction of primary clay in the eroded sediment is based upon results from a wide range of experiments, and is:

$$F_{clay} = 0.26\, O_{clay} \qquad (8)$$

where F_{clay} is the fraction of primary clay in the detached sediment and O_{clay} is the fraction of clay in the matrix soil. For small aggregates, when the matrix soil clay content is less than 25%:

$$F_{sagg} = 1.8\, O_{clay} \qquad (9)$$

when the matrix soil clay content is between 25 and 50%,

$$F_{sagg} = 0.45 - 0.6\,(O_{clay} - 0.25) \qquad (10)$$

and when the matrix soil clay content exceeds 50%,

$$F_{sagg} = 0.6\, O_{clay} \qquad (11)$$

The fraction of primary silt is computed as

$$F_{silt} = O_{silt} - F_{sagg} \qquad (12)$$

where O_{silt} is the fraction of silt in the matrix soil. If the equation results in a negative value for F_{silt}, F_{sagg} is set equal to O_{silt}, and F_{silt} is set equal to zero.

Primary sand fraction is calculated in Foster et al.'s (1985) procedure as:

$$F_{sand} = O_{sand}(1.0 - O_{clay})^5 \qquad (13)$$

where O_{sand} is the fraction of sand in the matrix soil. Fraction of large aggregates in the sediment is computed as:

$$F_{lagg} = 1.0 - F_{clay} - F_{silt} - F_{sagg} - F_{sand} \qquad (14)$$

The make-up of the small aggregate and large aggregate fractions must also be determined. For the small aggregates:

$$f_{clay:sagg} = O_{clay}/(O_{clay} + O_{silt}) \qquad (15)$$

$$f_{silt:sagg} = O_{silt}/(O_{clay} + O_{silt}) \qquad (16)$$

where $f_{clay:sagg}$ is the fraction of clay in the small aggregates and $f_{silt:sagg}$ is the fraction of silt in the small aggregates. Fraction of sand in the small aggregates is zero. For the large aggregates:

$$f_{clay:lagg} = [O_{clay} - F_{clay} - (F_{sagg} \times f_{clay:sagg})]/F_{lagg} \qquad (17)$$

$$f_{\text{silt:lagg}} = [O_{\text{silt}} - F_{\text{silt}} - (F_{\text{sagg}} \times f_{\text{silt:sagg}})]/F_{\text{lagg}} \tag{18}$$

$$f_{\text{sand:lagg}} = (O_{\text{sand}} - F_{\text{sand}})/F_{\text{lagg}} \tag{19}$$

where $f_{\text{clay:lagg}}$ is the fraction of clay in the large aggregates, $f_{\text{silt:lagg}}$ is the fraction of silt in the large aggregates, and $f_{\text{sand:lagg}}$ is the fraction of sand in the large aggregates. Foster et al. (1985) also assign specific gravities and particle diameters for each of the size classes. The equations presented here for estimation of soil particle fractions and characteristics are also utilized in the WEPP model (Flanagan and Nearing, 1995).

4.3. WEPP infiltration parameterization

The WEPP model was developed from 1985 to 1995 by a team of federal and university scientists within the U.S. WEPP is a physical process-based, distributed parameter, continuous simulation erosion prediction model. Daily climate inputs to the model drive the simulation of the rainfall-infiltration-runoff processes.

WEPP uses a Green–Ampt Mein–Larson (GAML) procedure (Mein and Larson, 1973), modified for unsteady rainfall (Chu, 1978) to compute infiltration during a rainstorm event. For the situation where there is ponding within a rainfall interval, the cumulative infiltration depth is computed using:

$$K_e t_c = F_i - \Psi \theta_d \ln\left[1 + \frac{F_i}{\Psi \theta_d}\right] \tag{20}$$

where K_e is the effective hydraulic conductivity (m s^{-1}), t_c is the corrected time to ponding (s), ψ is the average capillary potential (m), F is the cumulative infiltration depth at time i (m), and θ_d is the soil moisture deficit (m m^{-1}). Full details on the model procedures are presented in Stone et al. (1995). It is important to note that the effective hydraulic conductivity of a soil is not the same as the saturated hydraulic conductivity, nor equal in value to it (Alberts et al., 1995). Saturated hydraulic conductivity is a measure of a soil's ability to transmit water in a saturated state, a condition that very rarely occurs in the field.

The effective hydraulic conductivity is a critical parameter in *WEPP* model simulations. This value and any adjustments to it directly impact the amount and rates of infiltration and related runoff. WEPP can be applied using either constant or temporally varying values of conductivity. See Alberts et al. (1995) on methods to estimate time-invariant hydraulic conductivity for cropland and rangeland.

The power of the continuous simulation WEPP model is through the daily updating of soil, plant, and residue conditions – thus use of time-varying effective hydraulic conductivity is most often recommended. For cropland, baseline conditions are for a freshly tilled soil with no residue and no vegetation present. The conductivity of the soil in this state is called the baseline effective hydraulic conductivity. Adjustments to soil parameters are then made to the baseline values, as a function of the daily soil, plant and residue status.

For croplands, extensive model optimization runs on 43 soils were conducted using measured and curve number predictions for tilled fallow management (Alberts et al., 1995). The following equations are used within WEPP to predict baseline effective hydraulic conductivity. For soils with clay content less than or equal to 40%:

$$K_b = -0.265 + 0.0086\,\text{SAND}^{1.8} + 11.46\,\text{CEC}^{-0.75} \tag{21}$$

For soils having clay content greater than 40%:

$$K_b = 0.0066\, e^{(244/\text{CLAY})} \tag{22}$$

where K_b is baseline effective hydraulic conductivity (mm h^{-1}), SAND is percent sand content in the surface soil, CLAY is percent clay content in the surface soil, and CEC is the cation exchange capacity in the surface soil in meq per 100 g (Alberts et al., 1995; Flanagan and Livingston, 1995).

Data from natural rainfall studies on fallow, row-cropped, and perennial-cropped plots (Risse et al., 1994; Zhang et al., 1995a,b) were used to develop adjustment factors. A soil crusting and tillage adjustment can be computed based upon the amount of surface cover, the soil random roughness, the cumulative rainfall kinetic energy since the last tillage operation, a soil stability factor, and a crust factor. The adjustments to conductivity for row crops are a function of effective canopy cover, residue cover, and the storm rainfall amount. The following equations are used:

$$K_e = K_{\text{bare}}(1 - \text{scovef}) + (0.0534 + 0.01179 K_b)(\text{rain})(\text{scovef}) \tag{23}$$

$$K_{\text{bare}} = K_b[\text{CF} + (1 - \text{CF})e^{-C_{ss}\,E_a(1-\text{RR}/0.04)}] \tag{24}$$

where K_{bare} is the effective hydraulic conductivity for bare soil regions after adjustments for crusting and tillage (mm h^{-1}), CF is the crust factor (Rawls et al., 1990), C_{ss} is the soil stability factor (m^2 J^{-1}), RR is the random roughness of the soil surface (m), E_a is the cumulative kinetic energy of rainfall since the last tillage operation (J m^{-2}), scovef is the total effective surface cover, and rain is the storm rainfall amount (mm). Analysis of 88 plot-years of measured data under perennial crops found that on average the effective hydraulic conductivity is about 1.8 times greater than that from row crop conditions. See Alberts et al. (1995) for complete details on cropland hydraulic conductivity adjustments in WEPP.

Other sets of experiments were conducted to determine conductivity values for rangeland conditions. For rangelands, when rill surface cover is less than 45%, effective conductivity is predicted using:

$$\begin{aligned}K_{\text{erange}} = &\ 57.99 - 14.05\,\ln(\text{CEC}) + 6.20\,\ln(\text{ROOT10}) \\ &- 473.39\,\text{BASR}^2 + 4.78\,\text{RESI}\end{aligned} \tag{25}$$

while for rangelands when rill cover is greater than or equal to 45%

$$\begin{aligned}K_{\text{erange}} = &\ -14.29 - 3.40\,\ln(\text{ROOT10}) + 0.3783\,\text{SAND} + 2.0886\,\text{ORGMAT} \\ &+ 398.64\,\text{RR} - 27.39\,\text{RESI} + 64.14\,\text{BASI}\end{aligned} \tag{26}$$

where K_{erange} is effective rangeland hydraulic conductivity in mm h^{-1}, CEC is cation exchange capacity (meq per 100 g), ROOT10 is root biomass in the top 10 cm of the soil in kg m^{-2}, BASR is the product of the fraction of basal surface cover in rill areas and total basal surface cover, RESI is the product of the fraction of litter surface cover in inter-rill areas and the total litter surface cover, SAND is percent sand content of the surface soil, ORGMAT is percent organic matter in the surface soil and BASI is the product of the fraction of litter surface cover in interrill areas and the total basal surface cover (Alberts et al., 1995).

4.4. WEPP erodibility parameterization

WEPP uses four values from the hydrology component to estimate soil detachment: effective rainfall intensity, effective rainfall duration, peak runoff rate, and effective runoff duration. The equation in the model to estimate detachment on interrill areas is (Foster et al., 1995):

$$D_i = K_{iadj} I_e \sigma_{ir} SDR_{RR} F_{nozzle} \left(\frac{R_s}{w} \right) \qquad (27)$$

where D_i is the rate of interrill sediment delivery to the rills (kg s^{-1} m^{-2}), K_{iadj} is the adjusted interrill erodibility (kg s m^{-4}), I_e is the effective rainfall intensity (m s^{-1}), σ_{ir} is the interrill runoff rate (m s^{-1}), SDR$_{RR}$ is a sediment delivery ratio that is a function of random roughness, the row side-slope and the interrill sediment particle size distribution, F_{nozzle} is a nozzle energy factor for sprinkler irrigation, R_s is the spacing of the rills (m) and w is the rill width (m).

Detachment by flowing water at a point in a rill is computed using an excess flow shear stress equation (Foster et al., 1995):

$$D_f = K_{radj} (\tau_f - \tau_{cadj}) \left[1 - \frac{G}{T_c} \right] \qquad (28)$$

where D_f is the rill detachment rate (kg s^{-1} m^{-2}), K_{radj} is the adjusted rill erodibility (s m^{-1}), τ_f is the flow shear stress (Pa), τ_{cadj} is the adjusted soil critical shear stress (Pa), G is the sediment load (kg s^{-1} m^{-1}), and T_c is the sediment transport capacity at that point in the rill (kg s^{-1} m^{-1}).

Baseline conditions for cropland are for a freshly-tilled soil with no plant or residue cover. The adjusted interrill and rill erodibilities and critical shear stresses used in Equations (27) and (28) are obtained by multiplying the baseline values by a set of adjustment factors.

Field experiments conducted on 33 cropland soils (Elliot et al., 1989) and 18 rangeland sites (Simanton et al., 1987) provided information that allows the baseline K_i, K_r, and τ_c parameters to be estimated from site-specific soil properties.

For cropland soils with surface soil sand content of 30% or more, the WEPP erodibility estimation equations are (Flanagan and Livingston, 1995; Alberts et al., 1995):

$$K_{ib} = 2728000 + 192100 \, VFS \qquad (29)$$

$$K_{rb} = 0.00197 + 0.00030 \, VFS + 0.03863 \, e^{-1.84 \, ORGMAT} \qquad (30)$$

$$\tau_c = 2.67 + 0.065 \text{ CLAY} - 0.058 \text{ VFS} \tag{31}$$

and for cropland soils having less than 30% sand, the equations are:

$$K_{ib} = 6054000 - 55130 \text{ CLAY} \tag{32}$$

$$K_{rb} = 0.0069 + 0.134 \text{ e}^{-0.20 \text{ CLAY}} \tag{33}$$

$$\tau_c = 3.5 \tag{34}$$

where K_{ib} is baseline interrill erodibility (kg s m^{-4}), K_{rb} is baseline rill erodibility (s m^{-1}), τ_c is baseline critical shear stress (Pa), VFS is percent very fine sand in the surface soil (particle diameter size range 0.05–0.1 mm), CLAY is percent clay in the surface soil, and ORGMAT is the percent organic matter in the surface soil.

Baseline interrill erodibility on cropland is adjusted daily for a large number of factors. These include canopy cover, ground cover, roots, sealing and crusting, and freezing and thawing. Baseline rill erodibility is adjusted for incorporated residue, roots, sealing and crusting, and freezing and thawing effects. Baseline critical shear stress is adjusted daily for the effects of random roughness, sealing and crusting, and freezing and thawing.

For rangelands, interrill erodibility, rill erodibility and critical hydraulic shear stress are estimated using the following equations (Alberts et al., 1995; Flanagan and Livingston, 1995):

$$K_{irange} = 1810000 - 19100 \text{ SAND} - 63270 \text{ ORGMAT} - 8460000 \theta_{fc} \tag{35}$$

$$K_{rrange} = 0.0017 + 0.000024 \text{ CLAY} - 0.000088 \text{ ORGMAT} \\ - (0.00088 \text{ BD}_{dry}/1000) - 0.00048 \text{ ROOT10} \tag{36}$$

$$\tau_{crange} = 3.23 - 0.056 \text{ SAND} - 0.244 \text{ ORGMAT} + (0.9 \text{ BD}_{dry}/1000) \tag{37}$$

where K_{irange} is baseline interrill erodibility (kg s m^{-4}), K_{rrange} is baseline rill erodibility (s m^{-1}), τ_{crange} is baseline critical shear stress (Pa), BD$_{dry}$ is the dry soil bulk density (kg m^{-3}), and θ_{fc} is the volumetric water content of the soil at 0.033 MPa (m^3 m^{-3}).

Adjustments are made to rangeland interrill erodibility for ground cover and freezing and thawing effects. Freezing and thawing adjustments are also made to the rangeland rill erodibility and critical shear stress values (Alberts et al., 1995).

5. PROCEDURES TO DEVELOP EROSION MODEL PEDOTRANSFER FUNCTIONS

A range of techniques are available to parameterize soil erosion models and develop appropriate pedotransfer functions. Experimental techniques, simulation and statistical procedures will be discussed here.

5.1. Experimental techniques

Development of the USLE was based upon thousands of plot-years of field erosion experiment data, largely from long-term natural rainfall plots. A unit plot for the USLE was 72.6 ft long at a 9% slope and maintained in a continuous tilled-fallow. Plots with other lengths and gradients were also often used, and then observed soil loss values were adjusted back to those that would be expected from a unit plot. The experiment stations also had additional plots that would have different cropping and management practices as well as conservation practices. Runoff and sediment collected from the plots, as well as observed rain storm data on rainfall depths and intensities could then be used to develop the USLE factors. Given sufficient time and resources, these same approaches can still be used today (and are in some locations of the world), though it is often difficult to obtain sufficient resources to build, operate and maintain long-term natural rainfall erosion plots for a sufficient number of years to obtain meaningful data (Figure 1).

Figure 1. Natural rainfall erosion plots such as these at McCredie, Missouri have been used to collect data for empirical models such as the USLE (USDA-ARS photo courtesy of Don Meyer).

Alternative or supplemental experimental techniques to natural rainfall plots usually entail the use of rainfall simulators. Rainfall simulators are mechanical equipment developed to apply rainfall having intensities, depths, drop characteristics and energies that are similar to natural rainfall. Simulators allow the rapid generation of runoff and erosion data, often in weeks or months, compared to years or decades of natural rainfall experiments.

A rainfall simulator experiment should be designed to simulate the hydrologic and/or erosion processes of interest, and be able to adequately measure the desired variables. For process-based erosion models, most often the measurements of interest will be the rainfall rate, runoff rate, flow velocity, flow depth, sediment concentrations, and sediment particle size distributions. The compendium by Elliot et al. (1989) provides in-depth information on the procedures used in the WEPP cropland field rainfall simulator experiments on 33 U.S. soils. There, for each soil studied, they installed six long rill subplots that were 9 m long by 0.46 m wide, and six smaller interrill plots that were 0.75 m long by 0.5 m wide.

They also had four small plots 0.75 m long by 0.5 m wide for infiltration measurements. Soil pits were dug at each site, and extensive physical and chemical soil property analyses were conducted on the soil samples by the USDA-SCS National Soil Survey Laboratory in Lincoln, Nebraska.

5.2. Interrill erodibility

Rainfall simulation studies are most often conducted at uniform rainfall intensity rates, and water is applied so that runoff rates reach steady-state. Runoff samples are collected to analyze for sediment concentration throughout the event as well as to determine the size fractions of the sediment. With steady-state runoff rate and corresponding sediment discharge rate, the interrill erodibility value in Equation (27) can be back-calculated. A range of rainfall intensities can also be applied, to provide for a better average estimate of interrill erodibility. Often interrill erosion plots are set up so that they have relatively steep side-slopes to a collection trough, bare soil with no residue, and little roughness. In these cases, the adjusted interrill erodibility is approximately equal to the baseline interrill erodibility. Assuming that the plot conditions allow for all detached sediment on the interrill side-slopes to reach the rill and that the simulator nozzle produces rainfall equivalent to natural rainfall, Equation (27) can be rearranged to give:

$$K_{ib} = \frac{D_i}{I_e \sigma_{ir}} \tag{38}$$

If the interrill plot conditions are different from baseline, then appropriate adjustments to the calculation of K_{ib} need to be made to correct for residue, canopy, slope or roughness effects. For each soil studied, an average K_{ib} can be computed, and then using these values from a group of soils as well as measured soil properties, pedotransfer functions can be developed through statistical regression techniques.

5.3. Rill erodibility and critical shear stress

Rill erosion parameters are dependent upon some characteristic of the water flowing in the rills. This may be flow discharge, flow velocity, flow shear stress, or flow stream power. In the WEPP model, an excess flow shear stress equation (Equation (28)) is used. To develop rill erodibility and critical shear stress values, experiments can be conducted in which increasing levels of water flow are added to rills, and the corresponding runoff and sediment discharge measured. If clear water is used as the inflow, then the sediment load term in Equation (28) can be assumed to be approximately zero, and the equation becomes:

$$D_f = K_{radj}(\tau_f - \tau_{cadj}) \tag{39}$$

If the rill plot conditions are in a baseline state (bare, freshly tilled soil with no residue or canopy), then the adjusted rill erodibility and critical shear stress can be assumed to be the same as the baseline values. In a rainfall simulation study such as shown in Figure 2, under rainfall conditions the sediment discharge collected at the end of the plot must have an estimated interrill contribution subtracted from it in order to approximate the rill detachment rate. In the studies by Elliot et al. (1989), they took measurements on the rill

Figure 2. Overhead view of WEPP rainfall simulator experiment in progress at Cottonwood, South Dakota (USDA-ARS, photo by Tim McCabe).

plots under both rainfall and no rainfall conditions. In those experiments, measurements were also made of the rill geometry, flow depths and flow velocities so that values of flow shear stress could be estimated using:

$$\tau_f = \gamma RS \tag{40}$$

where γ is the density of water (N m^{-3}), R is the hydraulic radius of the rill flow (m), and S is the hydraulic gradient that was approximated with the slope of the rill bottom (m m^{-1}).

With values for flow shear stress for the various levels of inflow water and corresponding rill detachment rates, values of K_r and τ_c can be determined through linear regression. Some judgment may be necessary to identify and use only observed values where the flow shear stress has exceeded critical shear stress (non-zero detachment rates). Figure 3 shows an example plot of measured rill detachment rate (sediment discharge rate) vs. flow shear stress from a laboratory experiment. If these experiments are conducted over a range of soils and the soil physical and chemical properties measured, then pedotransfer functions for rill erodibility and critical shear stress can be developed through appropriate regression analyses.

5.4. Effective hydraulic conductivity

Information from rainfall simulation studies as well as from long-term natural rainfall erosion plots can be used to estimate hydraulic conductivity values for erosion models using Green–Ampt type equations. Several methods exist for these types of evaluations.

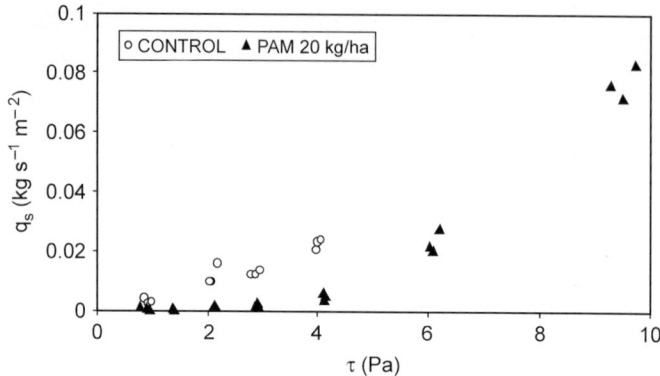

Figure 3. Example graph of measured sediment discharge (q_s) from a laboratory rill flume experiment vs. flow shear stress (τ_f) for two soil treatments (20 kg ha^{-1} anionic polyacrylamide; Control – untreated control).

In rainfall simulation studies, small plots in a baseline condition (for cropland – a freshly tilled, bare soil with no canopy or residue cover), can be covered with a porous material that will absorb the raindrop impact but still allow the water to infiltrate to the plot (e.g., fibrous furnace filter). Comparisons of runoff from these covered plots to uncovered plots exposed to raindrop impact and sealing can provide information on baseline conductivity as well as adjustments needed for soil sealing and crusting and consolidation. On both small and long plots that have been brought to steady-state runoff after considerable applications of rainfall, infiltration rate can be estimated and Equation (20) used to compute conductivity.

Another approach for estimation of the baseline effective hydraulic conductivity is to use the model as an optimization tool, and run multiple simulations until some determinant factor is minimized. For example with the WEPP model, input values for K_b can be set, the model run, then output values for total runoff depth over the period of simulation evaluated and compared to observed values. The minimization criteria may be to minimize the least square error, maximize the coefficient of determination, or maximize the model efficiency for individual storm event runoff, or monthly, annual, or average annual runoff values. These procedures can be automated to allow rapid multiple simulations and then estimation of the deviance of the model predictions from the observed runoff depths. For the WEPP model K_b estimations, a minimization procedure using the Nash–Sutcliffe model efficiency (Nash and Sutcliffe, 1970) on a storm-by-storm runoff basis (Alberts et al., 1995; Flanagan and Livingston, 1995) was used. Some of the large USLE database was used in these evaluations, and USLE data are available via the internet for other model parameterization efforts (http://topsoil.nserl.purdue.edu/nserlweb/usle).

6. SUMMARY

For wide-spread applicability, soil erosion prediction models often rely upon a variety of soil pedotransfer functions. These most often address soil erodibility and soil infiltration parameters. Development of these functions may be accomplished through experimental studies on a representative group of soils that cover the range of properties of the larger set of soils where the erosion model is to be applied. In some cases, existing long-term historical erosion plot data (e.g., the USLE database) may provide sufficient information to create erosion model pedotransfer equations.

REFERENCES

Alberts, G.A., Nearing, M.A., Weltz, M.A., Risse, L.M., Pierson, F.B., Zhang, X.C., Laflen, J.M., Simanton, J.R., 1995. Soil component. *In*: Flanagan, D.C., Nearing, M.A. (Eds.), USDA-Water Erosion Prediction Project: Hillslope Profile and Watershed Model Documentation, NSERL Report No. 10. USDA-ARS National Soil Erosion Research Laboratory, West Lafayette, IN, Chapter 7.

Browning, G.M., Parish, C.L., Glass, J.A., 1947. A method for determining the use and limitations of rotation and conservation practices in control of soil erosion in Iowa. J. Am. Soc. Agron. 39, 65-73.

Chu, S.T., 1978. Infiltration during an unsteady rain. Water Resour. Res. 14 (3), 461-466.

Elliot, W.J., Liebenow, A.M., Laflen, J.M., Kohl, K.D., 1989. A Compendium of Soil Erodibility Data from WEPP Cropland Soil Field Erodibility Experiments 1987 & 1988, NSERL Report No. 3. USDA-ARS National Soil Erosion Research Laboratory, West Lafayette, IN.

Elwell, H.A., 1978. Modeling soil losses in southern Africa. J. Agric. Engng Res. 23, 117-127.

Flanagan, D.C., Livingston, S.J. (Eds.), 1995. USDA-Water Erosion Prediction Project: WEPP User Summary, NSERL Report No. 11. USDA-ARS National Soil Erosion Research Laboratory, West Lafayette, IN.

Flanagan, D.C., Nearing, M.A. (Eds.), 1995. USDA-Water Erosion Prediction Project: Hillslope Profile and Watershed Model Documentation, NSERL Report No. 10. USDA-ARS National Soil Erosion Research Laboratory, West Lafayette, IN.

Foster, G.R., Lane, L.J., Nowlin, J.D., Laflen, J.M., Young, R.A., 1980. A model to estimate sediment yield from field-sized areas: development of model. *In*: Knisel, W.G. (Ed.), CREAMS: A Field-Scale Model for Chemicals, Runoff, and Erosion from Agricultural Management Systems, U.S. Department of Agriculture, Conserv. Res. Report No. 26, pp. 193-281, Chapter 3.

Foster, G.R., Young, R.A., Neibling, W.H., 1985. Sediment composition for nonpoint source pollution analyses. Trans. Am. Soc. Agric. Engng 28(1) 133-139, 146.

Foster, G.R., Flanagan, D.C., Nearing, M.A., Lane, L.J., Risse, L.M., Finkner, S.C., 1995. Hillslope erosion component. *In*: Flanagan, D.C., Nearing, M.A. (Eds.), USDA-Water Erosion Prediction Project: Hillslope Profile and Watershed Model Documentation, NSERL Report No. 10. USDA-ARS National Soil Erosion Research Laboratory, West Lafayette, IN, Chapter 11.

Hairsine, P.B., Rose, C.W., 1992a. Modeling water erosion due to overland flow using physical principles: 1. sheet flow. Water Resour. Res. 28 (1), 237-243.

Hairsine, P.B., Rose, C.W., 1992b. Modeling water erosion due to overland flow using physical principles: 2. rill flow. Water Resour. Res. 28 (1), 245-250.

Knisel, W.G. (Ed.), 1980. CREAMS: A Field-Scale Model for Chemicals, Runoff, and Erosion from Agricultural Management Systems, U.S. Department of Agriculture, Conserv. Res. Report No. 26, 640 pp.

Mein, R.G., Larson, C.L., 1973. Modeling infiltration during a steady rain. Water Resour. Res. 9 (2), 384-394.

Miller, M.F., 1926. Waste through soil erosion. J. Am. Soc. Agron. 27, 336-345.

Misra, R.K., Rose, C.W., 1996. Application and sensitivity analysis of process-based erosion model GUEST. Eur. J. Soil Sci. 47, 593-604.

Morgan, R.P.C., Quinton, J.N., Smith, R.E., Govers, G., Poessen, J.W.A., Auerswald, K., Chisci, G., Torri, D., Styczen, M.E., 1998. The European Soil Erosion Model (EUROSEM): a dynamic approach for predicting sediment transport from fields and small catchments. Earth Surf. Processes Landforms 23, 527-544.

Nash, J.E., Sutcliffe, V., 1970. River flow forecasting through conceptual models, I. A discussion of principles. J. Hydrol. 10 (3), 282-290.

Olson, T.C., Wischmeier, W.C., 1963. Soil-erodibility evaluations for soils on the runoff and erosion stations. Soil Sci. Soc. Am. Proc. 27 (5), 590-592.

Rawls, W.J., Brakensiek, D.L., Simanton, J.R., Kohl, K.D., 1990. Development of a crust factor for the Green Ampt model. Trans. Am. Soc. Agric. Engng 33 (4), 1224-1228.

Renard, K.G., Foster, G.R., Weesies, G.A., McCool, D.K., Yoder, D.C., 1997. Predicting Soil Erosion by Water: A Guide to Conservation Planning with the Revised Universal Soil Loss Equation (RUSLE), USDA-Agriculture Handbook, Vol. 703. U.S. Government Printing Office, Washington, DC, 404 pp.

Risse, L.M., Nearing, M.A., Savabi, M.R., 1994. Determining the Green–Ampt effective hydraulic conductivity from rainfall-runoff data for the WEPP model. Trans. Am. Soc. Agric. Engng 37 (2), 411-418.

Rosewell, C.J., Edwards, K., 1988. SOILOSS – a Program to Assist in the Selection of Management Practices to Reduce Erosion, Tech. Handbook No. 11. Soil Conservation Service of New South Wales, Sydney, NSW, 71 pp.

Sampson, A.W., Weyl, L.H., 1918. Range Preservation and Its Relation to Erosion Control on Western Grazing Lands, Bull. 675. U.S. Department of Agriculture, Washington, DC, 35 pp.

Simanton, J.R., West, L.T., Weltz, M.A., Wingate, G.D., 1987. Rangeland Experiments for Water Erosion Prediction Project, ASAE Paper No. 87-2545. American Society of Agricultural Engineering, St. Joseph, MI, 10 pp.

Smith, D.D., 1941. Interpretation of soil conservation data for field use. Agric. Engng 22, 173-175.

Stone, J.J., Lane, L.J., Shirley, E.D., Hernandez, M., 1995. Hillslope surface hydrology. In: Flanagan, D.C., Nearing, M.A. (Eds.), USDA-Water Erosion Prediction Project: Hillslope Profile and Watershed Model Documentation, NSERL Report No. 10. USDA-ARS National Soil Erosion Research Laboratory, West Lafayette, IN, Chapter 4.

Williams, J.R., Jones, C.A., Dyke, P.T., 1984. A modeling approach to determining the relationship between erosion and soil productivity. Trans. Am. Soc. Agric. Engng 27 (1), 129-144.

Wischmeier, W.H., Mannering, J.V., 1969. Relation of soil properties to its erodibility. Soil Sci. Soc. Am. Proc. 33 (1), 131-137.

Wischmeier, W.H., Smith, D.D., 1965. Predicting rainfall-erosion losses from cropland east of the Rocky Mountains. USDA-Agriculture Handbook 282. US Government Printing Office, Washington, DC, 47 pp.

Wischmeier, W.H., Smith, D.D., 1978. Predicting rainfall erosion losses: a guide to conservation planning. USDA-Agriculture Handbook 537. US Government Printing Office, Washington, DC, 58 pp.

Wischmeier, W.H., Johnson, C.B., Cross, B.V., 1971. A soil erodibility nomograph for farmland and construction sites. J. Soil Water Conserv. 26 (5), 189-193.

Woolhiser, D.A., Smith, R.E., Goodrich, D.C., 1990. KINEROS: A Kinematic Runoff and Erosion Model: Documentation and User Manual, ARS-77. USDA Agricultural Research Service, Washington, DC.

Zhang, X.C., Nearing, M.A., Risse, L.M., 1995a. Estimation of Green–Ampt conductivity parameters: part I. row crops. Trans. Am. Soc. Agric. Engng 38 (4), 1069-1077.

Zhang, X.C., Nearing, M.A., Risse, L.M., 1995b. Estimation of Green–Ampt conductivity parameters: part II. perennial crops. Trans. Am. Soc. Agric. Engng 38 (4), 1079-1087.

Zingg, A.W., 1940. Degree and length of land slope as it affects soil loss in runoff. Agric. Engng 21, 59-64.

Chapter 12

SOLUTE ADSORPTION AND TRANSPORT PARAMETERS

B. Minasny[1] and E. Perfect[2,*]

[1]Faculty of Agriculture, Food & Natural Resources, The University of Sydney, JRA McMillan Building A05, NSW 2006, Australia

[2]Department of Earth and Planetary Sciences, 1412 Circle Drive, University of Tennessee, Knoxville, TN 37996-1410, USA

*Corresponding author: Tel.: +1-865-974-6017

1. INTRODUCTION

Solute adsorption and transport in soil are important subjects as they deal with the movement of chemicals in environmental monitoring and the availability of nutrients for agronomic management. Pedotransfer functions (PTFs) have been developed mainly to predict soil water retention and hydraulic conductivity curves. In contrast, relatively less work has been done to develop, or to assess the usefulness of PTFs for predicting parameters of solute transport models. This chapter will review existing studies in this area and point to future research directions.

Many studies predicting solute movement in soil employ PTFs to predict soil hydraulic properties (Petach et al., 1991; Bouma et al., 1996; Gerke et al., 1999). In such studies, solute adsorption and transport parameters are usually arbitrarily defined or based on values from the literature. While databases of soil water retention curve and saturated hydraulic conductivity parameters are widely available, experiments on estimating solute adsorption and transport parameters are usually conducted on a case-by-case basis. Solute parameters are often difficult and tedious to measure, much more than soil hydraulic properties. Furthermore, they are often log-normally distributed with high spatial variation (Biggar and Nielsen, 1976). Solute velocities within a field can vary over an order of magnitude or more. Solute adsorption and transport parameters are also highly water content- and scale-dependent.

Previous studies have recognized relationships between solute parameters and basic soil properties, including some studies that developed empirical relations linking the geometry of soil aggregates to solute retention and movement (Passioura and Rose, 1971). However, most of the earlier works employed repacked columns of disturbed porous media (Passioura and Rose, 1971; Xu and Eckstein, 1997). A limited number of experiments on undisturbed columns have produced correlations between soil structure and solute parameters. Anderson and Bouma (1977) observed greater Cl^- dispersion in undisturbed soil samples with subangular blocky structure as compared to those with prismatic structure. Walker and Trudgill (1983) reported significant correlations between solute transport parameters and pore-geometry variables measured by image analysis of soil thin sections.

Several authors have investigated the relationship between pore structures revealed by dye-staining patterns and solute transport parameters. Seyfried and Rao (1987) and Vervoort et al. (1999) reported increasing solute dispersivity with decreasing percentage of dyed area. The percentage of dyed area can be thought of as a flow weighted measure of pore size and connectivity, with small values corresponding to high macropore connectivity and vice versa. Hatano et al. (1992) investigated the relationship between solute dispersion and the fractal geometry of dye-staining patterns. Booltink et al. (1993) predicted bypass flow in undisturbed soil columns from the fractal dimensions of the upper and lower parts of the column and the volume fraction of stained parts. Although these studies have shown good correlations between soil structure as measured by image analysis, their predictive functions are not efficient PTFs as it takes similar, or much more effort, to measure such staining parameters with current technology.

Recently, the prediction of solute adsorption and transport parameters from basic soil properties has gained more attention. We will review existing PTFs for predicting these parameters for both saturated and unsaturated porous media. Solute transformation, degradation, and mineralization are also important processes in soil. However, PTFs for predicting model parameters for these processes (Rasiah, 1995) are beyond the scope of this review.

The remainder of this chapter is organized as follows. First, PTFs for predicting solute adsorption will be presented (Section 2). Section 3 deals with PTFs for solute diffusion. Then PTFs for solute dispersion will be discussed (Section 4). This section includes parameters for the convective dispersion equation (CDE), the mobile–immobile model (MIM), as well as other physico-empirical models. Approaches to upscaling the predictions of solute PTFs are presented in Section 5. In the final section (Section 6), the state-of-the-science is summarized and some directions for future research are identified.

2. SOLUTE ADSORPTION

Most solutes undergo some form of chemical reaction during transport in the soil. Adsorption is an important process as it influences the movement of reactive solutes. This topic has become increasingly important as concerns have been raised about increased deposition of heavy metals and pesticides from the use of fertilizers and animal manures. Modeling the distribution and accumulation of solutes requires inputs of solute adsorption parameters, which can be accomplished by the use of PTFs. Examples of this approach include studies by Tiktak et al. (1999) and Kozak and Vacek (2000). Tiktak et al. (1999) used PTFs to assess the magnitude of subsurface accumulation of cadmium in the Netherlands, while Kozak and Vacek (2000) used them to map the spatial distribution of the herbicide atrazine across the Czech Republic. Three main groups of adsorbing chemicals are considered in this section: pesticides, heavy metals, and nutrients.

The adsorption process is usually summarized in the form of an adsorption isotherm. Measurements of adsorption isotherms are performed by equilibrating soil with solutions containing various concentrations of solutes. The simplest adsorption behavior observed can be described with a linear model:

$$S = K_d C \tag{1}$$

where S is the amount of solute adsorbed by soil [MM^{-1}], C is the concentration of solute in soil solution [ML^{-3}], and K_d is the distribution or adsorption coefficient [L^{3}M^{-1}][1]. Because of its simplicity, linear adsorption is usually used in models of solute movement in soils. The retardation coefficient (R) is a useful parameter describing the relative movement of solute to water defined by:

$$R = 1 + \frac{\rho}{\theta}\frac{\partial S}{\partial C} \qquad (2)$$

where ρ is the soil bulk density [ML^{-3}] and θ is the volumetric water content [L^{3}L^{-3}]. For linear adsorption R is given by:

$$R = 1 + \frac{\rho K_d}{\theta} \qquad (3)$$

The term $\rho K_d/\theta$ that represents the ratio of solute adsorbed to solute in solution, is also called the delay, a measure of the distance by which the solute lags behind water when solute moves one unit of length (Rowell, 1994). This relationship indicates that under partially saturated conditions, the R for a given solute increases as θ decreases. Another method for measuring the solute adsorption coefficient is by assuming linear adsorption and estimating the retardation factor from solute transport experiments by using an inverse solution, available in computer programs such as CXTFIT (Toride et al., 1995). The retardation factor is then treated as a fitting parameter along with other solute transport parameters (see Section 4).

When R is estimated using an inverse solution, its value in some cases may become less than unity, indicating that only a fraction of the liquid phase participates in the transport process (van Genuchten, 1981). For example, negatively charged ions are excluded from regions close to the surfaces of negatively charged clay minerals (Wagenet, 1983), e.g., chloride movement in fine-textured soil (Gonçalves et al., 2001). This phenomenon is referred to as anion exclusion.

When relatively small concentrations of solute are present in soil, the linear model usually fits the data sufficiently well. With increased concentration of solute C, the adsorption sites of the soil become saturated, and the incremental amount of solute adsorbed decreases. A more general adsorption curve called the Freundlich isotherm must then be used:

$$S = K_f C^m \qquad (4)$$

where K_f [L^{3}M^{-m}] is the distribution coefficient, which is related to the maximum adsorption, and m (typically has values <1) controls the shape of the adsorption curve that is related to the adsorption energy. Adsorption isotherm models based on different functions (with different underlying assumptions) are also available, as discussed by Hinz (2001). Adsorption of various elements by soil has been reviewed by Travis and Etneir (1981).

Buchter et al. (1989) studied the adsorption of 15 elements by 11 soil types. They found that the main chemical property controlling solute adsorption in soil, i.e., the K_f and m parameters in equation (4), is pH. The influence of soil pH on adsorption of heavy metals

---

[1] M stands for unit of mass, L for unit of length and T for unit of time.

was investigated by Harter (1983). Cation exchange capacity (CEC) also influences the K_f parameter for cation species, and the amount of amorphous iron oxides, aluminum oxides, and amorphous materials influence both cation and anion adsorption parameters. Buchter et al. (1989) found significant relationships between soil properties and adsorption parameters, and suggested developing predictive functions in the absence of experimental adsorption data.

PTFs can be used to predict adsorption parameters from information collected on other soil physico-chemical properties. For example, Shaw et al. (2000) found a strong correlation between clay content and the retardation factor for chloride:

$$R = 1.23 + (0.001)\text{Clay}^2 \tag{5}$$

Gonçalves et al. (2001) also developed a PTF for chloride retardation based on multiple linear regression with the van Genuchten water retention parameters as predictors. Unfortunately, these PTFs may not be very useful for general applications since they were derived using a narrow range of specific soil types.

The linear model is usually considered most appropriate for pesticide adsorption. Because organic matter is the main adsorbing component in soil, K_d values for pesticides have been shown to correlate well with soil organic matter content (Brouwer et al., 1990; Coquet and Barriuso, 2002). Adsorption coefficients for pesticides are usually expressed in terms of organic carbon content (K_{OC}). The K_{OC} for a particular pesticide can be calculated directly from its adsorption coefficient and the fraction of soil organic carbon content (OC, expressed in g C per 100 g soil):

$$K_{OC} = K_d/\text{OC} \tag{6}$$

The units of K_{OC} are normally expressed in $\text{cm}^3 \text{ g}^{-1}$, i.e.,

$$K_{OC} = \frac{\text{g adsorbed pesticide g}^{-1} \text{ organic carbon}}{\text{g pesticide in cm}^{-3} \text{ solution}} \tag{7}$$

K_{OC} values are used as a relative measure of the potential mobility of pesticides in soil. Values of K_{OC} for various pesticides have been published (Hamaker and Thompson, 1972; Rao and Davidson, 1980). Using the published K_{OC} value for a particular pesticide, the K_d value for other soils with known organic carbon contents can be calculated. This approach assumes that organic carbon is the main adsorbing component and that the adsorption behavior for one soil will be similar to others. This assumption is only approximately true, and K_{OC} values usually have a coefficient of variation between 40 and 60%. This topic is well reviewed by Wauchope et al. (2002). Many authors find that other soil properties (e.g., clay content) also play an important role in predicting the adsorption coefficient, and that the adsorption is nonlinear.

Kozak et al. (1992) provided predictions of the Freundlich parameter K_f for various pesticides using multiple regression analysis involving clay content, CEC, pH, and organic carbon content. Baskaran et al. (1996) predicted K_f for adsorption of atrazine from pH, soil organic matter content, clay content, and exchangeable cations for various soil types in New Zealand. Ebato et al. (2001) used a similar approach for soils in Japan.

Adsorption of heavy metals is now taken very seriously as a consequence of increasing applications of fertilizers, and agricultural and industrial wastes to soils. The soil properties most closely related to adsorption of heavy metals are organic matter content, pH, CEC, and clay content. Adsorption of arsenic (Livesey and Huang, 1981; Schug et al., 1999), cobalt, nickel, lead (Buchter et al., 1989), copper, zinc (Elzinga et al., 1997), and chromium (Stewart et al., 2003) has been investigated. Cadmium appears to be the most studied heavy metal for prediction of its sorption characteristics (Springob and Böttcher, 1998a,b; Schug et al., 1999; Tiktak et al., 1999; Springob et al., 2001; Holm et al., 2003).

Several PTFs have been developed to predict the Freundlich adsorption isotherm for heavy metals. These were derived using the log-transformed Freundlich equation, with predictor variables replacing the K_f and m parameters, i.e.,

$$\log(S) = [a_0 + a_1 v_1 + a_2 v_2 + ...] + [b_0 + b_1 v_1 + b_2 v_2 + ...]\log(C) \tag{8}$$

where a_i, b_i are parameters and v are the predictor variables. The soil properties most related to adsorption of heavy metals are: organic matter content, pH, CEC and clay content. Assuming constant m for a particular heavy metal, Elzinga et al. (1997) provided PTFs calibrated over a wide range of soil materials to predict Cd, Cu and Zn adsorption taking into account variation in pH, CEC, and other soil properties, i.e.,:

$$\log(S) = m \log(C) + a_0 + a_1 \text{ pH} + a_2 \log(\text{CEC}) + ... \tag{9}$$

Their PTFs are given in Table 1. These authors also attempted to use activity of metal in the solution, to correct the total concentration for ionic strength, but did not find any significant influence on the prediction.

Anderson and Christensen (1988) found that pH was the most influential factor in controlling the distribution of Cd in soils. Christensen (1989) correlated K_f values determined on 63 soil samples with soil properties such as pH, texture, CEC, humus content, extractable iron, manganese, and aluminum oxyhydroxides. Equilibrium pH was the dominant parameter according to stepwise regression analysis, accounting for 72% of the total variation in K_f. Sauvé et al. (2000) compiled data from more than 70 studies and found that approximately 50% of the overall variation in the K_f values for Cd could be explained by variations in soil solution pH. Including soil organic matter as additional predictor significantly improved the regression ($R^2 = 0.61$). Holm et al. (2003) experimentally eliminated the effect of pH in an attempt to identify key soil properties that control Cd sorption in soil. They found that organic carbon content and CEC were the only significant predictors; clay mineralogy did not appear to be an important factor in explaining variation in the Cd distribution coefficient.

Adsorption of nutrients (e.g., phosphate and potassium) is relevant to both agriculture and the environment. Nutrient leaching is detrimental to plant nutrition and leads to the contamination of surface and ground waters. Borggaard et al. (2004) derived a PTF to predict maximum phosphorous adsorption capacity from oxalate-extractable aluminum and iron and dithionite-extractable iron. They calibrated their model with non-calcareous soil samples from Denmark, Ghana and Tanzania. Studies of phosphorus adsorption using the standard Freundlich equation have shown a strong correlation between K_f and m and other soil properties (Freese et al., 1992).

Table 1
PTFs to predict adsorption of heavy metals by soil (Elzinga et al., 1997)

Metal	PTF equation	R^2	n
Cd	$\ln(S) = -2.01 + 0.692 \ln(C) + 0.627\ln(CEC) + 0.398pH$	0.59	1444
	$\ln(S) = -3.22 + 0.87 \ln(C) + 0.629 \ln(CEC) + 0.445pH - 0.471 \ln(Ca)$	0.78	1125
Cu	$\ln(S) = 0.228 + 0.65 \ln(C) + 0.6 \ln(CEC) + 0.163pH$	0.56	424
	$\ln(S) = 0.09 + 0.742 \ln(C) + 0.682 \ln(CEC) + 0.246pH + 0.32 \ln(OC) - 0.545 \ln(Clay)$	0.62	322
	$\ln(S) = -0.775 + 0.567 \ln(C) + 0.445 \ln(CEC) + 0.225pH - 0.625 \ln(SV)$	0.69	408
Zn	$\ln(S) = -1.07 + 0.697 \ln(C) + 0.675 \ln(CEC) + 0.282pH$	0.80	478
	$\ln(S) = -1.66 + 0.714 \ln(C) + 1.25 \ln(CEC) + 0.258pH + 0.117 \ln(OC) - 0.398 \ln(Clay)$	0.85	277

S is the adsorbed metal (mg kg^{-1}); C is the total concentration of the metal in solution (mg L^{-1}); Ca is concentration of calcium in solution (mol L^{-1}), SV is soil solution ratio (kg soil per kg solution); CEC in mmol$_c$ kg^{-1}, OC is mass of C (g per 100 g soil), Clay is the mass percentage of clay, n is the number of data points and ln refers to natural logarithm.

A modification of the Freundlich isotherm that takes into account the initial amount of solute present is often necessary for describing adsorption of nutrients in soil (Fitter and Sutton, 1975):

$$S = S_0 + K_f C^m \tag{10}$$

where S_0 [MM^{-1}] is the amount of solute initially adsorbed onto particle surfaces. Examples of PTFs developed using this model can be found in Scheinost and Schwertmann (1995) and Scheinost et al. (1997). Scheinost and Schwertmann (1995) working with phosphate found that the parameters in equation (10) could not be directly correlated to any basic soil properties, because they were highly correlated (implying a non-unique solution). Therefore, they proposed the following approach: set-up the expected relationship between the parameters of the model and soil properties, and then insert this relationship into the model and estimate the parameters of the relationship by fitting the extended model using nonlinear regression. Their prediction formula takes the form:

$$S = [a_0 + a_1 v_1 + \ldots](C^m - C_{eq}^m) \tag{11}$$

where a_i are parameters, v_i are the predictor variables, and C_{eq} is the solute concentration when $S = 0$. Using a similar approach, Scheinost et al. (1997) developed PTFs to describe the adsorption of potassium in soils.

3. DIFFUSIVE SOLUTE TRANSPORT

If a solute is moving in soil by diffusion alone, its transport can be expressed by Fick's first law:

$$J_D = -D_{eff} \frac{dC}{dx} \qquad (12)$$

where J_D is the flux density [$ML^{-2}T^{-1}$], D_{eff} is the effective diffusion coefficient of soil [L^2T^{-1}], C is the solute concentration [ML^{-3}], and x is the distance [L]. The effective diffusion coefficient summarizes many factors affecting solute diffusion in soil, and is highly dependent on water content. It is usually related to ionic diffusion in a pure water system, D_o, by the solute diffusivity (= ratio of the effective diffusion coefficient in soil to that for the solute in pure water), D_{eff}/D_o. Many studies also use the diffusion coefficient of the porous medium, D_p, which is related to D_{eff} by

$$D_p = D_{eff}\theta \qquad (13)$$

Further details on the theory of solute diffusion in porous media are given by Tinker and Nye (2000) and Flury and Gimmi (2002).

Determination of solute diffusivity is usually conducted with the so-called "half cell method." This method involves placing the soil sample in two cells: one source cell containing the solute and one recipient cell. The cells are incubated and then the soil cores are cut into slices and the spatial distribution of the solute is measured. This method of measurement is discussed by Flury and Gimmi (2002).

Olsen and Kemper (1968) and Bresler (1973) defined the following empirical relationship between D_p/D_o and water content,

$$\frac{D_p}{D_o} = p \exp(q\theta) \qquad (14)$$

where p and q are empirical constants with $q \approx 10$ and $0.005 < p < 0.001$. Concepts relating the diffusion coefficient to θ (and porosity, ϕ) that were developed to predict permeability and gas diffusivity, can be generalized and used to predict solute diffusion, i.e.,

Buckingam (1904): $D_p/D_o = \theta^2$
Penman (1940): $D_p/D_o = 0.66\theta$
Marshall (1959): $D_p/D_o = \theta^{1.5}$ \qquad (15)
Millington and Quirk (1961): $D_p/D_o = \theta^{10/3}/\phi^2$
Papendick and Campbell (1980): $D_p/D_o = 2.8\theta^3$
Sadeghi et al. (1989): $D_p/D_o = 0.18(\theta/\phi)^{2.98}$

These models predict solute or gas diffusivity well only in specific cases. For example, the often used Millington and Quirk (1961) model adequately predicts solute and gas diffusivity only for very coarse-textured soils (Moldrup et al., 2001, 2003). Therefore, Moldrup et al. (1997) suggested a general power function model called PMQ, which is a combination of the Penman (1940) and Millington and Quirk (1961) models:

$$\frac{D_p}{D_o} = 0.66\left(\frac{\theta}{\phi}\right)^{\frac{12-t}{3}} \tag{16}$$

where t is a tortuosity parameter. Fitting the model to solute diffusivity data for 20 soil types with a range of different textures (102 measurement points) yielded $t = 1$ (high tortuosity). Equation (16) can be used to predict solute diffusivity when only the soil–water content and total porosity are known. If further information (e.g., soil texture or the water retention curve) is available, soil-type dependent models (below) should be used instead (Moldrup et al., 2001; Moldrup, personal communication).

Transforming D_p into a dimensionless impedance factor (f_l) has been suggested as an alternative approach to predict solute diffusivity (Olesen et al., 1996, 1999). The impedance factor is defined as:

$$f_l = \frac{D_p/D_o}{\theta} \tag{17}$$

The impedance factor takes into account the liquid phase tortuosity, and decreases with increasing water content as the diffusion pathway becomes less tortuous. A linear relationship between the impedance factor and volumetric water content has been observed (Porter et al., 1960; Olesen et al., 1999). This relationship can be modeled as:

$$f_l = A + B\theta \tag{18}$$

Equation (18) can generate two parameters: the threshold water content, θ_{th}, and the impedance factor at saturation, f_s. The threshold water content corresponds to the soil water content where diffusion ceases due to a discontinuous diffusion pathway; it can be computed by extrapolating the linear model to the intercept of θ when $f_l = 0$. The impedance factor at saturation is found when θ is equal to the total porosity ϕ.

Olesen et al. (1996) suggested the following model to predict solute diffusivity based on the linear relationship between f_l and θ:

$$\begin{aligned}\frac{D_p}{D_o} &= 0.45\theta\left(\frac{\theta - \theta_{th}}{\phi - \theta_{th}}\right) &&\text{for } \theta > \theta_{th} \\ \frac{D_p}{D_o} &= 0 &&\text{for } \theta \leq \theta_{th}\end{aligned} \tag{19}$$

Olesen et al. (2001) simplified equation (19) based on their observation that the slope of the linear model does not vary much and can be set equal to 1.1. This approximation is known as the constant slope impedance factor, CSIF model:

$$\begin{aligned}f_l &= 1.1(\theta - \theta_{th}) &&\text{for } \theta > \theta_{th} \\ f_l &= 0 &&\text{for } \theta \leq \theta_{th}\end{aligned} \tag{20}$$

Based on the above equations, the diffusivity model becomes:

$$\frac{D_p}{D_o} = 1.1\theta(\theta - \theta_{th}) \text{ for } \theta > \theta_{th}$$
$$\frac{D_p}{D_o} = 0 \qquad \text{for } \theta \leq \theta_{th} \qquad (21)$$

Olesen et al. (2001) proposed the following relationship between θ_{th} and the slope of the soil water retention curve, b, in Campbell's model (Campbell, 1974):

$$\theta_{th} = 0.02b \qquad (22)$$

These authors also derived a PTF that predicts θ_{th} from basic soil information:

$$\theta_{th} = 0.81 P_{<2} - 0.90(P_{<2})^2 - 0.07(P_{2-20}) - 0.6\rho + 0.22(\rho)^2 + 0.42 \qquad (23)$$

where $P_{<2}$ is the mass percentage of particles <2 μm (clay), P_{2-20} is the mass percentage of particles between 2 and 20 μm (silt), and ρ is soil bulk density (in g cm^{-3}). The threshold water content can also be correlated with soil specific surface area expressed on a volumetric basis (Moldrup et al., 2001).

Although equation (21) is conceptually attractive for describing solute diffusion, the uncertainty in predicting the threshold water content from soil texture and/or soil water retention data limits its use as a predictive model. Moldrup et al. (2003) proposed a unified diffusivity model (UDM), which attempts to predict solute and gas diffusivity as a function of soil water content based on total porosity and pore-size distribution. Their model is given by:

$$\frac{D_p}{D_o} = \theta^T \left(\frac{\theta}{\phi}\right)^W \qquad (24)$$

where T is a tortuosity factor and W is a disconnectivity parameter. The term $(\theta/\phi)^W$ can be interpreted as the progressive loss of water film connectivity with decreasing water content, which acts to reduce solute diffusivity. Thus, liquid phase (solute) disconnectivity will decrease with increasing soil water content. For gas diffusivity, θ in equation (24) is replaced with the air-filled porosity, ε. For solute diffusion, the tortuosity factor is based on Buckingham's (1904) model ($T = 2$), with W given by $(b/3) - 1$ (Moldrup et al., 2003), giving:

$$\frac{D_p}{D_o} = \theta^2 \left(\frac{\theta}{\phi}\right)^{(b/3)-1} \qquad (25)$$

Based on data for 22 soil types, Moldrup et al. (2003) found that the UDM for solute diffusivity, equation (25) generally gives good predictions of D_p/D_o for sandy and clayey soils with a mean uncertainty of 0.03. Equation (25) needs the b parameter from the water retention curve. If water retention data are not available, this parameter can be estimated from PTFs that predict the parameters of Campbell's equation (Cosby et al., 1984); see Section 2 of this book for further discussion.

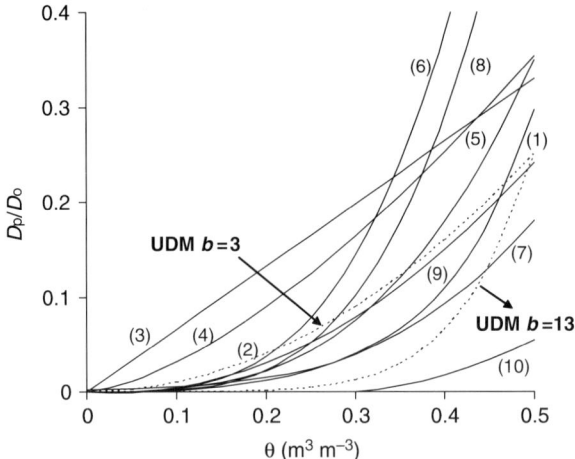

Figure 1. Graphical comparison of different solute diffusion models: (1) Olsen-Kemper with $p = 0.02$, $q = 10$, (2) Buckingham, (3) Penman, (4) Marshall, (5) Papendick–Campbell, (6) Millington–Quirk, (7) Sadeghi et al., (8) PMQ, (9) CSIF with $b = 3$, (10) CSIF with $b = 15$. Calculations were made assuming a total porosity 0.5 cm^3 cm^{-3}.

The predictions of various solute diffusion models are summarized in Figure 1, assuming soil with a porosity of 0.5 m^3 m^{-3}. As can be seen, the Penman and Marshall models (equation (15)) generally give higher predictions for the whole range of water content, while the Millington–Quirk, Papendick–Campbell and PMQ (equation (16)) give higher values with increasing moisture content. The CSIF model (equations (21 and 22)) gives low values of prediction for high b values (fine-textured soil). The predictions of the UDM (equation (25)) capture the range of diffusivities predicted by the other models through the variation of b values. The UDM with $b = 3$ (sandy soil) reduces to Buckingham's model.

It is noted that it is unnecessary to distinguish between solute diffusivity models for repacked and undisturbed soils (Moldrup, personal communication). Studies so far have not revealed any significant differences between solute diffusion measurements on repacked and undisturbed samples, as opposed to gas diffusivity where significant differences have been observed.

4. CONVECTIVE-DISPERSIVE SOLUTE TRANSPORT

4.1. Convection dispersion equation (CDE)

The one-dimensional flux of solute in the soil is a combination of diffusive and convective processes (Wagenet, 1983):

$$J_S = J_D + J_C \qquad (26)$$

where J is the mass of solute transported through a cross-sectional area over time

[ML^{-2}T^{-1}], and the subscripts S, D, and C refer to total solute, solute transported by diffusion, and solute transported by convection, respectively. The diffusive flux, J_D, is given by equation (12), while the convective flux, J_C, is given by:

$$J_C = -\theta D_m \frac{dC}{dx} + v\theta C \tag{27}$$

where D_m is the dispersion coefficient, which differs from that for molecular diffusion because water movement induces mechanical dispersion of solute molecules, and v is the average pore water velocity [LT^{-1}].

Combining equations (12), (26), and (27) gives:

$$J_S = -(\theta D_m + D_p)\frac{dC}{dx} + v\theta C \tag{28}$$

Combining equation (28) with the continuity equation (Wagenet, 1983), we can derive a general equation for solute transport in soil, i.e.,

$$\frac{\partial}{\partial t}(S + \theta C) = \frac{\partial}{\partial z}\left(\theta D \frac{\partial C}{\partial x}\right) - \frac{\partial(qC)}{\partial x} + \varphi \tag{29}$$

where S is the concentration of solute adsorbed by soil particles [ML^{-3}], D is the lumped apparent diffusion–dispersion coefficient [L^{2}T^{-1}], q is the Darcy flux $= v\theta$ [LT^{-1}], and φ is a solute source or sink term. D is a combination of mechanical dispersion and diffusion and its value depends on both θ and q.

Most laboratory studies are conducted under conditions of constant q and θ. For these conditions, and assuming linear adsorption or exchange reactions, i.e., equation (1), equation (29) becomes:

$$R\frac{\partial C}{\partial t} = D\frac{\partial^2 C}{\partial x^2} - v\frac{\partial C}{\partial x} \tag{30}$$

where R is the retardation coefficient (see Section 2). Equation (30) is usually referred to as the convective or advective dispersion equation (CDE or ADE). The theory of solute transport and assumptions behind this model are discussed by Skaggs and Leij (2002). Experimental methods for determining the parameters in equation (30) have been reviewed by Skaggs et al. (2002).

The critical parameters governing solute transport in the CDE are the pore water velocity, v, and the apparent diffusion–dispersion coefficient, D. Pore-water velocity is mainly controlled by external conditions (i.e., the pumping rate in laboratory experiments and the rainfall or infiltration rate under field conditions). If solute transport occurs under steady-state saturated flow conditions driven by gravity drainage (i.e., a unit hydraulic gradient in Darcy's law), v can be estimated by K_s/ϕ, where K_s is the saturated hydraulic conductivity. In the absence of additional information, this approach can be used to provide a simplistic approximation of the maximum v for a soil. However, in most field situations, partially saturated soil or transient conditions may hinder the use of such a simple steady-state approach.

In many laboratory studies, a linear relationship has been observed between D and v (Pfannkuch, 1962; Bresler, 1973; Shukla et al., 2003):

$$D = D_p + \lambda v \tag{31}$$

where D_p accounts for molecular diffusion, and λ relates mainly to the mixing of the solute by convection. Biggar and Nielsen (1976) measured field scale solute transport to ascertain the distribution of D and v at 8 depths to 1.8 m within 20 plots located in a 150 ha area of alluvial soil in California. They found that v and D were log-normally distributed, and that D (in cm^2 per day) was related to v by the following empirical expression:

$$D = 0.6 + 2.93 v^{1.11} \tag{32}$$

A general form of equations (31) and (32) can be written as:

$$D = D_o + \Lambda v^n \tag{33}$$

where Λ [L^{2-n}T^{n-1}] and n are constants. Assuming $n \approx 1$, and that the convective flux is much greater than the diffusive flux (i.e., $v \gg 0$), D_o can be neglected, and equations (31) and (33) are equivalent. Under these conditions, we can define the dispersivity, $\Lambda = \lambda$ [L] as:

$$\lambda = \frac{D}{v} \tag{34}$$

Dispersivity is a required input parameter in ground water contaminant transport models based on the CDE (Zheng and Bennett, 1995). In contrast, models for solute transport in the vadose zone such as SWAP (Van Dam et al., 1997), SWIM (Verburg et al., 1996), and Hydrus (Šimunek et al., 1999) make use of equations (31) and (33).

To date all attempts to predict parameters of the CDE from basic soil properties have focused on saturated conditions, and have employed the dispersivity concept. Xu and Eckstein (1997), working with mixtures of glass beads, showed that λ was positively correlated ($R = 0.84$) with the uniformity coefficient describing the mass-size distribution of glass beads. A limitation of this study was the choice of characteristics describing the solid phase as independent variables. Furthermore, packed beds of glass beads do not simulate the heterogeneity of natural porous media. Under saturated conditions, the magnitude of solute dispersion at any given flow rate is controlled by the pore-space geometry (Perfect and Sukop, 2001). Since it is the geometrical characteristics of the solids rather than the pore space that are measured in the packed bed approach, any resulting relationships developed for repacked or disturbed porous media are not directly applicable to undisturbed soil.

Pore-space geometry also determines the retention and transport of water in soil. Since hydraulic properties such as the saturated hydraulic conductivity and parameters describing the water retention curve are measured more frequently than pore characteristics, it seems logical to seek empirical relations between such properties. Perfect et al. (2002) analyzed 69 undisturbed soil columns from six soil types, ranging in texture from loamy sand to silty clay, under conventional-till and no-till management practices from Kentucky. Using Campbell's (1974) model for the water retention curve (ψ_b = air entry potential in cm and b = slope of the water retention curve), they developed the following PTF to predict dispersivity:

$$\lambda(\text{cm}) = -2.91 + 0.023\psi_b + 1.27b, \qquad R^2 = 0.47 \tag{35}$$

Equation (35) indicates that λ increases as ψ_b and b both increase, suggesting that fine-textured soils are more dispersive than coarse-textured ones.

Most solute transport experiments have been conducted using fully saturated materials. Under partially saturated conditions, D is theoretically related to both θ and v. However, the exact nature of this relationship is not well established, and its elucidation is an evolving area of research (Padilla et al., 1999; Toride et al., 2003). Very little experimental data are available showing relations between v, D and θ from dry to saturated conditions. Some researchers (De Smedt and Wierenga, 1984; Matsubayashi et al., 1997; Haga et al., 1999) have shown that the slope of the relationship between D and v increases with decreasing water content, implying that λ depends upon the degree of saturation. In contrast, the results of Yule and Gardner (1978) suggest that λ is independent of water content, and is solely a property of matrix geometry. If one assumes this to be the case it can easily be shown that for any given Darcy flux (q), the unsaturated diffusion–dispersion coefficient (D_u) is related to the saturated diffusion–dispersion coefficient (D_s) by:

$$D_u = D_s/S \tag{36}$$

where $S = \theta/\phi$ is the relative saturation.

Equation (36) indicates that the diffusion–dispersion coefficient for a partially saturated porous medium should increase as S decreases. A test of this relationship is shown in Figure 2 for steady-state unsaturated flow at a mean Darcy flux of 0.85 cm h^{-1} through an undisturbed block of Maury silt loam soil. The linear dependence of D_u on $1/S$ lends support to equation (36) and explains over 50% of the total variation. From the regression equation $D_s = 2.09$ cm^2 h^{-1}, as calculated from the sum of the slope and the intercept. Clearly, future attempts to develop PTFs for unsaturated solute transport parameters must take into account the functional dependencies of D, v, and even R, on water content.

4.2. Mobile–immobile model (MIM)

Studies of solute transport have shown that the asymmetry or tailing of breakthrough curves (early breakthrough) and also the deeper penetration of surface-applied chemicals in the field is not always as predicted by the CDE (Gaudet et al., 1977). These studies suggest that not all of the soil water appears to be actively and equally involved in solute transport. To account for this phenomenon, the soil water can be divided into two regions: a mobile fraction, θ_m, which is active in convective transport, and an immobile fraction, θ_{im}, the stagnant water, which is not involved in convective transport (Coats and Smith, 1964; van Genuchten and Wierenga, 1976):

$$\theta = \theta_m + \theta_{im} \tag{37}$$

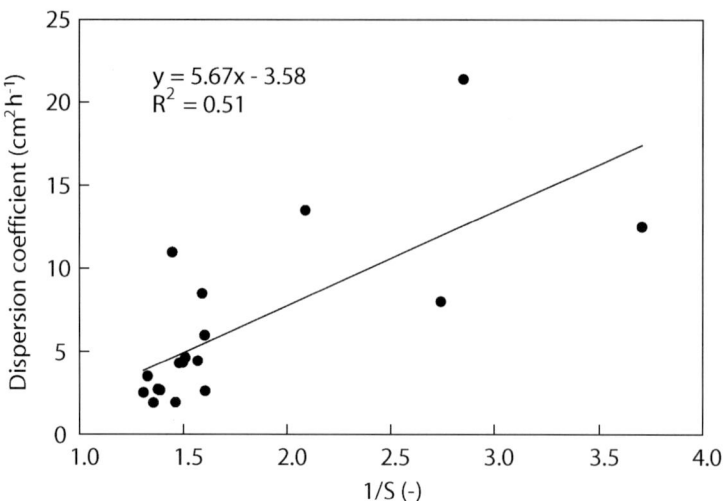

Figure 2. Relationship between diffusion–dispersion coefficient and relative saturation S. Data from Bejat et al. (2002).

where the subscript "m" refers to the mobile fraction, and "im" the immobile fraction. It follows that total solute concentration in the soil C^* is given by:

$$\theta C^* = \theta_m C_m + \theta_{im} C_{im} \quad (38)$$

This approach results in the MIM or two-region model, which under steady-state flow can be written as:

$$\theta_m \frac{\partial C_m}{\partial t} + \theta_{im} \frac{\partial C_{im}}{\partial t} = \theta_m D_m \frac{\partial^2 C_m}{\partial x^2} - \theta_m v_m \frac{\partial C_m}{\partial x} \quad (39)$$

with solute exchange between the two regions modeled as a first-order process, i.e.,

$$\theta_{im} \frac{\partial C_{im}}{\partial t} = \alpha^* (C_m - C_{im}) \quad (40)$$

where α^* [T^{-1}] is a mass transfer coefficient between the mobile and immobile regions.

Assuming linear adsorption, van Genuchten and Wierenga (1976) introduced the following dimensionless variables to facilitate the analytical solution of these partial differential equations:

$$T = \frac{vt}{L}; X = \frac{x}{L}; P = \frac{vL}{D}; \beta = \left(\frac{\theta_m + f\rho K}{\theta + \rho K} \right); \omega = \frac{\alpha^* L}{q} \quad (41)$$

where P = Peclet number, T = pore volume number, X = relative distance, β is a coefficient for partitioning between the mobile and immobile regions, L is the length of the column, f is the fraction of adsorption sites that equilibrate with the mobile liquid phase, and ω is a mass transfer coefficient governing the rate of solute exchange between the two liquid regions.

Attempts have recently been made to estimate D_m in equation (39) from basic soil properties (Shaw et al., 2000; Gonçalves et al., 2001). Shaw et al. (2000) analyzed 37 soil columns from argillic and kandic horizons in the upper coastal plain of Georgia, with clay contents ranging from 12 to 28% clay. They developed the following PTF to predict D_m from size of the structural units (size in mm) and clay content (clay, %):

$$D_m (\text{cm}^2 \text{ min}^{-1}) = 2.07 + 0.643(\text{size}^{1/2}) - 0.009 \text{ Clay}^2; R^2 = 0.30 \qquad (42)$$

Gonçalves et al. (2001) analyzed 24 fine-textured soils in an irrigation district in southern Portugal. They also used the MIM to parameterize their solute breakthrough curves, and PTFs were developed with linear regression and neural networks to predict the dispersion coefficient (D_m in cm^2 day^{-1}) from basic soil properties and parameters of the van Genuchten equation:

$$\log D_m = 2.524 - 0.389 \text{ OM} + 1.005 \log(K_s) - 2.005 \theta_r + 0.895 \log(\alpha); R^2 = 0.90 \qquad (43)$$

where OM is the organic matter content (% by weight), K_s the saturated conductivity (cm day^{-1}), θ_r (cm^3 cm^{-3}) and α (cm^{-1}) are parameters of the van Genuchten equation.

In our opinion, PTFs predicting D_m are not very helpful, because D_m depends upon the mean pore water velocity v, which is mainly controlled by experimental conditions. Most of the variability in D_m can be accounted for by differences in v (Table 2 in Gonçalves et al., 2001). As noted previously, the dispersivity is much more useful than D_m for characterizing the influence of pore-space geometry on solute mixing. It is possible to calculate an effective dispersivity (λ_{eff}) for the MIM model using the following equation (Vallochi, 1985):

$$\lambda_{\text{eff}} = \lambda_m \beta + \frac{v_m(1-\beta)^2}{\alpha^*} \qquad (44)$$

Shaw et al. (2000) found that λ_{eff} was higher in clayey than in sandy horizons.

We have re-analyzed the data for D_m in Tables 1 and 3 of Gonçalves et al. (2001) taking into account variations in v through the use of λ_m. Multiple regression analysis was used to explore relations between λ_m (in cm) and various predictor variables. This procedure resulted in the following PTF:

$$\lambda_m = 2.46 - 0.744 \ell - 0.04(1/\alpha); R^2 = 0.41, \text{RMSE} = 3.38 \text{ cm}, n = 23 \qquad (45)$$

where ℓ is a unit-less pore-connectivity factor in the Mualem–van Genuchten equation. The two predictor variables in equation (45) are both hydraulic parameters, which was also the case in the study by Perfect et al. (2002). The fact that water retention parameters (albeit from different models) were chosen in both analyses lends support to the idea that both dispersion and water retention are controlled by pore-size distribution and that one can be predicted from the other.

There are two main approaches in the literature for obtaining the other parameters in equation (39). The first is the inverse approach, which involves fitting calculated flux concentrations to observed flux concentrations from solute breakthrough experiments. This approach employs inverse solution of the MIM under prescribed initial and boundary conditions. Many studies use the computer program CXTFIT (Toride et al., 1995), which facilitates simultaneous estimation of v, D, R, β and ω via inverse solution. Problems can occur, however, when the solution is non-unique (more than one set of parameters can describe an experimental breakthrough curve). Then, the solute parameters given by the inverse program may have no physical meaning. This approach does not directly measure θ_m. Furthermore, in most studies, f is unknown and assumed to be equal to θ/θ_m (Nkedi-Kizza et al., 1983).

The second approach involves direct measurement. Ilseman et al. (2002) have recently proposed a laboratory method for direct measurement of the immobile soil water content and mass exchange coefficient. Clothier et al. (1992) introduced a simple and rapid means of measuring *in situ* the effective transport volume during near-saturated flow using a disc (or tension) infiltrometer. Water is applied to the soil using the infiltrometer until reaching steady-state flux. The disk is then removed and filled with a tracer such as bromide (Br) with concentration C_0. After the solution penetrates the soil to a specified depth, samples are extracted from underneath the disc and analyzed for concentration. Assuming that the mass transfer rate is negligible, the mobile water content can be deduced from:

$$\theta_m = \theta \frac{C^*}{C_0} \tag{46}$$

This approach provides a direct *in situ* method for measurement of θ_m from saturation to the limit of the tension infiltrometer (about 100 mm of suction).

Jaynes et al. (1995) extended the Clothier et al. (1992) method to allow for simultaneous estimation of α^*. Instead of a single conservative tracer, a series of fluorinated benzoate tracers are infiltrated, each for varying periods of time. After sufficient infiltration of the last tracer, soil samples are collected and analyzed, and the tracer concentration in the sample is subsequently used to calculate θ_{im} and α^*.

Various authors have observed different behavior of the mobile water fraction parameter β. Gaudet et al. (1977) reported increasing mobile water content with increasing θ and pore-water velocity. However, studies have also found that β appears to be constant with varying pore-water velocity (van Genuchten and Wierenga, 1977; Nkedi-Kizza et al., 1983). Bajracharya and Barry (1997) found that for structured porous media, β is often related to the volume fraction of aggregates. For a random distribution of hydraulic conductivities their results indicate that the rate parameter, α^*, increases linearly with pore-water velocity. The β parameter did not vary much with the pore-water velocity, and was a characteristic of the medium that varied with the standard deviation of the log of hydraulic conductivity. Griffoen et al. (1998) analyzed a database collated from published MIM experiments with emphasis on partially saturated and saturated soil materials. No general trend was found for the mobile water fraction, β. For saturated porous media,

β was found to be either constant or gradually decreasing with increasing mobile pore-water velocity v_m, while for partially saturated soil materials, β was either constant or increased with increasing θ.

The mass transfer rate is mainly governed by the mobile phase velocity rather than molecular diffusion. A strong linear relation is often observed between $\log(\alpha^*)$ and $\log(v)$, and the following expression has been proposed (Griffoen et al., 1998):

$$\alpha^* = v_m^a \tag{47}$$

where values of a range from 0.5 to 2.

Few attempts have been made to develop PTFs to predict the θ_m, θ_{im}, β, α^* and ω parameters of the MIM from basic soil properties. Based on a limited number of field measured θ_m values from Victoria, Australia ($n = 8$), Okom (1998) suggested that:

$$\theta_m = 0.2549(\rho/1.3)^{-0.2954} + 0.002\,\text{Silt} - 0.008\,\text{Clay}; R^2 = 0.67 \tag{48}$$

$$\theta_m/\theta = (\rho/1.3)^{-0.205} - 0.0089\,\text{Clay}; R^2 = 0.65 \tag{49}$$

Shaw et al. (2000) also provided equations to predict β and α^*:

$$\beta = 0.96 - 0.020\,\text{Clay}; R^2 = 0.43 \tag{50}$$

$$\ln(\alpha^*)(\text{min}^{-1}) = 4.51 - 0.065\,\text{Clay}^2 - 0.010\,\text{CEC}_{\text{Clay}}; R^2 = 0.57 \tag{51}$$

where CEC_{Clay} is CEC per gram of clay. Gonçalves et al. (2001) presented more complicated equations to predict β and ω involving water retention (van Genuchten equation) and particle-size (log-normal distribution) parameters:

$$\beta = -5.65 + 0.002\,\text{GSD} - 0.1104\,\text{OM} + 1.41\rho_b - 0.047\text{pH} + 0.061\phi - 3.13\theta_r$$
$$- 4.05\theta_s + 4.42\alpha + 2.097n - 0.048\ell; \text{RMSE} = 0.109, R^2 = 0.80 \tag{52}$$

$$\log(\omega) = -11.37 - 0.0199\,\text{FS} - 0.2126\,\text{Clay} + 0.0324\,\text{GSD}$$
$$+ 0.7136\,\text{pH} - 0.778\,\log(K_s) - 2.239\,\log(\alpha) + 4.27n$$
$$+ 1.988\,\log(K_o); \text{RMSE} = 0.488, R^2 = 0.46 \tag{53}$$

where K_o is the fitted saturated conductivity (cm day^{-1}), FS the fine sand content (0.2–0.02 mm) (% weight), and GSD is the geometric standard deviation of the particle-size distribution. Care must be taken when using these PTFs because they were calibrated from a small number of soil samples from particular soil types.

Using data mined from column studies in the literature (Griffoen et al., 1998; Schwartz, 1998; Shaw et al., 2000; Gonçalves et al., 2001), we found no apparent correlation between β and basic soil properties. This is probably because of the variable nature of the relationship between β and θ in the different studies. However, we were able to predict the mobile water content θ_m, as a function of clay/sand and water content, reasonably well. Using data for soil with clay contents between 0 and 60% and $\theta/\theta_m > 0.40$ (i.e., excluding bypass flow conditions), and assuming $\beta = \theta_m/\theta$, resulted in:

$$\theta_m = -0.1193 + 0.92\theta + 0.00115 \text{ Sand} \qquad R^2 = 0.44, \text{RMSE} = 0.07, n = 88 \qquad (54)$$

$$\log_{10}(\alpha^*) = -1.195 + 1.201 \log_{10}(v) - 0.0245 \text{ Clay} \qquad R^2 = 0.47, n = 93 \qquad (55)$$

where Sand and Clay are the weight percents of sand and clay, respectively.

We also compiled data that measured θ_m directly using a disc permeameter in the field (Jaynes et al., 1995; Angulo-Jamarillo et al., 1996; Casey et al., 1999; Okom et al., 2000). Using these data (28 soil samples with clay contents between 10 and 40%), we obtained the following PTF:

$$\theta_m = -0.1712 + 1.3148\theta - 0.00496 \text{ Clay}^2 \qquad R^2 = 0.47, \text{RMSE} = 0.08 \qquad (56)$$

Application of the PTF from column studies, equation (54), to the field data set used to derive equation (56) resulted in an $R^2 = 0.36$ and an RMSE = 0.09. Whereas application of the PTF derived from the field data set, equation (56), to a subset of the laboratory data with clay contents <30% resulted in predictions with an $R^2 = 0.59$ and an RMSE = 0.10. The general trend of these predictions is shown in Figure 3. It can be

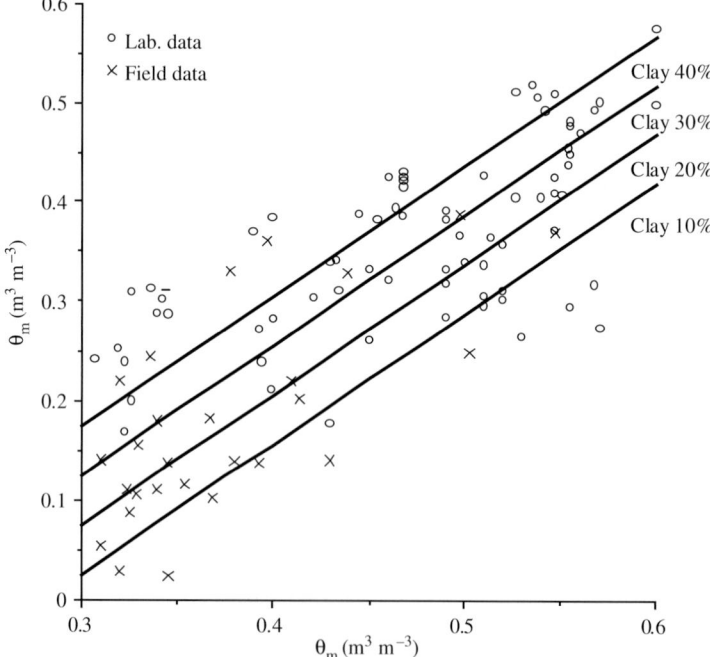

Figure 3. Mobile water content (θ_m) as a function of water content (θ) and clay content.

seen that equation (56) predicts the mobile water content reasonably well. Mobile water content increases with increasing water content and decreases with increasing clay content implying the presence of more stagnant water. These results suggest a constant value of θ_m/θ for any given clay content.

4.3. Other physico-empirical models

In addition to the CDE and MIM, several other physico-empirical solute transport models have been proposed, and are currently in use. These include the fractional CDE (Benson, 1998), mixing cell models (Frissel and Poelstra, 1967a,b; Bolt, 1979), kinematic wave models (Charbeneau, 1984; Germann et al., 1987), and transfer function models (Jury, 1982; Jury and Sposito, 1985). The key parameters of these models are normally obtained by calibration from laboratory breakthrough curves or field leaching experiments. We are unaware of any PTFs that have been developed to predict their input parameters from independent measurements.

Another class of models is that which relates solute dispersion to the distribution of pore water velocities in soil. These are quite attractive from a predictive stand point, since their parameters can be obtained from independent hydraulic properties, such as the water retention curve, $\theta(\psi)$ (Klinkenberg, 1957; Lindstrom and Boersma, 1971; Carbonell, 1979) or unsaturated hydraulic conductivity–water content function, $K(\theta)$ (Steenhuis et al., 1990; Durner and Flühler, 1996; Montas et al., 1997a). The concepts underlying these models are discussed by Perfect and Sukop (2001).

Lindstrom and Boersma (1971) considered the soil as a bundle of non-intersecting, parallel, and cylindrical capillary tubes of varying radii. Using this approach the size distribution of pore radii can be inferred from $\theta(\psi)$ and used to predict pore water velocities using Poiseuille's law. Solute dispersion is then related to the distribution of pore water velocities. Rao et al. (1976) evaluated this method against experimental data. Horton et al. (1987) predicted the initial breakthrough time of solutes moving through compacted soil materials. Scotter and Ross (1994) and Wang et al. (2002) developed simple models to predict solute breakthrough directly from $\theta(\psi)$. Steenhuis et al. (1990) related the solute breakthrough curve under steady-state saturated flow conditions to the empirical distribution of pore water velocities derived from $K(\theta)$. Montas et al. (1997b) accurately predicted 8 out of 10 solute breakthrough curves mined from the literature using independent parameters derived from $K(\theta)$.

The parameters of pore water velocity models can be obtained inversely from field or laboratory tracer tests, or independently from soil hydraulic properties. As a result, there is little motivation to develop PTFs to predict them from basic soil properties. In the absence of detailed hydrologic information, however, it is possible to estimate breakthrough from such properties by first predicting $\theta(\psi)$ or $K(\theta)$ using an established PTF (Rawls et al., 1991; Gonçalves et al., 1997), and then using these predictions as inputs to a pore water velocity model to predict solute dispersion. To our knowledge, this approach has not yet been explored.

5. UPSCALING PEDOTRANSFER FUNCTION PREDICTIONS

Many solute adsorption and transport parameters have been shown to be scale dependent. For example, recent theoretical advances in percolation theory and fractal geometry have demonstrated that the diffusion coefficient of a porous medium is not a constant when diffusion occurs in tortuous interconnected pores. The diffusing particle's trajectory is constrained by the geometry of the pore space. Within a 2D fractal porous medium, D_p becomes scale-dependent, as described by the following equation (Orbach, 1986; Crawford et al., 1993):

$$D_p(x) = D_p x^{2(d_m - d_t)/d_t} \tag{57}$$

where $D_p(x)$ is the scale-dependent D_p, d_t is the fractal dimension of the diffusing particles trajectory ($d_t = 1.5$ for classical Brownian motion), and d_m is the 2D mass fractal dimension of the porous medium within which the diffusing particle is confined. Anderson et al. (1996) have used equation (57) in conjunction with digitized images of thin sections, to investigate the influence of pore-space geometry on solute diffusion in a variety of soil types.

It is also well known that inverse estimates of dispersivity tend to increase as the volume of soil or aquifer material sampled in a solute transport experiment increases. Neuman (1990) statistically analyzed 131 longitudinal dispersivities from laboratory and field tracer studies conducted in porous media at scales ranging from <10 cm to >100 km. Regression analysis showed that although the data were widely scattered, a log–log relationship between dispersivity and apparent length scale, ℓ_a, was able to account for 74% of the total variation in λ. This relationship can be expressed as:

$$\lambda = 0.0175 \ell_a^{1.46} \tag{58}$$

Equation (58) holds for $\ell_a < 3500$ m. Similar relationships between λ and length scale have been reported by Gelhar et al. (1992) and Xu and Eckstein (1995). Several attempts have been made to explain the scale-dependency of λ theoretically based on fractals (Neuman, 1995; Kemblowski and Wen, 1993; Zhan and Wheatcraft, 1996; Hassan et al., 1997).

Rajaram (1997) investigated the scale dependency of the retardation factor in equation (30) and demonstrated its effect on solute transport. Local spatial variability in K_d values resulted in scale-dependent retardation at larger scales. Theoretical expressions were developed for calculating an effective or apparent retardation factor from knowledge of the statistical parameters of the $\ln(K_s)$ and K_d fields.

Because of the scale-dependency observed in solute adsorption and transport parameters, predictions made using PTFs are only valid for the scale at which the original measurements were made. The majority of solute PTFs were developed using soil columns at the laboratory-bench scale. To be useful, PTFs should also be able to predict the behavior of different sample volumes. Several techniques are available for upscaling PTF predictions, including fractal geometry, renormalization group theory, block averaging, and coarse graining (see Pachepsky et al. (2003) for a general introduction to these topics).

Perfect (2003) recently employed the fractal upscaling approach to test equation (35) against an independent data set derived from Schwartz (1998) and Vervoort et al. (1999). The raw predictions (applicable to 6-cm long columns) were upscaled to the dimensions of the columns used in the original studies based on the fractal model of

Neuman (1990, 1995), i.e., equation (58). It was assumed that parameters describing the water retention curve are scale-invariant.

For dispersivities up to ∼ 1 m, the upscaled predictions were positively related to the measured values ($R^2 = 0.62^{**}$), although there was a significant trend towards over estimation (Figure 4). Despite this tendency, the predictions still represent a significant

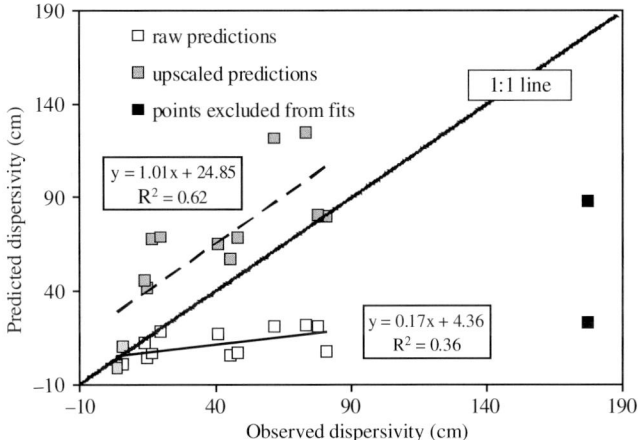

Figure 4. Raw and upscaled predicted dispersivities as related to the measured values (Perfect, 2003).

improvement over existing approaches for predicting λ. For example, longitudinal dispersivities are sometimes estimated as $1/10^{th}$ of the flow length of a transport experiment (Fetter, 1999). Applying this approach to the columns used by Schwartz (1998) and Vervoort et al. (1999) yields estimates of λ between 1.5 and 3.0 cm. Clearly, these values are much less accurate than the upscaled predictions in Figure 4. This is because equations (35) and (58) take into account differences in both soil hydraulic properties and flow length, whereas the $1/10^{th}$ rule-of-thumb is based exclusively on the scale effect.

Perfect (2003) also used the combined PTF and fractal upscaling approach, equations (35) and (58), to predict λ as a function of column length from Campbell model water retention parameters for 11 soil textural classes ranging from sand to clay. The predicted dispersivities are shown in Figure 5. There is a clear trend towards increasing dispersivity with increasing clay content, and this trend becomes more pronounced as the column length increases. The increase in λ with increasing clay content is probably a soil structural effect; equation (35) was developed using undisturbed columns, and in this condition fine-textured soils tend to have much a wider range of pore sizes than coarse-textured ones. In the absence of more detailed information for a particular soil, such an approach can be useful for estimating the dispersivity parameter required for modeling solute transport using the CDE. Further research along these lines may ultimately reduce our reliance on solute transport experiments and inverse procedures.

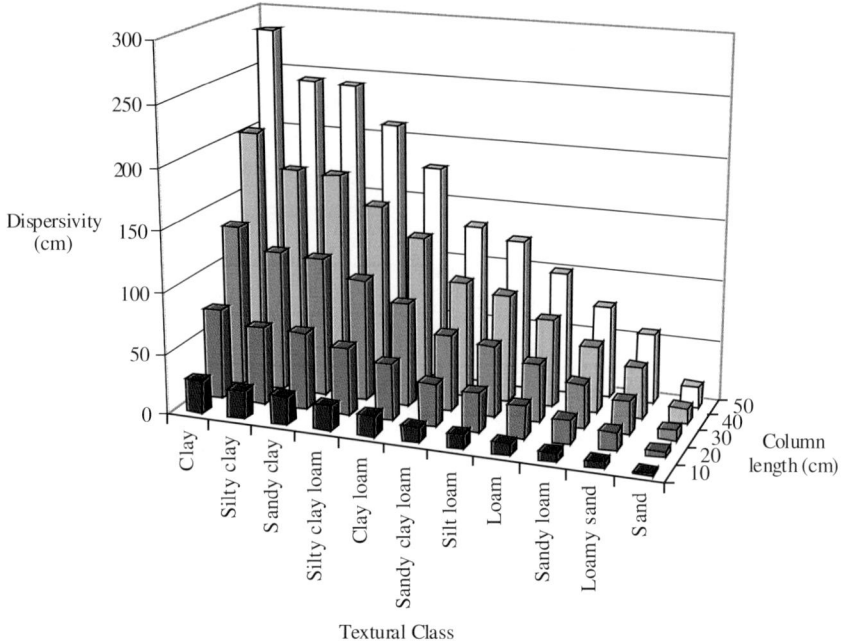

Figure 5. Dispersivities predicted from the water retention database of Cosby et al. (1984) as a function of soil textural class and column length (Perfect, 2003).

6. CONCLUSIONS AND FUTURE DIRECTIONS

This review suggests that input parameters required for solute adsorption and transport models can be predicted reasonably well from basic soil properties. Many studies have shown that the adsorption isotherm curve (parameterized using either a linear model or the Freundlich equation) can be predicted quite successfully in this way. The adsorption coefficient is mainly controlled by soil pH; clay content, CEC, and organic matter content have also proved to be good predictors. Recently, there have been several studies that have resulted in PTFs for predicting the retardation factor R or adsorption coefficient based on small sample sizes (Wang and Zhang, 2001). Care must be taken in the use of such functions because they are specifically calibrated on a specific soil type under certain laboratory conditions.

The UDM proposed by Moldrup et al. (2003) is a promising approach for predicting solute and gas diffusivities as a function of soil water content, based on total porosity and the slope of the water retention curve. This model has been validated against data for 22 soil types, and gives quite accurate predictions of D_p.

For the convective–dispersive model, there is a need to focus future studies on the prediction of dispersivity. PTFs for predicting v and D individually are not useful, because v is mainly controlled by experimental or field conditions, while D is directly related to v. Pore-size distribution appears to be the main predictor for dispersivity; the pore-size distribution can be represented by parameters of the water retention curve. There is also a need to increase our understanding of the physical meaning of the parameters of the MIM. In most studies, MIM parameters are derived from an inverse solution. If this solution is not unique, the parameters will not have any significant physical meaning. Techniques for the direct measurement of MIM parameters have recently been proposed and used in the field. Based on data collected using these methods, a PTF was presented that accurately predicts the mobile water content parameter from the total water content and percent clay content.

Overall, the main challenge for the future development of PTFs for solute transport parameters is the lack of databases containing large numbers of breakthrough curves for partially saturated conditions, collected and parameterized using relatively uniform methods. The hydrogeological community has made significant progress in compiling and evaluating dispersivity data for saturated geologic materials obtained from field and laboratory studies conducted at different spatial scales (Neuman, 1990; Gelhar et al., 1992). Similar efforts have begun for partially saturated materials (see Beven et al. (1993) for solute dispersion and Lorentz et al. (2001) for solute diffusion), but more are needed for calibration and testing of PTFs to predict solute movement in the vadose zone.

REFERENCES

Anderson, J.L., Bouma, J., 1977. Water movement through pedal soils. I. Saturated flow. Soil Sci. Soc. Am. J. 41, 413-418.

Anderson, P.R., Christensen, T.H., 1988. Distribution coefficients of Cd, Co, Ni, and Zn in soils. J. Soil Sci. 39, 15-22.

Anderson, A.N., McBratney, A.B., FitzPatrick, E.A., 1996. Soil mass, surface, and spectral fractal dimensions estimated from thin section photographs. Soil Sci. Am. J. 60, 962-969.

Angulo-Jaramillo, R., Gaudet, J., Thony, J., Vauclin, M., 1996. Measurement of hydraulic properties and mobile water content of a field soil. Soil Sci. Soc. Am. J. 60, 710-715.

Bajracharya, K., Barry, D.A., 1997. Nonequilibrium solute transport parameters and their physicl significance: numerical and experimental results. J. Contam. Hydrol. 24, 185-204.

Baskaran, S., Bolan, N.S., Rahman, A., Tillman, R.W., 1996. Pesticide sorption by allophanic and non-allophanic soils of New Zealand. New Zeal. J. Agric. Res. 39, 297-310.

Bejat, L., Perfect, E., Quisenberry, V.L., Coyne, M.S., Haszler, G.R., 2000. Solute transport as related to soil structure in unsaturated intact soil blocks. Soil Sci. Am. J. 64, 818-826.

Benson, D.A., 1998. The fractional advection–dispersion equation: development and application. Ph.D. Thesis, University of Nevada, Reno.

Beven, K.J., Henderson, D.E., Reeves, A.D., 1993. Dispersion parameters for undisturbed partially saturated soil. J. Hydrol. 143, 19-43.

Biggar, J.W., Nielsen, D.R., 1976. Spatial variability of leaching characteristics of a field soil. Water Resour. Res. 12, 78-84.

Bolt, G.H., 1979. Movement of solutes in soil: principles of adsorption/exchange chromatography. *In*: Bolt, G.H. (Ed.), Soil Chemistry. Elsevier Scientific, New York, pp. 285-348.

Booltink, H.W.G., Hatano, R., Bouma, J., 1993. Measurement and simulation of bypass flow in a structured clay soil: a physico-morphological approach. J. Hydrol. 148, 149-168.

Borggaard, O.K., Szilas, C., Gimsing, A.L., Rasmussen, L.H., 2004. Estimation of soil phosphate adsorption capacity by means of a pedotransfer function. Geoderma 118, 55-61.

Bouma, J., Booltink, H.W.G., Finke, P.A., 1996. Use of soil survey data for modeling solute transport in the vadose zone. J. Environ. Qual. 25, 519-526.

Bresler, E., 1973. Simultaneous transport of solutes and water under transient unsaturated flow conditions. Water Resour. Res. 9, 975-986.

Brouwer, W.W.M., Boesten, J.J.T.I., Siegers, W.G., 1990. Adsorption and transformation products of atrazine by soil. Weed Res. 30, 123-128.

Buchter, B., Davidoff, B., Amacher, M.C., Hinz, C., Iskandar, K., Selim, H.M., 1989. Correlation of Freundlich Kd and n retention parameters with soils and elements. Soil Sci. 148, 370-379.

Buckingham, E., 1904. Contribution to our knowledge of the aeration of soils. USDA Bur. Soil Bull., 25.

Campbell, G.S., 1974. A simple method for determining unsaturated conductivity from moisture retention data. Soil Sci. 117, 311-314.

Carbonell, R.G., 1979. Effect of pore distribution and flow segregation on dispersion in porous media. Chem. Engng Sci. 14, 1031-1039.

Casey, F.X.M., Jaynes, D.B., Horton, R., Logsdon, S.D., 1999. Comparing field methods that estimate mobile–immobile model parameters. Soil Sci. Soc. Am. J. 63, 800-806.

Charbeneau, R.J., 1984. Kinematic models for soil-moisture and solute transport. Water Resour. Res. 20, 699-706.

Christensen, T.H., 1989. Cadmium soil sorption at low concentrations: VII: Correlation with soil parameters. Water Air Soil Pollut. 44, 1-82.

Clothier, B.E., Kirkham, M.B., McLean, J.E., 1992. In situ measurement of the effective transport volume for solute moving through soil. Soil Sci. Soc. Am. J. 56, 733-736.

Coats, K.H., Smith, B.D., 1964. Dead-end pore volume and dispersion in porous media. Soc. Petrol. Engng J. 4, 73-84.

Coquet, Y., Barriuso, E., 2002. Spatial variability of pesticide adsorption within the topsoil of a small agricultural catchment. Agronomie 22, 389-398.

Cosby, B.J., Hornberger, G.M., Clapp, R.B., Ginn, T.R., 1984. A statistical exploration of the relationships of soil moisture characteristics to the physical properties of soils. Water Resour. Res. 20, 682-690.

Crawford, J.W., Ritz, K., Young, I.M., 1993. Quantification of fungal morphology, gaseous transport and microbial dynamics in soil: an integrated framework utilizing fractal geometry. Geoderma 56, 157-172.

De Smedt, F., Wierenga, P.J., 1984. Solute transfer through columns of glass beads. Water Resour. Res. 20, 225-232.

Durner, W., Flühler, H., 1996. Multi-domain model for pore-size dependent transport of solutes in soils. Geoderma 70, 281-297.

Ebato, M., Yonebayashi, K., Kosaki, T., 2001. Predicting Freundlich adsorption isotherm of atrazine on Japanese soils. Soil Sci. Plant Nutr. 47, 221-231.

Elzinga, E.J., Vissenberg, H.A., van den Berg, B., van Grinsven, J.J.M., Swartjes, E.A., 1997. Freundlich adsorptievergelijkingen voor cadmium, koper en zink op basis van literatuurgegevens. RIVM Report 711501001, Bilthoven, The Netherlands.

Fetter, C.W., 1999. Contaminant Hydrogeology, Second edition. Prentice Hall Inc., Upper Saddle River, NJ.

Fitter, A.H., Sutton, C.D., 1975. The use of Freundlich isotherm for soil phosphate sorption data. J. Soil Sci. 26, 241-246.

Flury, M., Gimmi, T.F., 2002. Solute diffusion. *In*: Dane, J.H., Topp, C.G. (Eds.), Methods of Soil Analyis Part 4. Physical Methods, SSSA Book Series No. 5. SSSA, Madison, pp. 1323-1351.

Freese, D., Van der Zee, S.E.A.T.M., Van Riemsdijk, W.H., 1992. Comparison of different models for phosphate sorption as a function of the iron and aluminium oxides of soils. J. Soil Sci. 43, 729-738.

Frissel, M.J., Poelstra, P., 1967a. Chromatographic transport through soils. I. Theoretical evaluations. Plant Soil 26, 285-302.

Frissel, M.J., Poelstra, P., 1967b. Chromatographic transport through soils. II. Column experiments with Sr- and Ca-isotopes. Plant Soil 27, 20-32.

Gaudet, J.P., Jegat, H., Vachaud, G., Wieranga, P.J., 1977. Solute transfer, with exchange between mobile and stagnant water, through unsaturated sand. Soil Sci. Soc. Am. J. 41, 665-670.

Gelhar, L.W., Welty, C., Rehfeldt, K.R., 1992. A critical review of data on field- scale dispersion in aquifers. Water Resour. Res. 28, 1955-1974.

Gerke, H.H., Arning, M., Stöppler-Zimmer, H., 1999. Modeling long-term compost application effects on nitrate leaching. Plant Soil 213, 75-92.

Germann, P.F., Smith, M.S., Thomas, G.W., 1987. Kinematic wave approximation to the transport of *Escherichia Coli* in the vadose zone. Water Resour. Res. 23, 1281-1287.

Gonçalves, M.C., Pereira, L.S., Leij, F.J., 1997. Pedo-transfer functions for estimating unsaturated hydraulic properties of Portuguese soils. Eur. J. Soil Sci. 48, 387-400.

Gonçalves, M.C., Leij, F.J., Schaap, M.G., 2001. Pedotransfer functions for solute transport parameters of Portuguese soils. Eur. J. Soil Sci. 52, 563-574.

Griffoen, J.W., Barry, D.A., Parlange, J.-Y., 1998. Interpretation of two-region model parameters. Water Resour. Res. 34, 373-384.

Haga, D., Niibori, Y., Chida, T., 1999. Hydrodynamic dispersion and mass transfer in unsaturated flow. Water Resour. Res. 35, 1065-1077.

Hamaker, J.W., Thompson, J.M., 1972. Adsorption. *In*: Hamaker, J.W., Goring, C.A.I. (Eds.), Organic Chemicals in the Soil Environment, Vol. 1. Marcel Dekker Inc., New York.

Harter, R.D., 1983. Effect of soil-pH on adsorption of lead, copper, zinc, and nickel. Soil Sci. Soc. Am. J. 47, 47-51.

Hassan, A.E., Cushman, J.H., Delleur, J.W., 1997. Monte Carlo studies of flow and transport in fractal conductivity fields: comparison with stochastic perturbation theory. Water Resour. Res. 33, 2519-2534.

Hatano, R., Kawamura, N., Ikeda, J., Sakuma, T., 1992. Evaluation of the effect of morphological features of flow paths on solute transport by using fractal dimensions of methylene blue staining pattern. Geoderma 53, 31-44.

Hinz, C., 2001. Description of sorption data with isotherm equations. Geoderma 99, 225-243.

Holm, P.E., Rootzen, H., Borggaard, O.K., Moberg, J.P., Christensen, T.H., 2003. Correlation of cadmium distribution coefficients to soil characteristics. J. Environ. Qual. 32, 138-145.

Horton, R., Thompson, M.L., McBride, J.F., 1987. Method of estimating the travel time of nonintersecting solutes through compacted soil material. Soil Sci. Soc. Am. J. 51, 48-53.

Ilsemann, J., van der Ploeg, R.R., Horton, R., Bachmann, J., 2002. Laboratory method for determining immobile soil water content and mass exchange coefficient. J. Plant Nutr. Soil Sci. 165, 332-338.

Jaynes, D.B., Logsdon, S.D., Horton, R., 1995. Field method for measuring mobile and immobile water content and solute transfer rate coefficient. Soil Sci. Soc. Am. J. 59, 352-356.

Jury, W.A., 1982. Simulation of solute transport using a transfer function. Water Resour. Res. 18, 363-368.

Jury, W.A., Sposito, G., 1985. Field calibration and validation of solute transport models for the unsaturated zone. Soil Sci. Soc. Am. J. 49, 1331-1341.

Kemblowski, M.W., Wen, J.-C., 1993. Contaminant spreading in stratified soils with fractal permeability distribution. Water Resour. Res. 29, 419-425.

Klinkenberg, I.J., 1957. Pore size distribution of porous media and displacement experiments with miscible liquids. Petrol. Trans. AIME 2, 366-369.

Kozak, J., Vacek, O., 2000. Pedotransfer functions as a tool for estimation of pesticides behavior in soils. Rostlinna Vyroba 46, 69-76.

Kozak, J., Valla, M., Prokopec, V., Vacek, O., 1992. Prediction of the role of soil organic matter and some other soil characteristics in herbicide adsorption. In: Kubat, J. (Ed.), Humus, Its Structure and Role in Agriculture and Environment. Elsevier, Amsterdam, pp. 165-169.

Lindstrom, F.T., Boersma, L., 1971. A theory on the mass transport of previously distributed chemicals in a water saturated sorbing porous medium. Soil Sci. 111, 192-199.

Livesey, N.T., Huang, P.M., 1981. Adsorption of arsenate by soils and its relation to selected chemical properties and anions. Soil Sci. 131, 88-94.

Lorentz, S., Goba, P., Pretorius, J., 2001. Hydrological process research: experiments and measurements of soil hydraulic characteristics. Water Research Commission Report No: 744/1/01, Pietermaritzburg, South Africa.

Marshall, T.J., 1959. The diffusion of gases through porous media. J. Soil Sci. 10, 79-82.

Matsubayashi, U., Devkota, L.P., Takagi, F., 1997. Characteristics of the dispersion coefficient in miscible displacement through a glass beads medium. J. Hydrol. 192, 51-64.

Millington, R.J., Quirk, J.M., 1961. Permeability of porous solids. Trans. Faraday Soc. 57, 1200-1207.

Moldrup, P., Olesen, T., Rolston, D.E., Yamaguchi, T., 1997. Modeling diffusion and reaction in soils.7. Predicting gas and ion diffusivity in undisturbed and sieved soils. Soil Sci. 162, 632-640.

Moldrup, P., Olesen, T., Komatsu, T., Schjønning, P., Rolston, D.E., 2001. Tortuosity, diffusivity, and permeability in the soil liquid and gaseous phases. Soil Sci. Soc. Am. J. 65, 613-623.

Moldrup, P., Olesen, T., Komatsu, T., Yoshikawa, S., Schjonning, P., Rolston, D.E., 2003. Modeling diffusion and reaction in soils: x. A unifying model for solute and gas diffusivity in unsaturated soil. Soil Sci. 168, 321-337.

Montas, H.J., Eigel, J.D., Engel, B.A., Haghighi, K., 1997a. Deterministic modeling of solute transport in soils with preferential flow pathways – Part 1. Model development. Trans. ASAE 40, 1245-1256.

Montas, H.J., Eigel, J.E., Engel, B.A., Haghighi, K., 1997b. Deterministic modeling of solute transport in soils with preferential flow pathways – Part 2. Model validation. Trans. ASAE 40, 1257-1265.

Nkedi-Kizza, P., Biggar, J.W., van Genuchten, M.T., Wierenga, P.J., Selim, H.M., Davidson, J.M., Nielsen, D.R., 1983. Modeling tritium and chloride 36 transport through an aggregated Oxisol. Water Resour. Res. 19, 691-700.

Neuman, S.P., 1990. Universal scaling of hydraulic conductivities and dispersivities in geologic media. Water Resour. Res. 26, 1749-1758.

Neuman, S.P., 1995. On advective transport in fractal permeability and velocity fields. Water Resour. Res. 31, 1455-1460.

Okom, A.E.A., 1998. Estimating soil hydraulic properties and the solute transport volume using surrogate variables. PhD Thesis, Environmental Horticulture and Resource Management, University of Melbourne, Melbourne.

Okom, A.E.A., White, R.E., Heng, L.K., 2000. Field measured mobile water fraction for soils of contrasting texture. Aust. J. Soil Res. 38, 1131-1142.

Olesen, T., Moldrup, P., Henriksen, K., Petersen, L.W., 1996. Modeling diffusion and reaction in soils: IV. New models for predicting ion diffusivity. Soil Sci. 161, 633-645.

Olesen, T., Moldrup, P., Gamst, J., 1999. Solute diffusion and adsorption in six soils along a soil texture gradient. Soil Sci. Soc. Am. J. 63, 519-524.

Olesen, T., Moldrup, P., Yamaguchi, T., Rolston, D.E., 2001. Constant slope impedance factor model for predicting the solute diffusion coefficient in unsaturated soil. Soil Sci. 166, 89-96.

Olsen, S.R., Kemper, W.D., 1968. Movement of nutrients to plant roots. Adv. Agron. 20, 91-151.

Orbach, R., 1986. Dynamics of fractal networks. Science 231, 814-819.

Pachepsky, Y., Radcliffe, D.E., Selim, H.M. (Eds.), 2003. Scaling Methods in Soil Physics. CRC Press, Boca Raton, FL.

Padilla, I.Y., Yeh, T.-C.J., Conklin, M.H., 1999. The effect of water content on solute transport in unsaturated porous media. Water Resour. Res. 35, 3303-3313.

Papendick, R.J., Campbell, G.S., 1980. Theory and measurement of water potential. In: Parr, J.F. (Ed.), Water Potential Relations to Soil Microbiology, Spec. Publ. No. 9. SSSA, Madison, WI, pp. 1-22.

Passioura, J.B., Rose, D.A., 1971. Hydrodynamic dispersion in aggregated media 2. Effects of velocity and aggregate size. Soil Sci. 111, 345-351.

Penman, H.L., 1940. Gas and vapour movements in soil: the diffusion of vapours through porous solids. J. Agric. Sci. 30, 437-462.

Perfect, E., 2003. A pedotransfer function for predicting solute dispersivity: model testing and upscaling. In: Pachepsky, Y., Radcliffe, D.E., Selim, H.M. (Eds.), Scaling Methods in Soil Physics. CRC Press, Boca Raton.

Perfect, E., Sukop, M.C., 2001. Models relating solute dispersion to pore space geometry in saturated media: a review. In: Selim, H.M., Sparks, D.L. (Eds.), Physical and Chemical Processes of Water and Solute Transport/Retention in Soils, SSSA Special Publ. 56. SSSA, Madison WI, pp. 77-146.

Perfect, E., Sukop, M.C., Haszler, G.R., 2002. Prediction of dispersivity for undisturbed soil Columns from water retention parameters. Soil Sci. Soc. Am. J. 66, 696-701.

Petach, M.C., Wagenet, R.J., DeGloria, S.D., 1991. Regional water-flow and pesticide leaching using simulations with spatially distributed data. Geoderma 48, 245-269.

Pfannkuch, H.O., 1962. Contribution à l'étude de déplacements de fluids miscibles dans un milieu poreux. Revue de l'Institut Francais du Pétrole 18, 215-270.

Porter, L.K., Kemper, W.D., Jackson, R.D., Stewart, B.A., 1960. Chloride diffusion in soils as influenced by moisture content. Soil Sci. Soc. Am. Proc. 24, 460-463.

Rajaram, H., 1997. Time and scale dependent effective retardation factors in heterogeneous aquifers. Adv. Water Resour. 20, 217-230.

Rao, P.S.C., Davidson, J.M., 1980. Estimation of pesticide retention and transformation parameters required in nonpoint source pollution models. In: Overcash, M.R., Davidson, J.M. (Eds.), Environmental Impact of Non-point Source Pollution. Ann Arbor Science Publ., Ann Arbor.

Rao, P.S.C., Green, R.E., Ahuja, L.R., Davidson, J.M., 1976. Evaluation of a capillary bundle model for describing solute dispersion in aggregated soils. Soil Sci. Soc. Am. J. 40, 815-820.

Rasiah, V., 1995. Comparison of pedotransfer functions to predict nitrogen-mineralization parameters of one- and two-pool models. Commun. Soil Sci. Plant Anal. 26, 1873-1884.

Rawls, W.J., Gish, T.J., Brakensiek, D.L., 1991. Estimating soil water retention from soil physical properties and characteristics. Adv. Agron. 16, 213-234.

Rowell, D.L., 1994. Soil Science: Methods and Applications. Longman Scientific and Technical, UK.

Sadeghi, A.M., Kissel, D.E., Cabrera, M.L., 1989. Estimating molecular diffusion coefficients of urea in unsaturated soil. Soil Sci. Soc. Am. J. 53, 15-18.

Sauvé, S., Hendershot, W., Allen, H.E., 2000. Solid-solution partitioning of metals in contaminated soils: dependence on pH, total metal burden, and organic matter. Crit. Rev. Environ. Sci. Technol. 34, 1125-1131.

Scheinost, A.C., Schwertmann, U., 1995. Predicting phosphate adsorption–desorption in a soilscape. Soil Sci. Soc. Am. J. 59, 1575-1580.

Scheinost, A.C., Sinowski, W., Aerswald, K., 1997. Regionalization of soil buffering functions: a new concept applied to K/Ca exchange curves. Adv. GeoEcology 30, 23-28.

Schug, B., Hoss, T., During, R.A., 1999. Regionalization of sorption capacities for arsenic and cadmium. Plant Soil 213, 181-187.

Schwartz, R.C., 1998. Reactive transport of tracers in a fine textured Ultisol. Ph.D Dissertation, Texas A&M University, College Station, TX.

Scotter, D.R., Ross, P.J., 1994. The upper limit of solute dispersion and soil hydraulic properties. Soil Sci. Soc. Am. J. 58, 659-663.

Seyfried, M.S., Rao, P.S.C., 1987. Solute transport in undisturbed columns of an aggregated tropical soil: preferential flow effects. Soil Sci. Soc. Am. J. 51, 1434-1444.

Shaw, J.N., West, L.T., Radcliffe, D.E., Bosch, D.D., 2000. Preferential flow and pedotransfer functions for transport properties in sandy Kandiudults. Soil Sci. Soc. Am. J. 64, 670-678.

Shukla, M.K., Ellsworth, T.R., Hudson, R.J., Nielsen, D.R., 2003. Effect of water flux on solute velocity and dispersion. Soil Sci. Soc. Am. J. 67, 449-457.

Šimunek, J., Šejna, M., Van Genuchten, M.Th., 1999. The HYDRUS-2D software package for simulating two-dimensonal movement of water, heat, and multiple solutes in variable saturated media. Version 2.0, IGWMC-TPS-53, International Ground Water Modeling Center, Colorado School of Mines, Golden, Colorado.

Skaggs, T.H., Leij, F.J., 2002. Solute transport: theoretical background. *In*: Dane, J.H., Topp, C.G. (Eds.), Methods of Soil Analysis Part 4. Physical Methods, SSSA Book Series, No. 5. SSSA, Madison, pp. 1353-1380.

Skaggs, T.H., Wilson, G.V., Shouse, P.J., Leij, F.J., 2002. Solute transport: experimental methods. *In*: Dane, J.H., Topp, C.G. (Eds.), Methods of Soil Analysis Part 4. Physical Methods, SSSA Book Series, No. 5. SSSA, Madison, pp. 1381-1434.

Springob, G., Böttcher, J., 1998a. Parameterization and regionalization of Cd sorption characteristics of sandy soils. I. Freundlich type parameters. Zeitschrift fur Pflanzenernahrung und Bodenkunde Z. Pflanzenernähr. Bodenkd. 161, 681-687.

Springob, G., Böttcher, J., 1998b. Parameterization and regionalization of Cd sorption characteristics of sandy soils. II. Regionalization: Freundlich k estimates by pedotransfer functions. Z. Pflanzenernähr. Bodenkd. 161, 689-696.

Springob, G., Tetzlaff, D., Schon, A., Bottcher, J., 2001. Quality of estimated Freundlich parameters of Cd sorption from pedotransfer functions to predict cadmium concentrations of soil solution. Trace elements in soil: bioavailability, flux and transfer. Lewis Publishers Inc., Boca Raton, USA, pp. 229-245.

Steenhuis, T.S., Parlange, J.-Y., Andreini, M.S., 1990. A numerical model for preferential solute movement in structured soils. Geoderma 46, 193-208.

Stewart, M.A., Jardine, M., Barnett, M.O., Mehlhorn, T.L., Hyder, L.K., McKay, L.D., 2003. Influence of soil geochemical and physical properties on the sorption and bioaccessibility of chromium(III). J. Environ. Qual. 32, 129-137.

Tiktak, A., Leijnse, A., Vissenberg, H., 1999. Uncertainty in a regional-scale assessment of cadmium accumulation in the Netherlands. J. Environ. Qual. 28, 461-470.

Tinker, P.B., Nye, P.H., 2000. Solute Movement in the Rizhosphere. Oxford University Press, Oxford.

Toride, N., Leij, F.J., van Genuchten, M.T., 1995. The CXTFIT code for estimating transport parameters from laboratory or field tracer experiments. Version 2.0. Research Report 137, US Salinity Laboratory, Riverside, CA.

Toride, N., Inoue, M., Leij, F.J., 2003. Hydrodynamic dispersion in an unsaturated dune sand. Soil Sci. Soc. Am. J. 67, 703-712.

Travis, C.C., Etneir, E.L., 1981. A survey of sorption relationships for reactive solutes in soil. J. Environ. Qual. 10, 8-17.

Vallochi, A., 1985. Validity of the local equilibrium assumption for modeling sorbing solute transport through homogeneous soils. Water Resour. Res. 21, 808-820.

van Dam, J.C., Huygen, J., Wesseling, J.G., Feddes, R.A., Kabat, P., van Walsum, R.E.V., Groenendijk, P., van Diepen, C.A., 1997. Theory of SWAP version 2.0. SC-DLO, Wageningen Agricultural University, Report 71, Department of Water Resources, The Netherlands.

van Genuchten, M.Th., 1981. Non-equilibrium transport parameters from miscible displacement experiments. Research Report No. 119, U.S. Salinity Laboratory, Riverside, California.

van Genuchten, M.T., Wierenga, P.J., 1976. Mass transfer studies in sorbing porous media: I. Analytical solutions. Soil Sci. Soc. Am. Proc. 40, 473-480.

van Genuchten, M.T., Wierenga, P.J., 1977. Mass transfer studies in sorbing porous media: II. Experimental evaluation with tritium. Soil Sci. Soc. Am. Proc. 41, 272-278.

Verburg, K, Ross, P.J., Bristow, K.L., 1996. SWIMv2.1 User Manual. Divisional Report No. 130, Division of Soils, CSIRO, Australia.

Vervoort, R.W., Radcliffe, D.E., West, L.T., 1999. Soil structure development and preferential solute flow. Water Resour. Res. 35, 913-928.

Wagenet, R.J., 1983. Principles of salt movement in soils. *In*: Nelson, D.W., Elrick, D.E., Tanji, K.K. (Eds.), Chemical Mobility and Reactivity in Soil Systems, SSSA Special Publication No. 11. ASA, SSSA, Madison, WI, pp. 123-140.

Walker, P.J.C., Trudgill, S.T., 1983. Quantitative image analysis of soil pore geometry: comparison with tracer breakthrough curves. Earth Surf. Processes Landforms 8, 465-472.

Wang, Y., Zhang, Y.P., 2001. Quantitative effect of soil texture composition on retardation factor of K^+ transport. Pedosphere 11, 377-382.

Wang, Q., Horton, R., Lee, J., 2002. A simple model relating soil water characteristic curve and soil solute breakthrough curve. Soil Sci. 167, 436-443.

Wauchope, R.D., Yeh, S., Linders, J.B.H.J., Kloskowski, R., Tanaka, K., Rubin, B., Katayama, A., Kördel, W., Gerstl, Z., Lane, M., Unsworth, J.B., 2002. Pesticide soil sorption parameters: theory, measurement, uses, limitations and reliability. Pest Manag. Sci. 58, 419-445.

Xu, M., Eckstein, Y., 1995. Use of weighted least-squares method in evaluation of the relationship between dispersivity and field scale. Ground water 33, 905-908.

Xu, M., Eckstein, Y., 1997. Statistical analysis of the relationships between dispersivity and other physical properties of porous media. Hydrogeol. J. 5, 4-20.

Yule, D.F., Gardner, W.R., 1978. Longitudinal and transverse dispersion coeffcients in unsaturated plainfield sand. Water Resour. Res. 14, 582-588.

Zhan, H., Wheatcraft, S.W., 1996. Macrodispersivity tensor for nonreactive solute transport in isotropic and anisotropic fractal porous media: analytical solutions. Water Resour. Res. 32, 3461-3474.

Zheng, C., Bennett, G.D., 1995. Applied Contaminant Transport Modeling. Van Nostrand Reinhold, New York.

Chapter 13

ESTIMATING SOIL SHRINKAGE PARAMETERS

E. Braudeau[1,*], R.H. Mohtar[2] and N. Chahinian[1]

[1]IRD/PRAM, Quartier Petit Morne, BP 214, 97285 Le Lamentin Cedex 2, Martinique

[2]Department of Agricultural and Biological Engineering, 1146 ABE Building, Purdue University, West Lafayette, IN 47907, USA

*Corresponding author: Tel.: +596-596-42-30-32; fax: +596-596-42-30-31;
e-mail: erik.braudeau@ird.fr

1. IMPORTANCE OF SHRINK–SWELL PROPERTIES

There are several models of soil water flow that consider the soil medium as an active site for chemical, physical and biological processes with a bimodal porous medium-, micro- and macro-pore systems (Jarvis, 1994; Tiktak, 2000; Van der Linden et al., 2001). Few of these models consider the soil medium as a structured medium with aggregates. However, the swelling–shrinkage behavior of these aggregates is almost never considered. Thus, literature on soil dynamics modeling describes soil hydraulic properties independently from the soil–water internal configuration. This leads to an empirical approach to represent and estimate soil–water dynamics and properties such as shrinkage, that induces cracks and fissures and play a major role in preferential flow, field capacity, wilting point, available water, air capacity, unsaturated hydraulic conductivity and water retention.

Most soil water models characterize soil medium by a constant bulk density and the dependencies of hydraulic conductivity k and soil matric potential h on volumetric water content θ. To estimate these properties, soil scientists often turn to pedotransfer function (PTF) using basic soil texture and constituents data. Several of these PTFs are available in the literature, as shown in the present book review. However, the internal soil structure and its volume change as a function of water content is not taken into consideration. This study presents a conceptual and functional model of structured soil–water medium in which the internal volumes change (aggregates, water pools, micro- and macro-pore systems) is governed by the shrinkage parameters that are easily determined using continuously measured soil shrinkage curve (SC) (Braudeau et al., 2004). This chapter also describes how to estimate these SC parameters using available soil survey data.

2. SOIL–WATER MEDIUM FUNCTIONAL MODEL

2.1. Soil–water medium hierarchy and functionality

The soil–water medium of the soil horizon is termed here "pedostructure." It consists of soil fabric with its variable amount of water. The pedostructure refers to the combination of the soil fabric (morphological aspect) and its hydraulic functioning as revealed by the SC (specific volume vs. water content). Braudeau et al. (2004) showed that the SC of the soil–water medium defines the specific volume of its basic component, the primary peds, V_{mi}. The primary peds have been first defined morphologically by Brewer (1964) as "the simplest peds occurring in a soil material: they cannot be divided into smaller peds, but they may be packed together to form compound peds of higher level of organization. S-matrix of a soil material is the material within the simplest (primary) peds, or composing apedal soil materials, in which the pedological features occur; it consists of plasma, skeleton grains, and voids that do not occur in pedological features other than plasma separations." Figure 1 shows a schematic representation of the pedostructure taking into consideration the four hierarchical structural levels of soil: the horizon, the pedostructure, the primary peds and the primary particles.

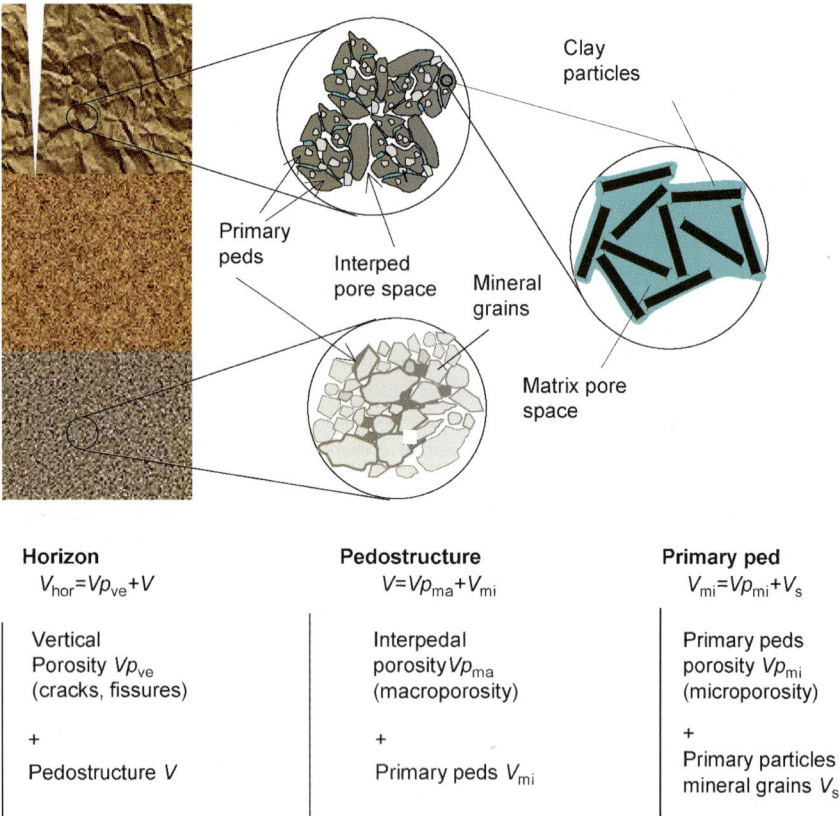

Horizon
$V_{hor}=Vp_{ve}+V$

Vertical
Porosity Vp_{ve}
(cracks, fissures)

+

Pedostructure V

Pedostructure
$V=Vp_{ma}+V_{mi}$

Interpedal
porosity Vp_{ma}
(macroporosity)

+

Primary peds V_{mi}

Primary ped
$V_{mi}=Vp_{mi}+V_s$

Primary peds
porosity Vp_{mi}
(microporosity)

+

Primary particles
mineral grains V_s

Figure 1. Schematic representation of the pedostructure, taking into consideration the structural levels of the soil horizon: the horizon, the pedostruture, the primary peds and the primary particles.

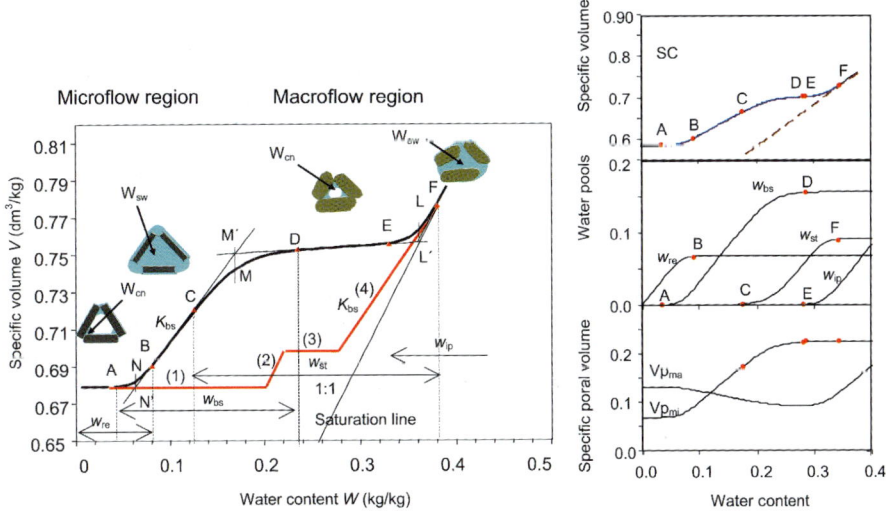

Figure 2. Various configurations of air and water partitioning into the two pore systems, inter- and intra-primary peds, related to the shrinkage phases of a standard SC. W_{cn} represents the condensed water lodged in interstitial pore site, and W_{sw} represents the swelling water lodged in interstitial pore site that can be interped macro-pores (ma) or matrix micro-pores (mi). The various pools of water, w_{re}, w_{bs}, w_{st}, w_{ip}, are represented with their water content curves. The linear and curvilinear shrinkage phases are delimited by the transition points (A–F). Points N', M', and L' are the intersection points of the tangents at those linear phases of the SC.

2.2. Characterization of the pedostructure using shrinkage curve

It was assumed that as soil dries, water leaves the soil from pore of gradually decreasing sizes. Starting from saturation (point F in Figure 2), the interpedal porosity (macro-pore volume, Vp_{ma}) empties up to point C while the primary peds porosity (micro-pore volume, Vp_{mi}) begins to shrink at point D, losing its water (W_{mi}) without air entry from point D up to point B.

For each pore system (inter-pedal and intra-primary peds), water removal is divided into two stages: a first stage during which water leaves the pore system without air intake, with the peds (macro-system) or clay particles (micro-system) approaching each other (shrinkage phases F–E and D–A, respectively). During the second stage water leaves the pore system while being replaced by air, and the peds or clay particles touch (shrinkage phases F–C or B–0, respectively). The ranges for covering the two phases are the curvilinear sections of the SC (phases F–E, C–D, and B–A). Thus, for each pore system, there are two water pools, "swelling" and "non-swelling." Removal of water from the swelling pool induces volume change (shrinkage), while removal of water from the non-swelling pool does not induce any significant volume change. The swelling and non-swelling water pools of the macro-pore system are w_{ip} and w_{st}, respectively, and those of the micro-pore system are w_{bs} and w_{re}. The subscripts *ip*, *st*, *bs*, and *re*, refer to as interpedal, structural, basic, and residual shrinkage phases, respectively.

Braudeau et al. (2004) showed that the volume change equation could be written as a linear combination of the water pool equations:

$$dV = K_{re}dw_{re} + K_{bs}dw_{bs} + K_{st}dw_{st} + K_{ip}dw_{ip} \qquad (1)$$

that gives, after integration:

$$V = V_o + K_{re}w_{re} + K_{bs}w_{bs} + K_{st}w_{st} + K_{ip}w_{ip} \qquad (2)$$

where $K(s)$ are the slopes of the linear shrinkage phases, $w(s)$ are the corresponding water pool, and V_o is the specific volume at dry state (Figure 2).

The authors showed also that the SC, continuously measured under laboratory conditions as in Braudeau et al. (1999) (30°C, unconfined sample, atmospheric pressure), can be assumed to represent a suite of equilibrium states between the water pools and the two pore systems, micro- and macro-pores. The water pools expressions as defined by the SC are:

$$w_{ip} = \frac{1}{k_L}\log[1 + \exp(k_L(W - W_L))]$$

$$w_{st} = -\frac{1}{k_M}\log[1 + \exp(-k_M(W - W_M))] - \frac{1}{k_L}\log[1 + \exp(k_L(W - W_L))]$$

$$w_{bs} = \frac{1}{k_N}\log[1 + \exp(k_N(W - W_N))] + \frac{1}{k_M}\log[1 + \exp(-k_M(W - W_M))]$$

$$w_{re} = -\frac{1}{k_N}\log[1 + \exp(-k_N(W - W_N))] + W_N$$

The parameters K_{re}, K_{bs}, K_{st}, K_{ip}, k_N, k_M, k_L, W_N, W_M, W_L, V_o used in these equations are parameters of the SC. They can be obtained graphically from the measured SC knowing points N′, M′, and L′, the intersection points of the straight lines tangent to the linear sections of the SC and N, M, and L, the corresponding points of the SC (Braudeau et al., 2004). Parameters k_N, k_M, k_L are determined using equations listed in Table 1.

Using the SC and water pool properties of Figure 2, the following relationships are defined:

$W_{mi} = w_{bs} + w_{re}$ the water content in primary peds

$W_{ma} = w_{ip} + w_{st}$

$Vp_{mi} = \min(Vp_{mi}) + w_{bs}/\rho_w$

$Vp_{ma} = V - Vp_{mi} - V_s$

Table 1
Relationships between pedostructure SC parameters (Braudeau et al., 2004) and the XP model transition points (Braudeau et al., 1999)

Shrinkage phases concerned	Relations between parameters of the PS model and with parameters of XP model
Basic and residual (N)	$k_N = (K_{bs} - K_{re})/(V_{N'} - V_N) \ln(2)$ $= 4.8 \ln(2)/(W_N - W_A)$ $= 3.46 \ln(2)/(W_B - W_N)$
Basic and structural (M)	$K_M = (K_{bs} - K_{st})/(V_{M'} - V_M) \ln(2)$ $= 4.8 \ln(2)/(W_M - W_D)$ $= 3.46 \ln(2)/(W_C - W_M)$
Interpedal and structural (L)	$K_L = (K_{ip} - K_{st})/(V_{L'} - V_L) \ln(2)$ $= 4.8 \ln(2)/(W_L - W_E)$ $= 3.46 \ln(2)/(W_F - W_L)$

PS, pedostructure; XP, exponential model of the SC.

$\min(Vp_{mi}) = \max(w_{re}/\rho_w) = W_N/\rho_w$

$\max(Vp_{mi}) = (\max(w_{re}) + \max(w_{bs}))/\rho_w = W_M/\rho_w$

$\max(Vp_{ma}) = Vp_{maSat} = (W_L - W_M)/\rho_w$

Figure 2 shows an example of the continuously measured SCs where the corresponding change in the water pools (w_{re}, w_{bs}, w_{st}, and w_{ip}) and the two specific pore volumes (Vp_{mi} and Vp_{ma}) are represented according to the above equations. Points A–F are the transition points of the shrinkage phases also shown on the corresponding curves of the water pools. These points mark the effective beginning or end of water removal from each pool. The water content at these points (W_A, W_B, ..., W_F) are characteristics of the soil and are determined from the continuously measured SC using a standard method developed by Braudeau et al. (2004) according to the equations listed in Table 1.

Figure 2 also shows a schematics of the swelling process when dry aggregated soil sample is immersed in water. Four events occur immediately before the primary peds swelling: (1) the entry of water into the soil medium through the interpedal voids; (2) the spacing of aggregates; (3) the water entry into the dry micro-pores of the primary peds, filling the residual micro-pore dry space in few seconds; and (4) swelling of the primary peds which takes up to 2 h (Braudeau, 1995). The length and slope of each phase are calculated according to the hydro-structural characteristics defined by the SC in the shrinkage cycle, assuming that they are the same for the swelling as for the shrinkage process, i.e.,: $K_{bs} = dV/dW_{mi}$; $\max(Vp_{mi})$ and $\min(Vp_{mi})$.

3. SEEKING PEDOTRANSFER FUNCTIONS FOR THE SC USING THE PEDOSTRUCTURE CHARACTERIZATION

The above section presents a conceptual pedostructure model that determine both the internal structural volume change as a function of water content and the corresponding SC.

This section demonstrates the use of available soil characteristics such as texture, COLE, water retention curve, field capacity and wilting point, to estimate the SC parameters according to their significance in the pedostructure model, and thus to construct an optimal approximation to the SC. Comparison of some SC parameters with existing PTFs will be presented.

The 11 parameters that characterize the SC described in the section above may not all be needed for modeling the *in situ* SC, therefore, we will examine the minimum set parameters to approximate the SC *in situ*, which will require additional hypothesis.

3.1. The required parameters for crossing scales from laboratory to the field

Table 2 summarizes the descriptive variables (specific volumes, water contents) of the soil horizon and of the soil medium (pedostructure, primary peds, and primary particles). The volume change relationship which was assumed above to calculate the variation of pedostructure volume according to the water pools can be used as a "scaling law" of the soil–water medium, relating specific structural volumes at different organizational levels. Starting from the micro-porosity of the primary peds, one can write:

$$dV_{mi} = k_{re}dw_{re} + (1/\rho_w)dw_{bs}$$

where $k_{re} = K_{re}/K_{bs}$. Then for the pedostructure level,

$$dV = K_{bs}\rho_w dV_{mi} + K_{st}dw_{st} + (1/\rho_w)dw_{ip}$$

To pass from the pedostructure to the horizon specific volume, V_{hor}, three weak assumptions can be made: the slopes K_{st} and K_{re} are negligible, $w_{ip} = 0$ due to overburden presssure, and the shrinkage of the pedostructure *in situ* is isotropic.

This latter assumption relates to the vertical porosity, Vp_{ve}, (which one can observe as cracks or fissures) of which the vertical direction is due to gravity and the opening to air due to the pedostructure shrinkage (Figure 3). Since the pedostructure volume change starts from point D, the opening of the vertical porosity which crosses the soil horizon appears only for $W < W_D$, while for $W > W_D$ the two specific volumes V_{hor} and V coincide: $V_{hor} = V$, and $V_D = V_{horD}$ can be taken as reference.

Table 2
Summary of parameters and variables for the soil–water medium at the pedeostructure and horizon scales, respectively. All variables are referred to mass of primary particles

Volume of concern	Specific volume	Specific pore volume	Water content	Condensed water pool	Swelling water pool
Horizon	V_{hor}	Vp_{hor}	W_{hor}		
Vertical porosity	Vp_{ve}		W_{ve}		
Pedostructure	V	Vp	W		
Interpedal porosity	Vp_{ma}		W_{ma}	w_{st}	w_{ip}
Primary peds	V_{mi}	Vp_{mi}	W_{mi}		
Primary peds porosity	Vp_{mi}		W_{mi}	w_{re}	w_{bs}
Primary soil particles	V_s				

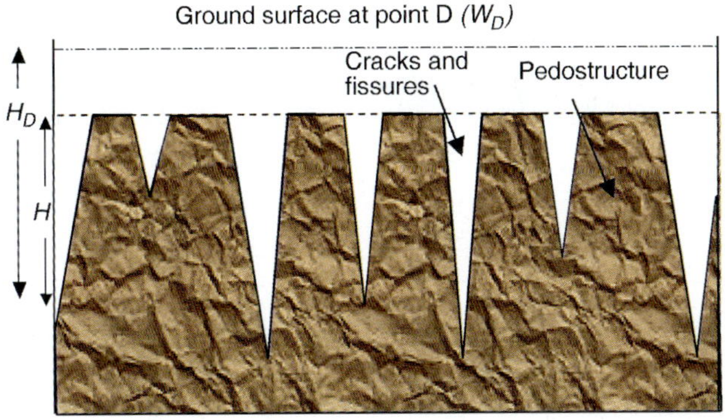

Figure 3. Schematic configuration of the soil horizon and a pedostructure showing the open cracks and fissures that opens to air as soil dries. H and H_D are the soil horizon depth at the desired moisture level and point D, respectively.

That leads to the following equations:

$$dV_{hor}/V_{hor} = dH/H = (1/3)dV/V$$

where H is the thickness of the soil horizon.

Integrating the above equation starting from point D, one obtains:

$$V_{hor}/V_{horD} = (H/H_D) = (\Delta V/V_D + 1)^{1/3} \approx (V - V_D)/3V_D + 1 = (V + 2V_D)/3V_D$$

$$V_{hor} = (V + 2V_D)/3 \text{ and } Vp_{ve} = V_{hor} - V = (2/3)(V_D - V)$$

Consequently, all of the variables listed in Table 2, from the particle level to the soil profile, are uniquely related to W at equilibrium state of the soil–water medium.

According to the assumptions posed above about *in situ* soil horizon shrinkage (W_D is the beginning of opening of the cracks, isotropic shrinkage for $W < W_D$ and vertical shrinkage for $W > W_D$), only the part of the curve corresponding to the water contents $< W_D$ has to be estimated and the number of independent parameters is reduced to six. Those are (while assuming $K_{st} = 0$ and $K_{re} = 0$) : $V_{N'} = V_A = V_o$; $V_{M'} = V_D = V_{L'}$; W_B, W_C, W_D, K_{bs}.

3.2. Significance of the SC parameters and its corresponding approximation

Table 3 presents the pedostructure shrinkage characteristics and its corresponding agronomic laboratory tests. The relationships between the phase transition points: A, B, ..., F, and the pedostructure SC parameters were shown in Table 1. Each shrinkage phase, delimited by these points corresponds to a particular configuration of the hydrated pedostructure which induces some particular physical soil agronomical property at the macroscopic scale. These properties are evaluated using standard laboratory tests such as the porous pressure plate apparatus to determine the available water capacity.

The permanent wilting point is defined as the soil moisture below which a plant can no longer extract water and wilts in an irreversible way. At that moisture level, water is held by

Table 3
Definition and mean of determination of some common soil physical properties according to the pedostructure model

Shrinkage characteristics and parameters	Definitions and significance with reference to the pedostructure		Correspondent agronomic laboratory tests
Particular soil water states			
w_A	Shrinkage limit of primary peds	w_{bs} vanishing	Shrinkage limit
w_B	Air entry point in primary peds	Effective decreasing of w_{re}	1500 kPa pressure plate test (approximate)
w_C	No more water in inter-ped pore space		Limit tensiometer reading (approximate)
w_D	Beginning shrinkage of primary peds	w_{st} vanishing	330 kPa pressure plate test (approximate)
w_E	Shrinkage limit between peds	Effective decreasing of w_{bs}	Higher limit of plasticity (approximate)
		w_{ip} vanishing	
Pedohydral parameters linked to structural properties			
W_M	Intersect of structural and basic linear shrinkage phases		Maximum value of Vp_{mi}
$W_{N'}$	Intersect of residual and basic linear shrinkage phases		Minimum value of Vp_{mi}
$W_{L'}$	Intersect of structural shrinkage phase and the load line		Total pore volume Vp at moist state (phase E-D)
$V_{M'} \approx V_{L'} \approx V_D$	Standard case, K_{re} and K_{st} negligible:	$V_D = K_{bs}(W_M - W_N) - V_A$	Specific volume of horizon and pedostructure at field capacity
$V_{N'} \approx V_o \approx V_A$			
Indicators of the soil swelling potential			
$CG^\mu = W_M - W_N/\rho_w$	Total shrinkage of primary peds, from wet to dry state		Swelling potential of soil
$CG = V_{M'} - V_{N'}$	Total shrinkage of pedostructure, from wet to dry state		Specific swelling potential of soil

the soil matrix at a very low potential which was found to correspond experimentally to near 1500 kPa of air pressure in the Richard's pressure membrane apparatus (Miller and Mazurak, 1955). Up to point B ($W > W_B$), the plasma of the primary peds ensures the flow of water in the soil and thus the renewal of the water extracted in contact with the membrane (or porous plate). Beyond point B, air entry breaks the capillary continuity of clay plasma and traps the water which no longer reaches the porous plate. This point is thus connected to the so-called permanent wilting point, or pF 4.2, currently given in soil databases.

Field capacity can be associated with the hydro-structural state at point D. At this point the gravitational drainage of the water from the inter-aggregate macro-porosity slows down because of the micro–macro-pore water exchange which takes place starting at this point.

In the same way than for the pressure membrane tests above, one can replace the various indicators of soil swelling such as the coefficient of linear extensibility (COLE) proposed by Grossman et al. (1968) with the pedostructure characterization. It consisted of measuring the 1D variation of an undisturbed soil sample between wet (or retention capacity) and dry state:

$$\text{COLE} = (L_{\text{moist}} - L_{\text{dry}})/L_{\text{dry}}$$

These indicators of soil swelling differ in the soil sample preparation and the initial water content. The COLE_{std} is measured on undisturbed soil cores while the COLE_{rod} uses water saturated soil paste rods. More recently, McKenzie et al. (1994) proposed the LS_{mod} test (modified Linear Shrinkage test) coming from a modification of the standard one LS_{std} that was habitually used in Australia (Standards Association of Australia, 1977). Table 4 gives their definition, the mode of preparation and the theoretical formulation of these indicators in terms of the SC parameters.

The indicator which corresponds best to the data provided by retractometry is the LS_{mod} because the soil sample preparation is the same (gently disaggregated and <2 mm sieved soil):

$$\text{LS}_{\text{mod}} = (1/3)(V_{M'} - V_{N'})/V_{M'}$$

The swelling capacity of the primary peds can be defined as:

$$CG^\mu(\max Vp_{mi} - \min Vp_{mi}) = (W_M - W_N)/\rho_w$$

The measurement of the COLE_{rod} or the LS_{std} on a sample would provide two essential soil properties for agronomic modelling (if the initial specific volume of the soil sample is known), namely the micro-swelling capacity and dry specific micro-pore volume:

$$CG^\mu = 3\text{LS}_{\text{std}}V_{M3} \text{ and } \min Vp_{mi} = W_N = V_{M3} - CG^\mu - V_s$$

where V_{M3} is the initial specific volume of the sample in the LS_{std} test (Table 4).

The COLE_{std} and LS_{mod} would provide the values of the swelling capacity of the aggregated soil which can be defined as:

$$CG = V_{M'} - V_{N'} = K_{bs}(W_M - W_N) = K_{bs}\rho_w CG^\mu$$

Table 4
Interpretation of the swelling indices according to the pedostructure soil water model

Index	Definition	Structure of sample Initial starting point	Formulation of index according to the pedostructure model
$COLE_{std}$	$(L_{-5\,kPa} - L_{dry})/L_{dry}$	Natural soil clods	$(V_{E1} - V_{A1})/3V_{A1} \approx (V_{M1} - V_{N1})/3V_{N1}$ $= K_{bs1}(W_M - W_N)/3(W_N + Vp_{ma1} + V_s)$
	$(L_{-33\,kPa} - L_{dry})/L_{dry}$	Near-saturation or -33 kPa matrix potential	$(V_{D1} - V_{A1})/3V_{A1} \approx (V_{M1} - V_{N1})/3V_{N1}$ $= K_{bs1}(W_M - W_N)/3(W_N + Vp_{ma1} + V_s)$
$COLE_{rod}$	$(L_{moist} - L_{dry})/L_{dry}$	Rod of remoulded soil paste	$(V_{L2} - V_{A2})/3V_{A2} \approx (V_{M2} - V_{N2})/3V_{N2}$ $= K_{bs2}(W_M - W_N)/3(W_N + Vp_{ma2} + V_s)$
	$(L_{-33\,kPa} - L_{dry})/L_{dry}$	Liquid-limit estimated at -33 kPa matrix potential	$(V_{D2} - V_{A2})/3V_{A2} \approx (V_{M2} - V_{N2})/3V_{N2}$ $= K_{bs2}(W_M - W_N)/3(W_N + Vp_{ma2} + V_s)$ (with $K_{bs2} \approx 1$ and $Vp_{ma2} \approx 0$)
LS_{std}	$(L_{LL} - L_{dry})/L_{LL}$	Remoulded soil paste	$(V_{L3} - V_{A3})/3V_{L3} \approx (V_{M3} - V_{N3})/3V_{M3}$ $= K_{bs3}(W_M - W_N)/3(W_M + Vp_{ma3} + V_s)$
		Liquid limit	(with $K_{bs3} \approx 1$ and $Vp_{ma3} \approx 0$)
LS_{mod}	$(L_{-5\,kPa} - L_{dry})/L_{-5\,kPa}$	Layer of <2 mm soil aggregates Near saturation (-5 kPa)	$(V_{E4} - V_{A4})/3V_{L4} \approx (V_{M4} - V_{N4})/3V_{M4}$ $= K_{bs4}(W_M - W_N)/3(W_M + Vp_{ma3} + V_s)$

The parameters in the last column have values different from one method to the other depending on the method of sample preparation.

and the specific volume at field capacity V_D:

$$CG = 3LS_{mod}V_{M4} \text{ and } V_D = V_{M4}$$

3.3. Construction of the SC from primary data of soil

These parameters, W_B, W_C and W_D are approximated by water contents at permanent wilting point (W_{PWP}); water content at ≈ 90 kPa, the tension pressure limit of the water column in a tensiometer, W_{90}; and water content at field capacity (W_{FC}) which is generally approximated by the water content at tensions varying between 5 and 40 kPa in the Richards pressure membrane test, depending on soil texture and mineralogy. These three points of the soil moisture characteristic curves can be obtained using porous pressure plate apparatus or evaluated from texture using PTFs (Donatelli et al., 1996). A comparison of W_B and W_D with the values of W_{PWP} and W_{FC} calculated by nine PTFs is given below.

The slope of the basic shrinkage phase (K_{bs}), which lies between 0.1 and 1.2, is a structural feature of the assembly of aggregates according to the relation:

$$K_{bs} = dV/dV_{mi}$$

which represents the volume change ratio between the aggregated soil sample and the primary peds. The parameter depends on the apparent sand content with respect to argillaceous plasma and thus on the clay/sand ratio. A PTF based on texture is needed for this parameter.

The remaining parameters V_o and V_D could not be obtained from texture but rather from indicators of the soil swelling potential such as COLE or LS (linear shrinkage index). For example, according to their formulation in Table 4, both measurements of $COLE_{rod}$ (on saturated paste) and LS_{mod} (on bed of aggregates <2 mm) would provide the following four SC parameters: $V_{N'}$, $V_{M'}$, W_M and W_N. Adding W_{FC} and W_{PFP} (e.g., W_D and W_B), the six parameters needed for modeling the *in situ* SC and consequently all the descriptive variables of the pedostructure can be obtained.

4. APPLICATION EXAMPLE

A pedological study in Tunisa of the irrigated perimeter of Cébala in the Low Valley of Majerda highlighted four types of alluvial soils, all located in the alluvial plain of the Delta of the Majerda River, at 10 km from the sea (Braudeau et al., 2001). These soils are differentiated by their behavior in four alluvial soil groups as: (a) vertic; (b) calcareous; (c) weakly saline; and (d) loamy. Table 5 shows the soil bulk density D_b at point D, $D_b = 1/V_D$. The water content at field capacity and permanent wilting point were estimated as $\theta_{FC} = W_D/\rho_w V_D$ and $\theta_{PWP} = W_B/\rho_w V_B$. The values in brackets refer to water contents of the horizon:

$$(\theta_{PWP}) = W_B/\rho_w V_B^{hor}.$$

4.1. Pedotransfer functions for calculating FC and PWP (W_D and W_B)

Nine different PTFs are proposed to calculate θ_{FC} and θ_{PWP} from soil texture, the results of which are given directly in volumetric water content (m^3 m^{-3}). They are

Table 5
Physical properties of the four alluvial soil

	Clay (%)	Fine silt (%)	Coarse silt (%)	Fine sand (%)	Coarse sand (%)	D_b (kg dm^{-3})	θ_{FC} (m^3 m^{-3})	θ_{PFP} (m^3 m^{-3})
Chalcareous	25	41.5	16.5	11	2	1.14	0.41	0.12 (0.13)
Weakly saline	28.6	32.4	16	18.4	1.4	1.32	0.38	0.12 (0.13)
Loam	23.6	18.4	19.4	34	1.6	1.4	0.34	0.11 (0.11)
Vertic	31.4	53.8	6.4	5	0.67	1.24	0.43	0.15 (0.16)

Values in brackets correspond to θ_{PFP} of the horizon at point B.

estimates of -33 kPa and -1500 kPa pressure plate water content, respectively. Results of these nine PTFs estimates for both moisture levels will be compared to the values of W_D and W_B derived by retractometry for the four soil types of Table 5. Results are shown in Figure 4. The volumetric water contents at field capacity (33 kPa) and wilting point (1500 kPa) for the four soil types using nine PTFs are presented along with the results given by retractometry (Table 5). The SOILPAR computer code (Acutis and Donatelli, 2003) was used to generate the soil moisture values with nine PTFs. In the figure, SC(1) represents the volumetric water content of the pedostructure (W_D/V_D and W_B/V_B); and SC(2) represents the volumetric water content associated with the volume of the horizon (W/V_{hor} with V_{hor} $(2V_D + V)/3$).

4.2. Values of LS$_{mod}$ for the four types of soil

LS$_{mod}$ was calculated using the SC measurement instead of the standard test described by McKenzie et al. (1994). The specific volume of the wet sample (V_{moist}) in the LS$_{mod}$ measurement is, in fact, the volume of the measuring cell divided by the weight of dry soil.

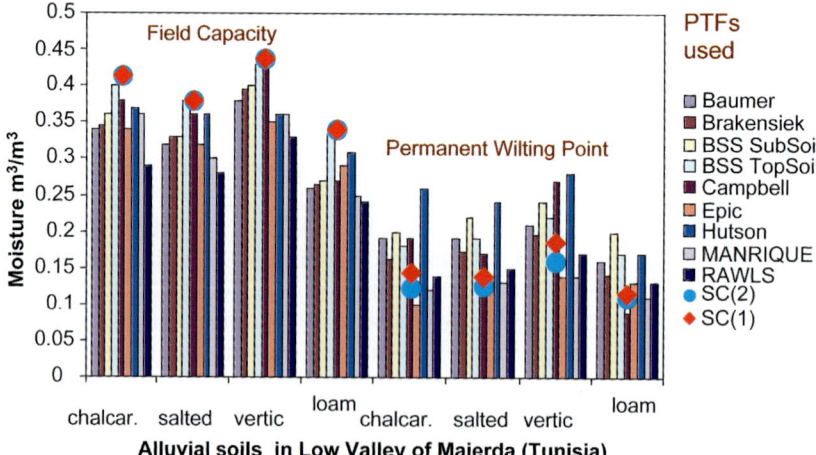

Figure 4. Comparison of field capacity and permanent wilting point values calculated using the nine PTFs proposed by SOILPAR (Acutis and Donatelli, 2003) with those determined by the SC (Braudeau and Donatelli, 2001).

Thus, two parameters, $V_o(=V_A=V_{N'})$, and $V_E(=V_D=V_{M'})$ are obtained as result of the test, instead of only $LS_{mod}=(L_{moist}-L_{dry})/L_{moist}$.

4.3. Value of K_{bs} as a function of texture

The multiple regression carried out in the pedological study of the irrigated perimeter of Cébala (Braudeau et al., 2001), between K_{bs} and clay (<2 mμ) and silt (<50 mμ) and sand (>50 mμ) contents yielded:

$$K_{bs} = 0.017(\text{Clay} + \text{silt}) + 0.006\text{sand} - 0.36$$

with a coefficient of regression $R^2 = 0.89$ for 26 observations. This relation can be used as a PTF for K_{bs}.

4.4. Equations used to build the shrinkage curve

Equations relating the parameters of the SC which make it possible to calculate the six basic parameters, $V_{N'}$, $V_{M'}$, W_N, W_M, k_N, and k_M, to model the SC are derived from the relationships between shrinkage characteristics given in Table 1 (Braudeau et al., 2004; Braudeau and Donatelli, 2001):

$$W_M = 0.42W_D + 0.58W_C \qquad W_N = 0.42W_A + 0.58W_B$$

$$k_M = 3.46 \log 2/(W_C - W_M) \qquad k_N = -4.8 \log 2/(W_A - W_N)$$

$$W_B/\rho_w = \theta_{PWP}V_B \qquad W_D/\rho_w = \theta_{FC}V_{M'} \qquad (V_{M'} - V_{N'})/(W_M - W_N) = K_{bs}$$

$$W_C = W_M + (V_C - V_{M'})/K_{bs} \qquad W_B = W_M + (V_B - V_{M'})K_{bs}$$

EAW = 1/3 UW (estimated using the agronomic relation used between the easily available soil water and the useful soil water) $(\theta_C - \theta_B) = 1/3(\theta_D - \theta_B)$ that

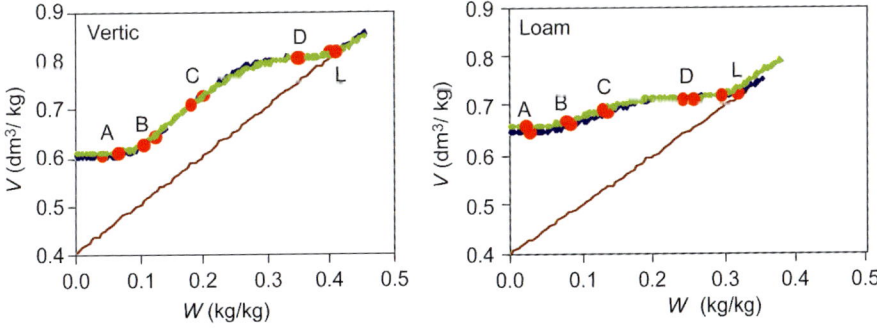

Figure 5. Superimposed observed (blue) and simulated (green) shrinkage curves from texture data, bulk density, and $COLE_{rod}$. The observed and calculated transition points are also shown.

corresponds to

$$W_C/\rho_w = \theta_C V_C = (W_D + 2W_B)/3$$

that gives 10 equations with 15 unknowns.

Five variables have to be estimated to determine the others. That can be θ_{PWP}, θ_{FC} and K_{bs} using texture, and the most and dry specific volumes ($V_{M'}$, $V_{N'}$) using the linear swelling index (LS_{mod}).

Figure 5 gives examples of the curves observed and simulated from primary data: the specific volume of the sample to V_D, at the holding capacity, $COLE_{rod}$ and texture; V_s is taken equal to 0.4 dm^3 kg^{-1}. The θ_{FC} and the θ_{PWP} were calculated by the most suitable methods defined on Figure 4 (BSS TopSoil and Rawls).

5. CONCLUSION

A systemic, process-based conceptual model of the soil SC for characterizing and parameterizing the soil–water medium is presented. The model demonstrated a unique link between soil water properties of non-rigid aggregated soil medium and its internal volume change. This leads to a physically based and functional model of the soil medium which can be used to accurately define and quantify commonly used field scale agronomic properties.

In this chapter, the required parameters for modeling the *in situ* soil hydro-structural properties are defined and calculated from the continuously measured SC. For cases where SC data is missing, approximate estimation of the soil–water medium parameters was developed and evaluated using texture, COLE, pF curve, field capacity and wilting point. Comparison of the SC estimates of agronomic parameters with some PTFs estimates of these parameters was presented.

ACKNOWLEDGEMENTS

This work was partially conducted while the first two authors were on leave at Cirad Cotton Decision Support group. The authors would like to thank the support and the encouragement provided by CIHEAM-IAM, IRD, Cirad, and Purdue University.

APPENDIX A. LIST OF PARAMETERS AND ABBREVIATIONS USED

The definition of the pedostructure variables and characteristics were presented in Braudeau et al. (2004), they are:

V	Pedostructure specific volume (dm^3 per kg of dry soil horizon)
V_{mi}	Primary peds specific volume (dm^3 per kg of dry soil horizon)
V_s	Solids (primary particles) specific volume (dm^3 per kg of dry soil horizon)
V_p	Pedostructure pore specific volume (dm^3 per kg of dry soil horizon)
$V_{p_{ma}}$	Macro-pore specific volume of pedostructure (dm^3 per kg of dry soil horizon)

Vp_{mi}	Micro-pore specific volume of pedostructure (dm³ per kg of dry soil horizon)
W	Pedostructure water content (soil moisture) (kg per kg of dry soil horizon)
W_{mi}	Primary peds water content (kg per kg of dry soil horizon)
W_{ma}	Interpedal water content (kg per kg of dry soil horizon)
w_{re}	Pedostructure residual water pool (kg per kg of dry soil horizon)
w_{bs}	Pedostructure basic water pool (kg per kg of dry soil horizon)
w_{st}	Pedostructure structural water pool (kg per kg of dry soil horizon)
w_{ip}	Pedostructure interpedal water pool (kg per kg of dry soil horizon)
A, B, C, D, E, F	Shrinkage transition points of the SC defined by the XP model
N', M', L'	Intersection points of the tangents to the SC at the linear phases
N, M, L	Characteristic points of the SC at the vertical (y-axis) of N', M', L'
I_{st}, I_{bs}	Inflection points of structural and basic shrinkage phases
W_A, W_B...W_F	Pedostructure water content (kg kg⁻¹) at points A, B...F
W_N, W_M, W_L	Pedostructure water content (kg kg⁻¹) at points N, M, L
V_A, V_B...V_F	Pedostructure specific volume (dm³ kg⁻¹) at points A, B...F
V_N, V_M, V_L	Pedostructure specific volume (dm³ kg⁻¹) at points N, M, L
$V_{N'}$, $V_{M'}$, $V_{L'}$	y-axis values (dm³ kg⁻¹) of points N', M', L' in the SC graph
K_{re}, K_{bs}, K_{st}, K_{ip}	Slopes of the SC linear phases (dm³ kg⁻¹)
k_N, k_M, k_L	Shape parameters (kg kg⁻¹) of the SC equation

REFERENCES

Acutis, M., Donatelli, M., 2003. SOILPAR 2.00: software to estimate soil hydrological parameters and functions. Eur. J. Agron. 18, 373-377.

Braudeau, E., 1995. Water uptake by swelling aggregates, Kearney Foundation Conference Vadose Zone Hydrology; Cutting Across Disciplines. University of California, Davis.

Braudeau, E., Donatelli, M., 2001. Parameters estimation of the soil characteristics shrinkage curve, Proceedings of Second International Symposium on Modelling Cropping Systems, 16–18 July, Florence, Italy, pp. 53–54.

Braudeau, E., Costantini, J.M., Bellier, G., Colleuille, H., 1999. New device and method for soil shrinkage curve measurement and characterization. Soil Sci. Soc. Am. J. 63, 525-535.

Braudeau, E., Zidi, C., Loukil, A., Derouiche, C., Decluseau, D., Hachicha, M., Mtimet, A., 2001. Un système d'information pédologique, le SIRS-Sols du périmètre irrigué de Cébala-Borj-Touil. (Basse Vallée de la Majerda). Bulletin Sols de Tunisie, numéro spécial 2001. Direction des sols (Ed.), Tunis, 134 pp.

Braudeau, E., Frangi, J.P., Mothar, R.H., 2004. Characterizing non-rigid dual porosity structured soil medium using its shrinkage curve. Soil Sci. Soc. Am. J. 68, 359–370.

Brewer, R., 1964. Fabric and Mineral Analysis of Soils. Wiley, New York.

Donatelli, M., Acutis, M., Laruccia, N., 1996. Evaluation of methods to estimate soil water content at field capacity and wilting point, Proceedings of Fourth European Society of Agronomy Congress, Veldhoven, The Netherlands, pp. 86–87.

Grossman, R.B., Brasher, B.R., Franzmeier, D.P., Walker, J.L., 1968. Linear extensibility as calculated from natural clod bulk density measurements. Soil Sci. Soc. Am. Proc. 32, 570-573.

Jarvis, N.J., 1994. The MACRO model (Version 3.1). Technical description and sample simulations. Reports and Dissertation, 19, Dept. Soil Sci., Swedish Univ. Agric. Sci., Uppsala, Sweden, 51pp.

McKenzie, N.J., Jacquier, D.J., Ringrose-Voase, A.J., 1994. A rapid method for estimating soil shrinkage. Aust. J. Soil Res. 32, 931-938.

Miller, S.A., Mazurak, A.P., 1955. An evaluation of permanent wilting percentage, 15-atmosphere moisture percentage, and hygroscopic coefficient of three soils in Eastern Nebraska. Soil Sci. Soc. Proc. 19, 260-263.

Standards Association of Australia, 1977. Determination of the linear shrinkage of a soil (Standard Method), Australian Standard 1289. Methods of Testing Soil for Engineering Purpose. Standards Association of Australia, North Sydney.

Tiktak, A., van den Berg, F., Boesten, J.J.T.I., Leistra, M., van der Linden, A.M.A., van Kraalingen, D., 2000. Pesticide Emission Assessment at Regional and Local Scales: User Manual of FOCUS Pearl version 1.1.1, RIVM Report 711401008, Alterra Report 28, RIVM, Bilthoven, 142pp.

Van der Linden, T., Boesten, J.J.T.I, Tiktak, A., van den Berg, F., 2001. PEARL model for pesticide behaviour and emissions in soil–plant systems. Description of processes. Alterra Report 13, RIVM Report 711401009, Alterra, Wageningen, 107pp.

Chapter 14

KEY SOIL WATER CONTENTS

E. Shein[1], A. Guber[2,*] and A. Dembovetsky[1]

[1]119992 Moscow, Leninskie Gory, Moscow State University, Soil Science Faculty, Russia

[2]Environmental Microbial Safety Laboratory, USDA-ARS-BA-ANRI-EMSL, Bldg. 173, Rm. 203, BARC-EAST, Powder Mill Road, Beltsville, MD 20705, USA

1. INTRODUCTION

In many applications, soil water retention data have to be presented in the form of the equation of the water retention curve. Several equations have been proposed to approximate typical S-shaped dependencies of soil water contents on logarithm of the capillary pressure. With the final goal to have such equation, there are two ways to develop a pedotransfer function for soil water retention. One can estimate coefficients, or parameters, of this equation from soil basic properties, or one can estimate water contents at some points on the water retention curve and then fit the equation to the estimated points (Pachepsky et al., 1999). Point estimates, at least in some cases, provided higher accuracy of the final water retention equation compared with parameter estimation (Pachepsky et al., 1996; Tomasella et al., 2003).

If the point estimation is the chosen way to develop PTF, then the selection of points on the water retention curve is the issue. Soil matric potentials of 0, −10, −33, and −1500 kPa have been often used to determine the characteristic points. These values of matric potentials are the result of general tacit agreement, and never have been proven to have a specific physical meaning. Attempts to relate water retention at −10 kPa or at −33 kPa to the field capacity of soils were not particularly successful (Haise et al., 1955; Cavazza et al., 1973; Peele and Beale, 1950; Ratcliffe et al., 1983).

A physics based approach to define several characteristic points on the water retention curve was developed by Voronin (1990). The idea was to find the segments of the water retention curve at which (a) the forces causing water retention have distinctly different physical nature, and (b) wet soil has distinctly different rheological behavior. The boundaries between these segments defined the key soil water contents. Voronin defined four key soil water contents that he termed maximum adsorbed water content (W_{maw}), maximum molecular water content (W_{mmw}), maximum capillary-sorptive water content (W_{mcsw}), maximum capillary water content (W_{mc}), all defined on mass basis (g g^{-1}). The methods were proposed to measure all four key water contents (Voronin, 1990; see also below). Measuring of the key water contents for a number of different soils gave a surprising result: there were simple relationships between the key water contents and soil matric potentials that had been held for soils of different origin and composition. Specifically, four regression equation have been found:

$$\log(p) = 5.20 + 3W_{maw} \qquad (1)$$

$$\log(p) = 2.17 + 3W_{mmw} \qquad (2)$$

$$\log(p) = 2.17 + 1W_{mcsw} \qquad (3)$$

$$\log(p) = 2.17 + 0W_{mc} \qquad (4)$$

where p is the soil matric potential taken with the opposite sign (cm). Water content W_{mc} corresponded to the constant value of $\log(p) = 2.17$. The location of those regression lines with respect to typical water retention curves is shown in Figure 1.

The existence of the relationships (1)–(4) should allow one to develop the water retention curve equation by (a) developing PTFs to estimate key water contents (Voronin et al., 1997), (b) determining values of soil matric potential from (1)–(3) for each of the key water contents, and thus obtaining points on the water retention curve as pairs of

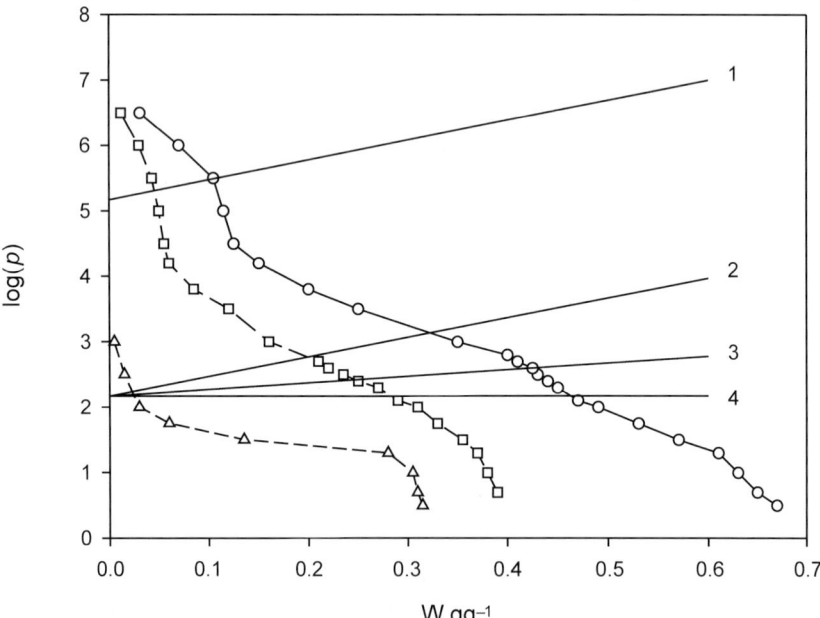

Figure 1. Definition of key water constants. Straight lines define water retention segments within which water retention is dominated by different forces and soil exhibits different rheological behavior. The number on line corresponds to the number of equation in the text. Symbols show measured water retention points.

values (W_{maw}, p_{maw}), (W_{mmw}, p_{mmw}), (W_{mcsw}, p_{mcsw}), and (W_{mc}, p_{mc}), (c) approximating these points with the equation of choice to find its coefficients (parameters).

The objective of this work was to test critical components of such approach. Specifically, we intended to test Equation (3) with data from a large database, and to define the minimum number of key water contents sufficient to estimate parameters in the Van Genuchten (1980) equation of the water retention curve.

2. MATERIALS AND METHODS

Data on 345 soil samples were extracted from the soil hydraulic database assembled at Soil Science Department in Lomonosov Moscow State University, Russian Federation. Soil samples originated from various regions of Russia, and encompassed wide range of textures and organic matter contents.

Soil water retention of undisturbed soil samples was measured (a) in the suction cell apparatus (Klute, 1986; Vadiunina and Korchagina, 1986) in the range of capillary potentials from 0 to 50–80 kPa, (b) in sand–kaolin boxes (Stakman et al., 1969; Varallyay and Mironenko, 1979) in the range of capillary potentials from 0 to 50–80 kPa, and (c) in the field using tensiometers and neutron probes for matric potentials within range from 0 to −50 kPa. Disturbed subsamples were used to measure water retention at −2.76, −20.5 and −81.6 MPa by equilibration with water vapor above saturated K_2SO_4, KCl and $Ca(NO_3)_2$ solutes. Sixth or eight matric potential levels were used in measurements with suction cells and sand–kaolin boxes.

The key water contents were measured as suggested by Voronin (1990). Specifically, the maximum capillary water content W_{mc} was measured as the liquid limit (ASTM, 1996). The maximum capillary-sorptive water content W_{mcsw} was measured as the soil field capacity after 6-h saturated infiltration and two days of free drainage in the field. The maximum capillary water content W_{mc} was measured as the soil plastic limit (ASTM, 1996). The maximum adsorbed water content W_{maw} was calculated from the equation: $W_{max} = W_a + (W_m)_e$, where W_a and $(W_m)_e$ – water contents at the adsorptive water film forms and water content forming water monomolecular film on the external surface of soil particles. These values are determined from the experimental curve of water vapor adsorption of soil (Voronin, 1990)

The Van Genuchten (1980) equation

$$\frac{W - W_r}{W_s - W_r} = \left(\frac{1}{1 + (\alpha p)^n}\right)^m \tag{5}$$

has been fitted to measured pairs of gravimetric water content and capillary pressure (W, p) to find intersections of the regression lines (1)–(4) with measured water retention curves.

3. ESTIMATING SOIL WATER CONTENTS AT FIELD CAPACITY

The comparison of the field capacity values estimated from the laboratory water retention curves and measured in the field is shown in Figure 2. The laboratory

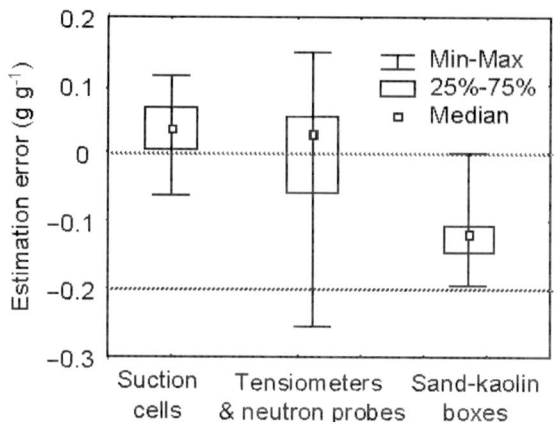

Figure 2. Errors in estimating field capacity from laboratory and field water retention measurements with three different methods. The error has been computed as the field-measured water content at field capacity minus the water content at intersection of the straight line from Equation (3) with the laboratory water retention curve.

suction cell method and the field measured water retention compares favorably with estimates from the Equation 3. More scatter is seen when the field water retention data are used. The sand–kaolin boxes have given systematic deviations. The laboratory-measured value was substantially smaller that the one predicted by Voronin's equation. The latter can be explained by the measurement procedure that includes the loss of contact between a sample and membrane each time when the sample is weighed and moved to the membrane with the next higher capillary pressure level.

We note that the Voronin's Equation (3) predicts the increase in the capillary pressure p as the field capacity increases. This corresponds to observations made by several authors Haise et al. (1955) compared field capacity with the water content at -33 kPa and found high correlation and significant differences between regression slopes and one. Field capacity was substantially higher than w_{33} in coarse-texture soils and substantially lower than w_{33} in fine-texture soils. A similar difference between regression slopes and one was demonstrated for wilting point and water content at 1500 kPa w_{1500}. Somewhat similar results were observed by Cavazza et al. (1973) who noted that the average values of field capacity were close to water contents at 20 kPa in loamy soils. However, the regression of water contents at 20 kPa vs. water content at field capacity produced slopes significantly less then one. Rivers and Shipp (1978); Cassel and Sweeney (1974); Bennet and Entz (1989) compared field water contents at field capacity with water content at -10 kPa w_{10} for coarse-texture soils and saw difference up to 100% in favor of the field capacity. The field capacity is often thought to be associated with a specific level of matric potential, i.e, with -33 kPa, -10 kPa (Da Silva et al., 1994), -5 kPa (Hollis et al., 1977; MacLean and Yager, 1972), etc. Voronin's theory and our results show that such association is not justified.

Table 1
Key soil water contents for two soils (g g^{-1})

Horizon	W_s	W_{mc}	W_{mcsw}	W_{mmv}	W_{mav}
	Typical Chernozem (Typic Haploboroll)				
A1	0.478	0.448	0.402	0.288	0.103
A2	0.446	0.419	0.343	0.247	0.097
B1	0.427	0.330	0.280	0.225	0.082
	Soddy-podzolic soil (Spodosol)				
A1	0.382	0.323	0.286	0.192	0.034
B1	0.317	0.205	0.181	0.118	0.013
B2	0.275	0.253	0.223	0.181	0.063

A comparison similar to the one described above has been done for other key water contents. We have concluded that the key water contents as measured with methods proposed by Voronin (1990) give reliable characteristic points on soil water retention curves.

4. SELECTION OF KEY WATER CONTENTS TO ESTIMATE VAN GENUCHTEN'S PARAMETERS

Four key water contents augmented by the water content at saturation present five points on water retention curve. It was suggested (Baumer and Brasher, 1982; Baumer and Rice, 1988) to use water content at saturation and two extra points on water retention curve to estimate parameters α and n in the van Genuchten equation. Specifically, Baumer and collaborators (Baumer et al., 1994) used water contents at saturation, at -34 kPa and at -1540 kPa. We proposed to estimate van Genuchten's parameters from water content at saturation and three points on water retention curve defined by the key water contents. A test study was carried out to decide which three points of the four available from Equation (1)–(4) should be selected. Soils in this study were Typical Chernozem (Typic Haploboroll) and Soddy-podzolic soil (Spodosol). Key water contents for those soils are summarized in Table 1.

We compared the accuracy of van Genuchten equation fits to the following sets of available soil water contents: (A) all key water contents are available, (B) W_s, W_{mc}, W_{mcsw} and W_{maw} available, (C) W_s, W_{mcsw}, W_{mmw}, and W_{maw} available, (D) W_s, W_{mc}, W_{mmw}, and W_{maw} available. Van Genuchten parameters were found using Marquardt-Levenberg algorithm. The accuracy of the approximation was evaluated using the root-mean-square error (RMSE) across all five points at the water retention curve.

Comparison of the fits (Figure 3 and Table 2) shows that data sets (c) and (d) provide an accuracy comparable with the five-point data set (a).The data set (b) consisting of W_s, W_{mc}, W_{mcsw} and W_{maw}) provides substantially lower accuracy (Table 2). We conclude that the set of key water contents to develop van Genuchten's equation should include W_s, W_{mmw}, W_{maw}, and either W_{mcsw} or W_{mc}.

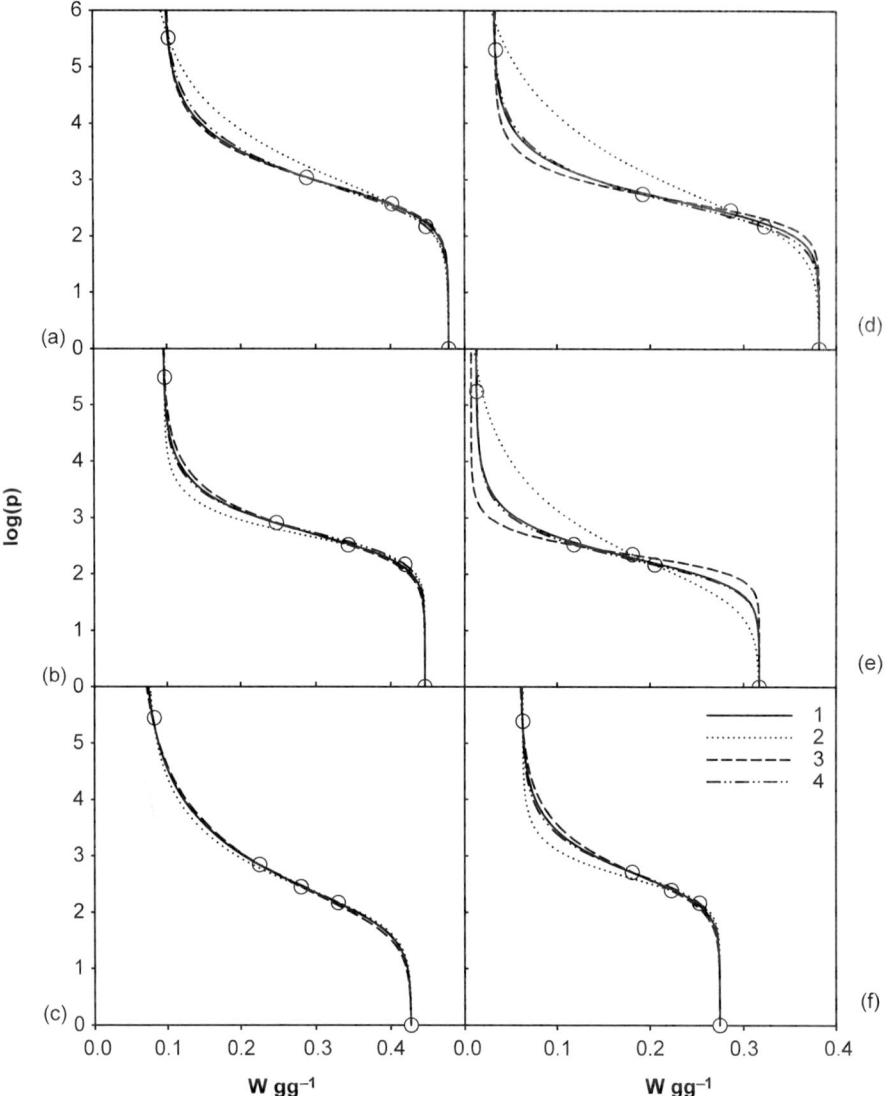

Figure 3. Water retention of Typical Chernozem: (a) horizon A1, (b) horizon B1, (c) horizon B2 and Soddy-podzolic soil: (d) horizon A1, (e) horizon A2, (f) horizon B1. Key water contents are shown with symbols. Lines show van Genuchten equation fitted to different sets of key water contents: (1) all key water contents are available, (2) W_s, W_{mc}, W_{mcsw} and W_{maw} available, (3) W_s, W_{mcsw}, W_{mmw}, and W_{maw} available, (4) W_s, W_{mc}, W_{mmw}, and W_{maw} available.

Table 2
Approximation coefficients of main hydrological characteristics for Typical Chernozem and Soddy-podzolic soil

Data set	Typical Chernozem (Typic Haploboroll)					Soddy-podzolic soil (Spodosol)				
	W_s (gg^{-1})	$\alpha \times 100$ (cm^{-1})	n	W_r (gg^{-1})	RMSE × 100 (gg^{-1})	W_s (gg^{-1})	$\alpha \times 100$ (cm^{-1})	n	W_r (gg^{-1})	RMSE × 100 (gg^{-1})
		Horizon A1						Horizon A1		
A	0.478	0.241	1.637	0.097	0.34	0.382	0.401	1.812	0.032	0.78
B	0.478	0.301	1.366	0.071	1.56	0.382	0.738	1.289	0.000	2.46
C	0.478	0.223	1.698	0.099	0.40	0.382	0.321	2.128	0.033	1.16
D	0.478	0.282	1.563	0.095	0.57	0.382	0.479	1.724	0.031	1.09
		Horizon B1						Horizon A2		
A	0.446	0.318	1.821	0.095	0.40	0.317	0.803	1.927	0.012	0.82
B	0.446	0.294	2.183	0.097	1.56	0.317	2.066	1.344	0.000	1.89
C	0.446	0.354	1.727	0.095	0.55	0.317	0.532	2.632	0.007	1.81
D	0.446	0.280	1.908	0.096	0.67	0.317	0.844	1.965	0.013	1.02
		Horizon B2						Horizon B1		
A	0.427	1.162	1.385	0.064	0.18	0.275	0.365	1.730	0.061	0.22
B	0.427	0.992	1.458	0.071	0.44	0.275	0.339	2.179	0.063	1.02
C	0.427	1.405	1.342	0.058	0.34	0.275	0.420	1.610	0.059	0.33
D	0.427	1.131	1.385	0.064	0.23	0.275	0.330	1.815	0.062	0.31

5. CONCLUDING REMARKS

Practical advantage of the key water content concept consists in the opportunity to use traditional measurements of field capacity, liquid limit, and plastic limit, and to couple those values with values of soil matric potential using the regression equations (1)–(4). Such pedotransfer approach relates easy-to-measure field values to water retention curves.

REFERENCES

ASTM, 1996. Annual Book of ASTM Standards. Am. Soc. For Testing and Materials, West Conshohocken, PA, Liquid Limit and Plastic Limit of Soils, D 4318-95.

Baumer, G.W., Brasher, B.R., 1982. Prediction of water contents at selected suctions, ASAE Paper No. 82-2590. American Society of Agricultural Engineers, St. Joseph, MI.

Baumer, O., Rice, J., 1988. Methods to predict soil input data for DRAINMOD, ASAE paper 88-2564. American Society of Agricultural Engineers, St. Joseph, Michigan.

Baumer, O., Kenyon, P., Bettis, J., 1994. MUUF v 2.14 User's Manual. USDA Natural Resource Conservation Service, Lincoln, Nebraska.

Bennet, D.R., Entz, T., 1989. Moisture-retention parameters for coarse-textured soils in Southern Alberta. Can. J. Soil Sci. 69, 263-272.

Cassel, D.K., Sweeney, M.D., 1974. In situ water holding capacities of selected North Dakota soils, Agric. Exper. Stn. Bull. 495. North Dakota State University, Fargo, ND.

Cavazza, L., Comenga, V., Linsalata, D., 1973. Correlation of field capacity between open field and laboratory determinations. *In*: Hadas, A., Swartzendruber, D., Rijtema, P.E., Fuchs, M., Yaron, B. (Eds.), Physical Aspects of Soil Water and Salts in Ecosystems. Springer, New York, pp. 87-193.

Da Silva, A.P., Kay, B.D., Perfect, E., 1994. Characterization of the least limiting water range of soils. Soil Sci. Soc. Am. J. 58, 1775-1784.

Haise, H.R., Haas, H.J., Jensen, L.R., 1955. Soil moisture of some Great Plains soils. II. Field capacity as related to 1/3-atmosphere precentage and minimum point as related to 15- and 26-atmosphere percentages. Soil Sci. Soc. Am. Proc. 19, 20-25.

Hollis, J.M., Jones, R.J.A., Palmer, R.C., 1977. The effect of organic matter and particle size on the water retention properties of some soils in the West Midlands of England. Geoderma 17, 225-238.

Klute, A., 1986. Water retention: Laboratory methods. *In*: Klute, A. (Ed.), Methods of soil analysis. Part 1. Physical and Mineralogical Methods. American Society of Agronomy, Soil Science Society of America, Madison, WI, USA, pp. 635-662.

MacLean, A.H., Yager, T.U., 1972. Available water capacities of Zambian soils in relation to pressure plate measurements and particle size analysis. Soil Sci. 113, 23-29.

Pachepsky, Ya. A., Timlin, D., Várallyay, G., 1996. Artificial neural networks to estimate soil water retention from easily measurable data. Soil Sci. Soc. Am. J. 60, 727-773.

Pachepsky, Ya.A., Rawls, W.J., Timlin, D.J., 1999. The current status of pedotransfer functions: their accuracy, reliability, and utility in field- and regional - scale modeling. *In*: Corwin, D.L., Loague, K., Ellsworth, T.R. (Eds.), Assessment Non-point Source Pollution in the Vadose Zone. Geophysical Monograph, 108. American Geophysical Union, Washington, DC, pp. 223-234.

Peele, T.C., Beale, O.W., 1950. Relation of moisture equivalent to field capacity and moisture retained at 15 atmospheres pressure to the wilting percentage. Agron. J. 42, 604-607.

Ratcliffe, L.F., Ritchie, J.T., Cassel, D.K., 1983. Field-measured limits of soil water availability as related to Laboratory-measured properties. Soil Sci. Soc. Am. J. 47, 770-775.

Rivers, E.D., Shipp, R.F., 1978. Soil water retention as related to particle size in selected sands and loamy soils. Soil Sci. 126, 94-100.

Stakman, W.P., Valk, G.A., van der Harst, G.G., 1969. Determination of soil moisture retention curves. I. Sand-box apparatus. II. Pressure membrane apparatus. ICW, Wageningen, The Netherlands.

Tomasella, J., Pachepsky, Y.A., Crestana, S., Rawls, W.J., 2003. Comparison of two approximation techniques to develop pedotransfer functions of water retention of Brazilian soils. Soil Sci. Soc. Am. J. 67, 1085-1092.

Vadiunina, A.F., Korchagina, Z.A., 1986. Methods of investigation soil physical properties. Moscow. Agropromizdat.

Van Genuchten, M.Th., 1980. A closed form equation for predicting the hydraulic conductivity of unsaturated soils. Soil Sci. Soc. Am. J. 44, 892-898.

Varallyay, Gy., Mironenko, E.V., 1979. Soil-water relationships in saline and alkali conditions. *In*: Kovda, V.A., Szabolcs, I. (Eds.), Modelling of soil salinization and alkalinization, Agrokemia es Talajtan (Suppl.), 28, pp. 33-82.

Voronin, A.D., 1990. An energy-based concept for the physical state of soils. Sov. Soil Sci. 22, 53-64.

Voronin, A.D., Dembovetskii, A.V., Shein, E.V., 1997. Analysis of main structural-functional relationships using databases on physical properties of soils. Pochovedenie 9, 1120-1123 (in Russian).

PART IV

SPATIAL COMPONENT IN PTF DEVELOPMENT

Chapter 15

DATA AVAILABILITY AND SCALE IN HYDROLOGIC APPLICATIONS

K. Smettem[1], G. Pracilio[1], Y. Oliver[1] and R. Harper[1]

[1]Centre for Water Research, The University of Western Australia, MO15, 35 Stirling Highway, Crawley, Western Australia, Australia 6009

1. INTRODUCTION

Describing the unsaturated soil water balance across a range of space and timescales and linking this information to surface runoff and groundwater flow remains a major research challenge in hydrology. Particular difficulties arise due to the non-linear nature of the fundamental hydraulic properties and extreme spatial and temporal heterogeneity encountered within the vadose zone in the field (Simunek et al., 1998). Indeed, many researchers believe that the lack of field-scale data remains a major limitation to the successful field application of physically-based models employing numerical solutions to Richards' equation (Mallants et al., 1998). Some researchers have expressed even more challenging views and consider that we are at a loss as to how to improve and validate distributed physically based models for hydrologic prediction (Beven, 1989; Grayson and Blöschl, 2000).

In describing the role of soils in the hydrologic cycle, Smettem (2002) pointed out that the need for highly detailed soils information in hydrologic modeling tended to decrease as the space and timescales under consideration increased. Milly (1994), e.g., was able to describe much of the annual variation in runoff from North American catchments by considering the soils within a catchment as little more than a lumped store for moisture.

The relative strengths and weaknesses of the reductionist (or "bottom-up") approach vs. the "top-down" data-driven approach (based on attempts to understand the controls on catchment response to rainfall) are reviewed in detail by Sivapalan et al. (2003).

We introduce these different perspectives to describing catchment hydrologic behavior in order to reflect on some critical questions: what soil information do hydrologists really need in order to model catchment processes? Is it advantageous to use a simpler indirect method for estimating hydraulic properties if it provides a continuous spatial coverage (as opposed to the usual discrete measurements)? Have we become too obsessed with establishing the best correlation between basic soil properties and measured hydraulic properties rather than considering the spatial and temporal dimensions of the processes we seek to study and the propagation of errors through the hydrologic models we use?

These questions are the subject of ongoing research and in this chapter we attempt to provide further background and insight into some of the key questions concerning pedotransfer functions, data availability and scale in hydrology.

As a point of departure, we commence by considering the issues associated with prediction of the soil water balance at a point in the landscape. This is the conventional basic step in the reductionist approach in order to test directly in the field the accuracy of the process descriptions and requirements for field parameter estimation.

2. DESCRIBING ONE-DIMENSIONAL FLOW

For one-dimensional flow, the Richards equation can be written as

$$\frac{\partial \theta}{\partial t} = \frac{\partial}{\partial z}\left(K \frac{\partial H}{\partial z}\right) + S \qquad (1)$$

where the hydraulic head, H, is the sum of the soil water pressure potential, h (L), and the gravitational potential, z (L), K is the hydraulic conductivity (Lt^{-1}), θ is the volumetric water content ($L^3 L^{-3}$), t is time and S is a source or sink strength ($L^3 L^{-3} t^{-1}$)

For some special cases, Equation (1) can be solved analytically (see Smith et al., 2002 for details) but in general, numerical solutions are required due to the highly non-linear nature of the fundamental parameter relations between the soil water pressure head and the soil water content and between the hydraulic conductivity and either the soil water pressure head or the soil water content. These parameter relations need to be known in order for a numerical solution to be implemented.

Numerous functions have been developed to describe the $\theta(h)$ relation. In our work we have routinely used the Brooks–Corey (1964) functions with smoothing to eliminate the original discontinuity (Hutson and Cass, 1987), and with no residual water content ($\theta_r = 0$). This approach has proved adequate for describing most Australian soils (Cresswell and Paydar, 1996; Bristow et al., 1999). The simplest unsmoothed version of the water retention function is written as (Campbell, 1985):

$$\frac{\theta}{\theta_s} = \left(\frac{h}{h_e}\right)^{-\lambda_w} \qquad h \leq h_e$$

$$= 1 \qquad 0 \geq h_e \qquad (2)$$

where θ_s is the saturated hydraulic water content, h_e is the air entry potential and λ_w is the pore size distribution index.

The problem of parameterization is also encountered in application of the widely used bucket-type models that usually estimate the flux by direct solution of the continuity equation:

$$\frac{\partial \theta}{\partial t} = -\frac{\partial q}{\partial z} \qquad (3)$$

where q is the water flux (Lt^{-1}).

Often the modeling procedure replaces θ with a depth increment of soil water (SW) over the depth of interest. ΔSW is then estimated from the water balance equation

$$\Delta SW = NI = P + SM - Q - E \tag{4}$$

where NI is net infiltration, P is precipitation, SM represents other forms of water input such as snowmelt or irrigation, Q is runoff and E is evapotranspiration.

In general, both physically-based and bucket-type water balance models have been developed and extensively tested at plot scale with detailed local parameterization. The tacit assumption of the reductionist paradigm when applied to larger field and catchment-scale problems is that the fundamental governing equations still apply and that upscaling is principally a matter of parameter acquisition.

From this reductionist perspective, upscaling of variably saturated flow is logically treated within a spatial framework as a gridded output of model realizations for each model cell or user defined region.

To reach this endpoint it is necessary to assign appropriate hydraulic properties to each cell or region. Much effort has been devoted to this property-gridding process and in the absence of measurements researchers often resort to pedotransfer functions (PTFs).

PTFs are defined as functional relationships that transfer properties available from routine soil surveys into missing soil properties, such as soil hydraulic characteristics (Bouma and van Lanen, 1987). The term *surrogates* has also been used to describe such auxillary data (Hamblin, 1991).

The procedure of embedding a physically based model in a Geographical Information System (GIS) framework is well described by Mohanty and van Genuchten (1996). Quantification of typical errors propagated in map algebra has received detailed treatment by Heuvelink (1998). The basic assumption that for each grid (over scales ranging from a few hectares to a few square kilometers) soil hydraulic parameters and vegetation parameters can be represented by single effective values is contrary to field observations of spatial variability. This assumption is addressed specifically by Vachaud and Chen (2002) and we shall return to it later.

Many physically based distributed hydrologic models also conceptualise the vadose zone transport processes as a sequence of vertical cells that do not interact laterally (Refsgaard and Storm, 1995) and thus again reduce the hydrodynamic modeling of the soil zone to a one-dimensional flow problem for each grid cell (e.g., Bui et al., 1996) or mapped regions representing a particular classified soil profile (Pracilio et al., 2003). Considerable additional computing power is required to link flow between cells in the unsaturated zone but some models have done this (Simunek, et al., 1996; Stolte et al., 1997). However, the question of anisotropy in hydraulic properties at field scale remains largely unresolved.

3. SOME ISSUES IN EXTRAPOLATING FROM POINT-BASED SOIL WATER BALANCE

Modeling the water balance at a particular site relies on the availability of accurate data. In the case of solutions to the Richards equation, the requisite hydraulic properties are the soil water retention relation, $\theta(h)$, and the unsaturated hydraulic conductivity relation, $K(h)$ or $K(\theta)$. For bucket-type models the required parameters are usually the saturated water content, θ_s, the water content at the "drained upper limit" or "field capacity", θ_{fc}, and the water content at lower limit, or "wilting point", θ_{wp}.

At any specific location, this usually equates to the estimation of a relatively large amount of model parameters because most soil profiles consist of layers with different physical properties. In general, if detailed field measurements of water balance are available then the most efficient approach is to derive the parameters directly from the measurements using an inverse optimization scheme (Lehmann and Ackerer, 1997; Simunek et al., 1998; Abbaspour et al., 1999, 2000). Clearly, the resulting parameter sets will be optimal for the conditions used to derive them. The physical consistency of the estimated parameters can only really be judged by using the calibrated model to predict the dynamics of the state variables for different combinations of initial and boundary conditions.

Although *in situ* field measurement of hydraulic parameters, by methods such as the instantaneous profile method, are generally preferred (Marion et al., 1994), limitations can arise due to their narrow measurement range and low spatial and temporal resolution (Tseng and Jury, 1993). This difficulty and the associated cost of obtaining the requisite hydraulic parameters, particularly at scales greater than specific experimental plots, has led to an ongoing interest in the utility of indirect methods such as PTFs.

It must be accepted that PTFs will give less precise model parameter estimates than direct measurement, particularly as scale increases. Wösten et al. (1990) investigated the effect of measurement scale on the soil water balance by comparison with neutron probe readings at 0.5 m depth. Not surprisingly, it was found that parameters measured directly gave better prediction of the measured water content than field averaged data on a regional scale, field averaged data on a national scale and data estimations obtained indirectly from basic soil properties.

PTFs have generally been divided into class PTFs, which use a designated soil descriptor such as soil type as the regression variable and continuous PTFs which use soil properties such as clay and organic matter contents as the regressed variables (Wösten, 1997). Recent advances include the use of Neural Networks (Pachepsky et al., 1996; Schaap and Leij, 1998), which are essentially multivariate nonlinear regression tools.

In soil hydrology, continuous PTFs have historically been employed to predict certain points on the soil water retention curve (Rawls et al., 1993). Water retention points such as "wilting point" or "lower limit" of plant available water and the drained upper limit (the so-called "field capacity") have particular application in bucket-type hydrological models such as PAWC (Bristow et al., 1999) and plant growth models such as CERES-Maize (Jones and Kiniri, 1986). More recently, attention has focused on predicting the parameters of specific analytical functions describing the water retention and hydraulic conductivity relations that can be incorporated directly into numerical solutions of the Richards equation (Rawls and Brakensiek, 1989; Vereecken et al., 1989; Smettem et al., 1994). This approach is computationally more efficient than estimating particular points on the water retention curve and then fitting an analytical model through the predicted points. The term "physical–empirical model" has been used to describe this estimation procedure, which attempts to retain a rational physical basis for the PTF.

We have shown in previous work (Smettem et al., 1999; Smettem and Bristow, 1999) that a relatively simple physico-empirical PTF can be derived from clay content and bulk density. The motivation for this development arose because of the paucity of quantitative physical information in existing Australian National and State soil databases. There is no database in Australia with the equivalent resolution of STATSGO (Soil Survey Staff, 1994) and presently no development of databases such as CONUS-SOIL (Miller and White, 1998), specifically tailored for environmental modeling applications. Most maps and reports that have been published in Australia for land resource assessment purposes contain only qualitative or semi-quantitative spatial information about soil physical properties, usually based on hand texture determination and qualitative soil structure descriptors. Although the data were adequate for some land evaluation purposes it was clear to us that for hydrologic applications a different line of attack would be required.

Our approach has therefore been to develop and test tools for rapid spatial appraisal of hydrologically significant soil physical properties in individual catchments. We introduce and discuss some of these specific approaches in this chapter but commence with a brief review of some simple physico-empirical PTFs that are suited to the available data.

3.1. Background to a simple physico-empirical pedotransfer function

Particle size distribution data is the most readily available data associated with soil surveys, and has a long history of use in the derivation of surrogates for estimating water retention data (Arya and Paris, 1981; Haverkamp and Parlange, 1986; Wu et al., 1990; Chang and Uehara, 1992; Smettem et al., 1994; Smettem and Gregory, 1996; Smettem et al., 1999). The underlying assumption is that there is some form of shape similarity between the cumulative particle size distribution (on a mass or particle number basis) and the water retention curve. Many of the aforementioned authors have shown that the power function particle size model has sufficient shape similarity to the Brooks–Corey water retention function for practical application. Assuming that the degree of saturation W is equal to the cumulative PSD function $F(d)$, we can write

$$W = F(d) = \frac{M(r < R)}{M_t} \tag{5}$$

where M is the mass fraction, M_t is the total mass, r is the particle size radius, and R is the sieve radius. Because soil survey data rarely contain more than four or five particle-size intervals, it is appropriate to represent $F(d)$ by a power function (Smettem et al., 1994; Bui et al., 1996)

$$F(d) = \frac{M(r < R)}{M_t} = \left(\frac{R}{R_1}\right)^{\lambda_p} \tag{6}$$

where R_1 is an upper size limit and λ_p is a particle size distribution index obtained from a log–log relation between $F(d)$ and R/R_1. If we fix R_1 at the conventional soil survey upper limit for sieve analysis (1 mm) then the relation between λ_p and the clay fraction C_f is

$$C_f = \frac{M_c}{M_t} = \left(\frac{0.001}{1}\right)^{\lambda_p} \tag{7}$$

Using % clay rather than the clay fraction allows us to write

$$\lambda_p = -0.1448 \ln(\% \text{ clay}) + 0.667 \tag{8}$$

If shape similarity between the PSD and the water retention curve WRC is assumed we can use Equations (2) and (6) to write

$$\left(\frac{h}{h_e}\right)^{-\lambda_w} = \left(\frac{R}{R_1}\right)^{-\lambda_p} \tag{9}$$

and λ_w and λ_p are equivalent if h_e is scaled to R_1. Note that R_1 is fixed (at 1 mm in our case) but that in practice h_e will vary with factors such as texture, structure and bulk density, so the coefficients relating λ_w and clay content will most likely differ from those relating λ_p and clay content. In practice it is therefore more useful to write λ_w in general terms as

$$\lambda_w = -a \ln(\% \text{ clay}) + b \tag{10}$$

where a and b are coefficients obtained from analysis of existing data sets.

Smettem et al. (1999) found that for the Australian datasets they studied the coefficients were: $a = 0.1087$, $b = 0.5$.

Working independently on one of the three New Zealand regional datasets used by Smettem et al. (1999) and augmenting it with additional data to create a second dataset, Grewal et al. (1998) reached the same conclusion that clay content was the most significant variable for estimating the slope of the power function water retention relation. These authors showed that for their datasets, clay content alone accounted for 66% of the variation in the pore size distribution index. Including bulk density into a multiple regression improved the correlation by a further 6%. However, bulk density information is not at present routinely available to us in Australian State land resource datasets.

Written in terms of %clay, the pore size distribution index calculated by Grewal et al. (1998) is

$$\lambda_w = (-1.71 + 0.558 \text{ \%clay})^{-1} \tag{11}$$

Earlier, Cosby et al. (1984) reported a similar regression equation

$$\lambda_w = (3.1 + 0.157 \text{\%clay})^{-1} \tag{12}$$

An intercomparison of these three relationships is shown in Figure 1. The original PTF of Smettem et al. (1994) gives the highest λ_w values for a given %clay but was derived entirely without calibration. The parameters used by Smettem et al. (1999) return very similar values of λ_w to the PTF of Cosby. The relation established by Buchan and Grewal (1990) gives λ_w values that are a little lower than the other PTFs for clay contents $>5\%$. At very low clay contents ($<5\%$), their relation gives high λ_w values compared to the Smettem and Cospy PTFs (data not shown in Figure 1).

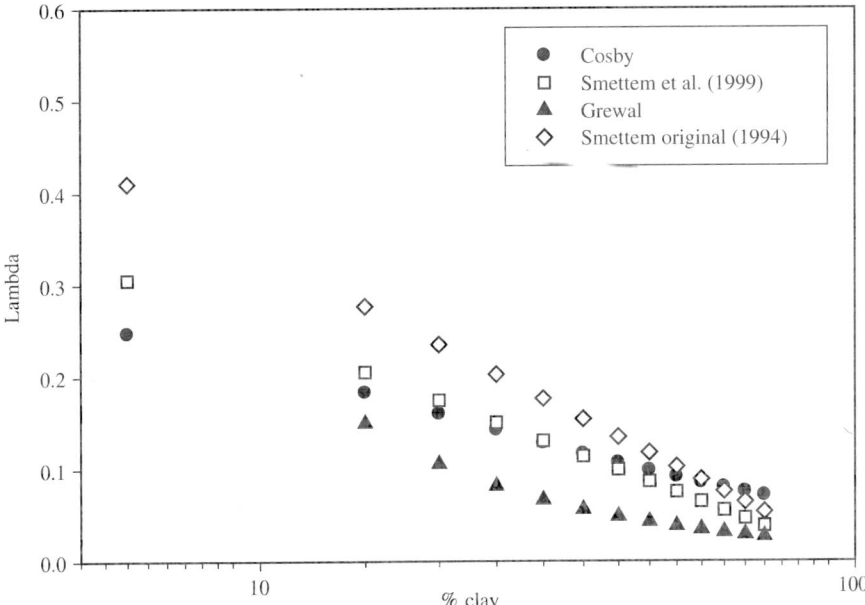

Figure 1. Comparison of four simple PTF relations for estimating the Brooks–Corey Lambda value from %clay.

Apart from λ_w, we also require the air-entry potential, h_e and the field-saturated water content θ_s to define the WRC. This is the "minimum set" of parameters required to describe the WRC (Buchan and Grewal, 1990).

The air-entry potential (cm) can be approximated using particle size data (Campbell, 1985)

$$h_e = 5\, d_g^{0.5} \qquad (13)$$

Here d_g is the geometric mean particle size (mm) which is related to λ_p. Smettem et al. (1999) give

$$\lambda_p = \gamma d_g^{0.5} \qquad (14)$$

Using six particle size classes, Smettem et al. (1999) evaluated γ as 0.87. With this value of γ in Equation (14), then introducing Equation (14) into Equation (13) and simplifying gives

$$h_e(\text{cm}) = 4.35 \lambda_p^{-1} \qquad (15)$$

Introducing Equation (15) into Equation (10) gives

$$h_e = 4.35/[-a\ln(\%\text{clay}) + b] \qquad (16)$$

If the field saturated water content is known, or estimated from bulk density (Cresswell and Paydar, 1996), then description of the WRC is completed by estimation from clay content alone using Equations (10) and (16) to obtain the parameters in Equation (2).

3.2. Difficulties with estimation of the "air entry" point

There are a number of problems associated with estimation of the so-called air entry point, h_e, that require some comment. First, in our experience any non-linear least squares fitting routine used to obtain water retention parameters usually returns very wide confidence intervals on the estimate of h_e regardless of the number of datapoints used in the fitting procedure.

From an analysis of Australian and UK soils, Gregson (1987) found a mean value of h_e to be zero. Although setting h_e to zero simplifies the $h(\theta)$ relation to a "one-parameter" model (assuming θ_s is known) it precludes the development of any physico-empirical $K(h)$ relation. Working with their two datasets, Grewal et al. (1998) reached a similar conclusion that h_e could be set to a constant but that the "best" constant was zero for one dataset and -10 cm for another dataset.

The derivation of both λ_w and h_e from %clay also infers an interrelation between these parameters but after log transformation of Equation (2) we have (Buchan and Grewal, 1990)

$$\ln h = a + b \ln(\theta/\theta_s) \tag{17}$$

where $a = \ln h_e$ and $b = 1/\lambda_w$.

The parameters a and b are now uncoupled and independent.

3.3. An intercomparison of three simple PTFs

We have recently completed collection and analysis of water retention data from 60 undisturbed soil cores (3 cm deep and 5.3 cm diameter) obtained from surveys across three catchments in Western Australia. Draining water retention data $h(\theta)$ were obtained for field saturation and nine different negative soil water potentials using Tempe cells and pressure plate techniques to cover the range from 0 to $-15,000$ cm. The pressure increments were $-10, -20, -30, -50, -70$ and -90 cm H_2O. Each core was then split into three 1 cm deep sections and pressure plate measurements were performed at -1000, $-10,000$ and $-15,000$ cm H_2O. Bulk density was determined following completion of the water retention measurements and oven drying of the samples at 105 °C.

Soil from the actual cores used for the water retention measurements was used for determining the particle-size distribution (PSD) using sieving and the pipette method described by Coventry and Fett (1979). Size classes included gravel (>2 mm diameter), coarse sand (0.2–2 mm), fine sand (0.02–0.2 mm), silt (0.002–0.2 mm) and clay (<0.002 mm). The fine sand was also divided into the 0.02–0.05 mm fraction to provide an estimate of the USDA silt fraction (0.002–0.05 mm).

The new data are used here to perform an intercomparison of three simple PTFs and to demonstrate the additional error associated with using these PTFs compared to the direct non-linear least squares fitting of Equation (2) to the water retention data. The PTFs evaluated are: (1) Smettem et al. (1999), who give the parameters in Equation (2) using Equations (10) and (16); (2) Cosby et al. (1984), who give λ_w in Equation (2) from the relation in Equation (12) and h_e (cm) from

$$h_e = 10(-0.0095\%\text{sand} + 1.54) \tag{18}$$

and (3) Saxton et al. (1986), who represented the WRC with a three part equation to account for the double inflection in the curve. The equations of Saxton et al. (1986) cover the ranges: (1) saturation to air entry; (2) air entry to -100 cm; and (3) -100 to $-15,000$ cm and are given by

$$\theta = \left(\frac{h}{10A}\right)^{1/B} \quad h \text{ from } 100 \text{ to } 15,000 \text{ cm} \tag{19}$$

$$\theta = \left[10 - \left(\frac{h}{10}\right)\right] \bigg/ \left[(10 - h_e)(\theta_s - \theta_{100})\right] \quad h \text{ from } h_e \text{ to } 100 \text{ cm} \tag{20}$$

$$\theta = \theta_s \quad h < h_e \tag{21}$$

The parameters in Equations (20) and (21) are

$$\theta_{100} = \exp(\ln 10 - \ln A)/B \tag{22}$$

$$A = 100 \exp[-4.396 + (-0.0715 \%\text{clay}) + (-0.000488 \%\text{sand}^2)$$
$$+ (-0.00004285 \%\text{sand}^2 \times \%\text{clay}) \tag{23}$$

$$B = -3.14 + (-0.00222 \%\text{clay}^2) + (-0.00003484 \%\text{sand}^2 \times \%\text{clay}) \tag{24}$$

Both Cosby et al. (1984) and Saxton et al. (1986) also give equations for estimating θ_s but for this intercomparison we use the measured field saturated water contents. The lower limit to the sand class in Equations (18), (23) and (24) is 0.5 mm.

Measured against predicted water contents are shown in Figure 2a–d. It is visually apparent that the greatest scatter is observed with the Saxton PTF and that all the PTFs have considerably more scatter than the direct estimation (Figure 2a) using Equation (2).

Two useful statistics for evaluating model performance are the residual mean square error (RMSE) and the mean bias error (MBE). The RMSE for an individual core is given by

$$\text{RMSE} = \left[\frac{1}{n_o}\sum_{i=1}^{n_o}(P_i - O_i)^2\right]^{1/2} \tag{25}$$

where P_i and O_i are the predicted and observed values and n_o is the number of observations.

The RMSE has the dimensions of the observations and so the smaller the RMSE the better the model prediction. This statistic does not however differ between under- and

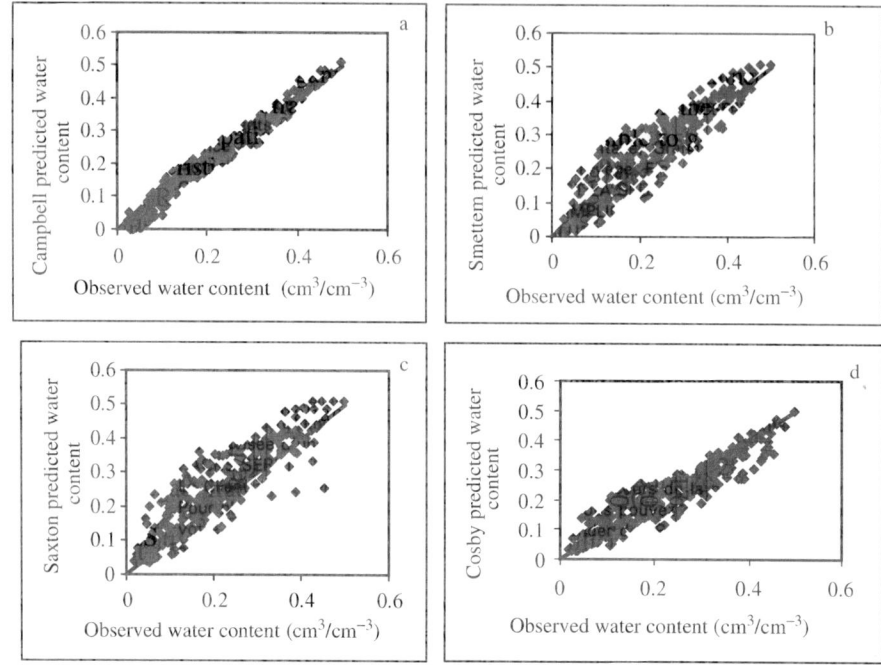

Figure 2. Comparison of observed v's predicted water contents obtained by (a) fitting the Campbell form of the Brooks–Corey water retention model to the water retention data, with estimation from three simple PTFs developed by (b) Smettem et al. (1999), (c) Saxton et al. (1986) and (d) Cosby et al. (1984).

over-prediction and so for this purpose we use the MBE defined by

$$\text{MBE} = \frac{1}{n_o} \sum_{i=1}^{n_o} (P_i - O_i) \qquad (26)$$

As expected, the RMSEs of the PTF models are greater than the RMSEs obtained from the fitted Water Retention Curve data. Average RMSEs are: Equation (2) fitted 0.0172; Saxton PTF 0.0558; Smettem PTF 0.0482; Cosby PTF 0.0392. For both the Smettem and Cosby PTFs 75% of the MBEs were within ± 0.04 but this fell to 50% for the Saxton PTF.

The MBEs for the three PTFs are shown in Figure 3. The Cosby PTF has the smallest bias, with all estimates $< \pm 0.06$ and 60% $< \pm 0.04$. The Smettem et al. (1999) PTF has 63% of estimates $< \pm 0.04$ but 10% of estimates > 0.06, whereas the Saxton PTF has 20% of estimates > 0.06.

Kern (1995) performed an intensive intercomparison of five PTFs, using data from the USDA-SCS National Soil Survey Laboratory Pedon Database. The PTFs included those of Cosby and Saxton in the study. The focus was on model performance at three particular negative soil water potentials (-100, -330 and -15000 cm) and results showed that the Cosby model performed well at predicting the -15000 cm water contents but Saxton model gave better predictions of water content at -100 and -330 cm.

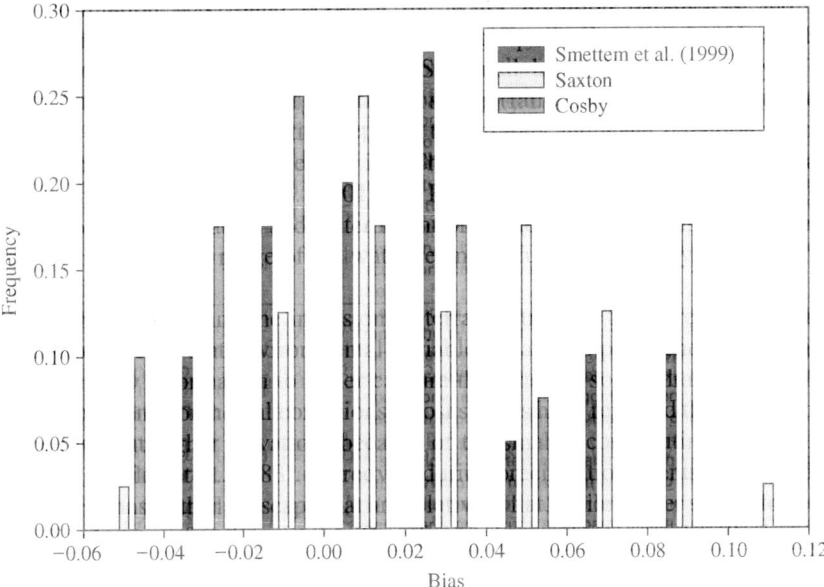

Figure 3. Comparison of bias for three simple PTFs used to describe water retention data from three catchments in Western Australia vs. bias obtained by directly fitting the Campbell water retention model to the water retention data.

From this analysis we conclude that there is little to choose between the Smettem and Cosby PTFs. Either PTF would be a reasonable first choice for estimating WRCs across regions where data is scarce.

3.4. Estimating the hydraulic conductivity "matching point" in the Brooks–Corey $K(h)$ or $K(\theta)$ relation

A number of researchers have developed pedotransfer function relationships to estimate the saturated hydraulic conductivity from basic data (Puckett et al., 1985; Campbell, 1985; Dane and Puckett, 1992). Some researchers have separated "macropore" and "matrix" components (Rawls et al., 1993; Smettem and Bristow, 1999) on the basis that K_{sat} often leads to overestimation of the unsaturated hydraulic conductivity–water content relation or unsaturated hydraulic conductivity–negative soil water pressure head relation, $K(h)$. (Smettem and Clothier, 1989; Jarvis and Messing, 1995; Mohanty et al., 1997).

Smettem and Bristow (1999) showed a general relation between the model of Campbell (1985) and the model of Rawls et al. (1993) and then used the simple model of Campbell to estimate the matrix K_{sat} (denoted K_m).

The Campbell (1985) equation is

$$K_m(\text{mm h}^{-1}) = C d_g \tag{27}$$

where d_g is the geometric mean particle size (mm) and C is a constant that was originally evaluated as 144 for the dataset of Hall et al. (1977).

Introducing Equation (14) into Equation (27) gives

$$K_m = (h_e/50)^{-2} = 2500 C\, h_e^{-2} \tag{28}$$

Accepting the original value of C gives K_m solely as a function of h_e, which is evaluated as a function of clay content from Equation (16).

Puckett et al. (1985) used regression to derive K_{sat} (mm h^{-1}) as a function of clay content. Their empirical equation is

$$K_{sat} = 157 \exp(-0.1975\ \%\text{clay}) \tag{29}$$

Dane and Puckett (1992) also report that clay content was the only significant variable for estimating saturated hydraulic conductivity of over 755 soils in Florida. Therefore, although it may be difficult to justify a direct relation between "matrix" hydraulic conductivity and air entry there are still useful empirical relations between K_m and %clay.

We emphasize that development of simple physico-empirical PTFs from minimal input data was originally motivated by the routinely available data in Australian soil databases.

4. SPATIAL MAPPING OF CLAY CONTENT USING ANCILLARY DATA

From the foregoing discussion it is clear that clay content is a critical soil texture parameter in development of simple PTFs. It is, therefore, necessary to adequately map the spatial distribution of clay content at the scale of interest.

At this stage, we do not advocate one ancillary method in preference to another but do recommend the use of any method that provides a continuous surface rather than relying on extrapolation from point measurements.

4.1. Gamma radiometric techniques

Gamma-rays are attenuated as they pass through matter due to interaction with electrons and atomic nuclei (Parasnis 1997). The degree of attenuation depends on the electron density of the matter through which the gamma-ray passes and is estimated from mass attenuation coefficients of the constituent elements and the density of source material (Lide, 1999). Duval et al. (1971) suggest that the attenuation coefficient is independent of material because oxygen and nitrogen are the dominant elements per unit volume in most earth materials, and that this is valid for air and most rock types. The percentage of the gamma-ray signal (Y) from a particular depth in centimeters (T) can be obtained from

$$Y(T) = (1 - e^{0.046 p_e T}) \times 100 \tag{30}$$

where p_e is the dry bulk density (g cm^{-3}).

At a bulk density of 1.6, about 50% of the gamma-ray signal would be derived from the top 10 cm and about 90% from the top 30 cm of soil.

Gamma-ray attenuation increases with soil moisture content and for this reason it is preferable to make measurements during the dry summer months. The percentage of signal attenuation increases by approximately 1% for each 1% volumetric increase in water content (Cook et al., 1996).

Data collected by radiometrics can be used to produce images of the spatial concentrations of radionuclides through the landscape. These are related primarily to the mineralogy and geochemistry of parent material and secondly to weathering processes leading to soil formation (Wilford et al., 1997). Radiometric data therefore provides an opportunity to remotely sense information relevant to the distribution of surface soil texture in the landscape.

4.2. High resolution airborne radiometric systems

Airborne radiometry has significant advantages over other soil mapping techniques, as it is able to map soil variables continuously at a high spatial resolution across the landscape rather than relying on the traditional method of extrapolating from point sources. An important advantage compared to other remote sensing techniques is that dense vegetation will only reduce gamma-ray detection by about 15% (Aspin and Bierwirth, 1997).

In earlier work (Smettem et al., 1994), radiometric imagery had been used as one line of "evidence" to establish prior Bayesian probabilities for the most likely clay class at any spatial location within a catchment. At the time, the available radiometric imagery was in the form of the standard ternary map. This map is generally enhanced in processing to provide the greatest discrimination between the red, green and blue color lines and between bright and dark areas. A typical "expert" interpretation of the ternary image for granitic parent material is shown in Table 1.

Table 1
Geochemical interpretation of radiometric data for granitic parent material in Western Australia

Radiometric data	Mineral	Landscape information
Potassium (K)	K-feldspar	Granite
Uranium (U), Thorium (Th)	Monazite	Laterite cap or ironstone gravels
Low total count	Silica	Sand
High total count	Adsorbed onto clay	Clay

More recently, Taylor et al. (2002) performed a detailed analysis of both airborne and ground-based radiometric measurements from a 7 by 11 km area portion of the Elashgin Catchment, 206 km north-east of Perth and 10 km south-east of Wyalkatchem, in the central eastern wheat belt of Western Australia.

In this study, high-resolution airborne radiometric measurements were made at 20 m intervals, along flight-lines 25 m apart, with a very low flying height of 20 m. Therefore, the majority of the signal (66%) emanated from a "footprint" with a width of approximately 20 m and length of 40 m based on approximate calculations for a moving detector by Ward (1981). A 32 L NaI crystal was used to detect gamma-ray radiation and 256 channels were monitored for a sampling period of half a second. These were then windowed into counts in four primary channels K, U, Th and TC (total count), converted to percentage for the former and ppm for the remaining three channels. Corrections were made for dead time, variations in terrain clearance and background radiation. A spectral stripping algorithm was then applied to the data before it was interpolated to a pixel size of 6 m. Digital elevation measurements were made simultaneously to the radiometric survey.

Taylor et al. (2002) reported good correlations between %clay and total count, TC, ($r^2 = 0.71$) and used this to map the distribution of surface clay content across the catchment. Using Equation (10), with the a and b coefficients for the Australian soil datasets reported by Smettem et al. (1999) allows the PTF derived distribution of λ_w to be mapped. A map for the Elashgin catchment, with the PTF derived λ_w draped on a digital elevation model (DEM) is shown in Figure 4. Because we are mapping from clay content as the primary variable it is possible to qualitatively visualize the

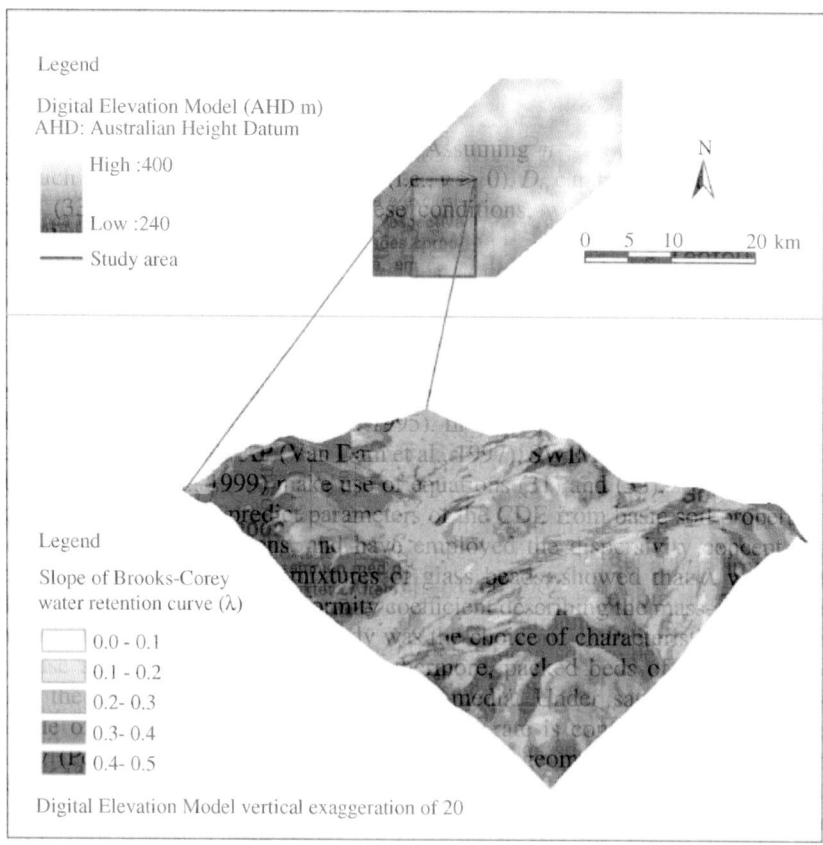

Figure 4. Spatial distribution of the Brooks–Corey pore size distribution index, λ, for the Elasghin catchment in Western Australia derived from ancillary radiometric data using the simple PTF of Smettem et al. (1999).

catenary sequence from the map. In this landscape we often find deep sands on the catchment divides (high λ values), alluvial and colluvial material on slopes and material with higher clay content in the valley floors. Geomorphologists report these catenary patterns to be repetitive throughout catchments in this region, so it may be possible to identify class PTFs from DEM derivatives augmented with limited soils information.

5. REDUNDANCY OF SOIL TEXTURAL CLASSES AND THE INTERRELATION WITH CLIMATE

Vachaud and Chen (2002) have shown that for long term modeling (using the ANSWERS model in their case), and depending upon the conditions of the soil, there exists a domain in the textural triangle for which classes of soil can be defined by a single set of textural parameters (sand, silt and clay content at the barycentre of the class) This "insensitive" area covering the classes: silt loam, silt, loam, sandy loam and sandy clay loam (the sand class was excluded from the analysis but would also be insensitive under their model analysis) has important consequences for large-scale distributed hydrological models as a simple "look-up" table could be used to define the average class properties and there is no need to account for the spatial variability of textural parameters within a class located in this insensitive area. In their study, the control appears to be rainfall intensity and the insensitive area covers those classes that have not run off and therefore exert no direct control on the soil water balance. Clearly, insensitive classes could be model, climate and application dependent. More work is needed in this area to determine how sensitive various physically based models are to soil input parameters. As more efficient solutions to the Richards equation are becoming available (e.g., Ross, 2003) it is critical that modelers and experimentalists work on identifying and adequately parameterising the governing hydrological processes at particular space and timescales of interest.

6. CONCLUDING REMARKS

The development of pedotransfer functions has catalyzed an important dialogue between soil scientists and hydrologists that may ultimately enrich our understanding of catchment hydrologic behavior. Runoff and redistribution processes are complex, and detailed spatial information at hydrological scales of interest is often lacking. This has led to a reliance on model calibration, rather than a more general conceptual understanding of the processes that govern the spatial and temporal distributions of water within catchments. Pedotransfer functions, together with other auxillary methods for sensing the response of catchments to rainfall are not simply land evaluation methodologies, but offer the hydrologist important tools for exploring the spatial response of catchments to runoff.

REFERENCES

Abbaspour, K., Sonnleitner, M., Schulin, R., 1999. Uncertainty in estimation of soil hydraulic parameter estimation by inverse modeling: example lysimeter experiments. Soil Sci. Soc. Am. J. 63, 501-509.

Abbaspour, K., Kasteel, R., Schulin, R., 2000. Inverse parameter estimation in a layered unsaturated field soil. Soil Sci. 165, 109-123.

Arya, L.M., Paris, J.F., 1981. A physicoempirical model to predict soil moisture characteristics from particle-size distribution and bulk density data. Soil Sci. Soc. Am. J. 45, 1023-1030.

Aspin, S.J., Bierwirth, P.N., 1997. GIS analysis of the effects of forest biomass on gamma-radiometric images. Proceedings of the 3rd National Forum on GIS in the Geosciences, Australian Geological Survey Organisation Report 1997/36.

Beven, K.J., 1989. Changing ideas in hydrology—the case for physically-based models. J. Hydrol. 105, 157-172.

Bouma, J., van Lanen, J.A.J., 1987. Transfer functions and threshold values: from soil characteristics to land qualities. *In*: Beck, K.J. (Ed.), Quantified Land Evaluation, Proceedings of an International Society of Soil Science and Soil Science Society of America Workshop, Washington D.C., International Aerospace Survey Earth Science Publication No 6. ITC Publication, Enshede, The Netherlands, pp. 106-110.

Bristow, K.L., Smettem, K.R.J., Ross, P.J., Ford, E.J., Roth, C.H., Verburg, K., 1999. Obtaining hydraulic properties for soil water balance models: some pedotransfer functions for tropical Australia. *In*: van Genuchten, M.Th., Leij, F.J., Wu, L. (Eds.), Proceedings of the International Workshop on Characterization and Measurement of the Hydraulic Properties of Unsaturated Porous Media. University of California, Riverside, USA, pp. 1103-1120.

Brooks, R.H., Corey, A.T., 1964. Hydraulic Properties of Porous Media, Hydrol. Pap. No 3. Colorado State University, Fort Collins, Colorado, USA.

Buchan, G.D., Grewal, K.S., 1990. The power-function model for the soil moisture characteristic. J. Soil Sci. 41, 111-117.

Bui, E.N., Smettem, K.R.J., Moran, C.J., Williams, J., 1996. Use of soil survey information to assess regional salinization risk using geographical information systems. J. Environ. Qual. 25, 433-439.

Campbell, G.S., 1985. Soil Physics with BASIC. Elsevier, New York.

Chang, H., Uehara, G., 1992. Application of fractal geometry to estimate soil hydraulic properties from the particle-size distribution. *In*: van Genuchten, M.Th., Leij, F.J., Lund, L.J. (Eds.), Proceedings of the International Workshop on Estimating the Hydraulic Properties of Unsaturated Soils. University of California, Riverside, USA, pp. 125-138.

Cook, S.E., Corner, R.J., Groves, P.R., Grealish, G.J., 1996. Use of gamma radiometric data for soil mapping. Aust. J. Soil Res. 34, 183-194.

Cosby, B.J., Hornberger, G.M., Clapp, R.B., Ginn, T.R., 1984. A statistical exploration of the soil moisture characteristic to the physical properties of soils. Water Resour. Res. 20, 682-690.

Coventry, R.J., Fett, D.E.R., 1979. A pipette and sieve method of particle-size analysis and some observations on its efficacy. Divisional Report No 38, CSIRO Division of Soils.

Cresswell, H., Paydar, Z., 1996. Water retention in Australian soils I. Description and prediction using parametric functions. Aust. J. Soil Res. 34, 265-283.
Dane, J.H., Puckett, W.E., 1992. Field soil hydraulic properties based on physical and mineralogical information. In: van Genuchten, M.Th., Leij, F.J., Lund, L.J. (Eds.), Proceedings of the International Workshop on Indirect Methods for Estimating the Hydraulic Properties of Unsaturated Soils. University of California, Riverside, USA, pp. 389-403.
Duval, J.S., Cook, B., Adams, J.A.S., 1971. Circle of investigation of an airborne gamma-ray spectrometer. J. Geophys. Res. 76, 8466-8470.
Grayson, R.B., Blöschl, G., 2000. Spatial Patterns in Hydrological Processes: Observations and Modeling. Cambridge University Press, Cambridge, p. 406.
Gregson, K., 1987. A one-parameter model for the soil water characteristic. J. Soil Sci. 38, 483-487.
Grewal, K.S., Buchan, G.D., Metha, S.C., 1998. Relationships between soil water retention and soil compaction. J. Ind. Soc. Soil Sci. 46, 165-171.
Hall, D.G.M., Reeve, M.J., Thomasson, A.I., Wright, V.F., 1977. Water retention, porosity and density of field soils, Soil Surv. Tecch. Monogr. 9. Harpenden:Rothamsted Experimental Station, UK, 74 pp.
Hamblin, A., 1991. Sustainable agricultural systems: what are the appropriate measures for soil structure? Aust. J. Soil Res. 29, 709-715.
Haverkamp, R., Parlange, J-Y., 1986. Predicting the water retention curve from the particle size distribution 1. Sandy soils without organic matter. Soil Sci. 142, 325-329.
Heuvelink, G.B.M., 1998. Error Propagation in Environmental Modelling with GIS. Taylor and Francis, London, 144 p.
Hutson, J.L., Cass, A., 1987. A retentivity function for use in soil–water simulation models. J. Soil Sci. 38, 105-113.
Jarvis, N.J., Messing, I., 1995. Near saturated hydraulic conductivity in soils of contrasting texture measured by tension infiltrometers. Soil Sci. Soc. Am. J. 59, 27-34.
Jones, C.A., Kiniri, J.R., 1986. CERES-MAIZE Model: A Simulation Model of Maize Growth and Development. A&T University Press, Texas, 194 pp.
Kern, J.S., 1995. Evaluation of soil water retention models based on basic soil physical properties. Soil Sci. Soc. Am. J. 59, 1134-1141.
Lehmann, F., Ackerer, P., 1997. Determining the soil hydraulic properties by a inverse method in one-dimensional unsaturated flow. J. Environ. Qual. 26, 76-81.
Lide, D.R., 1999. CRC Handbook of Chemisty and Physics. CRC Press, Boca Raton.
Mallants, D., Tseng, P.H., Vanclooster, M., Feyen, J., 1998. Predicted drainage for a sandy loam soil: sensitivity to hydraulic property description. J. Hydrol. 206, 136-148.
Marion, J.M., Or, D., Rolston, D.E., Kavvas, M.L., Biggar, J.W., 1994. Evaluation of methods for determining soil–water retentivity and unsaturated hydraulic conductivity. Soil Sci. 158, 1-13.
Miller, D.A., White, R.A., 1998. A Conterminious United States multi-layer soil characteristics data set for regional climate and hydrology modelling, Earth Interactions, online at http://EarthInteractions.org
Milly, P.C.D., 1994. Climate, soil water storage, and the average annual water balance. Water Resour. Res. 30, 2143-2156.

Mohanty, B.P., van Genuchten, M.T., 1996. An integrated approach for modelling water flow and solute transport in the vadose zone. Application of GIS to the Modelling of Non-Point Source Pollutants in the Vadose Zone, Special Publication, 48. Soil Science Society of America, Madison, WI, USA, pp. 217-233.

Mohanty, B.P., Bowman, R.S., Hendrickx, J.M.H., van Genuchten, M.T., 1997. New piecewise-continuous hydraulic functions for modelling preferential flow in an intermittent flood-irrigated field. Water Resour. Res. 33, 2049-2063.

Pachepsky, Y.A., Timlin, D., Varallyay, G., 1996. Artificial neural networks to estimate soil water retention from easily measurable data. Soil Sci. Soc. Am. J. 60, 727-733.

Parasnis, D.S., 1997. Principles of applied geophysics, Fifth edition. Chapman & Hall, London.

Pracilio, G., Asseng, S., Cook, S.E., Hodgson, G., Wong, M.T.F., Adams, M.L., Hatton, T.J., 2003. Estimating spatially variable deep drainage across a central-eastern wheatbelt catchment, Western Australia. Aust. J. Agric. Res. 54, 789-802.

Puckett, W.E., Dane, J.H., Hajek, B.F., 1985. Physical and mineralogical data to determine soil hydraulic properties. Soil Sci. Soc. Am. J. 49, 831-836.

Rawls, W.J., Brakensiek, D.L., 1989. Estimation of soil water retention and hydraulic properties. In: Morel-Seytoux, H.J. (Ed.), Unsaturated Flow in Hydrologic Modeling: Theory and Practice, NATO ASI Series C. Kluwer, Dordrecht.

Rawls, W.J., Brakensiek, D.L., Logsdon, S.D., 1993. Predicting saturated hydraulic conductivity utilizing fractal principles. Soil Sci. Soc. Am. J. 52, 26-29.

Refsgaard, J.C., Storm, B., 1995. Mike-She. In: Singh, V.J.P. (Ed.), Computer Models of Watershed Hydrology. Water Resour. Pub. Co., USA, pp. 809-846.

Ross, P.J., 2003. Modelling soil water and solute transport – Fast, simplified numerical solutions. Agron. J. 95, 1352–1361.

Saxton, K.E., Rawls, W.J., Romberger, J.S., Papendick, R.I., 1986. Estimating generalized soil–water characteristics from texture. Soil Sci. Soc. Am. J. 50, 1031-1036.

Schaap, M.G., Leij, F.J., 1998. Using neural networks to predict soil water retention and soil hydraulic conductivity. Soil Till. Res. 47, 37-42.

Simunek, J., Sejna, M., van Genuchten, M.T., 1996. The HYDRUS-2D Software Package for Simulating Water Flow and Solute Transport in Two-Dimensional Variably Saturated Media, Version 1.0, IGWMC-TPS-56. International Ground Water Modeling Center, Colorado School of Mines, Golden, CO.

Simunek, J.M., van Genuchten, M.T., Gribb, M., Hopman, J.W., 1998. Parameter estimation of unsaturated soil hydraulic properties from transient flow process. Soil Till. Res. 47, 27-36.

Sivapalan, M., Blöschl, G., Zhang, L., Vertessy, R., 2003. Downward approach to hydrological prediction. Hydrol. Process. 17, 2101-2112.

Smettem, K.R.J., 2002. The role of soil in the hydrologic cycle. In: Lal, R. (Ed.), Encyclopedia of Soil Science. Dekker, New York, pp. 671-673.

Smettem, K.R.J., Bristow, K.L., 1999. Obtaining soil hydraulic properties for soil water transport and leaching models from survey data 2. Hydraulic conductivity. Aust. J. Agric. Res. 50, 1259-1262.

Smettem, K.R.J., Clothier, B.E., 1989. Measuring unsaturated sorptivity and hydraulic conductivity using multiple disc permeameters. J. Soil Sci. 40, 563-568.

Smettem, K.R.J., Gregory, P.J., 1996. The relation between soil water retention and particle size distribution parameters for some predominantly sandy Western Australian soils. Aust. J. Soil Res. 34, 695-705.

Smettem, K.R.J., Bristow, K.L., Ross, P.J., Haverkamp, R., Cook, S., Johnson, A.K.L., 1994. Trends in water balance modelling at field scale using Richards' equation. Trends Hydrol. 1, 383-402.

Smettem, K.R.J., Oliver, Y.M., Heng, L.K., Bristow, K.L., Ford, E.J., 1999. Obtaining soil hydraulic properties for soil water transport and leaching models from survey data I. Water Retention. Aust. J. Agric. Res. 50, 283-289.

Smith, R.E., Smettem, K.R.J., Broadbridge, P., Woolhiser, D.A., 2002. Infiltration theory for hydrologic applications, Water Resource Monograph 15. American Geophysical Union, Washington DC.

Soil Survey Staff, 1994. State Soil Geographic Database (STATSGO) Data Users Guide, USDA Natural Resources Conservation Service Misc Pub 1492. US Government Printing Office, Washington DC, p. 88.

Stolte, W.J., McFarlane, D.J., George, R.J., 1997. Flow systems, tree plantations, and salinisation in a Western Australian catchment. Aust. J. Soil Res. 35, 1213-1229.

Taylor, M., Smettem, K.R.J., Pracilio, G., Verboom, W., 2002. Relationships between soil properties and high-resolution radiometrics, central eastern Wheatbelt, Western Australia. Exploration Geophys. 33, 95-102.

Tseng, P.H., Jury, W.A., 1993. Simulation of field measurement of hydraulic conductivity in unsaturated heterogeneous soil. Water Resour. Res. 29, 2087-2099.

Vachaud, G., Chen, T., 2002. Sensitivity of computed values of water balance and nitrate leaching to within class variability of transport parameters. J. Hydrol. 264, 87-100.

Vereecken, H., Maes, J., Feyen, J., Darius, P., 1989. Estimating the soil moisture retention characteristics from texture, bulk density and carbon content. Soil Sci. 148, 389-403.

Ward, S.H., 1981. Gamma-ray spectrometry in geological mapping and uranium exploration. Econ. Geol., 75th Anniversary Volume, 840-849.

Wilford, J.R., Bierwirth, P.N., Craig, M.A., 1997. Application of airborne gamma-ray spectrometry in soil/regolith mapping and applied geomorphology. AGSO J. 17, 201-216.

Wösten, J.H.M., 1997. Pedotransfer functions to evaluate soil quality. In: Gregorich, E.G., Carter, M.R. (Eds.), Soil Quality for Crop Production and Ecosystem Health, Developments in Soil Science, Vol. 25. Elsevier, The Netherlands, pp. 221-245.

Wösten, J.H.M., Schuren, C.H.J.E., Bouma, J., Stein, A., 1990. Functional sensitivity analysis of four methods to generate soil hydraulic functions. Soil Sci. Soc. Am. J. 54, 832-836.

Wu, L., Vomicil, J.A., Childs, S.W., 1990. Pore size, particle size, aggregate size and water retention. Soil Sci. Soc. Am. J. 54, 95.

Chapter 16

THE ROLE OF TERRAIN ANALYSIS IN USING AND DEVELOPING PEDOTRANSFER FUNCTIONS

N. Romano[*] and G.B. Chirico

Department of Agricultural Engineering, Division for Land and Water Resources Management, University of Naples "Federico II", Via Università, 100, 80055 Portici (Naples), Italy

[*]Corresponding author: Tel.: +39-081-2539421; fax: +39-081-2539412

1. INTRODUCTION

The influence of topography on soil properties and processes has long been recognized (Freeze, 1974; Ruhe, 1956; Walker et al., 1968), but only recently we are witnessing an increase in investigations and more quantitative analyses on this research topic (Florinksy et al., 2002; Gessler et al., 2000; Heuvelink and Webster, 2001; Lark, 1999). This is partly due to the improvements in existing techniques of modeling soil spatial variation, such as geostatistical interpolations or stochastic simulations, along with the availability of new tools for landscape analysis, such as digital elevation models (DEMs) of terrain. This chapter discusses how terrain features can be employed to efficiently parameterize the soil hydraulic behavior in land-surface models via pedotransfer functions (PTFs) and to attempt to improve the prediction performance of PTFs at various scales.

There is an increasing need in simplified approaches that should not only be able to capture the significant spatial patterns of soil hydraulic properties at larger scales, but also to estimate those more sensitive hydraulic parameters at fine scales. Crop models in precision farming and contaminant transport models are typical applications where such methods are required. As amply illustrated in previous chapters, PTFs are being developed for estimating soil hydraulic properties as alternative methods to direct measurements. PTFs express soil water retention and hydraulic conductivity characteristics as function of basic soil properties, such as texture, oven-dry bulk density, porosity, and organic carbon content that are easier to measure or available in existing soil databases (Elsenbeer, 2001; Tietje and Hennings, 1996; Tietje and Tapkenhinrichs, 1993). Wösten et al. (2001) and Minasny and McBratney (2002) discussed the matter of reliability of PTFs, but more work is needed to understand merits and shortcomings of PTFs and to explore the opportunities of enhancing PTF predictive abilities.

Basic soil physical and chemical properties exhibit significant spatial variations over a wide range of scales, from the local scale to the regional scale, with generally greater spatial correlation lengths compared to those of soil hydraulic properties (Blöschl and Sivapalan, 1995). Direct characterization of these properties over large spatial scales requires sampling that can be prohibitively expensive with the current measurement techniques. Therefore, PTF application may appear infeasible when soil hydraulic properties are needed over large land areas.

An efficient and cost-effective approach to the soil hydraulic characterization over larger land areas is to employ strategies for spatial interpolation to make predictions at unsampled locations. There are two alternatives: interpolating the soil hydraulic data after applying PTFs in the few observational points or interpolating the basic soil properties prior to applying the PTFs. Sinowski et al. (1997) and Heuvelink and Pebesma (1999) have evaluated these alternatives. They concluded that interpolating before using PTFs is the best option, as this appears the only way to ensure that the spatial structure of the basic soil properties (that can be very different from each other) is translated into the derived hydraulic property.

Several interpolation techniques can be found in literature. In particular, there are deterministic approaches, such as thin plate splines (Hutchinson, 1993) or statistical approaches, such as kriging and stochastic simulations that treat the target variable as a random variable. The performance of any interpolation strategy is significantly affected by the amount of data available and in particular the spacing between data (Hutchinson, 1993; Western and Blöschl, 1999). An advantage of kriging over deterministic techniques is that the former: (a) accounts for the spatial variability through the semivariogram of the variable of interest; (b) provides a measure of associated uncertainty and (c) enables ancillary information to be included through co-kriging. Conditional stochastic simulations, sampling either randomly or with the Latin–Hypercube technique, retain the advantages of kriging, but can be preferred for some applications as they take into account the tails of the relevant frequency distributions that are usually considered as outliers. This in turn may be important when making predictions with comprehensive models, i.e., in modeling solute transport in unsaturated zone.

To improve the interpolation, soil data should be complemented with auxiliary variables that exhibit a certain degree of correlation with the target soil properties and can be effectively measured at large spatial scales. These auxiliary variables can be related to the hydrological, geomorphological, geochemical or biological processes contributing to the soil formation and functioning in the area of interest (Jenny, 1980).

Terrain features are the environmental variables most commonly employed to complement soil data for two fundamental reasons. First, topography is an important pedogenetic variable and is strongly related with other pedogenetic variables, which are parent material, climate, the activity of the living organisms (biotic factors), and the age of soil (Jenny, 1980). Terrain affects lateral surface and subsurface flow, both through macropores and the soil matrix, and processes controlling lateral soil movements and deposition. Terrain also affects local climatic conditions, in particular, temperature regime and incident solar radiation, which in turn influence the vegetation dynamics and the soil moisture temporal dynamics associated with the evapotranspiration. These hydrological and geomorphological processes are the engines of important processes contributing to the soil formation, such as mineral weathering, leaching, erosion, sedimentation, decomposition, horizonation, etc. Moreover, terrain features can reflect spatial variability in the underlying parent material and differences in age of the corresponding soil profiles. Therefore, topography presents efficient, albeit ancillary information for assessing the spatial distribution and organization of soil properties, at least assuming that the effects of other environmental state variables controlling the soil formation are uniform within the area of interest.

The second reason for using terrain attributes as ancillary data is that nowadays they are widely available in a digital form at relatively low cost with high spatial resolution. Very often, terrain is the only information available for a spatial analysis. New and powerful techniques have been developed for automatic surveying and a wide array of efficient software tools are available to support work in digital storage, presentation and analysis of terrain data (Tarboton, 1997; Wilson and Gallant, 2000). To address soil-landscape analysis, terrain data is aggregated into simple terrain attributes or terrain indices, automatically computed from the elevation data. Such terrain indices have been introduced for assessing the spatial variability of other terrain-dependent and soil-related variables, such as soil moisture, erosion and shallow landslide hazard, with underlying processes similar to those affecting soil formation, albeit having much finer temporal scale than soil formation. It is important to note here that using the ancillary information for interpolating soil physical and chemical properties does not improve the predictive performance of the PTFs themselves. Improving the predictive performance of PTFs by incorporating ancillary environmental attributes as independent variables in PTFs is a completely different issue that has been recently addressed (Leij et al., 2004; Pachepsky et al., 2001; Rawls and Pachepsky, 2002; Romano and Palladino, 2002). The majority of PTFs are regression equations that are derived from data collected during site-specific field campaigns. Such PTFs have demonstrated their ability to estimate soil hydraulic data across the study region with acceptable precision compared to the costs of investigation. However, as all empirical relationships, PTFs can be effective only within soil types and environmental conditions for which they have been developed (Bastet et al., 1999; Tomasella and Hodnett, 1998). Terrain attributes can contribute as a source of site-specific information that is not included in PTFs, which have been developed with soil databases collected at continental scale.

The problems of better assessing the spatial variability of soil hydraulic properties and improving in the local estimation of these properties can be translated into two main issues: (i) more efficient interpolations between available input predictor variables and (ii) the possibility to locally calibrate a PTF in a cost-effective way. Therefore, a specific objective of this chapter is to present a framework in which terrain attributes can help to attack both these problems when a PTF is employed to estimate soil hydraulic properties. Techniques developed for computing terrain attributes are illustrated in Section 2, while the issue of interpolating spatially sparse data of soil properties using terrain attributes as an ancillary data is discussed in Section 3. Section 4 discusses the opportunities in improvement of predictive capabilities of PTFs via incorporation of topographic information.

2. TERRAIN ANALYSIS FOR LANDSCAPE DESCRIPTION

The basic topographic information for terrain analysis is given in the form of a DEM. A DEM is an ordered array of numbers that represents the spatial distribution of elevations above some arbitrary datum in a landscape (Moore et al., 1991). There are three major ways of structuring a DEM: elevation data on a regular grid; irregularly spaced (x, y, z) data; and digitized contour lines stored in the form of (x, y) coordinate pairs along each contour line of specified elevation. A DEM can be analyzed to derive specific terrain attributes, representing properties of terrain surface that are significant to describe environmental properties and processes active in the landscape. Terrain attributes generally used for soil-landscape analysis are presented in Table 1. Elevation, slope angle, slope orientation (aspect), plan, profile and tangential curvatures, and specific catchment area (SCA) are classified as primary terrain attributes, since they can be computed independently from the elevation data. The other attributes are known as secondary attributes or compound indices, since they are computed by combining primary attributes (Wilson and Gallant, 2000). Secondary terrain attributes comprise, in particular, topographic wetness indices, stream-power indices, the sediment transport index, the direct solar radiation index, and temperature indices. An ordered array of numbers representing the spatial distribution of terrain attributes computed from a DEM is formally defined as

Table 1
Main landform attributes used in soil spatial characterization (Moore et al, 1991; Shary, 1995; Wilson and Gallant, 2000; Florinksy et al., 2002)

Attribute	Definition	Process or physical variable to which can be correlated
Elevation (m)	Elevation above a reference point	Local climatic gradients, vegetation patterns
Slope, β (radiants)	Angle between the horizontal plane and a plane tangent to the terrain surface	Velocity of lateral fluxes
Aspect (radiants)	Angle clockwise from north of the projection of the normal vector to the terrain surface on the horizontal plane	Direction of lateral fluxes, relative intensity of the solar radiation
Profile curvature (m^{-1})	Curvature of the terrain surface in the direction identified by the aspect angle	Downstream acceleration of lateral fluxes, rate of erosion processes
Tangential curvature (m^{-1})	Curvature of the terrain surface in the direction orthogonal to that identified by the aspect angle	Convergence of lateral fluxes, rate of lateral accumulation
Specific catchment area, a (m)	Upslope catchment area per unit contour length	Rate of lateral fluxes
Wetness index	$\ln(a/\tan \beta)$	Local flow accumulation
Potential solar radiation index	Ratio of the incident solar extraterrestrial radiation on the local sloping surface to that on the horizontal surface at a given latitude and longitude	Relative variability of evapotranspiration rate
Stream power index (m)	$a \tan \beta$	Potential flow erosion

Digital Terrain Model, or DTM (Moore et al., 1991). Terrain attributes can be computed with any of the three types of elevation data structures mentioned above. These three structures present, however, different advantages and limitations with respect to the analysis of terrain.

Local terrain attributes, such as slope, aspect, profile and tangential curvatures can be computed after fitting a surface to the point elevation data within the DEM. Several methods have been proposed for fitting a surface to the elevation data, such as kriging (Matheron, 1973) or spline interpolation (Hutchinson and Gessler, 1994). Slope and aspect are expressed using the first-order derivatives of the fitted surface, whereas profile and tangential curvatures are expressed using the second-order derivatives. The derivatives can be computed by means of various finite-difference schemes (Moore et al., 1993). Local terrain attributes can be also computed directly from gridded DEMs without any surface fitting, although this approach is thought to produce poorer results (Moore et al., 1991). SCA patterns are computed with special terrain analysis algorithms that generally produce an intermediate data sets including: (1) the spatial discretization of the terrain in a network of elemental units (element network); (2) the connectivity among the elements; and (3) the element terrain attributes.

There are three basic types of element networks, corresponding to the three basic methods of structuring elevation data, namely grid networks, triangulated irregular networks (TINs), and contour-based networks (Figure 1).

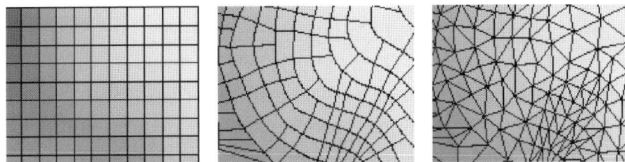

Figure 1. Different types of element networks. From left to right: grid, contour and TIN networks.

Contour-based networks employ the set of contour lines to subdivide the surface in irregular polygons. Each polygon consists of an upper and lower contour segment and two flow lines connecting the upper and lower contours. Contour lines are assumed to be equipotential lines; pairs of orthogonals to the equipotential lines, called "flow lines", form the "stream tubes". The major advantage of contour-based DEMs is that their structure is based "on the physics of water movement on the land surface" (Moore et al., 1991). In other words, the element network is structured according to the flow paths defined by the terrain. Thus once the flownet is constructed, there is no ambiguity in the lateral distribution of the flow. A contour-based terrain analysis, if feasible, has to be preferred to other methods. Contour-based methods, however, are not robust. The construction of the contour-based element networks is fairly complicated, time-consuming and prone to error. Complex terrains often require extensive user involvement, with fine-tuning, adding and deleting contours in order to avoid invalid flow lines (Maunder, 1999). Contour-based networks also do not offer an efficient structure for creating and storing spatial data. For all these reasons, contour-based networks are not commonly used in terrain-based applications.

Procedures based on grids or on TINs do not require much effort in the spatial discretization, as they produce a network structured according to the spatial distribution of the point elevation data. On the other hand, these procedures are more approximated in the computation of the element connectivity and attributes because of the inconsistencies between the element network structure and the underlying hypotheses concerning the lateral flow response of the system.

The TIN-based models usually sample characteristic points, such as ridges, peaks and breaks in slope, forming an irregular network of points stored as a set of 3D spatial co-ordinates along with the pointers to neighbors in the network. TINs are very efficient methods for representing terrain, because the density of points can be varied to suit the complexity of the surface. Points can be dense where elevation is changing rapidly and sparse in flatter areas. The elementary area is the facet, defined as the triangle having three adjacent points in the network as vertices. The problem with TINs for distributed modeling is that flow paths are difficult to represent. The irregularity of the TIN makes the computation of terrain attributes much more difficult than by grid modeling. There can also be difficulties in determining the upslope connection of a facet.

Overall, grid-based procedures are the most popular. Various algorithms have been proposed to analyze lateral flow paths with gridded elevation data. These algorithms can be divided into two main categories: single-flow-direction (SFD) algorithms and multiple-flow-direction (MFD) algorithms. The SFD algorithms connect each element with a single downslope element. MFD algorithms connect each element with one or more downslope elements. In the latter case, the proportion of flow allocated to each downslope element is defined by the connectivity.

SFD algorithms in use are D8 (O'Callaghan and Mark, 1984), Rho8 (Fairfield and Leymarie, 1991) and Lea's method (Lea, 1992). MFD methods in use are the multiple slope methods proposed by Quinn et al. (1991) and Freeman (1991), and the D_∞ method (Tarboton, 1997). All these algorithms produce terrain discretization with rectangular pixels with sizes defined by the spacing of the elevation data. The center of the pixel is considered as point source of the outgoing lateral flow and the flow is treated as 1D. Different criteria have been adopted to define the downslope elements and the proportions of the outgoing flow to be apportioned to each downslope elements. Basically, these criteria stem from different approaches to definitions of the local slopes and flow directions according to the relative differences in elevation of each pixel to the other eight surrounding pixels. MFD algorithms have been generally recognized as performing better than SFD algorithms, at least at the hillslope scale.

The connectivity of an element network defines how the elementary units are laterally linked. In particular, for each element the connectivity defines: (1) the downslope elements and (2) the proportion of outgoing lateral fluxes allocated to each downslope element. Attributes of an element network define the terrain attributes of the elements that are useful in the description of the lateral and vertical processes. The elements are generally assumed to be trapezoidal. The element attributes are computed by averaging in space the 3D properties of the element. Examples of element attributes are the spatial dimensions of the element, the slope and the element aspect. It is worth noting that element attributes such as slope, aspect and curvature derived from terrain element networks generally differ from the corresponding local attributes derived by surface fitting. The former are in fact properties of the elementary units in which the terrain surface is discretized; the latter are point properties of the fitted terrain surface. When primary attributes are compounded into terrain indices (also known as secondary terrain attributes), it is important that all primary attributes are derived by the same method of terrain surface analysis characterization.

Another approach to define SCA is taken in the DEMON algorithm (Costa-Cabral and Burges, 1994). Unlike other grid methods, this algorithm treats the flow originated by each pixel as 2D and computes the contributing area patterns by tracking downslope flow tubes starting from the corners of each pixel. It also defines the flow width to be associated to each pixel by projecting the pixel to the orthogonal of the flow direction. The DEMON is, probably, the most precise in revealing the SCA patterns on grids. However, it is not a robust method and it cannot be used in all practical circumstances (Tarboton, 1997).

2.1. Primary terrain attributes

Elevation, slope, profile and tangent curvatures are local properties of the terrain surface analyzed with respect to the gravitational field, while aspect is a local property of the terrain surface analyzed with respect to both the solar radiation field and the gravitational field. SCA is a nonlocal property as it characterizes the entire upslope area. Elevation, aspect and slope can represent together the spatial variability of the local microclimatic conditions from the hillslope scale (i.e., variability within a hillslope) to the catchment scale (i.e., among adjacent hillslopes). Elevation can be also directly correlated with precipitation and temperature patterns. Slope and aspect together influence the solar incident radiation and its variability among adjacent hillslopes (Western et al., 1999). These microclimatic conditions characterize the soil moisture regime and the biological activity over seasonal temporal scales and influence the variability of soil properties among adjacent hillslopes. Elevation can be also a surrogate of the variability of the parent material.

The computed primary terrain attributes are sensitive to the spacing of the original terrain data (Figure 2). Differences in computed terrain attributes owing to different spatial resolutions of the original elevation data are to be expected and several studies have explored the sensitivity of computed terrain attributes to the grid-DEM spatial resolution and the flow-direction algorithm (Quinn et al., 1991; Wilson et al., 2000; Zhang and Montgomery, 1994). It has been shown that in grid-DEMs the percentage of the catchment

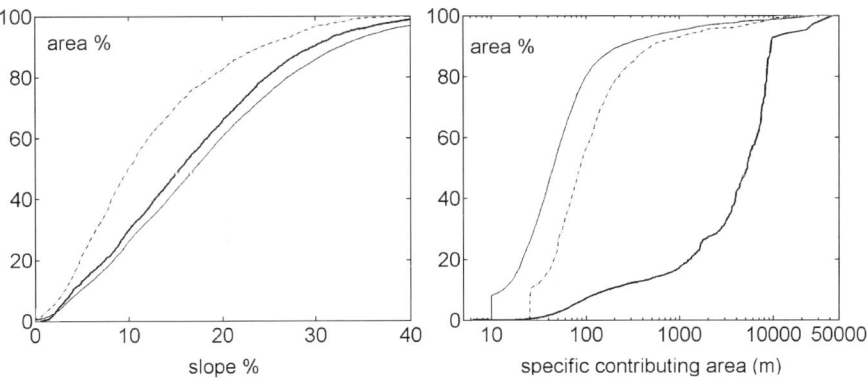

Figure 2. Effects of the DEM scale and element network type on the computed terrain attributes. Sample distributions of the slope (left) and of the SCA (right) computed on a 1 km^2-catchment using: a 10-m grid DEM (thin continuous line), a 25-m grid DEM (dashed line) and a contour network (thick continuous line).

steeper than a given slope systematically decreases as the grid size increases and the largest effect is for the steepest portions of the catchment (Wilson et al., 2000; Zhang and Montgomery, 1994). SCA is also sensitive to grid-DEM resolution. Larger grid sizes produce specific contributing area distributions biased toward larger values (Zhang and Montgomery, 1994). Larger pixel size in fact increases the minimum contributing area that can be represented (particularly in the ridge areas) and in the valleys distributes the contributing area on large areas around the channels.

This sensitivity of the terrain attributes to the spacing of the original elevation data can be significant in soil-landscape modeling. The resolution of the elevation data ideally should account for the nature of the modeled soil spatial variability and the data availability for model testing. However, there are several practical constraints that often make the choice of the scale a "pragmatic decision" (Grayson and Blöschl, 2000), and the resolution of the available topographic data is often one of those.

2.2. Secondary terrain attributes

Secondary terrain attributes are derived empirically or by an analytical synthesis of the processes controlling the spatial dynamics of the variable of interest, under simplifying hypotheses concerning the actual processes. Table 1 lists three secondary attributes that are generally used in soil-landscape analysis: the topographic wetness index (Beven and Kirkby, 1979), the potential solar radiation index and the stream power index (Lee, 1978; Moore et al., 1988). The topographic wetness index aims at describing the spatial variability of soil moisture, and it has been derived under the following main hypotheses: (1) uniform upslope recharge rate; (2) equilibrium condition of the upslope catchment response (steady-state assumption); (3) lateral sub-surface flow parallel to the surface slope and, therefore, unit contour length equivalent to the unit flow width; (4) hydraulic gradient of lateral sub-surface flow equal to the terrain slope. The potential solar radiation ratio, which is the ratio of the potential solar radiation on a sloping surface to that on a horizontal surface, has been suggested as an approximate method for examining the spatial distribution of solar radiation across a catchment, assuming that topography is the only source of variability of the incident radiation and neglecting the atmospheric effects. It is a function of the latitude, the declination of the sun (i.e., the day of the year) and local topographic attributes of slope and aspect. The stream power index is expressed as the product of SCA, which is assumed proportional to the local lateral discharge rate, and the local slope, representative of the local energy gradient and it is a measure of the erosive power of the surface lateral flow.

Other compound indices have been suggested in literature. For example, Park et al. (2001) proposed the terrain characterization index, which is given by the product of the surface curvature and the logarithm of the SCA.

3. TERRAIN ATTRIBUTES AS AUXILIARY DATA FOR INTERPOLATING SOIL PROPERTIES

As discussed in the "Introduction" section, the spatial interpolation of soil physical and chemical properties can be an effective strategy for predicting soil hydraulic properties over large spatial scales by means of PTFs, when only sparse soil sample observations are available.

The quality of the interpolated pattern obviously depends not only on the accuracy, the spacing and the extent of the observations, but also on the interpolation strategy employed. Any interpolation procedure implicitly or explicitly assumes some degree of spatial organization for the interpolated variable. The issue is then whether the spatial structure introduced by the interpolation procedure is realistic (Blöschl and Grayson, 2000).

Soil properties can exhibit different degrees of spatial organization, ranging from the absence spatial organization (white noise) or limited organization, to the case of highly structured patterns over different spatial scales (Blöschl, 1996; Western et al., 2001). Limited spatial organization is referred to the degree of continuity in soil properties, which can be expressed by semivariograms. The semivariograms can be easily defined from sample data and used for spatial interpolation with geostatistical techniques (Goovaerts, 1999; Western et al., 1999).

Highly structured patterns are more difficult to identify from sample data alone. Examples are soil with dominant textural class along convergent features, such as drainage lines or hollows, or connected bands of higher hydraulic conductivity along depositional landforms. In these cases, auxiliary information concerning the physical processes underlying the development of this spatial structure is required to support the information obtained from soil sampling.

Terrain attributes provide auxiliary information about the spatial organization of the soil properties caused by the influences of terrain on soil-forming processes. The scale resolved by the DEM from which terrain attributes are derived represents a lower limit. An upper limit is represented by the spatial scales of the physical processes pertinent to the terrain attributes. Attributes such as slope, concavity, wetness index, representing the conditions of lateral flow processes, are able to describe mainly the source of variability at the hillslope scale. Other terrain attributes, such as the potential solar radiation index, are able to describe the source of variability among adjacent hillslopes (Western et al., 1999).

There are several spatial interpolation methods employing auxiliary environmental data. These methods can be divided into two categories (Blöschl and Grayson, 2000): single-step and dual-step methods.

Single-step methods optimize the interpolation by employing jointly the original sample data and the auxiliary data. Examples of single-step methods are kriging with external drift and the co-kriging. A review of these methods can be found in Odeh et al. (1994, 1995). Kriging with external drift postulates the existence of a relationship between the original data and the auxiliary data. This relationship represents a trend in the spatial pattern of the analyzed variable that is added to its stochastic component. The parameters of this relationship are estimated simultaneously with the parameters of the semivariogram of the spatially dependent component. Co-kriging is instead a multivariate extension of the kriging, which exploits simultaneously semivariograms of both auxiliary and original data and their covariance. Co-kriging can be either isotopic or heterotopic (Odeh et al., 1995). Heterotopic co-kriging is more generalized and more appropriate in soil-terrain analysis, since it recognizes that the target soil variable can be undersampled compared to terrain attributes, whose resolution is generally determined by the original DEM resolution. Co-kriging generally introduces less spatial dependence between auxiliary and original data and therefore the interpolated patterns tend to be smoother (Blöschl and Grayson, 2000). Dual-step methods quantify the relationship between soil properties and terrain features separately from the spatial interpolation. The most popular technique is the multi-linear regression approach. In a first step, a regression relationship is developed between terrain attributes and soil properties, and this relationship is then used to estimate the soil properties across study area where soil sample data are not available. Moore et al. (1993) explored the performance of this technique in a 5.4 ha study site. They found significant correlation between terrain attributes and soil properties. In particular, slope and wetness index accounted for about one-half of the variability of the depth of A horizon, organic matter content, pH, extractable phosphorous, silt and sand contents. Similar studies have been carried out then by several authors (Brubacker et al., 1994; Florinksy and Arlashina, 1998; Florinksy et al., 2002; McKenzie and Ryan, 1999; Park et al., 2001).

A limitation of the multilinear regression methods is that while accounting for the deterministic relation between original data and auxiliary variables, they do not represent the spatial structure of the soil properties, e.g., the limited degree of spatial structure represented by the semivariograms. Analyzing the regression residuals can reveal this small-scale variability. Residuals represent the component of soil spatial variability that cannot be captured by the terrain attributes. In particular, the residuals can reflect: (a) the source of soil variability characterized by the spatial scales that are finer than the scale resolved by the DEM and (b) the source of variability that are related to other environmental variables that are independent from the topography (Western et al., 1999). In order to integrate this small-scale variability in the interpolated pattern, regression residuals can be interpolated by ordinary kriging and the final patterns given by the sum of the regressed values and the interpolated residuals. This approach is also known as regression–kriging method (McBratney et al., 2000; Odeh et al., 1995; Western and Grayson, 2000). Another possible hybrid method, combining regression models and geostatistical model is "kriging with uncertain data," which involves a simple regression followed by ordinary kriging of the regressed values (Odeh et al., 1994, 1995).

An optimal statistical method valid for all soil variables and study cases does not exist. Odeh et al. (1994, 1995) have compared the performance of multi-linear regression, co-kriging (both isotopic and heterotopic), kriging with external drift, the combined regression–kriging technique and kriging with uncertainty in interpolating soil properties using terrain attributes in experimental fields. The performance of each method differs significantly, depending on the target soil variable. Heterotopic co-kriging and regression–kriging methods appear the most promising.

The methods illustrated above always presume that a significant correlation between terrain attributes and soil properties can be established across the entire study area. This condition can be a strong limitation for their applicability, especially when the analysis is extended to larger spatial scales, beyond the spatial scales of variability that can be captured by the terrain attributes. The relationship between soil properties and terrain attributes can be very complex at these large spatial scales, since it is the result of the interaction between several soil forming factors that are variable in space.

In order to account for the nonlinearity in the relationships between terrain attributes and soil properties, methods alternative to the multilinear-regression have been suggested, such as the generalized linear models and regression tree models (McKenzie and Ryan, 1999), conditional probability approaches (Cook et al., 1996; Lagacherie and Voltz, 2000), artificial neural networks (ANNs) (Zhu et al., 1997). Another procedure that has been suggested is to characterize terrain attributes by fuzzy c-means classification procedure (also known as continuous classification) and then employ the corresponding membership grades (rather than the actual terrain attributes) as predictor variables in regression models or auxiliary variables in co-kriging in order to interpolate the observed soil properties in sample points (De Bruin and Stein, 1998; Lark, 1999).

Ultimately, all the above-mentioned methods are multivariate analysis techniques aiming at exploring the correlation between soil physical and chemical properties and terrain variables that are easier to measure in the field. This correlation is exploited for interpolating soil physical and chemical data measured in a few sample data.

4. TERRAIN ATTRIBUTES AS INPUT PARAMETERS IN PTFs

The predictor variables of the PTFs in the literature comprised particle-size distribution parameters (e.g., geometric mean diameter and its standard deviation) or contents of textural components (usually, percent of sand, silt, and clay contents according to the USDA classification system), oven-dry bulk density, ρ_b, and soil organic matter (Briggs and Shantz, 1912; Gupta and Larson, 1979). Measured values of water content, θ, at matric pressured heads, h, of -3.3 and -150 m (say, $\theta_{-3.3}$ and θ_{-150}) were then added in a few PTFs to enhance estimation of the soil water retention function, $\theta(h)$ (Rawls and Pachepsky, 2002). Note that the two retention values $\theta_{-3.3}$ and θ_{-150} are often found in available databases because of their use in some crop and environmental models to identify the conditions of field capacity and permanent wilting in a field soil. Those input variables were almost the only predictors used in PTFs developed to infer any soil hydraulic property for a long period of time. Williams et al. (1993) recognized the importance of incorporating information on soil structure and clay mineralogy for an indirect estimation of the soil water retention. By the late 1990s, Lin et al. (1999) had carried out a systematic study on quantification of soil morphological properties for their use into PTFs of both soil water retention and hydraulic conductivity characteristics. Surprisingly, despite various efforts aimed at developing new types of PTFs (Minasny et al., 1999; Wösten et al., 1999), little or no information on terrain features was incorporated into PTFs for estimating soil hydraulic properties in the previous century. We would also point out that most investigations were undertaken with the major objective of developing empirical relationships from data measured in particular environments and specific types of soils, but no information was available on how, if at all, an existing PTF might be calibrated to give reliable results in another environment. Only during the early 21st century, Palladino et al. (2000) and Pachepsky et al. (2001) published results on an explicit use topographic variables in indirect estimation of soil hydraulic properties.

Within a study aimed at examining the spatial variability of soil properties across a gently sloping 3.7-ha field, Pachepsky et al. (2001) evaluated correlations between soil water retention and terrain attributes determined using a 30-m DEM obtained from aerial photography data. Thirty-nine soil samples were collected from 4 to 10 cm depth along four 30-m transects and for each of them texture, oven-dry bulk density, and several $\theta(h)$ values were measured. The topographical variables of slopes, profile curvature and tangential curvature were computed in the DEM grid nodes and interpolated to the soil sampling locations. Among the measured soil water retention data points, it was found that θ at h of -1.0 and -3.3 m correlated reasonably well with the terrain attributes considered, and the related regression model explained about 67 and 60% of variation in soil water content, respectively. A weaker correlation was found for water contents at $h = -10.0$ m, with the regression model relating θ_{-10} to the terrain attributes now explaining only about 20% of variation. A decrease in soil water retention was reported for steeper slopes at intermediate pressure heads. It was found that θ_{-1} and $\theta_{-3.3}$ were negatively correlated with profile curvature, while positively correlated with tangential curvature. These results show the potential offered by the topographic attributes for the interpretation of the field-scale variability of soil hydraulic properties.

Romano and Palladino (2002) extended the early work by Palladino et al. (2000) and pointed to the idea that topography can represent suitable ancillary information for calibrating a published PTF across a specific environment under study. They evaluated the prediction performances of three most-used PTFs for determining the soil water retention function, namely the PTFs proposed by Rawls and Brakensiek (1989) and by Vereecken et al. (1989), and that developed on the basis of the European database of soil hydraulic properties, HYPRES (Wösten et al., 1999). The soil-landscape map has been developed for a catchment that covered a hilly area of about 32 km² in southern Italy, and 88 undisturbed soil cores were collected, 50 m apart and from depth 5–12 cm, along two transects located on opposite hillsides with respect to the main river. All cores were subjected to standard measurement techniques to determine soil physical and chemical properties (texture, oven-dry bulk density, organic carbon content) and hydraulic properties (soil water retention function and saturated hydraulic conductivity). Primary (slope, aspect, surface curvatures, downward and upward flow-path length, and specific contributing area) and secondary (wetness index and solar radiation index) terrain attributes at the sampling locations were obtained using a 25-m DEM. The topographic variables were employed to adjust van Genuchten's (vG) $\theta(h)$ parameters of the published PTFs by an optimization procedure that minimized the deviations between measured and PTF-estimated water retention functions. To obtain a soil water retention function, $\theta(h)_\tau$, that incorporates also information from terrain analysis, the soil water retention predictions of a published PTF, $\theta(h)_{\text{PTF}}$, are modified by calibrating a linear polynomial model used to interpret the residual $\theta(h)_\tau - \theta(h)_{\text{PTF}}$. Specifically, describing the observed and PTF-predicted water retention data by vG relation $\theta(h)$, each of the four vG parameters is adjusted according to:

$$\pi_{i\tau} = \pi_{i\text{PTF}} + \sum_{j=1}^{t} a_{ij}\tau_j \tag{1}$$

where π_i represents one of the parameters featuring in the vG relation (i.e., θ_s, θ_r, α, or n), τ_j is a terrain attribute (for example, slope or aspect) and a_{ij} is the coefficient of variable τ_j. The subscript PTF means that the specific retention parameter π_i is calculated using the pedotransfer algorithm as originally proposed by the authors. The unknown coefficient values a_{ij} in equation (1) are simultaneously determined by minimizing the following objective function OF(**a**):

$$\text{OF}(a) = \sum_{s=1}^{N} \left\{ \frac{1}{(\xi_u - \xi_l)} \int_{\xi_l}^{\xi_u} [\theta(\xi)_{\text{vG}} - \theta(\xi; a)_\tau]^2 d\xi \right\}_s \tag{2}$$

that accounts for the discrepancies between the observed, $\theta(\xi)_{\text{vG}}$, and the predicted, $\theta(\xi;a)_\tau$ soil water retention function, with $\xi = \log(|h|)$ and ***a*** being the vector of unknown polynomial coefficients a_{ij}. The summation term is extended to the whole set of N

available water retention functions, with index s representing the sth soil samples collected along the transects. The integral in the right-hand side of equation (2) is computed in the interval from $\xi_l = \log(|-1 \text{ cm}|) = 0.00$ and $\xi_u = \log(|-16,000 \text{ cm}|) = 4.20$. Comparisons between estimation performances of original and adjusted PTFs demonstrated the benefit of incorporating terrain attributes, especially slope, aspect, and profile curvature, to improve the indirect predictions of PTFs. Moreover, indices of prediction efficiency show that the terrain-adjusted PTFs have a more similar prediction behavior among the PTFs considered, at least for the data set examined. For reliable application to a certain environment of any indirect method, especially if it pertains to an empirical one (empirical is the vG's water retention relation and empirical are the PTF retention coefficients obtained by multiple regressions using national or regional databases), additional hydraulic data should be measured in the specific area of interest for verification and recalibration purposes. With respect to this, Romano and Palladino (2002) advanced a technique that calibrates an already developed PTF by making use of topographic variables to limit the number of the additional measured soil hydraulic functions and reduce the bias in the indirect predictions. Figure 3 depicts a result of the calibration of the PTF by Rawls and Brakensiek (1989) by including also topographic information on slope gradient and orientation as input predictors. This procedure has also shown the advantage of yielding a more realistic spatial portrayal of the soil hydraulic behavior along the study transects, especially by lowering the bias introduced by PTFs which were developed in other environments, e.g., the PTF of Rawls and Brakensiek (1989) for USA, or for particular soil types, e.g., the PTF of Vereecken et al. (1989) for Belgian soils.

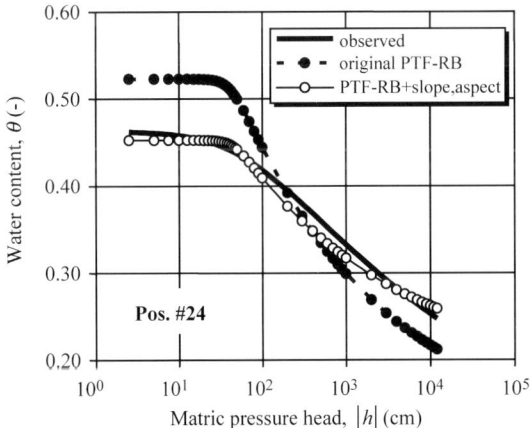

Figure 3. Local refinement of the original water retention prediction by the PTF of Rawls and Brakensiek (1989), PTF-RB, using information on slope and aspect (from Romano and Palladino, 2002; reproduced with permission).

Rawls and Pachepsky (2002) utilized a relatively large number of soil data from the Natural Resources Conservation Service (NRCS) database for the indirect estimation of the soil water contents at -3.3 and -150 m matric pressure head. The classic input predictors of soil textural fractions were supplemented by classes, rather than by numbers, describing slope position and landscape surface shape. Information on soil horizons was also added. The class PTFs were developed using the regression-tree method, which is a recursive data-partitioning algorithm that at a first stage divides the available set of data into two subsets by minimizing the variance in the predictions. Then, this operation is carried out recursively on each subset, and so on. For volumetric water content at -3.3 m, Figure 4 shows the tree with the related branches and terminal nodes. It is interesting to note that this technique enabled soils to be partitioned into homogeneous groups called topotextural groups. The authors observed that by coupling terrain features with the texture information it was possible to estimate water retention values in the A-horizon more accurately than from laboratory texture only.

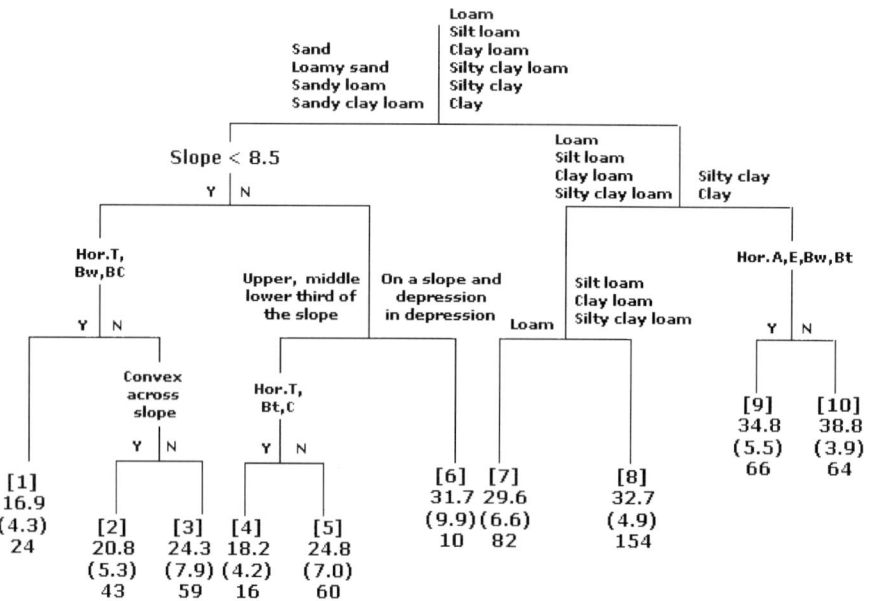

Figure 4. Output of the regression tree procedure for estimating soil water content at a pressure head of -3.3 m, i.e., water retention $\theta_{-3.3}$ (redrawn from Rawls and Pachepsky, 2002).

Leij et al. (2004) carried out a comprehensive study on the indirect estimation of both soil water retention and hydraulic conductivity characteristics with PTFs generated with the artificial neural networks (ANNs) method. One specific aim of this investigation was to examine whether, and to what extent, topographic variables can improve the prediction performances or could be used in certain circumstances to replace some missing basic soil properties. Soil physical, chemical, and unsaturated hydraulic properties were measured in one hundred soil cores collected at 50-m intervals along a 5-km hillslope transect. The Spearman rank analysis has shown that the parameters featuring in the water retention function are correlated with elevation, slope, aspect, and potential solar radiation, although these correlations tend to be smaller than those with the basic soil properties.

A weak correlation was found between the water retention parameters and terrain curvatures, presumably because of the relatively large spacing between samples and errors involved in the computation of curvature. Saturated hydraulic conductivity, K_s, was correlated with slope and potential radiation in spring (fall) and winter. To a certain extent, the relatively low level of correlation between topography and water retention parameters should have been expected (see, Section 2 of Chapter 4 in this book). On the other hand, as already shown by Pachepsky et al. (2001) and Romano and Palladino (2002), correlations between topographic variables and water retention values appear in general more evident and stronger in the wetter range of the water retention function, namely for higher matric pressure heads.

The rationale underlying the inclusion of terrain attributes to indirectly estimate soil hydraulic properties is that environmental variables related to pedogenetic dynamics can represent ancillary information to better capture the inherent soil hydraulic behavior and account for local environmental conditions. Most studies have reported that water contents tend to be lower at higher elevations because of the smaller contributing area and lower clay content (Hawley et al., 1983). Moreover, during precipitation events, infiltrated water persists in locations of the landscape that are relatively flat, while higher slopes foster rapid lateral redistribution of soil moisture. Therefore, whereas texture and bulk density can hardly help in understanding the internal spatial architecture of a soil, elevation and slope gradient can add some useful information about the weathering or about the fact that a certain soil horizon could be affected by material transport from one position in the landscape to another. Areas with steeper slopes will experience less infiltration and crust formation and more surface runoff; the surface water content is likely to be lower (Moore et al., 1988) and the hydraulic conductivity higher (Casanova et al., 2000). Although soils in some landscape positions can have very similar percentages of clay content, the type of clay minerals in those positions can be different because of erosional processes (Lee et al., 2003). The type of clay mineral, e.g., smectitic or kaolinitic clay minerals, affects the aggregation of primary particles (i.e., sand, silt, and clay) into secondary units, and subsequently the formation of intra-aggregate pores and inter-aggregate pores. The overall porosity of a soil as determined by the oven-dry bulk density does not distinguish these two categories of soil pores. Differences between water retention of smectitic (montmorillonitic) and kaolinitic soil have long been recognized (Hodnett and Tomasella, 2002).

The orientation of a slope influences the incident solar radiation and evapotranspiration, thereby providing valuable information about interactions between the clay fraction of soils and organic substances, which in turn may strongly affect the water retention capacity of soils. Hanna et al. (1982) observed higher available water contents and Cerdà (1997) reported higher infiltration rates for north-facing slopes, while in Chile (southern hemisphere) Casanova et al. (2000) found a higher hydraulic conductivity at south-facing slopes compared with north-facing ones. The higher organic matter and clay contents for the south-facing slope presumably lead to more stable macropores. The difference in aspect values can effectively account for soil organic matter quality Therefore, a combination of slope and aspect attributes may be useful to provide information on structure of soil. Slope and aspect can also give some indication of soil thickness since, for example, in the northern hemisphere the north-facing (south-facing in the southern hemisphere), milder slopes will tend to have wetter soil moisture conditions and show better developed soil profiles than south-facing, steeper slopes. That may help in setting the lower boundary conditions for hydrologic modeling.

Land surface curvature is an important moderator of lateral flow pattern, as depressed concave areas will be wetter than convex or planar areas with corresponding differences in sedimentation patterns and biological activity. Solar radiation, which is determined from the slope, aspect and solar position, is a key driving force for evapotranspiration; the water content will be lower for surfaces receiving more solar radiation (Western et al., 1999). A correlation between solar radiation and hydraulic properties is thus conceivable because moisture regime, temperature and vegetation affect soil formation. Contributing area and slope determine the flux and gradient for overland flow. A compound topographic index, such as the wetness index, has a potential to provide a PTF-based prediction method in a cost-effective way with ancillary information on downward movement of water and soil particles. One will readily have a useful, albeit rough, map of the zones where soil particles are most likely to be removed, transported and deposited. The wetness index can also provide information on positions of wetter areas, where deeper and organic-rich soil profiles are most likely to be developed, and dryer areas, where thin soil profiles with lower organic C contents usually develop. Other topographical attributes, such as the compound stream power and sediment transport capacity indices (Moore et al., 1993) and the elevation-relief ratio for skewness of elevation, have also been used in erosion studies. It remains to be seen whether and how those attributes can contribute in PTF development.

Overall, the reported studies and the related discussion above confirm the usefulness of topographic attributes as ancillary data to indirectly estimate soil hydraulic properties. One should note that the extent of such utility is site specific and relies on integration of local environmental conditions in certain primary and secondary terrain variables. Information from terrain analysis coupled with the use of efficient interpolation schemes are useful in reducing the spatial uncertainty of the PTF estimates of soil hydraulic properties over a certain area, whereas further investigations are needed to evaluate under which circumstances certain topographical variables can better infer the soil hydraulic properties at one specific location.

5. CONCLUDING REMARKS AND FUTURE DEVELOPMENTS

Terrain attributes can be evaluated at various scales. Although specific analysis is yet to be performed, we conjecture that the best results for soil hydrological applications will be obtained with those terrain analysis methods that better represent the terrain features at the hillslope scale, such as contour-based methods or grid-based multiple-flow direction methods, since the hillslope scale is the dominant spatial extent controlling the soil formation. Another important factor to be considered is the original DEM resolution, since this affects the minimum spatial scales at which soil variability can be potentially resolved. Higher DEM resolution is to be preferred given the short spatial correlation lengths observed in soil properties.

As outlined in this chapter, terrain attributes can be employed in conjunction with PTFs in order to address two fundamental issues: the first is to assess soil hydraulic properties over large spatial scales when PTF input variables are available in a limited amount of observation points; the second is to improve PTF predictions accounting for local soil-forming factors not represented by the traditional PTF predictor variables.

In the first case, the spatial correlation between soil physical and chemical properties and terrain attributes is exploited for interpolating soil physical and chemical data measured in few samples. Various interpolation techniques are available, some have been shown to perform better than others in predicting the actual soil properties. The choice of the interpolation technique should depend on the amount of field data available and on the specific purpose of soil characterization. The optimal approach is to evaluate several interpolation techniques from a functional point of view, considering the processes of interest, the spatial and temporal scale of analysis and the modeling approach.

In the second case, terrain features are employed as ancillary information to calibrate PTFs for a specific environment under study. It has been pointed out that PTFs generate evident systematic errors (biases) when they are applied to types of soils and environments that differ from those used for their development (Romano and Palladino, 2002; Romano and Santini, 1997). Under such conditions, PTFs can basically drive modelers and end-users in recognizing the most significant changes in soil hydraulic properties occurring across a domain of interest. The inclusion of terrain attributes as input variables has shown the potential ability of reducing the biases in PTF predictions, and thus appears a good tool to improve the pedotransfer methods when describing the spatial structure of soil hydraulic property variability within a smaller domain of interest.

There exists a pressing need to develop an efficient PTF for deriving the unsaturated hydraulic conductivity relationship between unsaturated hydraulic conductivity, K, and soil water content, θ, or matric pressure head, h. Up to now, almost all applications of process-based hydrologic models basically take soil water retention parameters from measurements, but treat soil hydraulic conductivity parameters as calibrating parameters. As pointed out by Childs (1969), efforts to statistically correlate the scale parameter K_s only to particle-size distribution are bound to fail or will be of scant practical utility. Specifically, the estimation of saturated hydraulic conductivity, K_s, close to the soil surface is rather difficult as this parameter is often highly affected by preferential flow or the presence of coarse fragments and stones in the soil matrix. Soil structure has to be mainly associated with inter-aggregate pores, decayed root channels, earthworm holes, and cracks of varying dimension because of drying/wetting cycles in soil. In addition to the traditional input variables of textural fractions, organic matter content and oven-dry bulk density, indicators of soil structure and probably vegetation cover condition should feature more explicitly in an enhanced pedotransfer prediction of K_s. The results by Leij et al. (2004) have shown that including terrain attributes as ancillary information in PTFs is beneficial to gain better PTF-predictions of K_s, but it seemed not sufficient to explain observed values. In an attempt to improve the indirect prediction of hydraulic conductivity parameters, studies are under way to explore the feasibility of adding information about land-use and vegetation (cover, distribution, density, etc.), and hence to account somehow for mechanical and biological macroporosity. This kind of information can be conveniently retrieved from spatial and temporal variations in the normalized difference vegetation index (NDVI) through interpretation of satellite images. Remotely sensed data are usually acquired at a coarse resolution of about 1 km^2 with the advanced very high resolution radiometer (AVHRR) from the National Oceanic and Atmospheric Administration (NOAA) satellites. Information of this kind is expected to be more effective when one is interested in gaining an average portrayal of soil hydraulic conductivity behavior and modeling hydrologic processes for a relatively large catchment extent.

More interestingly, integrating microwave measurements from space or airborne platforms with terrain analysis will enable remotely sensed near-surface soil water contents and their spatial organization in the landscape to be obtained and possibly incorporated into the pedotransfer approach. A combination of these tools could also increase the working spatial resolution, thus making it possible to get predictions of soil hydraulic characteristics over smaller areas.

ACKNOWLEDGEMENTS

The study present in this paper has been in part financially supported by P.O.N. project "AQUATEC—New technologies of control, treatment, and maintenance for the of water emergency", funded by the Italian Ministry of University.

REFERENCES

Bastet, G., Bruand, A., Voltz, M., Bornand, M., Quétin, P., 1999. Performance of available pedotransfer functions for predicting the water retention properties of French soils. *In*: Genuchten, v., Leij, F.J., Wu, L. (Eds.), Characterisation and Measurement of the Hydraulic Properties of Unsaturated Porous Media. University of California, Riverside, USA, pp. 981-991.

Beven, K.J., Kirkby, M.J., 1979. A physically based, variable contributing area model of basin hydrology. Hydrol. Sci. Bull. 24, 43-69.

Blöschl, G., 1996. Scale and Scaling in Hydrology. Wiener Mitteilungen, Band 132, Technische Universität Wien, Institut für Hydraulik, Gewässerkunds und Wasserwirtschaft, Wien.

Blöschl, G., Grayson, R.B., 2000. Spatial observations and interpolation. *In*: Grayson, R.B., Blöschl, G. (Eds.), Spatial Patterns in Catchment Hydrology – Observations and Modelling. Cambridge University Press, Cambridge, pp. 51-81.

Blöschl, G., Sivapalan, M., 1995. Scale issues in hydrological modelling – a review. Hydrol. Process. 9, 251-290.

Briggs, L.J., Shantz, H.L., 1912. The Wilting Coefficient for Different Plants and Its Indirect Determination. Government Printing Office, Washington, 83 pp.

Brubacker, S.C., Jones, A.J., Frank, K., Lewis, D.T., 1994. Regression models for estimating soil properties by landscape position. Soil Sci. Soc. Am. J. 58, 1763-1767.

Casanova, M., Messing, I., Joel, A., 2000. Influence of aspect and slope gradient on hydraulic conductivity measured by tension infiltrometer. Hydrol. Process. 14, 155-164.

Cerdà, A., 1997. Seasonal changes of infiltration rates in a Mediterranean scrubland on limestone. J. Hydrol. 198, 209-225.

Childs, E.C., 1969. An Introduction to the Physical Basis of Soil Water Phenomena. Wiley, London.

Cook, S.E., Corner, R.J., Grealish, G., Gessler, P.E., Chartres, C.J., 1996. A rule-based system to map soil properties. Soil Sci. Soc. Am. J. 60, 1893-1900.

Costa-Cabral, M.C., Burges, S.J., 1994. Digital Elevation Model Networks (DEMON) – a model of flow over hillslopes for computation of contributing and dispersal areas. Water Resour. Res. 30 (6), 1681-1692.
De Bruin, S., Stein, A., 1998. Soil-landscape modelling using fuzzy c-means clustering of attribute data derived from a Digital Elevation Model (DEM). Geoderma 83, 17-33.
Elsenbeer, H., 2001. Preface of the special issue on pedotransfer functions in hydrology. J. Hydrol. 251, 121-122.
Fairfield, J., Leymarie, P., 1991. Drainage networks from digital elevation models. Water Resour. Res. 27 (5), 709-717.
Florinksy, I.V., Arlashina, H.A., 1998. Quantitative topographic analysis of gilgai soil morphology. Geoderma 82, 359-380.
Florinksy, I.V., Eilers, R.G., Manning, G.R., Fuller, L.G., 2002. Prediction of soil properties by digital terrain modelling. Environ. Model. Software 17, 295-311.
Freeman, T.G., 1991. Calculating catchment area with divergent flow based on a regular grid. Computers Geosci. 17 (3), 413-422.
Freeze, R.A., 1974. Streamflow generation. Rev. Geophys. Space Phys. 12, 627-647.
Gessler, P.E., Chadwick, O.A., Chamran, F., Althouse, L., Holmes, K., 2000. Modeling soil-landscape and ecosystem properties using terrain attributes. Soil Sci. Soc. Am. J. 64, 2046-2056.
Goovaerts, P., 1999. Geostatistics in soil science: state-of-art and perspectives. Geoderma 89, 1–45.
Grayson, R.B., Blöschl, G., 2000. Spatial modelling of catchment dynamics. In: Grayson, R.B., Blöschl, G. (Eds.), Spatial Patterns in Catchment Hydrology – Observations and Modelling. Cambridge University Press, Cambridge, pp. 51-81.
Gupta, S.C., Larson, W.E., 1979. Estimating soil water retention characteristics from particle size distribution, organic matter percent and bulk density. Water Resour. Res. 15 (6), 1633-1635.
Hanna, Y.A., Harlan, P.W., Lewis, D.T., 1982. Soil available water as influenced by landscape position and aspect. Agron. J. 74, 999-1004.
Hawley, M.E., Jackson, T.J., McCuen, R.H., 1983. Surface soil moisture variation on small agricultural watersheds. J. Hydrol. 62, 179-200.
Heuvelink, G.B.M., Pebesma, E.J., 1999. Spatial aggregation and soil process modelling. Geoderma 89, 47-65.
Heuvelink, G.B.M., Webster, R., 2001. Modelling soil variation: past, present, future. Geoderma 100, 269-301.
Hodnett, M.G., Tomasella, J., 2002. Marked differences between van Genuchten soil water-retention parameters for temperate and tropical soils: a new water-retention pedo-transfer function developed for tropical soils. Geoderma 108, 155-180.
Hutchinson, M.F., 1993. Development of a continent-wide DEM with applications to terrain and climate analysis. In: Goodchild, M.F., Parks, B.O., Steyaert, L.T. (Eds.), Environmental Modelling with GIS. Oxford University Press, New York.
Hutchinson, M.F., Gessler, P.E., 1994. Splines – more than just a smooth interpolator. Geoderma 62, 45-67.
Jenny, H., 1980. The soil resource, origin and behaviour. Springer-Verlag, New York.

Lagacherie, P., Voltz, M., 2000. Predicting soil properties over a region using sample information from a mapped reference area and digital elevation data: a conditional probability approach. Geoderma 97, 187-208.

Lark, R.M., 1999. Soil-landform relationships at within-field scales: an investigation using continuous classification. Geoderma 92, 141-165.

Lea, N.L., 1992. An aspect driven kinematic routing algorithm. *In*: Parsons, A.J., Abrahams, A.D. (Eds.), Overland Flow: Hydraulics and Erosion Mechanics. Chapman & Hall, New York.

Lee, R., 1978. Forest Micrometeorology. Columbia University Press, New York, 276 pp.

Lee, B.D., Sears, S.K., Graham, R.C., Amrhein, C., Vali, H., 2003. Secondary mineral genesis from chlorite and serpentine in an ultramafic soil toposequence. Soil Sci. Soc. Am. J. 67, 1309-1317.

Leij, F.J., Romano, N., Palladino, M., Schaap, M.G., Coppola, A., 2004. Topographical attributes to predict soil hydraulic properties along a hillslope transect. Water Resour. Res. 40, W02407, doi:10.1029/2002WR001641.

Lin, H.S., McInnes, K.J., Wilding, L.P., Hallmark, C.T., 1999. Effects of soil morphology on hydraulic properties: II. Hydraulic pedotransfer functions. Soil Sci. Soc. Am. J. 63, 955-961.

Matheron, G., 1973. The intrinsic random functions and their applications. Adv. Appl. Prob. 5, 438-468.

Maunder, C.J., 1999. An automated method for constructing contour-based digital elevation models. Water Resour. Res. 35 (12), 3931-3940.

McBratney, A.B., Odeh, I.O.A., Bishop, T.F.A., Dunbar, M.S., Shatar, T.M., 2000. An overview of pedometric techniques for use in soil survey. Geoderma 97, 293-327.

McKenzie, N.J., Ryan, P.J., 1999. Spatial prediction of soil properties using environmental correlations. Geoderma 89, 67-94.

Minasny, B., McBratney, A.B., 2002. Uncertainty analysis for pedotransfer functions. Eur. J. Soil Sci. 53, 417-429.

Minasny, B., Bell, M.A., van Keulen, H., 1999. Comparison of different approaches to the development of pedotransfer functions for water-retention curves. Geoderma 93, 225-253.

Moore, I.D., Burch, G.J., McKenzie, D.H., 1988. Topographic effects on the distribution of surface water and the location of ephemeral gullies. Trans. ASAE 31, 1098-1107.

Moore, I.D., Gessler, P.E., Nielsen, G.A., Peterson, G.A., 1993. Soil attribute prediction using terrain analysis. Soil Sci. Soc. Am. J. 57, 443-452.

Moore, I.D., Grayson, R.B., Ladson, A.R., 1991. Digital terrain modeling – a review of hydrological, geomorphological, and biological applications. Hydrol. Process. 5 (1), 3-30.

O'Callaghan, J.F., Mark, D.M., 1984. The extraction of drainage networks from digital elevation data. Computer Vision, Graphics and Image Processing 28, 323-344.

Odeh, I.O.A., McBratney, A.B., Chittleborough, D.J., 1994. Spatial prediction of soil properties from landform attributes derived from a digital elevation model. Geoderma 63, 197-214.

Odeh, I.O.A., McBratney, A.B., Chittleborough, D.J., 1995. Further results on prediction of soil properties from terrain attributes: heterotopic cokriging and regression-kriging. Geoderma 67, 215-226.

Pachepsky, Y.A., Timlin, D.J., Rawls, W.J., 2001. Soil water retention as related to topographic variables. Soil Sci. Soc. Am. J. 65, 1787-1795.

Palladino, M., Romano, N., Santini, A., 2000. Use of pedotransfer functions and DEM data to upscale soil hydraulic properties from the pedon to the catchment scale. In: Maione, U. (Ed.), New Trends in Water and Environmental Engineering for Safety and Life: Eco-compatible Solutions for Aquatic Environments. A.A. Balkema, Rotterdam, pp. 102.1-102.11.

Park, S.J., McSweeney, K., Lowery, B., 2001. Identification of the spatial distribution of soils using a process-based terrain characterization. Geoderma 103, 249-272.

Quinn, P., Beven, K., Chevallier, P., Planchon, O., 1991. The prediction of hillslope flow paths for distributed hydrological modelling using digital terrain models. Hydrol. Process. 5, 59-80.

Rawls, W.J., Brakensiek, D.L., 1989. Estimation of soil water retention and hydraulic properties. In: Morel-Seytoux, H.J. (Ed.), Unsaturated Flow in Hydrologic Modeling – Theory and Practices, NATO ASI Series. Kluwer, Dordrecht.

Rawls, W.J., Pachepsky, Y.A., 2002. Using field topographic descriptors to estimate soil water retention. Soil Sci. 167 (7), 423-435.

Romano, N., Palladino, M., 2002. Prediction of soil water retention using soil physical data and terrain attributes. J. Hydrol. 265, 56-75.

Romano, N., Santini, A., 1997. Effectiveness of using pedo-tranfer functions to quantify the spatial variability of soil water retention characteristics. J. Hydrol. 202, 137-157.

Ruhe, R.V., 1956. Geomorphic surfaces and the nature of soils. Soil Sci. 82, 441-455.

Shary, P.A., 1995. Land surface in gravity points classification by a complete system of curvatures. Math. Geol. 27 (3), 373-390.

Sinowski, W., Scheinost, A.C., Auerswald, K., 1997. Regionalisation of soil water retention curves in a highly variable soilscape, II. Comparison of regionalisation procedures using a pedotransfer function. Geoderma 78, 145-159.

Tarboton, D.G., 1997. A new method for the determination of flow directions and upslope areas in grid digital elevation models. Water Resour. Res. 33 (2), 309-319.

Tietje, O., Hennings, V., 1996. Accuracy of saturated hydraulic conductivity prediction by pedo-transfer functions compared to the variability within FAO textural classes. Geoderma 69, 71-84.

Tietje, O., Tapkenhinrichs, M., 1993. Evaluation of pedo-transfer functions. Soil Sci. Soc. Am. J. 57, 1088-1095.

Tomasella, J., Hodnett, M.G., 1998. Estimating soil water retention characteristics from limited data in Brazilian Amazonia. Soil Sci. 163, 190-202.

Vereecken, H., Maes, J., Darius, P., 1989. Estimating the soil moisture retention characteristics from texture, bulk density and carbon content. Soil Sci. 148, 389-403.

Walker, P.H., Hall, F.F., Protz, R., 1968. Relation between landform parameters and soil properties. Soil Sci. Soc. Am. Proc. 32, 101-104.

Western, A.W., Blöschl, G., 1999. On the spatial scaling of soil moisture. J. Hydrol. 217 (3-4), 203-224.

Western, A.W., Grayson, R.B., 2000. Soil moisture and runoff processes at Tarrawarra. In: Grayson, R.B., Blöschl, G. (Eds.), Spatial Patterns in Catchment Hydrology – Observations and Modelling. Cambridge University Press, Cambridge, pp. 209-246, Chapter 9.

Western, A.W., Grayson, R.B., Blöschl, G., Willgoose, G.R., McMahon, T.A., 1999. Observed spatial organization of soil moisture and its relation to terrain indices. Water Resour. Res. 35 (3), 797-810.

Western, A.W., Blöschl, G., Grayson, R.B., 2001. Toward capturing hydrologically significant connectivity in spatial patterns. Water Resour. Res. 37 (1), 83-97.

Williams, J., Prebble, J.E., Williams, W.T., Hignett, C.T., 1993. The influence of texture, structure and clay mineralogy on the soil moisture characteristic. Aust. J. Soil Res. 21, 15-32.

Wilson, J.P., Gallant, J.C., 2000. Digital terrain analysis. *In*: Wilson, J.P., Gallant, J.C. (Eds.), Terrain Analysis: Principles and Applications. Wiley, New York, pp. 1-27.

Wilson, J.P., Repetto, P.L., Snyder, R.D., 2000. Effect of data source, grid resolution, and flow routing method on computed topographic attributes. *In*: Wilson, J.P., Gallant, J.C. (Eds.), Terrain Analysis: Principles and Applications. Wiley, New York, pp. 133-161.

Wösten, J.H.M., Lilly, A., Nemes, A., Le Bas, C., 1999. Development and use of a database of hydraulic properties of European soils. Geoderma 90, 169-185.

Wösten, A.W., Pachepsky, Y.A., Rawls, W.J., 2001. Pedotransfer functions: bridging the gap between available basic soil data and missing soil hydraulic characteristics. J. Hydrol. 251, 123-150.

Zhang, W., Montgomery, D.R., 1994. Digital elevation model grid size, landscape representation, and hydrologic simulations. Water Resour. Res. 30 (4), 1019-1028.

Zhu, A.X., Band, L., Vertessy, R., Dutton, B., 1997. Derivation of soil properties using a Soil Land Inference Model (SoLiM). Soil Sci. Soc. Am. J. 61, 523-533.

Chapter 17

SPATIAL STRUCTURE OF PTF ESTIMATES

N. Romano

Department of Agricultural Engineering, Division for Land and Water Resources Management, University of Naples "Federico II", Via Università, 100, 80055 Portici (Naples), Italy
Tel.: +39-081-2539421; fax: +39-081-2539412

1. BACKGROUND AND JUSTIFICATION

Soils are highly heterogeneous porous media whose unsaturated hydraulic properties are being determined with substantial investment of labor and time resources. This is chiefly due to the nonlinear dependence among the relevant variables and the need to use relatively expensive facilities with skilled people to carry out specifically designed hydraulic tests either on undisturbed soil cores or *in-situ* on selected plots. The difficulties are exacerbated in part by the spatial and temporal variability of the properties. The variability is significant at all scales of interest, making it extremely difficult to capture flow behavior at one particular scale, whether it is the scale of single soil pores or the catchment scale. In most practical cases, spatial variability has been shown to dominate over temporal variability and this chapter focuses attention on soil hydraulic characterization at relatively large spatial scale via pedotransfer functions (PTFs). The leitmotiv of this chapter will be the use of methods to determine the soil hydraulic properties for field- and large-scale modeling that are more practical and also prove to be effective at detecting the spatial correlation structure of these properties.

Soil is commonly characterized from the hydraulic point of view using nonlinear relationships between the volumetric soil water content, θ, the matric pressure head, h, and the hydraulic conductivity, K, in terms of the water retention, $\theta(h)$, and hydraulic conductivity, $K(\theta)$, functions. Knowledge of these functions is essential when water flow in soil is modeled using the Richards equation (Kutílek and Nielsen, 1994). Sometimes the hydraulic response of a field soil to boundary conditions is evaluated by its ability to store water between the two distinct values of water contents at the conditions of *field capacity*, θ_{fc}, and *permanent wilting*, θ_{pw}. Accordingly, soil hydraulic behavior is parameterized by introducing the concept of available water, $AW = \theta_{fc} - \theta_{pw}$, without accounting for the soil water retention and hydraulic conductivity functions of the soil profile. This approximate view may generate a misconception, as it happens in the case of a layered soil profile with contrasting soil hydraulic properties of each layer, and may lead to erroneous conclusions when applied for example to precision farming or modeling the hydrologic response of hillslopes or a small catchment. A discussion on that matter is far beyond the scope of this chapter, but few details are addressed below as they can be of interest for the estimation of soil water retention data via PTFs (Reynolds et al., 2000). The customary practice of calculating θ_{pw} as the water retention value $\theta(h = -150$ m) and identifying the condition of *field capacity* as the value of θ at $h = -3.30$ m($\approx -1/3$ bar) is questionable and represents only a rough generalization of different, and sometimes contrasting experimental results (Romano and Santini, 2002; Minasny and McBratney, 2003). Even though θ_{pw} could be evaluated with some confidence from knowledge of the soil water retention characteristic, it should be noted that the PTF estimates become very uncertain when approaching the dry end of the function $\theta(h)$, i.e., at very low matric pressure heads. On the other hand, θ_{fc} should represent the average water content when the hydraulic conductivity is so low that the soil profile does not virtually drain water any longer under a nearly unit gradient of the total hydraulic head at the lower boundary of the flow domain (Romano and Santini, 2002). Both water retention and hydraulic conductivity affect the drainage rate at the unit gradient, and it may thus be difficult to identify a PTF to estimate θ_{fc}. Because of soil–vegetation–atmosphere interactions, the modern view of soil water storage in a field soil calls for considerations not only about unsaturated soil hydraulic properties but also of growth status of plants and their root system as well as meteorological conditions. Overall, we suggest to use PTFs that estimate the unsaturated soil hydraulic functions $\theta(h)$ and $K(\theta)$, rather than to rely on operationally defined soil water contents such as field capacity and wilting point.

The measurement of soil hydraulic properties with either direct methods or parameter optimization approaches is in general not easy and usually restricted to small-scale studies (Romano and Santini, 1997). This is especially true for the measurements of soil hydraulic conductivity that not only shows a complex nonlinear dependence on matric pressure head, but also spans several orders of magnitude as soil conditions vary from saturation to dryness. Some predictive methods have been proposed that estimate the soil hydraulic conductivity characteristic from the measured water retention, but they require: (i) a large number of data points to adequately describe the soil water retention characteristic with parametric relationships and (ii) an accurate value of hydraulic conductivity at saturation. The availability of an unsaturated value of K but near saturation improves the prediction of the unsaturated conductivity function (Van Genuchten and Nielsen, 1985; Durner et al., 1999; Schaap and Leij, 2000). If the hydraulic characterization of soil has to be done to relatively large areas, one needs to resort to simplified techniques such as the PTFs that estimate "what we need", namely the soil hydraulic properties, on the basis of "what we have" or "what can be easily obtained", e.g., basic soil physical and chemical properties (Bouma, 1989). Exhaustive information on the pedotransfer concept can be found in other chapters of this book.

Difficulties in determining soil hydraulic properties at a certain scale, the influence of heterogeneity on water flow and solute transport in the vadose zone, and issues relating to the extrapolation of information from known measurement sites represent sources of uncertainties which any modeling effort has to cope with. There are various environmental and management problems that require characterization of unsaturated soil hydraulic behavior with a trade-off between reasonable burdens and good accuracy (i.e., small systematic errors) and precision (i.e., a small dispersion around the mean) (Holt et al., 2002). Through analysis of the results from differently developed PTFs, one objective of this paper is to provide highlights about the potential offered by this technique together with some limitations in assessing an average portrayal of soil hydraulic behavior at field scale. Yet spatial variability remains a primary issue to be addressed adequately if model predictions should be of any practical use. Too often practical application of process-based distributed models is hampered by the lack of information concerning key input parameters (soil hydraulic properties; boundary conditions, especially that one at the lower boundary, etc.) as well as their fluctuations in space (and in time, of course).

This chapter would also offer due consideration about the fact that an indirect method for soil hydraulic characterization is successful if it also enables the relevant spatial patterns to be identified adequately. In this way, soil hydraulic spatial response can be effectively interpreted via distributed modeling. Recent advances in parameterization of hydrologic processes at hillslope or catchment scale, for example, has recognized the importance not only of describing the spatial variations in data and parameters, but also of detecting the possible presence of organization in the spatial patterns (Grayson and Blöschl, 2000). A question can be whether, and to what extent, indirect soil hydraulic characterization based on PTFs can meet this latter requirement. Precision agriculture, as another example, now faces more with the issue of reliably predicting the impact of management strategies that tend to maximize crop production while minimizing environmental damage, and it seems that there is no real need to produce only very detailed maps of soil properties or at least this will not be the major aim any longer (Batchelor et al., 2002). Although precision agriculture benefits from several advanced information technologies (Bouma et al., 1999) that combine the new generation of soil–vegetation–atmosphere models with local monitoring and remote sensing techniques, mapping of land-use, and geographical information systems (D'Urso et al., 1999), the start-up costs for the community and the costs of updating information are still relatively high. Therefore, to evaluate different management scenarios for precision agriculture and deal with within-field spatial variability in a cost-effective way one should account for a hydrologically functional behavior of the soil system. Can PTFs constitute valuable tools to estimate soil hydraulic behavior from a functional point of view? If this were the case, which functional variable or specific aspect of soil behavior can PTFs describe best? Moreover, public agencies, such as the US Environmental Protection Agency or the European Environment Agency, recommend the use of computer simulation models to predict contamination from chemical substances or design remediation strategies, for example in case of nonpoint-source (diffuse) pollution problems. Usually these models pertain to the category of rate-based mechanistic models that use the Richards and convective–dispersion equations (Addiscott and Wagenet, 1989), such as leaching estimation and chemistry model (LEACHM, Wagenet and Hutson, 1986), root zone water quality model (RZWQM, Starks et al., 2003), or HYDRUS-2D (Šimunek et al., 1999). However, model assessment of leaching risks in the case of practical situations becomes reliable only if one accounts properly for field-observed spatial variability (Rao and Wagenet, 1985; Mulla et al., 1999).

What seems to be lacking in the literature is the developing of suitable techniques that characterize soil hydraulic behavior under field conditions, a task that should always comprise both cost-effective determination of the soil hydraulic properties and reliable assessment of the structure of their spatial variability. Therefore, another objective of this paper is to offer a more coherent view of these two apparently separate questions. After some general statements on describing quantitatively soil spatial variability, in subsequent sections the determination of the spatial structure using hydraulic data estimated from existing PTFs is reviewed and discussed. It is then shown that PTF estimates can be improved by including ancillary information from terrain analysis and an evaluation is performed to examine whether, and to what extent, the proposed technique is effective also in terms of better capturing spatial variability patterns. Finally, comments are made on some practical aspects relating to possible biases in PTF estimates and scale issues.

2. SOIL HYDRAULIC PROPERTY VARIATIONS AND THE ROLE OF SIMPLIFIED PREDICTIVE METHODS

Several studies have shown that soil properties such as physical and chemical variables (e.g., texture, dry bulk density, organic matter content, cation exchange capacity, etc.) or hydraulic functions (essentially, the soil water retention, $\theta(h)$, and hydraulic conductivity, $K(\theta)$, functions), are randomly distributed over a landscape and only very intensive surveys allow one to describe this variability accurately. Such intensive surveys are notoriously labor- and time-consuming. Pedological and geomorphological studies are carried out to describe and classify pedons and characterize landscapes with the major aim to define soil maps. These maps are useful for many applications, but in general are based on factors that are weakly correlated to soil hydraulic properties (McBratney and Odeh, 1997). Within a soil map unit, it is common to find variations in the soil water retention and hydraulic conductivity functions that should be accounted for to avoid unacceptable errors in the predictions of simulation models.

Soil spatial variability has been investigated mainly qualitatively since the early decades of the last century, while more quantitative studies appeared only in the late 20th century (Burrough, 1993). In recent years, an extensive research effort has been devoted to ascertain the effect of natural and human-induced variability of soil properties on the behavior of ecosystems and on the simulation modeling results. Natural variations in soil behavior are generally referred to as intrinsic variations because of, for example, changes in soil-forming factors. Human activity, such as tillage practices, different types of irrigation management, and agrochemical applications, represent the so-called extrinsic sources of spatial variability. Variations in space exhibited by soil hydraulic properties are governed by several interactive soil-forming factors and processes that may act across different scales and are not currently well understood. Therefore, a deterministic description of the spatial variability of soil properties is not currently feasible. Other tools exist to analyze spatial data, including basic statistics and cross-correlation, stochastic and spectral approaches, geostatistics, state-space modeling, and fractal theory.

We will be using geostatistics in this section. One should note that geostatistics is generally considered to be the best method of spatial interpolation that also includes information on uncertainty (Goovaerts, 1997). Under the assumption of ergodicity and the validity of the intrinsic hypothesis, the key tool of geostatistics is the semivariogram function, $\gamma(u)$, that provides an average measure of dissimilarity between measured values as a function of separation distance, u, and, if necessary because of the anisotropy, direction. For the sake of simplicity, we limit our discussion to using the one-dimensional semivariogram. Such an approach is sufficient, for example, when one should describe the spatial variability along a transect with uniformly spaced observations. In that case, the experimental semivariogram can be computed from the observations as follows

(Goovaerts, 1997):

$$\gamma(u) = \frac{1}{2N(u)} \sum_{i=1}^{N(u)} [v(\xi_i) - v(\xi_{i+u})]^2 \qquad (1)$$

where $v(\xi)$ denotes the value of the variable v at location ξ, and $N(u)$ is the number of data pairs for the separation distance, u. Notice, however, that estimation of semivariance γ using equation (1) can be sensitive to the presence of outliers in the spatial series (see Lark (2000) for a discussion on the robust estimation of a semivariogram). The shape of a semivariogram is commonly identified by characteristic parameters referred to as the *sill*, *range*, and *nugget*. In general, semivariance γ increases initially as u increases. In case of an *intrinsically stationary* process, at larger u the semivariogram $\gamma(u)$ tends to level off and fluctuate around a value called the sill. The range is the distance at which the semivariance approaches the sill and represents the separation distance beyond which two values of the variable can be considered as statistically independent. The nugget is a measure of the discontinuity at the origin of the semivariogram (i.e., the extrapolation of γ at $u = 0$) and mainly arises from the existence of spatial variations at distances smaller than the shortest sampling interval. It is important to point out that the nugget effect depends on the selected sampling scale. In addition, the nugget effect may also arise from measurement errors, and it can be associated with the lower bound for precision (i.e., small random errors) of the measuring technique employed if this discontinuity at the origin persists when considering sampling distances of different sizes.

Information on the role of the semivariogram in the analysis of spatial data can be found in many papers and treatises (Goovaerts, 1997). It is important to point out here that the shape of the semivariogram enables one to obtain qualitative and quantitative information on the structure of spatial variability, as well as carry out an efficient subsequent interpolation phase (kriging) between the available observations and develop optimal sampling schemes. The problem of estimating the value of a variable at any unsampled location of the domain of interest is crucially important. Predicting input parameters and data at unsampled locations always results in a certain degree of uncertainty, which propagates then into the simulation of the processes under study. When estimating soil hydraulic properties by PTFs, the interpolation can be done either before or after applying PTFs. Sinowski et al. (1997) compared one procedure that interpolates first the basic soil properties, and then applies the selected PTF to the interpolated data for estimating the soil water retention functions, $\theta(h)$, versus another procedure that estimates first the $\theta(h)$ functions at sampled locations by using the PTF, and then interpolates the PTF-estimated $\theta(h)$. They concluded that the former method is superior. Further discussion on the interpolation problem using PTFs can be found in the chapter by Romano and Chirico (2005) of this book.

With a view to the overall aim of this paper, it is worth making some preliminary comments about the sources of uncertainty and sensitivity of a tool employed to describe soil spatial variations. Systematic and/or random errors always affect observations and the use of different instruments and techniques of measurement yields different values of a variable of interest. Acknowledging this fact, the question that we pose now is to examine the potential impact of using a certain method for determining the soil hydraulic properties on the subsequent analysis of spatial variability. This is an issue that is sometimes overlooked or even ignored, especially when a choice is made between using a method that is known to be accurate, albeit cumbersome ("rigorous method"), instead of using a method that relies on various simplifications ("simplified method"). Romano (1993) carried out a transient drainage experiment on a 50-m long transect of an Andosol fairly uniform to a depth of 90 cm from the surface. At 50 locations, 1-m apart, measurements of matric pressure head, with a tensiometer positioned at the depth $z = 30$ cm, and soil water content, with a neutron probe positioned at $z = 30$, 45, 60, and 75 cm, were collected at frequent but irregular time intervals during the field experiment. This set of measurements was used as an input information in a parameter optimization method. The proposed inverse method, among other things, had the novelty of optimizing not only the data, but also the process since it enabled the lower boundary condition to be set as an additional unknown of the inverse problem. The available data were also employed in a simplified procedure suggested by Sisson and van Genuchten (1991). For comparison purposes, both methods used the following closed form parametric expressions of van Genuchten–Mualem (or VGM relations) to describe soil hydraulic properties:

$$\theta(h) = \theta_r + (\theta_s - \theta_r)[1 + |\alpha h|^{1/1(1-m)}]^{-m} \qquad (2a)$$

$$K(\theta) = K_s S_e^\lambda [1 - (1 - S_e^{1/m})^m]^2 \qquad (2b)$$

where $S_e = (\theta - \theta_r)/(\theta_s - \theta_r)$ is the effective saturation, calculated from the saturated, θ_s, and residual, θ_r, values of water content θ, and K_s is the saturated hydraulic conductivity. The symbols α, m, and λ represent empirical parameters. If one interprets the soil water retention function as an equivalent pore-size distribution of the porous medium, the scale parameter α corresponds to the modal pore size, whereas the shape parameter m can be related to the second moment of that distribution, i.e., its standard deviation. This general remark can be of interest in the context of inferring these water retention parameters via PTFs. Parameter λ represents a tortuosity factor that generates an always monotonic function (2b) if greater than -2. For soils with small values of parameter m, namely soils showing relatively wider pore-size distributions, the condition $dK/dS_e > 0$ can be satisfied even for λ smaller than the threshold value reported above (Durner et al., 1999).

A general idea about spatial variability in the study area can be retrieved from Figure 1 showing values of K at $\theta = 0.30$, denoted as $K_{0.30}$. The reader is directed to the paper by Romano (1993) for additional details on the field test and the two methods employed. Differences exist between values of $K_{0.30}$ as determined by the two methods at the various locations and in a few cases they are also well pronounced. Nevertheless, it is interesting to note that the two methods result in spatial series that have close mean values along the transect, namely $\underline{K}_{0.30}^R = 0.022$ cm/h for the optimization method considered as the rigorous ("R") one and $\underline{K}_{0.30}^S = 0.019$ cm/h for the simplified ("S") method. Therefore, these data show that simplified methods can represent cost-effective options if the purpose of the investigation is mainly to obtain the average hydraulic properties of the study area. However, a simplified approach may prove highly inefficient at detecting the structure exhibited in space by a variable of interest. To facilitate comparisons, for variable $\log(K_{0.30})$, Figure 2 shows the experimental semivariogram scaled to the sample variance, $\gamma(u)/\sigma^2$. In these graphs, the semivariance γ was computed up to a separation distance, u, of 12 m to comply with the empirical rule to stop the calculation of $\gamma(u)$ when u becomes

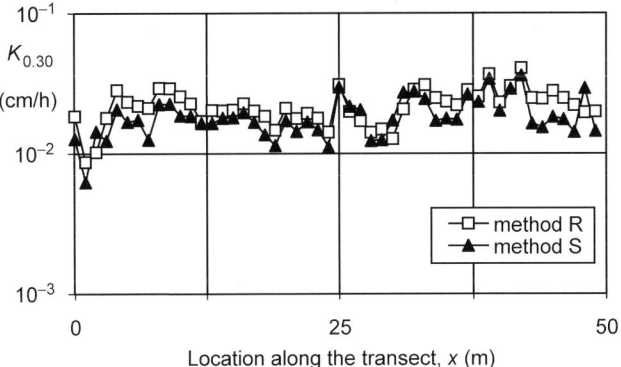

Figure 1. Hydraulic conductivity $K_{0.30}$ (cm h^{-1}) at water content of 0.30 m^3 m^{-3} along a 50-m transect measured at 1-m intervals. Squares show values found using the inverse method by Romano (1993), denoted as method R, and triangles show values estimated with the method of Sisson and van Genuchten (1991), denoted as method S. $K_{0.30}$ corresponds to unsaturated hydraulic conductivity across the transect when the effective soil water saturation is on average equal to 75%.

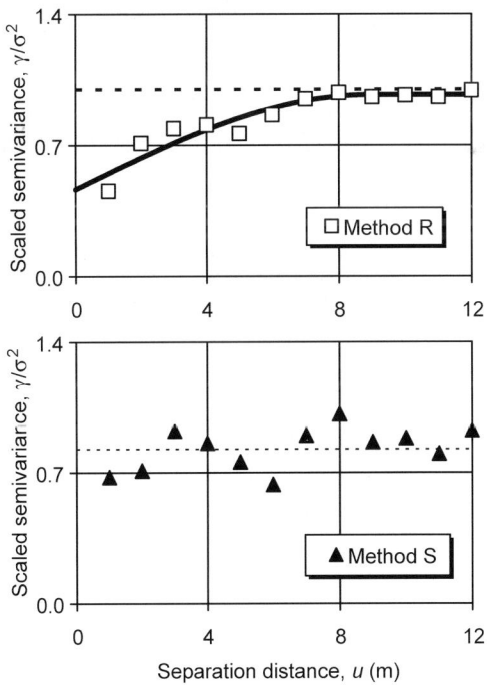

Figure 2. Scaled semivariograms of the log($K_{0.30}$) with the spherical models fitted obtained from method R (top) and method S (bottom). Dashed lines indicate sample variances. $K_{0.30}$ (in cm h^{-1}) is the hydraulic conductivity at water content of 0.30 m^3 m^{-3}.

greater than about N/4, with N being the size of the sample data ($N = 50$, in this case). The semivariogram for variable $\log(K_{0.30})$ as computed by method R shows the existence of the spatial structure with a range approximately equal to 8 m. Similar patterns of the spatial variability were found for the same soil in a nearby area (Ciollaro and Romano, 1995). Instead, the semivariogram computed from the values of method S is a pure nugget effect; that is, this method says that the decimal logarithm of hydraulic conductivity $K(\theta = 0.30)$ varies randomly across the transect. This can be in part due to the fact that method S generates values of $K_{0.30}$ that fluctuate more over shorter distances and captures with difficulty the spatial connectivity among the actual data. Therefore, a simplifying assumption may have distorted the available data set in such a way that while the basic

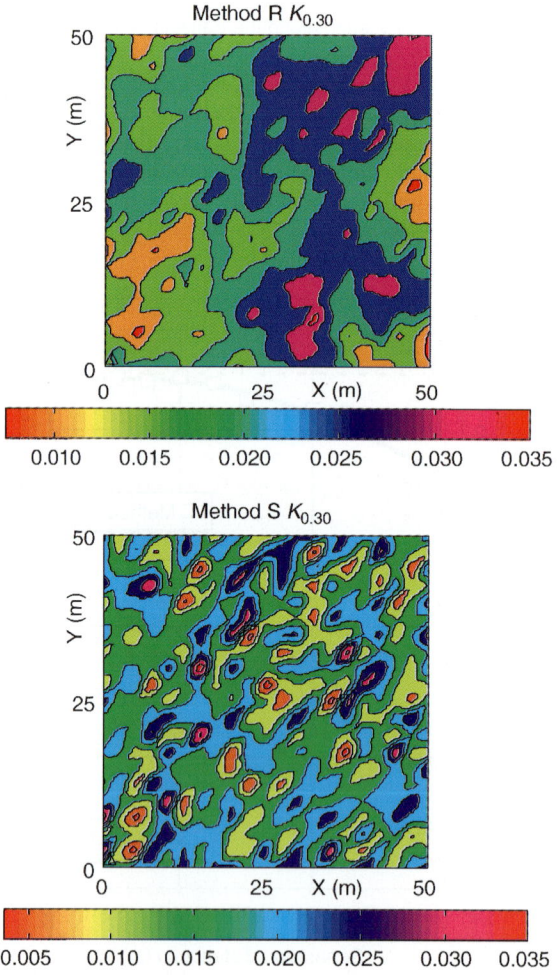

Figure 3. Random fields of variable $K_{0.30}$ generated using the spatial structure determined with method R (top) and method S (bottom). $K_{0.30}$ (in cm h^{-1}) is the hydraulic conductivity at water content of 0.30 m^3 m^{-3}.

statistics is preserved somewhat, the assessment of the spatial structure becomes very unreliable. That may dramatically hamper our ability to predict the soil hydraulic behavior.

To summarize this interesting finding, 2D random fields have been generated under the hypothesis of isotropy using the HYDRO-GEN code developed by Bellin and Rubin (1996). For an area of 50×50 m^2, different hypothetical realizations of K were computed using the semivariograms obtained from methods R and S, respectively, and Figure 3 shows the maps of the resulting fields of variable $K_{0.30}$. The simulated fields are markedly different and comparison between them gives an overall idea of the magnitude of the bias when information gained from an inaccurate measuring method is transferred to analyze the structure of spatial variations. As pointed out by Romano and Palladino (2002), a simplified method may produce information with some loss of contents that tends to obscure the actual spatial dependence and results in unsound maps of soil hydraulic variables.

3. CASE STUDY AND DISCUSSION

A transect case study will be presented with the major aim of providing a critical account of the parameterization of field-scale soil water flow models using PTFs. The problem of an adequate description of the spatial structure of soil hydraulic properties is addressed here with reference to two different pedotransfer methods pertaining to the statistical regression equations or the artificial neural networks (ANNs), respectively.

Data were collected from a hillslope of the Sauro River catchment area near the village of Guardia Perticara in Basilicata, Italy. The hydrology of the catchment is affected by the seasonality of the precipitation. The streamflow regime depends strongly on seasonal variations, with low or no flow during most of the year and high discharge peaks of short duration around autumn or early winter. Santini et al. (1999) provided a soil-landscape description of the study hillslope. The soil classification for this research was based on the US soil taxonomy (Soil Survey Staff, 1998). From the summit to shoulder hillslope position, soils ranged from shallow soils developing on the steep and stable slope (representative soil – *Lithic Haplustoll*) to moderately deep calcareous soils (representative soil – *Calcic Ustochrept*), respectively. Soils along upper and lower backslope are mainly deep soils developing on gentle slope (representative soil in the upper backslope – *Chromic Calciustert*; representative soils in the lower backslope – *Typic Haplustert* and *Calcic Ustochrepts*). In the footslope, deep and moderately deep soils of steeper slope were found (representative soils – *Chromic Calciustert* and *Calcic Ustochrept*). The toeslope mainly had soils of the alluvial plain (representative soil – *Typic Ustifluvent*). Vegetation mainly consisted of deciduous/perennial trees and shrubs, the latter being partly present on land that was used for cattle grazing and horticulture prior to 1955. Land uses were mainly wheat crops, olive trees, or abandoned lands.

A total of 100 points were sampled at 50-m intervals along a 5-km transect that run approximately in the N–S direction, from an elevation of 1072 m down to 472 m above sea level. Undisturbed soil samples were taken at 5-cm soil depth using 7-cm long steel cylinders with an inner diameter of 7.2 cm to determine basic soil properties, water retention functions and saturated hydraulic conductivity in the laboratory using standard techniques. Additional details can be found in the chapter by Romano and Santini (1997).

The collected samples comprised a relatively wide range of soil texture classes, with variations from about 3 to 58% for sand contents and from about 14 to 53% for clay contents, while silt contents were within a narrower range from about 24 to 54%. Values of the saturated hydraulic conductivity, K_s, were determined with the falling head method. Soil water retention data points were measured up to a matric pressure head of about -2.50 m using a recently designed suction table apparatus (Romano et al., 2002). The membrane plate apparatus was used to measure water retention points at pressure heads of -30, -60, and -120 m. To allow more objective comparisons between the observed and estimated water retention characteristics, for each soil sample the measured $\theta(h)$-values were fitted with van Genuchten's closed-form relation (2). The VGM retention parameter set $\{\theta_s, \theta_r, \alpha, n = 1 - 1/m\}$ was optimized using the Downhill Simplex Method (Press et al., 1992). Selected PTFs were evaluated with the fitted water retention data sets as a reference for comparisons. Unsaturated hydraulic conductivity functions for each sample were obtained from the VGM relation (2b) using the measured K_s-values and setting parameter λ at -1.0 (Schaap and Leij, 2000). For all variables, normality at 95% significance level was tested using the κ^2 statistic of D'Agostino et al. (1990), accounting for the combined effects of skewness and kurtosis.

3.1. Potential and limitations of using PTF estimates to capture the spatial structure of soil hydraulic parameters

In this section, the data presented by Romano and Santini (1997) are revisited using the PTF-HYPRES that was recently developed from a database of soils at European scale (Wösten et al., 1999). Readers are directed to the cited publication for details on the PTF-HYPRES regression equations to calculate the hydraulic parameter set $\{\theta_s, \theta_r, \alpha, n = 1/(1 - m), K_s, \lambda\}$ featuring in equation (2).

Figure 4 depicts scatter plots of van Genuchten's parameters α and n as computed by fitting the measured data with van Genuchten's relation (2) (denoted α_{fitted} and n_{fitted}) or estimated by PTF-HYPRES (denoted α_{HYPRES} and n_{HYPRES}). In Figure 4 (top), data pairs $\alpha_{fitted}-\alpha_{HYPRES}$ are mostly contained in a relatively narrow band, though very different from the 1:1 straight line. This can be attributed to the fact that the water retention scale parameter α should be mostly related to soil structural properties, which are actually not input descriptors for PTF-HYPRES. At larger α-values, the points spread in a typically fan-shaped way. For the examined transect data, the PTF-HYPRES shows a systematic overestimation of parameter α_{fitted}. Figure 4 (bottom) depicts a poorly defined correspondence between the n_{fitted} and n_{HYPRES} data points. For values of n_{fitted} greater than about 1.5, parameter n_{HYPRES} is essentially constant. Therefore, PTF-HYPRES seems virtually unable to identify the shape of the particle-size distribution, especially for soils with a predominant sand fraction. This outcome is rather difficult to interpret, as the water retention shape parameter, n, should probably depend chiefly on soil textural properties. A poor correspondence between computed and PTF-estimated retention parameters α and n has also been observed when using other PTFs with the datasets collected by Romano and Palladino (2002). Nonetheless, it does not seem important to seek a good correlation between the computed and predicted shape or scale parameters as recent PTFs are able to provide relatively good estimates of the water retention variables all the same. The other side of the coin, however, is that statistics and spatial analysis based on the PTF-estimated shape or scale parameters should be evaluated with great care as the relevant results might be misleading or meaningless. At least for the

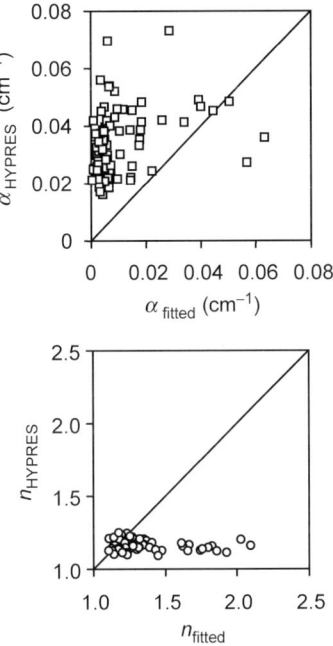

Figure 4. Scatter plots of VGM water retention parameters α (top) and n (bottom). Parameters α_{fitted} and n_{fitted} are obtained by fitting the available row data to van Genuchten's relation (2); parameters α_{HYPRES} and n_{HYPRES} are obtained from the PTF-HYPRES regression equations.

examined transect data, scale parameter α seems more sensitive to spatial variations than the shape parameter n. This result was also found by Ciollaro and Romano (1995) and Meyer and Gee (1999).

Fitted and HYPRES-estimated soil water retention data were statistically analyzed and Table 1 compares descriptive statistics for $\theta(h)$ values at matric pressure head, h, of -10, -100, and -1000 cm. Standard deviation, σ, for the HYPRES-estimated water contents is always smaller than that for the corresponding fitted water contents, indicating a smoothing effect that is more pronounced in both wetter (for $h = -10$ cm) and drier (for $h = -1000$ cm) soil conditions. This seems a feature of the indirect water retention estimates using present-day PTFs. To illustrate this latter finding and also to show the spatial variations of water contents along the study transect, Figure 5 compares the series of fitted and HYPRES-estimated water content θ at $h = -10$ cm.

To obtain basic quantitative information on similarity between the data sets, first the F-test was performed at 5% level of significance and for all of the three cases the null hypothesis of equality of variances could be accepted at the level of significance adopted. Student's t-tests at 10% level of significance (5% in each tail) were then performed to test the null hypothesis that two sample means are equal (or, using a better statistical jargon, that it cannot be excluded that two samples come from populations with different means). This hypothesis should be rejected in all three cases, namely the means of fitted

Table 1
Statistical properties of some fitted, HYPRES, and $\text{ANN}_{\text{basic+terrain}}$-estimated soil water retention variables for the study transect

Variable	Summary statistics[a]						MXE	ME%	RMSE%
	Min	Max	$\bar{x}$	σ	CV	κ_c^2			
From VGM Eq. (2a) fitted to measured retention values									
θ_{-10}[b]	0.313	0.526	0.418	0.0412	9.85	5.21	nd	nd	nd
θ_{-100}	0.286	0.493	0.390	0.0463	11.9	0.151	nd	nd	nd
θ_{-1000}	0.157	0.427	0.286	0.0482	16.9	0.580	nd	nd	nd
HYPRES-estimated water retention data									
θ_{-10}	0.353	0.523	0.430	0.0359	8.36	4.15	0.119	−1.13	3.28
θ_{-100}	0.257	0.465	0.359	0.0407	11.4	0.754	0.114	3.10	4.78
θ_{-1000}	0.152	0.349	0.258	0.0401	15.5	0.314	0.134	2.72	5.15
$\text{ANN}_{\text{basic+terrain}}$-*estimated water retention data*									
θ_{-10}	0.358	0.482	0.416	0.0233	5.61	0.298	0.0793	0.232	2.54
θ_{-100}	0.312	0.440	0.388	0.0283	7.28	2.64	0.0761	0.148	2.86
θ_{-1000}	0.200	0.312	0.266	0.0212	7.99	4.23	0.139	1.97	4.45

[a]Min, minimum value; Max, maximum value; $\bar{x}$, mean; σ, standard deviation; CV, coefficient of variation (in percent); κ_c^2, κ^2 statistic for normality test computed from the data (critical value $\kappa_{\alpha=0.05}^2 = 5.99$). MXE is the maximum error; ME is the mean error (in percent); RMSE is the root mean square error (in percent).

[b]θ_{-10}, θ_{-100}, and θ_{-1000} are volumetric water contents at matric pressure heads of −10, −100, and −1000 cm, respectively.

and HYPRES-estimated values should be considered different. Therefore, if the soil water retention characteristics are estimated with PTF-HYPRES and the specific problem at hand or model to be used is sensitive to the average water retention behavior, then the actual data give a better description of the process. However, a general

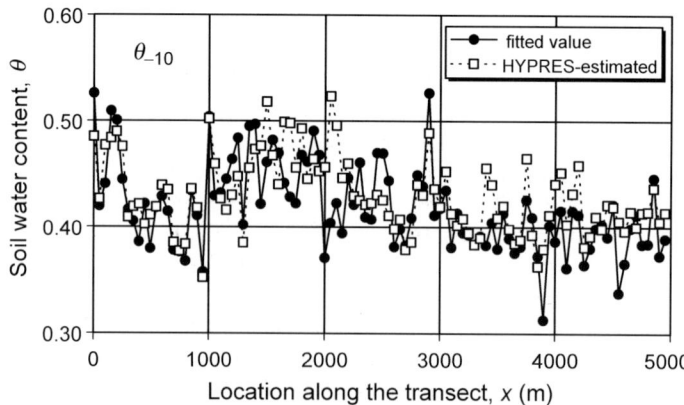

Figure 5. Values of soil water content θ at $h = -10$ cm as fitted from measured data or estimated by PTF-HYPRES along a 5-km transect with 50-m sampling intervals.

similarity in variability of soil water contents is observed in terms of coefficient of variation, $CV = \sigma/\underline{x}$, (the ratio of the standard deviation to its mean). The CVs of HYPRES-estimated water contents are relatively close to the CVs calculated from the fitted water contents. Even though PTF-HYPRES show low accuracy in generating reliable mean water retention values along the study transect, the deviations from the means are estimated fairly well. This was pointed out by Romano and Santini (1997) with reference to other recent PTFs. Moreover, using PTF-HYPRES preserves the finding that CV increases as soil becomes drier. The κ_c^2 values reported in Table 1 show that the HYPRES-estimated spatial series are normally distributed (i.e., $\kappa_c^2 < \kappa_{0.05}^2$). Closer inspection of frequency histograms, as well as of the skewness and kurtosis statistics, reveals that the probability functions of the water retention variables estimated by PTF-HYPRES are somewhat skewed (i.e., less symmetric) than the probability function of the variables obtained from the observed data, while the respective coefficients of kurtosis are closer, indicating similarity in peakedness. Similar statistical evaluations are reported in Table 2 for spatial series of the unsaturated hydraulic conductivity functions. The D'Agostino et al. (1990) normality test (at the 0.05 confidence level) shows that the decimal log transform applied to the VGM-fitted conductivity data is successful in generating series that follow the normal distribution. Neither of the log-transformed HYPRES-estimated hydraulic conductivity variables follows the normal distribution of probability. In general, these findings should be taken in due account when statistical results are then used for stochastic simulations (Goovaerts, 2000).

Table 2
Statistical properties of some fitted and HYPRES-estimated unsaturated hydraulic conductivity variables for the study transect

Variable	Summary statistics[a]						MXE	ME%	RMSE%
	Min	Max	$\underline{x}$	σ	CV	κ_c^2			
From VGM Eq. (2b) using measured K_s									
$\log(K_{-10})$[b]	−3.53	0.834	−1.23	0.962	78.2	2.60	nd	nd	nd
$\log(K_{-100})$	−4.72	0.343	−2.03	1.15	56.7	2.68	nd	nd	nd
$\log(K_{-1000})$	−6.71	−1.25	−3.72	1.29	34.7	2.95	nd	nd	nd
HYPRES-estimated hydraulic conductivity data									
$\log(K_{-10})$	−2.61	−0.840	−1.39	0.372	26.7	19.9	2.54	16.5	108
$\log(K_{-100})$	−4.12	−2.15	−2.80	0.437	15.6	24.8	3.52	76.8	146
$\log(K_{-1000})$	−7.87	−4.84	−5.50	0.539	9.81	40.2	3.95	98.4	168

[a]Min, minimum value; Max, maximum value; $\underline{x}$, mean; σ, standard deviation; CV, coefficient of variation (in percent); κ_c^2, κ^2 statistic for normality test computed from the data (critical value $\kappa_{\alpha=0.05}^2 = 5.99$). MXE is the maximum error; ME is the mean error (in percent); RMSE is the root mean square error (in percent).
[b]$\log(K_{-10})$, $\log(K_{-100})$, and $\log(K_{-1000})$ are decimal logarithm of hydraulic conductivity at matric pressure heads of −10, −100, and −1000 cm, respectively.

Residual error analysis was carried out to highlight the general performance of a PTF using the following statistical indicators MXE, ME%, and RMSE%:

$$\text{MXE} = \max_{i=1}^{N}[\text{abs}(\text{err}_i)] \tag{3}$$

$$\text{ME\%} = \frac{\sum_{i=1}^{N} \text{err}_i}{N} \times 100 \tag{4}$$

$$\text{RMSE\%} = \left[\frac{\sum_{i=1}^{N} (\text{err}_i)^2}{N}\right]^{1/2} \times 100 \tag{5}$$

where N is the number of observations, and $\text{err}_i = (f_i - e_i)$ is the error between fitted value, f_i, and PTF-estimated value, e_i. The MXE statistic is a local indicator of the goodness of the PTF estimates. Accuracy and precision of a PTF are evaluated quantitatively using the ME% and RMSE% statistics. ME% reveals the presence of biases (i.e., a systematic overestimation, if negative, or underestimation, if positive, of the fitted values), whereas RMSE% quantifies the scatter around the regression line in a plot depicting fitted and PTF-estimated quantities. If compared with the results of Romano and Santini (1997) obtained for the same data set using other PTFs, the PTF-HYPRES gives similar RMSEs, but generates water contents that underestimate more the fitted water contents.

To describe the reliability in the HYPRES water retention estimates from saturation to dryness, Figure 6 shows the mean relative error, MRE%, and prediction efficiency, PEf%, statistics:

$$\text{MRE\%}(h) = \frac{1}{N}\sum_{i=1}^{N}\left(1 - \frac{\theta(h)_{\text{PTF},i}}{\theta(h)_{\text{VGM},i}}\right) \times 100 \tag{6}$$

$$\text{PEf\%}(h) = \left\{1 - \frac{\sum_{i=1}^{N}[\theta(h)_{\text{VGM},i} - \theta(h)_{\text{PTF},i}]^2}{\sum_{i=1}^{N}[\theta(h)_{\text{VGM},i} - \overline{\theta(h)}_{\text{VGM}}]^2}\right\} \times 100 \tag{7}$$

as a function of matric pressure head, h. In these equations, N is the size of the total available data and $\overline{\theta(h)}_{\text{VGM}}$ denotes the arithmetic mean of the individual fitted values $\theta(h)_{\text{VGM}}$. The $\overline{\text{MRE}\%}$ provides a measure of the dispersion of HYPRES-estimations, which overall seems fairly acceptable as values of MRE% less than 5% occur for 5.0 cm $< |h| <$ 50 cm and for 2500 cm $< |h| <$ 5000 cm. The PEf% statistic provides an indication of the HYPRES ability to reproduce the 1:1 line. High values of PEf% (near 100%) indicate a very good agreement between fitted values and PTF-HYPRES estimates,

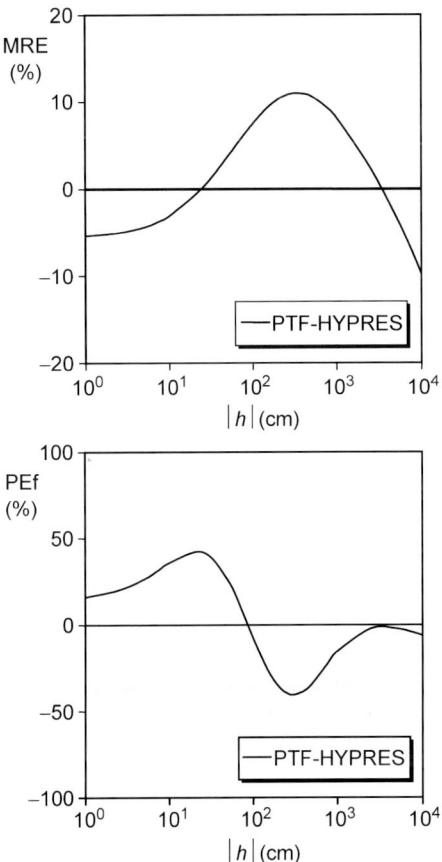

Figure 6. MRE (top) and PEf (bottom) statistics (in percent) as a function of matric pressure head, $|h|$ (absolute value), for PTF-HYPRES water retention estimates along a 5-km transect with 50-m sampling intervals.

whereas negative values would suggest that only the mean of the observations yields an acceptable picture of the average patterns of the variable. Note that the lower limit for PEf% is $-\infty$. The PEf% indicator takes on values greater than zero only for $|h|$ less than about 100 cm, with a peak of 42.1% for $|h| = 25$ cm. Therefore, soil water retention behavior along the transect is reasonably estimated only under wetter conditions in soil. Romano and Palladino (2002) have recently shown that this performance can be improved if a PTF is calibrated using terrain attributes of the area of interest.

Semivariograms based on the fitted and HYPRES-estimated spatial data were computed. For comparisons purposes only, Tables 3 and 4 report the parameters that identify the spherical semivariogram models. The ratio between nugget and total semivariance can be viewed as an indicator of the strength of spatial dependence with the sampling scale adopted: *strong* if that ratio is ≤ 25, *moderate* if the ratio is between 25 and 75%, and *weak* if the ratio is ≥ 75. The three-fitted water retention variables of Table 3 are

Table 3
Parameters of the spherical semivariograms for fitted, HYPRES-, ANN$_{basic}$-, and ANN$_{basic+terrain}$-estimated soil water retention

	Variance ($\times 10^{-3}$)	Semivariance $\times 10^{-3}$			Range (m)
		Nugget	Structural	Total	
From VGM Eq. (2a) fitted to measured values					
θ_{-10}[a]	1.70	0.968	0.499	1.47	824.7
θ_{-100}	2.14	1.36	0.685	2.04	1284
θ_{-1000}	2.32	1.33	1.12	2.45	380.5
HYPRES-estimated water retention data					
θ_{-10}	1.29	0.672	0.703	1.38	951.2
θ_{-100}	1.66	0.682	1.29	1.97	1160
θ_{-1000}	1.61	0.672	1.35	2.02	1173
ANN$_{basic}$-estimated water retention data					
θ_{-10}	0.303	0.133	0.115	0.249	1498
θ_{-100}	0.635	0.304	0.327	0.631	1677
θ_{-1000}	0.489	0.242	0.276	0.517	1192
ANN$_{basic+terrain}$-estimated water retention data					
θ_{-10}	0.544	0.221	0.205	0.426	1120
θ_{-100}	0.799	0.276	0.453	0.729	1574
θ_{-1000}	0.451	0.139	0.305	0.444	1213

[a] θ_{-10}, θ_{-100}, and θ_{-1000} are volumetric water contents at matric pressure heads of -10, -100, and -1000 cm, respectively.

Table 4
Parameters of the spherical semivariograms for fitted and HYPRES-estimated unsaturated hydraulic conductivity

Variable	Variance	Semivariance			Range (m)
		Nugget	Structural	Total	
From VGM Eq. (2b) using measured K_s					
$\log(K_{-10})$[a]	0.925	0.589	0.246	0.835	521.9
$\log(K_{-100})$	1.33	0.792	0.365	1.16	530.5
$\log(K_{-1000})$	1.67	1.02	0.437	1.45	517.2
HYPRES-estimated soil hydraulic conductivity data					
$\log(K_{-10})$	0.134	0.0336	0.0795	0.113	710.6
$\log(K_{-100})$	0.191	0.0498	0.109	0.159	769.5
$\log(K_{-1000})$	0.257	0.0735	0.141	0.214	824.2

[a] $\log(K_{-10})$, $\log(K_{-100})$, and $\log(K_{-1000})$ are decimal logarithm of hydraulic conductivity at matric pressure heads of -10, -100, and -1000 cm, respectively.

moderately spatially dependent with the ratio of nugget semivariance to total semivariance ranging from about 54 to 67%. For the HYPRES-estimated water retentions, the corresponding nugget semivariances expressed as a percentage of the total semivariance range from about 33 to 49%. Therefore, HYPRES-estimated retention variables show a moderate spatial dependence that is more pronounced than that of the fitted retention variables. This feature can be a result of the dominant control of textural properties that basically represent intrinsic sources of variation, on the spatial variability exhibited by PTF-estimated retention variables. The retention variables obtained from the measurements are also very sensitive to tillage practices and changes in land use, which represent extrinsic sources of variability, and thus their structure of variability is less moderate and tends toward the class of weak spatial dependence. This is also confirmed by the nugget values of the fitted retention variables (Table 3) that are greater than those of the HYPRES-estimated retention variables, indicating significant variations of the fitted retention variables at scales smaller than those of the sampling interval used along the transect ($\Delta x = 50$ m). Note that fitted water retention variable θ_{-1000} has a much smaller range (range = 380 m) than those of the other two fitted retention variables. As soil dries out, the character of spatial correlation in soil water retention can change even significantly, but the observed range for θ_{-1000} can be also due to a spurious effect of some larger measurement errors of θ at smaller h that generate rapid fluctuations between the spatial data. Results for the water retention are summarized in the plot of Figure 7 that presents the comparison between the scaled spherical semivariograms for water retention variable θ_{-10}. The two-scaled semivariograms have similar shape and show typical decreases in spatial correlation as separation distances increase. It is interesting to note that the semivariograms reach the respective sills at very similar values of the *range* of

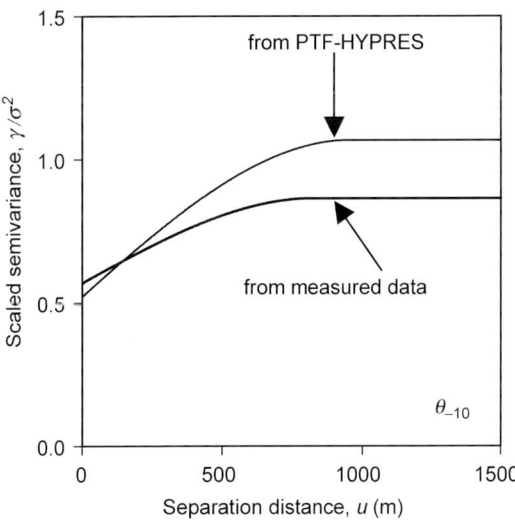

Figure 7. Scaled semivariograms of soil water content at the matric pressure head of -10 cm, θ_{-10}, calculated from the observations (thick line) or PTF-HYPRES estimates (thin line) along a 5-km transect with 50-m sampling intervals.

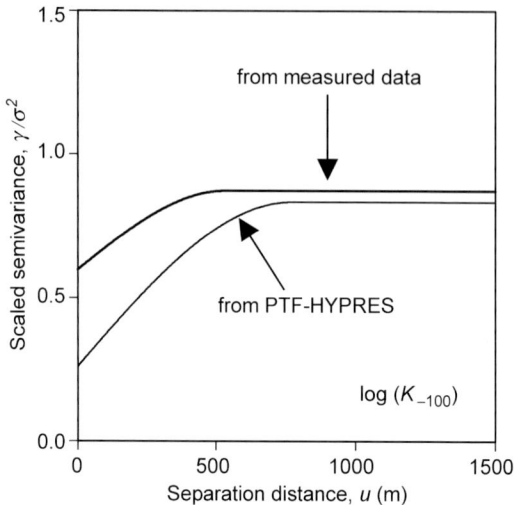

Figure 8. Scaled semivariograms of hydraulic conductivity log(K_{-100}) calculated from the observations (thick line) or PTF-HYPRES estimates (thin line) along a 5-km transect with 50-m sampling intervals.

correlation. However, as it may occur, the sill of a semivariogram is not always equal to the sample variance, i.e., at larger u-values the ratio γ/σ can differ from unity. A discussion on this latter issue is beyond the scope of this chapter, but can be found in the paper by Barnes (1991). The semivariograms in Figure 7 have similar intercepts but different slopes indicating that structural semivariance should be related to slightly different factors: intrinsic factors dominate the semivariograms of the HYPRES-estimates, whereas extrinsic factors also come into play for the semivariograms of the fitted retention variables. Overall, the semivariograms depicted in Figure 7 confirms the relatively good efficiency of a PTF in detecting the spatial dependency of a soil water retention variable. As for the unsaturated hydraulic conductivity characteristic, Figure 8 shows the scaled semivariograms for variable log(K_{-100}). This plot highlights the smoothing effect of the PTF with respect to the actual variability exhibited by the observations. This is reflected in the nugget values. For the semivariogram of log(K_{-100}) computed from K_s measurements, the large nugget with respect to the sill shows that most of the hydraulic conductivity variations along the transect occur at relatively short distances. Therefore, while observations and PTF estimates lead to semivariograms with similar range and sill values, hence comparable long-range spatial variations, the pattern of variation for observed hydraulic conductivity exhibits more complexity than that estimated by the PTF as it is affected to a greater extent by processes acting at different scales.

All these results illustrate that PTFs may be a good tool to estimate changes in soil hydraulic properties at coarse scale and hence to provide a good description of spatial variation at that scale. This may also be due in part to the fact that the shape of a semivariogram is less affected by the presence of systematic errors in the data. Usually semivariograms show different patterns if the relevant data fluctuate differently around the means. The problem remains as to how to reduce the bias when describing the average soil hydraulic behavior with a PTF. This will be addressed in the next section.

3.2. Assessment of soil hydraulic spatial variability using ANNs and terrain attributes

In the previous section, we have seen that PTFs estimate the soil hydraulic properties with some bias, but are able to provide a reasonable portrayal of the spatial variations exhibited by soil hydraulic properties. Romano and Palladino (2002) have put forward the idea of producing unbiased PTF-output by making an existing PTF more efficient with ancillary information from topography. In this section, the potential benefit of including terrain attributes in the input set of descriptors is evaluated within the framework of the ANNs approach. Increasingly, ANNs are being used to identify complex input–output relationships.

Topographical variables for the study transect were obtained from a 30 × 30-m grid size Digital Elevation Model using the geographic resources analysis support system (GRASS) software for geographic information systems. Selected terrain attributes at each sample location were interpolated with the nearest neighbor procedure using Surfer® version 7.00 (Golden Software, Inc., Golden, CO). The ANNs consisted of feed-forward back-propagation networks with a hidden layer containing six hidden nodes and sigmoidal transfer functions, and were developed to predict the hydraulic parameters $\{\theta_s, \theta_r, \alpha, m, K_s\}$ from different sets of predictors, partly selected based on preliminary analysis using Spearman ranking. Spearman coefficients showed that elevation above mean sea level, z, was positively correlated with organic carbon content (OC) and silt content, and negatively with oven-dry bulk density, ρ_b, and sand content. The soil water retention parameters of equation (2) were somewhat correlated with the topographical attributes of elevation, z, maximum slope, β, orientation of the slope (or aspect, $\cos \phi$, with ϕ in degrees clockwise from north), and potential solar radiation, but not as strongly as with the soil physical properties. Correlation was also investigated for soil water contents at specific matric pressure heads. As for the correlation between soil water contents and topography, elevation was mainly correlated with water contents in the wet range of water retention ($h > -250$ cm). Slope has a slight negative correlation with water contents at lower h, and this may be due to the lower clay content of the uppermost soil horizons positioned on steeper slopes. Slope orientation shows a modest correlation with water contents in the entire matric pressure range investigated. As the aspect was found to be fairly independent of soil attributes, it can represent a valuable predictor for soil water retentions. No correlation was found between K_s and basic soil properties, and no obvious explanation was found for this lack of correlation. The parameter λ was set at -1.0 (Schaap and Leij, 2000). Optimization of the ANNs was performed by minimizing the squared residuals of the hydraulic parameters using the Levenberg–Marquardt algorithm. The bootstrap method was employed for randomized sampling with replacement to divide the 100-sample set in calibration and validation parts (containing roughly 64 and 36% of the samples, respectively). The calibration and bootstrap procedure was repeated 30 times. This procedure reduces the risk of bias and allows estimation of the estimation uncertainty. The bootstrap method was combined with the TRAINLM routine of the neural network toolbox of MATLAB® version 4.0 (MathWorks Inc., Natick, MA). Further details can be found in Leij et al. (2004).

Several ANNs have been developed using different combinations of predictors, such as sand, silt, and clay contents (SSC), oven-dry bulk density (ρ_b), organic carbon content (OC), elevation (z), slope (β), aspect ($\cos \phi$), profile and tangential curvatures, and potential solar radiation. Compound topographical variables, such as the wetness index, sediment transport capacity indices, and elevation relief ratio, were also considered. To facilitate comparisons with the results of the previous section, Table 1 shows statistical evaluations for the ANN$_{basic+terrain}$ obtained with the following predictors: sand, silt, and clay contents, oven-dry bulk density, organic carbon content, elevation, slope, and aspect. These input predictors comprise those of a classic PTF (SSC, ρ_b, OC) together with three additional terrain attributes (z, β, $\cos \phi$). Student's t-tests show that differences between the means of fitted and ANN-estimated water retention variables are not significant, but the standard deviations show pronounced differences. In general, the residual error analysis shows the benefit of having incorporate terrain attributes into the pedotransfer predictions. The MXE, ME%, and RMSE% have improved, specifically bias (i.e., ME%) and the measure of variation around the mean (i.e., RMSE%) are now considerably smaller. Results obtained with the classic physico-chemical input descriptors (SSC, ρ_b, OC), denoted as ANN$_{basic}$, are not shown in this table for brevity.

To further demonstrate the effectiveness of including terrain variables in the pedotransfer method, Figure 9 compares PEf%(h) curves for ANN$_{basic}$ and ANN$_{basic+terrain}$. Values of PEf% for ANN$_{basic+terrain}$ are greater than zero for $|h|$ up to about 2000 cm. Moreover, PEf% is nearly constant at 61% for $|h|$ values ranging from 1 to approximately 150 cm and then approaches zero for $|h|$ of about 2500 cm. The average description of the water retention characteristics offered by ANN$_{basic+terrain}$ along the transect appears fairly efficient.

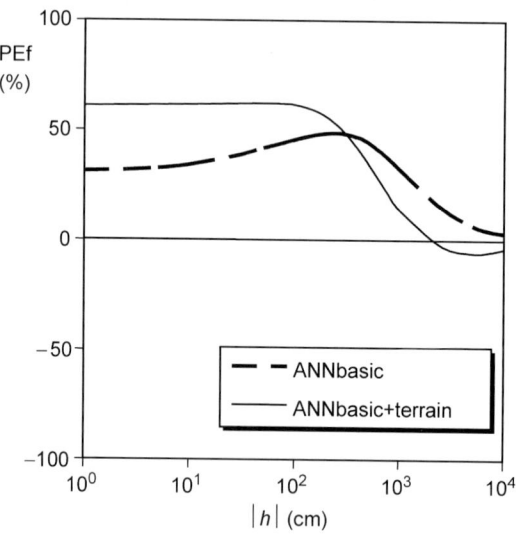

Figure 9. PEf statistics (in percent) as a function of matric pressure head, $|h|$ (absolute value), for ANNs water retention estimates along a 5-km transect with 50-m sampling intervals. Dashed line refers to ANN$_{basic}$ with SSC, ρ_b, OC as input descriptors, whereas solid line refers to ANN$_{basic+terrain}$ which includes as additional descriptors the terrain attributes of elevation, z, slope, β, and aspect, $\cos \phi$.

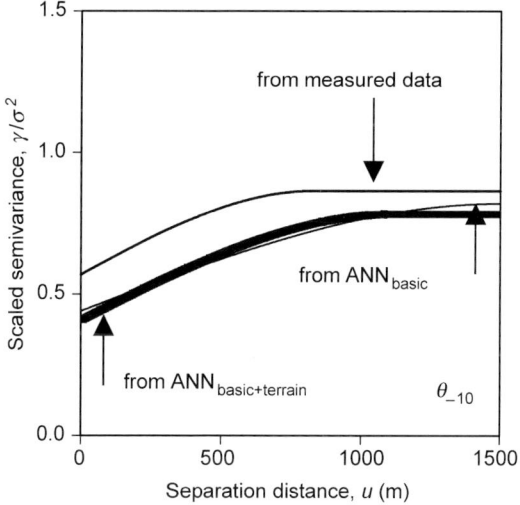

Figure 10. Scaled semivariograms of soil water content at the matric pressure head of -10 cm, θ_{-10}, calculated from the observations (thick line) or ANNs estimates (thin and bold lines) along a 5-km transect with 50-m sampling intervals.

Figure 10 presents spherical semivariograms of the two ANNs for water retention variable θ_{-10}. There are relatively small differences between the ANN semivariograms that are both close to the semivariogram of the observations. Note that the slope of the ANN$_{basic}$ semivariogram is slightly different than that of the semivariogram of the observation. This situation was already encountered in the case of the PTF-HYPRES (Figure 7), which uses the same input predictors of ANN$_{basic}$. The semivariograms from the observations and ANN$_{basic+terrain}$ have more similar slopes. While including topographical attributes as ancillary information does help to increase accuracy in PTF estimation of soil hydraulic properties, it does not appear to alter the spatial fluctuations of PTF estimates around the average and hence the detection of their spatial correlation structure.

4. CONCLUDING REMARKS WITH AN EYE ON SCALE ISSUES

It has been observed that the availability of spatially distributed information is usually the most critical factor in land surface modeling and this applies to different types of parameters and data of the physical system under consideration. The examples discussed in this paper with reference to soil hydraulic properties demonstrate the potential usefulness of employing the pedotransfer approach to provide simulation models, especially distributed parameter models, with a good average portrayal of soil hydraulic behavior. Hopefully this outcome would contribute to the increase of the use of distributed models, with a fruitful link between soil or soil-landscape mapping units and soil hydraulic functions, instead of the bucket-type models essentially based only on a simplistic view of the water-holding capacity of soil. Beven (2000, p. 172) questioned the representation capability offered by PTFs. However, one should make a distinction between point and average estimations. In concurrence with the recent findings by Soet and Stricker (2003) and Nemes et al. (2003), the results presented in this paper demonstrate that PTFs show great potential when employed to identify the functional behavior of a system, as in the case discussed here of describing the spatial variations of soil hydraulic properties. One should go beyond the fact that PTFs are strictly valid only within the environmental niches (e.g., field, regional, or continental scale) where they were developed. A PTF seems rather flexible with respect to its local calibration and this opens new opportunities in soil hydraulic characterization using indirect methods. Romano and Palladino (2002) have shown that PTF estimates can be suitably calibrated using ancillary information from digital elevation models. Our results reinforce the need for detailed, but limited in number, local measurements that reflect soil hydraulic behavior and would provide an information base for calibration of PTFs.

The selection of a mathematical model to understand natural phenomena or to predict the behavior of a particular physical system is never simple. This task is unavoidably conditioned by economic considerations, personal biases and the modeler's experience and expertise, site-specific hydrological issues, the scientific rigor to be applied, and the available data (Bouma and Hoosbeek, 1996). In practice distributed models do not use long-time series of meteorological and hydrological data for their calibrations, yet they demand an enormous amount of information to describe the spatial distributions of system parameters and input data, such as topographic characteristics, vegetation coverage, land use, and soil properties. Some parameters of the system under consideration should be determined in the field irrespective of the hydrologic model being used. However, as the area of interest increases, field campaigns become very expensive to carry out and model parameters may begin to lose their physical significance (Woolhiser, 1996).

Two basic problems affect the practical implementation of a hydrologic model: *spatial variability* and *scale* (Grayson and Blöschl, 2000). The interaction between spatial variability and scale is particularly evident for unsaturated flow processes, mainly because of the significant nonlinearity involved. Determination of soil hydraulic properties, either directly or indirectly via PTFs, is carried out at the laboratory scale, which is much smaller than the field or catchment scales representing the typical environments where computer models are required to predict the evolutions of fundamental processes. A mismatch does exist between the scales of physical measurements and the scales of model predictions (Beven, 1995; Blöschl, 2001). Therefore, the problem arises of embedding the small-scale variability into the larger scale represented by the numerical grid of the model. This is the upscaling problem consisting in the identification of a suitable equivalent parameter, which is a single value or function that relates spatially averaged variables (Kabat et al., 1997; Jhorar et al., 2002). For example, the equivalent hydraulic conductivity by definition relates the spatially averaged flux to the spatially averaged hydraulic gradient. A method that provides sufficiently reliable spatial patterns of estimated soil hydraulic properties can be a good candidate for the subsequent task of determining a soil hydraulic property over an extent (e.g., a model grid square or a certain region) that is different from the measurement support (e.g., a soil core).

Assessment of the spatial structure of variability exhibited by soil hydraulic properties that are estimated by simplified indirect method, such as PTFs, is in its early days, but the research framework is appearing to take shape. An emerging challenge is the determination of equivalent hydraulic parameters using PTF estimates and information on their variations in space. We have a long way to go yet.

ACKNOWLEDGEMENTS

I would like to gratefully acknowledge Alessandro Santini, who took me under his wing and introduced me into the world of Soil Hydrology. He is a model for me, giving me an example to follow not only in my academic activities, but also in my personal life.

The study presented in this paper has been in part financially supported by P.O.N. project *AQUATEC*—New technologies of control, treatment, and maintenance for the solution of water emergency", funded by the Italian Ministry of University.

REFERENCES

Addiscott, T.M., Wagenet, R.J., 1989. Concepts of solute leaching in soils: a review of modelling approaches. J. Soil Sci. 36, 411-424.
Barnes, R.J., 1991. The variogram sill and the sample variance. Math. Geol. 23, 673-678.
Batchelor, W.D., Basso, B., Paz, J.O., 2002. Examples of strategies to analyze spatial and temporal yield variability using crop models. Eur. J. Agron. 18, 141-158.
Bellin, A., Rubin, Y., 1996. HYDRO-GEN: a spatially distributed random field generator for correlated properties. Stoch. Hydrol. Hydraul. 10, 253-278.
Beven, K.J., 1995. Linking parameters across scale: subgrid parameterizations and scale dependent hydrological models. Hydrol. Process. 9, 507-525.
Beven, K.J., 2000. Ranfall – Runoff Modelling: The Primer. John Wiley & Sons Ltd, Chichester, UK.
Blöschl, G., 2001. Scaling in hydrology. Hydrol. Process. 15, 709-711.
Bouma, J., 1989. Using soil survey data for quantitative land evaluation. Adv. Soil Sci. 9, 177-213.
Bouma, J., Hoosbeek, M.R., 1996. The contribution and importance of soil scientists in interdisciplinary studies dealing with land. *In*: Wagenet, R.J., Bouma, J. (Eds.), The Role of Soil Science in Interdisciplinary Research, pp. 1-15, SSSA Special Publ. No. 45. Soil Society of America, Inc., Madison, WI.
Bouma, J., Stoorvogel, J., van Alphen, B.J., Booltink, H.W.G., 1999. Pedology, precision agriculture, and the changing paradigm of agricultural research. Soil Sci. Soc. Am. J. 63, 1763-1768.
Burrough, P.A., 1993. Soil variability: a late 20th century view. Soil Fertil. 56, 529-562.
Ciollaro, G., Romano, N., 1995. Spatial variability of the hydraulic properties of a volcanic soil. Geoderma 65, 263-282.
D'Agostino, R.B., Belanger, A., D'Agostino, R.B., 1990. A suggestion for using powerful and informative tests of normality. Am. Stat. 44, 316-321.
D'Urso, G., Menenti, M., Santini, A., 1999. Regional application of one-dimensional water flow models for irrigation management. Agric. Water Manag. 40, 291-302.

Durner, W., Schultze, B., Zurmühl, T., 1999. State-of-the-art in inverse modeling of inflow/outflow experiments. In: van Genuchten, M.T., Leij, F.J., Wu, L. (Eds.), Characterization and Measurement of the Hydraulic Properties of Unsaturated Porous Media. University of California, Riverside, CA, pp. 661-681.

Goovaerts, G., 1997. Geostatistics for Natural Resources Evaluation. Oxford University Press, New York.

Goovaerts, G., 2000. Estimation or simulation of soil properties? An optimization problem with conflicting criteria. Geoderma 97, 165-186.

Grayson, R.B., Blöschl, G., 2000. Spatial processes: organization and patterns. In: Grayson, R.B., Blöschl, G. (Eds.), Spatial Patterns in Catchment Hydrology. Cambridge University Press, Cambridge, UK, pp. 3-16.

Holt, R.M., Wilson, J.L., Glass, R.J., 2002. Spatial bias in field-estimated unsaturated hydraulic properties. Water Resour. Res. 38 (12), 1311, doi:10.1029/2002WR001336.

Jhorar, R.K., Bastiaanssen, W.G.M., Feddes, R.A., van Dam, J.C., 2002. Inversely estimating soil hydraulic functions using evapotranspiration fluxes. J. Hydrol. 258, 198-213.

Kabat, P., Hutjes, R.W.A., Feddes, R.A., 1997. The scaling characteristics of soil parameters: from plot scale heterogeneity to subgrid parameterization. J. Hydrol. 190, 363-396.

Kutílek, M., Nielsen, D.R., 1994. Soil Hydrology. GeoEcology Textbook. Catena Verlag, Cremlingen-Destedt, Germany.

Lark, R.M., 2000. A comparison of some robust estimators of the variogram for use in soil survey. Eur. J. Soil Sci. 51, 137-157.

Leij, F.J., Romano, N., Palladino, M., Schaap, M.G., Coppola, A., 2004. Topographical attributes to predict soil hydraulic properties along a hillslope transect. Water Resour. Res. 40, W02407, doi:10.1029/2002WR001641.

McBratney, A.B., Odeh, I.O.A., 1997. Application of fuzzy sets in soil science: fuzzy logic, fuzzy measurements and fuzzy decisions. Geoderma 77, 85-113.

Meyer, P.D., Gee, G.W., 1999. Information on hydrologic conceptual models, parameters, uncertainty analysis, and data sources for dose assessments at decommissioning sites, NUREG/CR-6656, PNNL-13091, Washington, DC, pp. 1-34.

Minasny, B., McBratney, A.B., 2003. Integral energy as a measure of soil-water availability. Plant Soil 249, 253-262.

Mulla, D.J., Mallawatantri, A.P., Wendroth, O., Joschko, M., Rogasik, H., Koszinski, S., 1999. Site-specific management of flow and transport in homogeneous and structured soils. In: Parlange, M.B., Hopmans, J.W. (Eds.), Vadose Zone Hydrology: Cutting Across Disciplines. Oxford University Press, New York, USA, pp. 396-417.

Nemes, A., Schaap, M.G., Wösten, J.H.M., 2003. Functional evaluation of pedotransfer functions derived from different scales of data collection. Soil Sci. Soc. Am. J. 67, 1093-1102.

Press, W.H., Teukolsky, S.A., Vetterling, W.T., Flannery, B.P., 1992, Numerical Recipes in FORTRAN: The Art of Scientific Computing, 2nd Ed. Cambridge University Press, New York, NY.

Rao, P.S.C., Wagenet, R.J., 1985. Spatial variability of pesticides in field soils: methods of data analysis and consequences. Weed Sci. 33, 18-24.

Reynolds, C.A., Jackson, T.J., Rawls, W.J., 2000. Estimating soil water-holding capacities by linking the Food and Agriculture Organization soil map of the world with global

pedon databases and continuous pedotransfer functions. Water Resour. Res. 36, 3653-3662.
Romano, N., 1993. Use of an inverse method and geostatistics to estimate soil hydraulic conductivity for spatial variability analysis. Geoderma 60, 169-186.
Romano, N., Chirico, G.B., 2005. The role of terrain analysis in using and developing pedotransfer functions. In: Pachepsky, Y.A., Rawls, W.J. (Eds.), Development of Pedotransfer Functions in Soil Hydrology. Elsevier, Amsterdam.
Romano, N., Palladino, M., 2002. Prediction of soil water retention using soil physical data and terrain attributes. J. Hydrol. 265, 56-75.
Romano, N., Santini, A., 1997. Effectiveness of using pedo-transfer functions to quantify the spatial variability of soil water retention characteristics. J. Hydrol. 202, 137-157.
Romano, N., Santini, A., 2002. Water retention and storage: field. In: Dane, J.H., Topp, G.C. (Eds.), Methods of Soil Analysis, Part 4, Physical Methods, SSSA Book Series N.5. Soil Society of America, Inc., Madison, WI, pp. 721-738.
Romano, N., Hopmans, J.W., Dane, J.H., 2002. Water retention and storage: suction table. In: Dane, J.H., Topp, G.C. (Eds.), Methods of Soil Analysis, Part 4, Physical Methods, SSSA Book Series N.5. Soil Society of America, Inc., Madison, WI, pp. 692-698.
Santini, A., Coppola, A., Romano, N., Terribile, F., 1999. Interpretation of the spatial variability of soil hydraulic properties using a land system analysis. In: Feyen, J., Wiyo, K. (Eds.), Modelling of Transport Processes in Soils at Various Scales in Time and Space. Wageningen Pers, Wageningen, The Netherlands, pp. 491-500.
Schaap, M.G., Leij, F.L., 2000. Improved prediction of unsaturated hydraulic conductivity with the Mualem–van Genuchten model. Soil Sci. Soc. Am J. 64, 843-851.
Šimunek, J., Sejna, M., van Genuchten, M.T., 1999. HYDRUS-2D, simulating water flow, heat, and solute transport in two-dimensional variably saturated media. Version 2.0, US Salinity Laboratory, ARS/USDA, Riverside, California and International Ground Water Modeling Center, IGWMC – TPS 53, Colorado School of Mines, Golden, Colorado.
Sinowski, W., Scheinost, A.C., Auerswald, K., 1997. Regionalization of soil water retention curves in a highly variable soilscape: II. Comparison of regionalization procedures using a pedotranfer function. Geoderma 78, 145-160.
Sisson, J.B., van Genuchten, M.T., 1991. An improved analysis of gravity drainage experiments for estimating the unsaturated soil hydraulic functions. Water Resour. Res. 27, 569-575.
Soet, M., Stricker, J.N.M., 2003. Functional behaviour of pedotransfer functions in soil water flow simulations. Hydrol. Process. 17, 1659-1670.
Soil Survey Staff, 1998. Keys to Soil Taxonomy, Eighth edition. USDA-NRCS, Washington, DC.
Starks, P.J., Heathman, G.C., Ahuja, L.R., Ma, L., 2003. Use of limited soil property data and modeling to estimate root zone soil water content. J. Hydrol. 272, 131-147.
Van Genuchten, M.T., Nielsen, D.R., 1985. On describing and predicting the hydraulic properties of unsaturated soils. Ann. Geophys. 3, 615-628.
Wagenet, R.J., Hutson, J.L., 1986. Predicting the fate on nonvolatile pesticides in the unsaturated zone. J. Environ. Qual. 15, 315-322.
Woolhiser, D.A., 1996. Search for physically based runoff model – a hydrologic El Dorado. ASCE J. Hydraul. Engng. 122, 122-129.
Wösten, J.H.M., Lilly, A., Nemes, A., Le Bas, C., 1999. Development and use of a database of hydraulic properties of European soils. Geoderma 90, 169-185.

PART V

USER-ORIENTED TECHNIQUES AND SOFTWARE

Chapter 18

SOIL INFERENCE SYSTEMS

A. McBratney and B. Minasny

Faculty of Agriculture, Food & Natural Resources, The University of Sydney, JRA McMillan Building A05, NSW 2006, Australia

1. SOFTWARE FOR PEDOTRANSFER FUNCTIONS

There are many water-retention pedotransfer functions generated using new or existing datasets. Pedotransfer functions have been used to translate available soil data to other data that are not available, or are more difficult to measure (Figure 1). There seems to be a lack of effort in gathering and using the available pedotransfer functions. Several software products have been made available, mainly to predict soil hydraulic properties (Table 1). The software either uses functions developed by the respective authors, or is a compilation of published PTFs.

Mishra et al. (1989) devised a program SOILPAR for predicting the soil water retention curve, and saturated hydraulic conductivity along with their uncertainty, from particle-size distribution data. Water retention is predicted from particle-size distribution data according to the model of Arya and Paris (1981), and saturated conductivity is predicted using a modified Kozeny-Carman equation. Although the program is now regarded as obsolete, they emphasized quantifying the uncertainty of PTFs prediction based on a first-order Taylor analysis.

Many programs simulating soil–water flow now have pedotransfer functions embedded in them to predict the soil hydraulic properties, e.g., SWAP (van Dam et al., 1997) uses the PTFs of Wösten et al. (1999); and HYDRUS (Simunek et al., 1999) uses the PTFs of Schaap et al. (2001).

These PTF programs (Table 1) simply predict hydraulic properties given the required inputs. Attempts have been made to use an "expert system" to select the available functions given the inputs we have. Program SH-Pro made such an attempt by providing a system that is able to pick a suite of published PTFs that suits the amount of available inputs. However, the possible (minimum data) is governed by the available functions and the user still has to decide which functions to use. Hubrecht and Feyen (1996) derived various pedotransfer functions predicting soil thermal properties based on the availability of the input data. Schaap et al. (1998) called this a hierarchical system based on the availability of data, specifically:

- texture class,
- particle size analysis (or PSA: sand, silt and clay content),
- PSA and bulk density,

Figure 1. Current use of pedotransfer functions.

- PSA, bulk density and θ at -33 kPa,
- PSA, bulk density and θ at -33 and -1500 kPa

In reality there will be lots of combinations of available data, e.g., the most common variable missing in general soil survey data is bulk density. The available data could be particle-size analysis, gravimetric water content at -10 kPa; or particle size analysis with organic carbon or cation exchange capacity (CEC). A system should be able to utilise whatever available soil information, and select the best functions that will produce a prediction with some optimality criterion, such as minimum variance. Furthermore, a system should be able to provide other useful soil physical and chemical properties rather than just hydraulic properties.

The following section will discuss on the concept of soil inference systems and will illustrate it for the prediction of various soil physical and chemical properties.

2. SOIL INFERENCE SYSTEMS

McBratney et al. (2002) introduced the concept of a soil inference system, where pedotransfer functions are the knowledge rules for soil inference engines. A soil inference system takes measurements we know with a given level of (un)certainty, and infers data that we do not know with minimal inaccuracy, by means of logically conjoined pedotransfer functions.

Table 1
Available PTF software

Pedon-SEI (Ungaro and Calzolari, 2001; Ungaro et al., 2001):
http://www.fi.cnr.it/irpi/pedone/Pedon_introd.htm
Prediction: Water retention (Brooks–Corey model), Italian soil;
Availability: Web-based, Freeware

Rosetta (Schaap et al., 2001):
http://www.ussl.ars.usda.gov/MODELS/rosetta/rosetta.htm
Prediction: Water retention and hydraulic conductivity curves (Mualem-Van Genuchten model);
Availability: Freeware

SH-Pro (Cresswell et al., 2000):
Prediction: Various published PTFs for water retention and saturated hydraulic conductivity;
Availability: Freeware, contact author: cresswell@csiro.au

SOILPROP (Mishra et al., 1989; Environmental System and Technologies, 1990):
Prediction: Water retention (Arya-Paris model) and saturated hydraulic conductivity with uncertainty analysis;
Availability: Commercial, Obsolete

SoilPar (Acutis and Donatelli, 2003):
http://www.sipeaa.it/ASP/ASP2/SOILPAR.asp
Prediction: Various published PTFs for water retention and saturated hydraulic conductivity;
Availability: Freeware

SoilVision (Fredlund et al., 1999):
Prediction: Water retention and saturated hydraulic conductivity;
http://www.soilvision.com/
Availability: Commercial

Soil water Characteristics, Hydraulic properties Calculator (Saxton et al., 1986):
http://www.bsyse.wsu.edu/saxton/soilwater/
Prediction: Water retention and saturated hydraulic conductivity;
Availability: Freeware

SWLIMITS (Ritchie et al., 1999; Suleiman and Ritchie, 2001):
http://nowlin.css.msu.edu/software/swlimit_form.html
Prediction: Soil water limits and saturated hydraulic conductivity;
Availability: Web-based, Freeware

Neuro θ, Theta (Minasny and McBratney, 2003a):
http://www.usyd.edu.au/su/agric/acpa/software
Prediction: Water retention (Van Genuchten model) and saturated hydraulic conductivity curves for Australian soil;
Availability: Freeware

Neuro Multistep (Minasny et al., 2004):
http://www.usyd.edu.au/su/agric/acpa/software
Prediction: Water retention and unsaturated hydraulic conductivity (Mualem-Van Genuchten model) for Alluvial soils from California derived from Multistep outflow experiment;
Availability: Freeware

Dale et al. (1989) discussed the role of expert systems in soil classification, and similar principles can be applied to the proposed inference system. This is shown in Figure 2, where the system has a source, an organizer and a predictor. The sources of knowledge to predict soil properties are collections of pedotransfer functions and soil databases. The organizer arranges and categorizes the pedotransfer functions with respect to their required inputs and the soil types from which they were generated. The inference engine is a collection of logical rules selecting the pedotransfer functions with the minimum variance. The rules can be a collection of "if-then" statements, or based on probabilistic Bayesian inference. Uncertainty of the prediction can be assessed using Monte-Carlo simulations. The inference system operates through a user interface that will return the predictions of soil physical and chemical properties with their uncertainties based on the information provided.

McBratney et al. (2002) demonstrated a rudimentary soil inference system (SINFERS) in the form of a specially adapted spreadsheet. It has two essentially new features; first, it contains a suite of published pedotransfer functions. The output of one PTF can act as the input to other functions (if no measured data are available). Secondly, the uncertainties in estimates are inputs and the uncertainties of subsequent calculations are performed. The input consists of the essential soil properties. The inference engine will work in the following manner:

1. Predict all the soil properties using all possible combinations of inputs and PTFs.
2. Select the combination that leads to a prediction with the minimum variance.

The first column of the spreadsheet consists of the essential soil properties, followed by their estimates. Typically, hydraulic PTFs require four basic soil properties: clay and sand content, bulk density and organic carbon content. The steps in SINFERS are:

1. Make sure that there are estimates for clay and sand content, bulk density and organic carbon content.
2. If one or more of these properties are not available, then estimate them from the available information using PTFs. For example, clay and sand content can be estimated

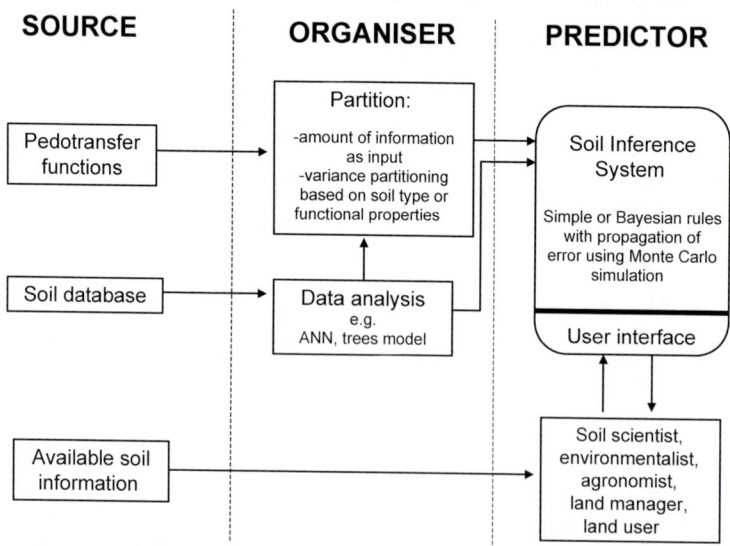

Figure 2. A soil inference system (modified and updated from Dale et al., 1989).

from the average values of a texture class, and bulk density can be estimated from sand and clay content.
3. Once these values have been obtained, other derived values can be calculated arithmetically, e.g., the geometric mean of particle diameter, sand-sized particles according to different classification.
4. The next step is to predict more difficult-to-measure soil properties such as the water retention curve, hydraulic conductivity, soil strength, gas diffusion coefficient, solute transport parameters, CEC, P buffering capacity and pH of buffering capacity.
5. These values in turn can be used to predict various soil physical and chemical parameters, e.g., nonlimiting water range (NLWR), optimal water content for tillage, risk for soil acidification, etc.

Uncertainty can be quantified in terms of the model (PTFs) uncertainty and input uncertainty. If the PTFs have a large training set and were trained/calibrated adequately, the uncertainty in the model is usually smaller than the uncertainty in the inputs. In the case where a predictor in a PTF is not measured but estimated from other PTFs, the overall uncertainty is important. Using the risk-analysis software @RISK (Palisade, 2000), which is an "add-in" to the MS Excel spreadsheet, the overall uncertainties in the prediction can be calculated. The user can define the distribution for each input variable and their correlations. We use a normal error distribution for the input variables, but some distributions need to be truncated to avoid negative values (such as clay and organic matter content). The program @RISK uses Latin hypercube sampling to sample the multivariate joint distribution of the input variable and a Monte Carlo simulation is performed to calculate the distribution of the prediction. This is achieved by sampling repeatedly from the assumed probability distribution of the input variables and evaluating the result of the PTFs for each sample. The distribution of the results, along with the mean, standard deviation and other statistical measures can then be estimated.

3. A SCHEME FOR DEFINING UNCERTAINTIES OF DATA INSIDE/OUTSIDE THE TRAINING SET

In order to assess whether a sample is within or outside the training set, to avoid extrapolation, a method that penalizes the prediction uncertainty when the sample is outside the training set should be developed. McBratney et al. (2002) use a simplification of fuzzy k-means with extragrades (De Gruijter and McBratney, 1988) to achieve this, where they define a class, within the class is the training set, while samples outside the class boundary are considered as extragrades (outliers).

The membership is controlled by a fuzziness exponent ϕ, which defines the smoothness of the membership, and the parameter α which defines the distance from the data center (means) for a data point to be considered as an outlier, ($0 < \alpha < 1$). The smaller the value, the further the distance before a point is classified as an outlier. The membership of a sample in the defined dataset is defined as

$$m = \frac{d^{-2/(\phi-1)}}{d^{-2/(\phi-1)} + \left(\frac{1-\alpha}{\alpha} d^{-2/(\phi-1)}\right)^{-1/(\phi-1)}} \tag{1}$$

while the membership defining a sample as being outside of the dataset is:

$$m^* = \frac{\left(\frac{1-\alpha}{\alpha}d^{-2}\right)^{-1/(\phi-1)}}{d^{-2/(\phi-1)} + \left(\frac{1-\alpha}{\alpha}d^{-2/(\phi-1)}\right)^{-1/(\phi-1)}} \quad (2)$$

where d is the Mahalanobis distance between the sample x and the center (mean) of the data c:

$$d = \sqrt{(x-c)^T \mathbf{V}^{-1}(x-c)} \quad (3)$$

Here $\mathbf{V}$ is the variance covariance matrix of the data. This distance takes into account the correlation among variables.

The standard deviation of the prediction σ_a can be calculated as:

$$\sigma_a = \sqrt{\frac{\sigma_p^2}{(1-m^*) + (0.0001\sigma_p^2)}} \quad (4)$$

where σ_p^2 is the variance of the PTF. The first term of the denominator in the above equation calculates the belongingness of a new sample in the training set, while the second term avoids infinite calculation when $m^* = 1$. When a sample is within the training set, $\sigma_a = \sigma_p$; but the prediction error will increase if the data is outside the training set.

We seek the values for the ϕ and α parameters such that say 95% of the training data is considered as "within the data set" and 5% is considered outside (outliers). We found that a ϕ value of 1.3 provides a good representation of the membership. The value α can be found by optimization such that it provides an average m^* value of 0.05. We found that values around 0.01–0.02 satisfy this criterion.

As an example, we have a dataset of sandy materials and we wish to develop a PTF to predict field capacity from bulk density and clay and sand content (mean values = 1.5 Mg m^{-3}; 16%, 70%). We use the value of $\phi = 1.3$ and $\alpha = 0.014$. The extragrade membership m^* for different clay contents at mean values of sand and bulk density is given in Figure 3a. As the clay content moves away from the mean value, m^* increases rapidly. The resulting prediction error is shown in Figure 3b. The standard deviation of prediction, as calculated from the least-squares estimates, only has a value of 0.013 m^3 m^{-3} for a clay content of 40%; the suggested modification accounts for the extrapolation ($m^* = 0.906$) and penalizes the prediction error as it is outside the boundary of training set (the standard deviation of prediction becomes $= 0.042$ m^3 m^{-3}).

4. EXAMPLE OF SINFERS

We present an example of the use of a rudimentary soil inference system, SINFERS, to predict some important soil physical and chemical properties In this example, we wish to predict soil physical and chemical properties of a kaolinitic light clay. The data available

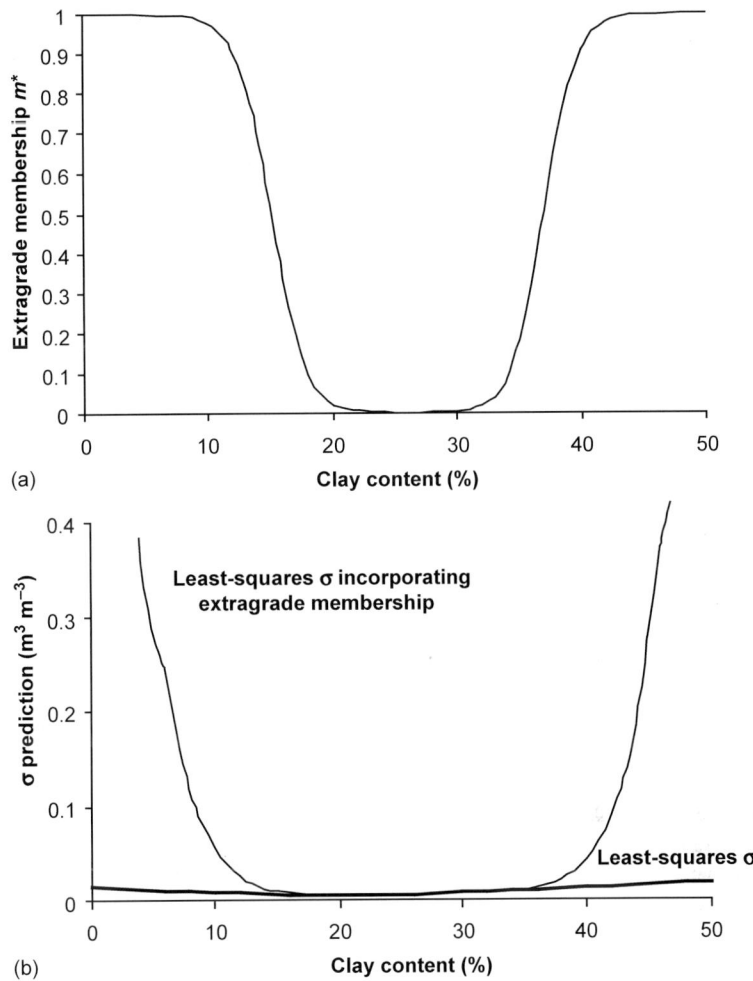

Figure 3. (a) Extragrade membership as a function of clay defining the outliers from the training data-set; (b) uncertainty in the prediction as a function of clay content.

from laboratory analysis for this soil are particle-size distribution (sand, silt, and clay content), bulk density, organic carbon content and pH in water (Table 2). These values were placed in the "Input" column along with their standard deviation SD.

The program first predicts water retention curve using a parametric form of van Genuchten (van Genuchten, 1980) using a neural networks PTF (Minasny and McBratney, 2002). The predicted parameters θ_r, θ_s, α, and n can be used to derive many soil physical parameters related to water transport, plant water availability, solute transport and mechanical properties.

We would also like to predict parameters of the Campbell's equation for water retention (Campbell, 1974): air entry potential h_b, and slope of the water retention curve b,

Table 2
Predicted soil properties of a clay using the soil inference system

Soil properties	Units	Inputs			Prediction		
		External condition	Measurement/Estimate	SD	Estimate	SD model	SD input
Basic properties							
Clay <2 μm	%		44	1	44.0		1.00
Silt 2–20 μm	%		19	2	19.0		1.83
Silt 2–50 μm	%				35.6	0.48	2.42
Fine sand 20–200 μm	%		27	1			
Coarse sand 200–2000 μm	%		10	1			
Sand 20–2000 μm	%		37	1	37.0		1.40
Sand 50–2000 μm	%				20.4	0.48	1.83
ln (d_g)	ln (mm)				−3.89		0.06
Mean geometric diameter (d_g)	mm				0.02		0.00
Field bulk density	g cm^{-3}		1.32	0.05	1.32		0.05
Particle density	g cm^{-3}		2.65	0.05	2.65		
Total porosity	cm^3 cm^{-3}				0.502		0.02
Water retention							
Van Genuchten							
θ_r	cm^3 cm^{-3}				0.010		0.006
θ_s	cm^3 cm^{-3}				0.481		0.005
α	kPa^{-1}				0.252		0.004
n	–				1.119		0.006
θ Inflection point	cm^3 cm^{-3}				0.377		0.004
h Inflection point	cm				293.5		59.949
Slope at inflection point	1/ln (cm)				−0.039		0.001

Parameter	Units	Value	Value
Campbell			
h_b	cm	18.0	0.000
b		9.1	0.330
Mass fractal dimension D_m		2.89	0.004
Available water capacity			
ψ at FC	kPa	−40	
θ at FC	$cm^3\,cm^{-3}$	0.365	0.007
θ −10 kPa	$cm^3\,cm^{-3}$	0.418	0.005
θ −1500 kPa (WP)	$cm^3\,cm^{-3}$	0.242	0.008
Average water capacity	$cm^3\,cm^{-3}$	0.123	0.001
Integral energy	$J\,kg^{-1}$	44.8	0.169
Gas diffusion			
Critical ODR	$\mu g\,cm^{-2}\,s^{-1}$	$3.3\,10^{-3}$	
Required aerobic depth	cm	30	
Fraction of oxygen in air	$cm^3\,cm^{-3}$	0.21	
O_2 concentration at surface, C_o	$\mu g\,cm^{-3}$	0.37	
O_2 Gas diffusivity	$cm^2\,s^{-1}$	0.0027	
Air-filled porosity at required ODR	$cm^3\,cm^{-3}$	0.134	0.001
Diffusivity at ODR	$cm^2\,s^{-1}$	0.0027	0.000
θ at ODR	$cm^3\,cm^{-3}$	0.367	0.020
θ at 10% air filled porosity	$cm^3\,cm^{-3}$	0.402	
Soil resistance			
R_s at −1500 kPa	MPa	0.00	10000	0.000
θ at 2 MPa	$cm^3\,cm^{-3}$	0.26	10000	0.019
LLWR			
Upper limit	$cm^3\,cm^{-3}$	0.365	0.010
Lower limit	$cm^3\,cm^{-3}$	0.262	0.018
LLWR	$cm^3\,cm^{-3}$	0.103	0.023

Table 2. Continued

Tillage θ				
Optimum θ	cm^3 cm^{-3}	0.377		0.004
Optimum ψ	kPa	29.35		6.0
θ Lower tillage limit	cm^3 cm^{-3}	0.347		0.010
θ Upper tillage limit	cm^3 cm^{-3}	0.418		0.004
Infiltration at saturation				
Effective porosity ϕ_e	cm^3 cm^{-3}	0.084		0.026
ln K_s (Equation (20))	ln (mm h^{-1})	1.288	0.188	0.217
K_s (Equation (20))	mm h^{-1}	3.6	1.207	0.831
ln K_s (Equation (21))	ln (mm h^{-1})	1.163		1.179
K_s (Equation (21))	mm h^{-1}	3.2		5.597
Initial condition at FC				
Δ theta		0.116		0.010
Sorptivity	mm h$^{-0.5}$	9.7		1.469
Microscopic capillary length λ_c	mm	130.7		1.902
Mean pore size λ_m	mm	0.057		0.001
Irrigation rate r	mm h^{-1}	10		
Time to ponding t_p	h	0.65		0.250
Initial condition at WP				
Δ theta		0.239		0.011
Sorptivity	mm h$^{-0.5}$	14.7		1.842
λ_c	mm	135.4		1.061
λ_m	mm	0.055		0.001
Irrigation rate r	mm h^{-1}	10		
t_p	h	1.23		0.321

Solute transport				
Dispersivity	cm			0.419
θ mobile	cm^3 cm^{-3}			0.004
Travel time to reach 1 m	day			0.831
K_{oc} Diuron	cm^3 g^{-1}	351		31.6
K Diuron	cm^3 g^{-1}			1.802
R Diuron				6.539
Thermal properties at field capacity				
Thermal conductivity λ_h	W m^{-1} K^{-1}			0.044
Thermal capacity C_h	MJ m^{-3} K^{-1}			0.063
Thermal diffusivity D_h	mm^2 s^{-1}			0.010
Electrical properties at FC				
Frequency	GHz	1.2		
Real part permittivity ε'				0.535
Imaginary part permittivity ε''				0.063
Bulk ECσ	mS m^{-1}			0.673
Chemical properties				
Organic C (Walkley & Black)	%	1.1	0.5	0.500
Total C	%			0.65
Organic matter	%			1.15
CEC		6	0.1	9.69
pH 1:5 water				6.00
pH 1:5 CaCl$_2$				5.3
pH buffering capacity	mmol H kg^{-1} pH^{-1}			21.66
Net acid addition rate	kmol H$^+$ ha^{-1} yr^{-1}	2.5		3.46
Time for pH to drop by 1 unit	years			2.91

Additional values visible in upper rows: 9.12, 0.346, 4.0, 351, 3.86, 14.97, 0.74, 2.49, 0.30, 20.67, 2.80, 29.68, 1.1, 1.43, 2.51, 341.1, 17.15, with 72.1, 2.29, 0.009, 2.33 in adjacent columns.

because many PTFs use these parameters as inputs (Perfect et al., 2002; Moldrup et al., 2000). We can predict h_b and b using the PTFs of Campbell (1985) or Williams et al. (1992), however, the water retention will not be the same as those with van Genuchten's parameters predicted from neural networks. For example, if we use the PTFs of Williams et al. (1992), the predicted parameters are $h_b = 45.9$ cm, and $b = 18$, and the resulting water retention is quite different the one predicted using neural networks. Since we have more confidence in our PTFs and would like the prediction to be uniform, we predict Campbell's parameters from the water retention curve predicted using neural networks. We can extract points along the curve and fit them with Campbell's equation using nonlinear least-squares. A simpler approach that can be easily implemented in a spreadsheet is by approximating h_b with $1/\alpha$ then with log transformation:

$$\log\left(\frac{\theta}{\theta_s}\right) = -\frac{1}{b}[\log(h) - \log(h_b)] \tag{5}$$

b can be predicted using linear regression. Value of h_b can be refined by calculating its value from Equation (5), this step can be repeated until the difference of the parameters at successive steps is smaller than a critical value. Using this approach, we estimated $h_b = 18$ cm and $b = 9.1$, these values give a water retention curve that is close to the one predicted using neural networks (Figure 4).

The b parameter can be related to the mass fractal dimension of soil (Anderson and McBratney, 1995):

$$D_m = d_e - 1/b \tag{6}$$

where d_e is the embedding dimension ($= 3$). The fractal dimension of this soil is 2.9, which can be used to calculate the scale-variant bulk density (Anderson and McBratney, 1995).

The predicted van Genuchten parameters can be used to estimate other useful properties. The slope of the water retention curve at its inflection point when the pressure head (h) is plotted on the log scale (see Figure 4), is suggested as a measure of soil microstructure, thus a good measure of soil physical quality (Dexter, 2004):

$$\frac{d\theta}{d\ln(h_{inf})} = -mn(\theta_s - \theta_r)\alpha^n h_{inf}^n [1 + (\alpha h_{inf})^n]^{-m-1} \tag{7}$$

where θ_r, and θ_s is the residual and saturated water content, α is a scaling parameter, n and m are empirical parameters, and h_{inf} is the pressure head at the inflection point:

$$h_{inf} = \frac{1}{\alpha}\left(\frac{1}{m}\right)^{1/n} \tag{8}$$

The inflection point occurs at the potential of -29 kPa with the slope of 0.04 $\ln(cm^{-1})$. The value of the slope is greater than the suggested critical value of 0.035 (Dexter, 2004), suggesting a good structural stability. The optimum water content for tillage can be derived from Equation (7) (Dexter and Bird, 2001), where the optimum θ is defined at the

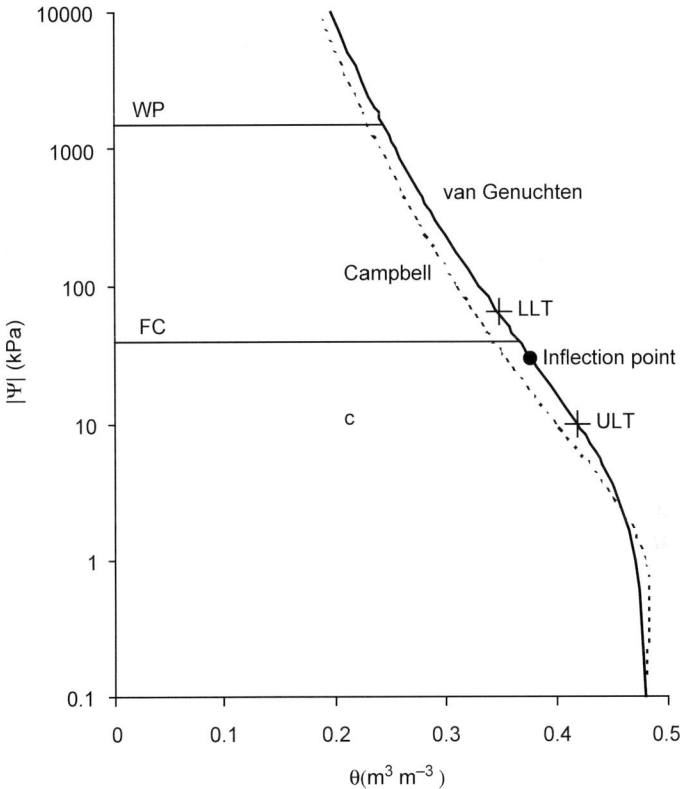

Figure 4. Predicted water retention curve according to the van Genuchten (solid line) and Campbell function (dashed line). The inflection point of the curve is marked by black dots on the curve. LLT and ULT are the upper and lower limit for tillage, FC is field capacity and WP is wilting point.

inflection point:

$$\theta_{opt} = \theta_r + (\theta_s - \theta_r)\left(1 + \frac{1}{m}\right)^{-m} \qquad (9)$$

The upper (wet) tillage limit or plastic limit is approximated by:

$$\theta_{wet} = \theta_{opt} + 0.4(\theta_s - \theta_{opt}) \qquad (10)$$

The lower (dry) tillage limit is calculated under conditions that

$$\theta_{dry} h_{dry} = 2\theta_{opt} h_{opt} \qquad (11)$$

Given the known optimal water content and water potential $\theta_{opt} h_{opt}$, we need to solve for two unknowns: water content and potential at dry limit θ_{dry} h_{dry}. Because θ_{dry} is related to h_{dry} by the van Genuchten equation, we can calculate them iteratively with the iteration function embedded in the spreadsheet. The optimal water content for tillage calculated for this soil is at 0.38 m^3 m^{-3}, and with range between 0.35 and 0.42 m^3 m^{-3} at water potentials of -64 and -10 kPa.

The water retention curve is used to derive soil available water capacity. According to the classical concept of field capacity, water potential at field capacity is around -10 to -40 kPa and can vary between soil types (Romano and Santini, 2003). We made an approximation (Figure 5) where water potential at field capacity varies with sand and clay content. For coarse textured soil, the value is around -5 kPa and for medium textured soil around -10 to -30 kPa and for clay soil around -50 kPa.

Wilting point is well established to correspond to potential of -1500 kPa, although it may depend on plant species. The water potential at field capacity of this soil is at -40 kPa, with water content 0.36 m^3 m^{-3}. Thus the available water capacity is 0.12 m^3 m^{-3}.

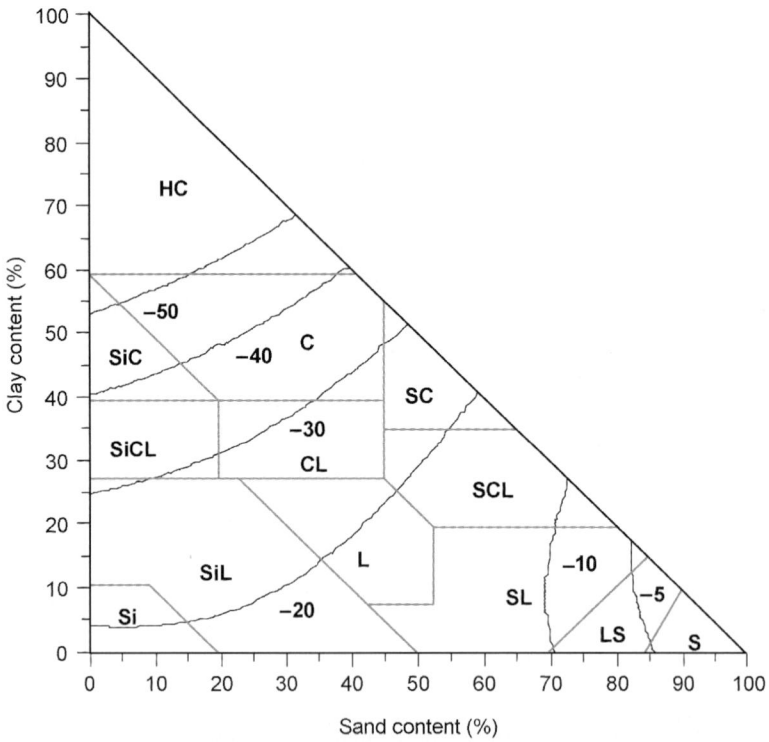

Figure 5. Water potential at field capacity as a function of sand and clay content (S, sand; Si, silt; L, loam; C, clay; HC, heavy clay).

We can calculate the integral energy (Minasny and McBratney, 2003b) which is an estimate of the total energy required to extract water from field capacity to wilting point:

$$E_1[\theta_{FC} - \theta_{WP}] = \frac{1}{\theta_{FC} - \theta_{WP}} \int_{\theta_{WP}}^{\theta_{FC}} \psi(\theta)\, d\theta \tag{12}$$

The amount of energy required for the plants to extract water from an initial condition at field capacity to its maximum limit at wilting point is: 45 J kg^{-1}.

The NLWR (Letey, 1985) is the range in soil water content in which water, oxygen and mechanical resistance are least limiting to plants. The limit at the upper end is defined by the soil water content at a value of nonlimiting oxygen diffusion (oxygen diffusion rate (ODR), as measured by platinum electrode of 3.3×10^{-3} g cm^{-2} s^{-1}) or at field capacity, whichever is the smaller, and at the lower end by the water content at a value of nonlimiting soil strength (2 MPa) or the water content at wilting point (-1500 kPa), whichever is the larger (Wu et al., 2003).

Water content at the critical value of ODR, θ_{ODR}, is found by the help of the gas diffusivity PTF of Moldrup et al. (2000):

$$\frac{D_g}{D_{0g}} = \varepsilon^2 \left(\frac{\varepsilon}{\phi}\right)^{3/b} \tag{13}$$

where D_g is gas diffusion coefficient (cm^2 s^{-1}) and D_{0g} is gas diffusion in pure air, ε is air-filled porosity, ϕ is total porosity and b is Campbell's parameter. Applying Fick's law, we can calculate oxygen flux f_g (g cm^{-2} s^{-1}) to a depth L from its concentration C_0 (g/m^3) at the soil surface:

$$f_g = D_g \frac{C_0}{L} \tag{14}$$

We use an aerobic depth L of 30 cm, and by assuming 21% of O_2 in the air, employing the ideal gas equation, we define the soil surface O_2 concentration as $C_0 = \varepsilon(2.7 \times 10^{-4})$ g O_2 cm^{-3} soil. Substituting PTFs of Moldrup for D_g into the above equation with D_{0g} for oxygen = 0.23 cm^2 s^{-1}, we seek a value of ε_{ODR} that satisfy $f_g = 3.3 \times 10^{-3}$ µg cm^{-2} s^{-1}. Water content at the critical value of ODR is calculated from:

$$\theta_{ODR} = \phi - \varepsilon_{ODR} \tag{15}$$

The critical water content for achieving this ODR at depth of 30 cm is 0.37 cm^3 cm^{-3} with an air-filled porosity of 0.13 cm^3 cm^{-3}. The air-filled porosity is higher than the conventional value quoted as critical value for root growth of 10% (Collis-George, 1959). The value of θ at air filled porosity of 0.1 (0.4 cm^3 cm^{-3}) is only able to supply oxygen at the critical ODR to a depth of 10 cm.

Soil strength at wilting point was evaluated using the soil mechanical resistance PTF of Da Silva and Kay (1997):

$$R_s = c_1 \theta^{c_2} \rho^{c_3} \tag{16}$$

with following values for c_1, c_2, c_3:

$$c_1 = \exp(-3.67 + 0.765 \text{ OC} - 0.145 \text{ Clay})$$
$$c_2 = -0.481 + 0.208 \text{ OC} - 0.124 \text{ Clay} \tag{17}$$
$$c_3 = 3.85 + 0.0963 \text{ Clay}$$

where R_s is the mechanical resistance (MPa), ρ is bulk density (g cm^{-3}), clay is the clay content (%) and OC is the organic carbon content (%). However, this PTF was developed in Canada using soil with small to medium clay content (9–39%) and small to large sand content (8.6–81%). Using the technique described in Section 3, this sample was found to be well outside of training set with an outlier membership $m^* = 1$; hence this has large uncertainty, and the prediction is inaccurate, and probably not at all useful.

The upper limit of the NLWR is calculated as:

$$\min \theta \{ \text{ODR} = 3.3 \times 10^{-3} \text{ g cm}^{-2} \text{ s}^{-1}, \text{ field capacity} \}. \tag{18}$$

θ satisfying this condition is at 0.365 cm^3 cm^{-3} (field capacity). The lower limit is calculated from:

$$\max \theta \{ \text{mechanical resistance} = 2 \text{ MPa, wilting point} \}. \tag{19}$$

Therefore, the NLWR is 0.10 cm^3 cm^{-3}, which is 2% less than the value calculated from classical available water capacity.

We calculated the transport of water under saturated conditions. The saturated conductivity K_s can be calculated using different PTFs, e.g., Minasny and McBratney (2000) provided the following functions:

$$\ln K_s = 18.29 - 14.496 \rho_b + 3.98 \rho_b^2 - 0.0668 \text{ Clay} + 0.478 \ln(d_g) \tag{20}$$

$$\ln K_s = 10.8731 + 3.9140 \ln(\varepsilon_{10}) \tag{21}$$

where K_s is in mm h^{-1}, d_g is the geometric mean particle diameter in ln(mm), and ε_{10} is the air-filled porosity at -10 kPa. The value predicted according to Equation (20) is 3.6 mm h^{-1}, while Equation (21) gave a similar value of 3.2 mm h^{-1}. However, the uncertainty of Equation (21) is quite large, about 100%. This will be a problem if we wish to use the calculated K_s for other predictions. Therefore, we choose the prediction using Equation (20). Assuming the Campbell's soil hydraulic model, the soil–water diffusivity is:

$$D(\theta) = -\frac{h_b b K_s}{\theta_s} \left(\frac{\theta}{\theta_s} \right)^{2+b} \tag{22}$$

Sorptivity, S is a measure of the capacity of the soil to absorb water by capillarity, is calculated according to Parlange (1975):

$$S^2 = \int_{\theta_i}^{\theta_s} (\theta_s + \theta - 2\theta_i) D(\theta) d\theta \tag{23}$$

where θ_i initial water content, and θ_s is water content at saturation. We calculated sorptivity from initial condition of field capacity and wilting point to saturation. Sorptivity from initial condition at field capacity to saturation is 9.7 mm h$^{-0.5}$, if the initial condition is at wilting point, S is larger = 14.7 mm h$^{-0.5}$. The importance of capillary sorption to water flow can be expressed by a parameter called the macroscopic capillary length λ_c (White and Sully, 1987):

$$\lambda_c = \frac{1}{K_s} \int_{\theta_i}^{\theta_s} D(\theta) d\theta \tag{24}$$

It is a flow-weighted mean soil–water potential, which can be used for scaling of soil–water potential. The parameter λ_c can be used to derive a measure of soil structure through its relationship with a characteristic mean pore size λ_m by means of the capillary rise equation (White and Sully, 1987):

$$\lambda_m = \frac{\sigma_w}{\rho_w g \lambda_c} \tag{25}$$

where σ_w [MT^{-2}] is the soil–water surface tension, ρ_w is the density of water and g [LT^{-2}] is the gravity acceleration. For pure water at 20 °C, Equation (25) reduces to:

$$\lambda_m = 7.4/\lambda_c \tag{26}$$

where λ_m and λ_c is in mm. The mean pore size for this soil is 0.06 mm. White et al. (1989) define an estimate of time for ponding t_p for rainfall/irrigation at rate of r:

$$t_p = \frac{0.55}{r} \left(\frac{S^2}{K_s} \right) \ln \left(\frac{r}{r - K_s} \right) \tag{27}$$

For an irrigation of 10 mm h^{-1}, the soil is going to pond within half an hour if its initial condition is at field capacity, and it will take twice as long when it is at wilting point.

Solute transport can be predicted using parameters of Campbell's. A dispersivity of 11 cm was estimated from air entry value h_b and the slope of the water retention curve b, according to the PTF of Perfect et al. (2002). Mobile water content θ_m can calculated (see Chapter 12), it appears that about 72% of the water content at saturation is available for solute transport. Using this value, we can calculate the approximate travel time for a nonadsorbing solute from the surface to depth of 1 m: $t = \theta_m L/q_s$, where L is the depth, and q_s is the water flux under saturated condition, taken as saturated hydraulic conductivity K_s (assuming a unit potential gradient). Because this is a clay soil with relatively low K_s(3.6 mm h^{-1}) it will take about 4 days for a nonreactive solute applied on the surface to reach a depth of 1 m. We are also interested in the movement of a pesticide Diuron in the soil under field capacity conditions. The organic carbon partition coefficient K_{oc} for Diuron is 351 cm^3 g^{-1} (Hamaker and Thompson, 1972) with a 9% coefficient of variation. The distribution coefficient for the soil, K_d, is calculated as:

$$K_d = K_{oc} \times \text{fraction of organic carbon} \tag{28}$$

The retardation factor at field capacity can be calculated (Chapter 12):

$$R = 1 + \frac{\rho K}{\theta_{FC}} \tag{29}$$

The distribution coefficient in the soil is 3.9 cm^3 g^{-1}. The value of R obtained is 15, meaning that the pesticide is strongly bound in the soil, and the flux will be about 15 times slower compared to a nonadsorbing solute. However the uncertainty of the estimate of K_d and R is quite high (uncertainty of 45%), which is due to the uncertainty of the K_{oc} value.

Thermal properties of the soil can be calculated from basic soil properties (De Vries, 1966; Campbell, 1985; Farouki, 1986; Hubrechts and Feyen, 1996). The mixture model of De Vries (1963) was used to calculate the soil heat capacity:

$$C_h = \rho_s c_s \phi_s + \rho_w c_w \theta \tag{30}$$

where C_h is the volumetric heat capacity (J m^{-3} K^{-1}), subscripts s and w refer to soil solids and water, ρ is the density (kg m^{-3}), and c is the specific heat capacity: 0.73 kJ kg^{-1} K^{-1} for soil and 4.18 kJ kg^{-1} K^{-1} for water. The PTFs of Campbell (1985) were used to predict soil thermal conductivity λ_h (in W m^{-1} K^{-1}). They used clay content and bulk density ρ_b as predictors:

$$\lambda_h = A + B\theta - (A - D)\exp[-(C\theta)^4] \tag{31}$$

where:

$$\begin{aligned} A &= 0.65 - 0.78\rho_b + 0.6\rho_b^2 \\ B &= 1.06\rho_b \\ C &= 1 + 2.6\left(\frac{\text{Clay}}{100}\right)^{-0.5} \\ D &= 0.03 + 0.1\rho^2 \end{aligned} \tag{32}$$

The thermal diffusivity D_h (m^2 s^{-1}) is found as the ratio between thermal conductivity and capacity. The predicted thermal properties as a function of water content are given in Figure 6. As can be seen this model has quite a large uncertainty given a small uncertainty in the input variables, with bulk density as the largest contribution to the variation.

We are also interested in the electrical properties of the soil, in particular the dielectric constant and electrical conductivity. Permittivity (ε) is a measure of the relative ability of a material to store a charge at a given electrical field or frequency. The dielectric constant of a material is the ratio between the permittivity and permittivity of free space

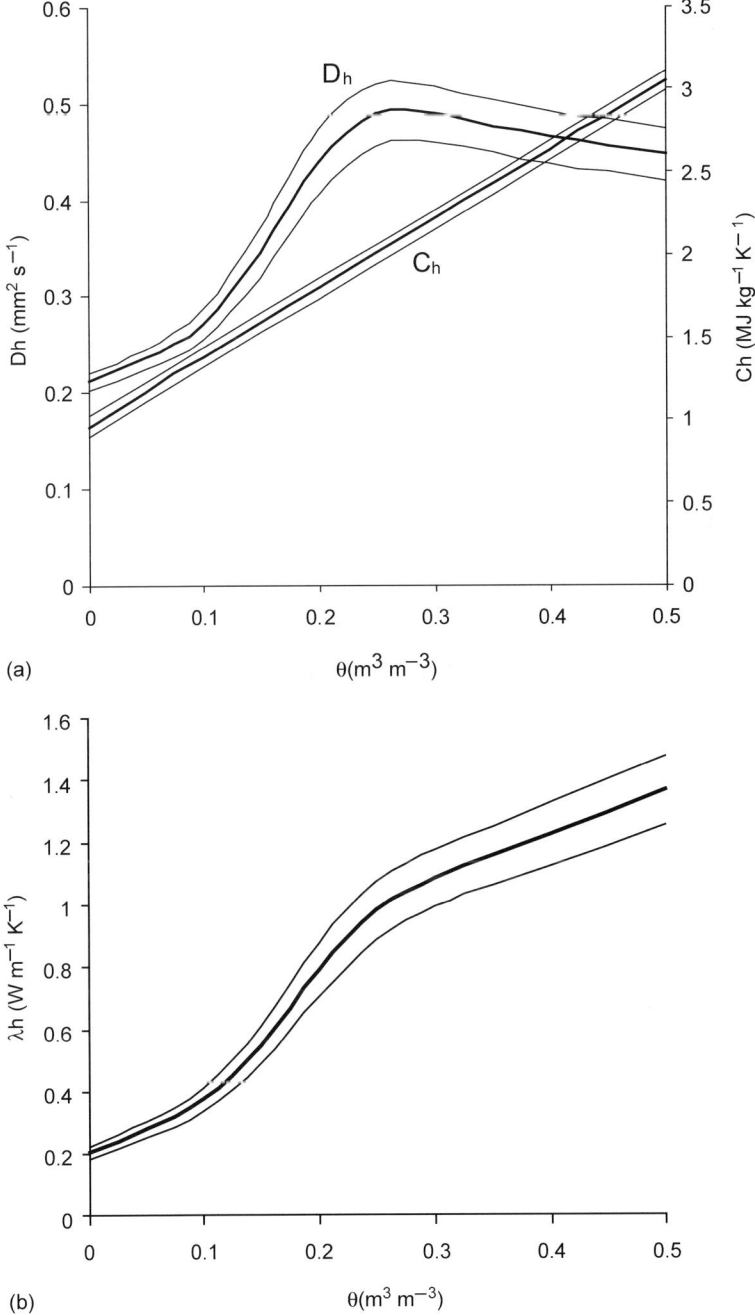

Figure 6. (a) Soil thermal conductivity λ_h, (b) soil thermal diffusivity (D_h) and capacity (C_h). Solid lines represent prediction from pedotransfer functions, while dashed lines enveloping the prediction curves are the 95% confidence interval of prediction.

ε_0 $(1/(36\pi) \times 10^{-9} \text{ F m}^{-1})$ $\varepsilon^* = \varepsilon/\varepsilon_0$. The dielectric constant can be expressed as:

$$\varepsilon^* = \varepsilon' + i\varepsilon'' \tag{33}$$

where ε^* is the relative complex dielectric constant of soil–water mixture, ε' is the real, ε'' is the imaginary part and i is $(-1)^{1/2}$. The imaginary part is the dielectric loss, a measure of the proportion of the charge transferred in conduction and stored in polarization (Saarenketo, 1998). The dielectric constant values are calculated based on the semi-empirical dielectrical mixing model (Peplinski et al., 1995; Dobson et al., 1985; Hendrickx et al., 2003) (Figure 7):

$$\varepsilon' = \left[1 + \frac{\rho_b}{\rho_s}(\varepsilon_s^\varphi) + \theta^{\beta'}\varepsilon_{fw}^{\prime\varphi} - \theta\right]^{1/\varphi}, \quad \varepsilon'' = \left[\theta^{\beta''}\varepsilon_{fw}^{\prime\prime\varphi}\right]^{1/\varphi} \tag{34}$$

where ρ_b is bulk density (in g/cm^3), φ is an empirical constant $= 0.65$; β' and β'' are empirical constants:

$$\begin{aligned}\beta' &= 1.2748 - 0.519\frac{\text{Sand}}{100} - 0.152\frac{\text{Clay}}{100}, \\ \beta'' &= 1.33797 - 0.603\frac{\text{Sand}}{100} - 0.166\frac{\text{Clay}}{100}\end{aligned} \tag{35}$$

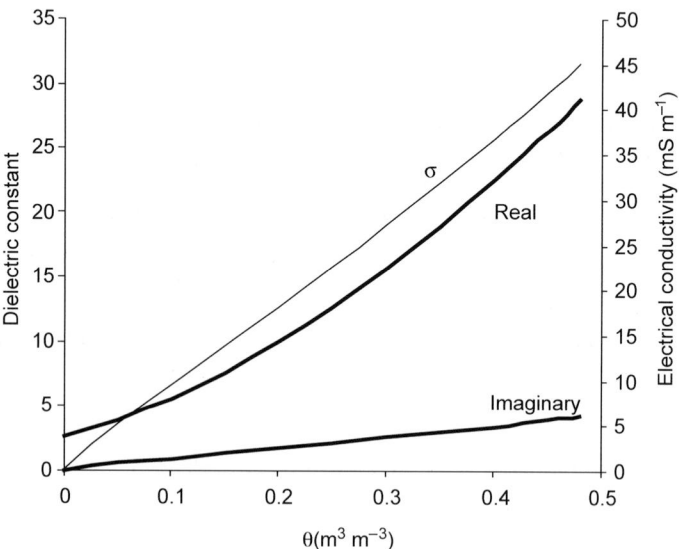

Figure 7. Predicted real ε and imaginary part of permittivity (bold lines) and electrical conductivity (σ) (solid line) of the soil as a function of water content.

where Sand and Clay are percent weight of sand and clay content, ε'_{fw} and ε''_{fw} are the real and imaginary parts of the relative dielectric constants of free water, given by:

$$\varepsilon'_{fw} = \varepsilon_{w\infty} + \frac{\varepsilon_{w0} - \varepsilon_{w\infty}}{1 + (2\pi f \tau_w)^2}, \quad \varepsilon''_{fw} = \frac{2\pi f \tau_w (\varepsilon_{w0} - \varepsilon_{w\infty})}{1 + (2\pi f \tau_w)^2} + \frac{\sigma_{eff}}{2\pi \varepsilon_0 f} \frac{(\rho_s - \rho_b)}{\rho_s \theta} \quad (36)$$

τ_w is the relaxation time for water, f is the frequency (Hz), ε_{w0} is the static dielectric constant for water, and $\varepsilon_{w\infty}$ is the high frequency limit of $\varepsilon'_{fw} = 4.9$. At room temperature $2\pi\tau_w = 0.58 \times 10^{-10}$ s and $\varepsilon_{w0} = 80.1$. The effective conductivity σ_{eff} in the range of 0.3–1.3 GHz is calculated as:

$$\sigma_{eff} = 0.0467 + 0.2204\rho_b - 0.4111\frac{Sand}{100} + 0.6614\frac{Clay}{100} \quad (37)$$

For a GPR with frequency of 1.2 GHz we obtain the dielectric constants of the soil as a function of soil water content (Figure 7). The relationship between electrical conductivity σ and ε'' is given by (Saarenketo, 1998):

$$\sigma = \varepsilon'' \varepsilon_0 f \quad (38)$$

Using this relationship, the bulk soil conductivity as a function of water content can be plotted (Figure 7). This estimate clearly does not take into account any information about the electrolyte concentration of the soil and therefore information on the EC of soil solution is necessary.

We also predict several important chemical properties, namely CEC with the PTF of McBratney et al. (2002):

$$CEC(mmol(+)kg^{-1}) = -29.250 + 8.139\,Clay + 0.253\,Clay \times OC \quad (39)$$

where Clay is the percent by weight of clay content and OC is organic carbon content in percent by weight. A CEC of 341 mmol(+)kg^{-1} is predicted, and the error in the model is quite small (2.3 mmol(+)kg^{-1}) compared to the error in the input variable (10 mmol(+)kg^{-1}). This is because the model is well calibrated on a wide range of soil materials (McBratney et al., 2002). We can transform the pH measured in water (1:5 soil to water ratio) into pH as measured in CaCl$_2$ (1:5) according to Henderson and Bui (2002). Other chemical properties that can be predicted are soil pH buffering capacity (pHBC; Noble et al., 1997). From pHBC, if we assume a net input of 2.5 kmol H$^+$ per hectare every year, then the time needed for the pH to drop by 1 unit can be calculated according to Helyar et al. (1990):

$$T(year) = \frac{pHBC(kmol\,H^+\,kg^{-1}\,pH^{-1}) \times \rho(kg\,m^{-3}) \times 1500(m^3)}{NAAR(kmol\,H^+\,ha^{-1}\,year^{-1})} \quad (40)$$

where 1500 is the volume of soil in the 0–0.15 m depth interval for 1 ha; and NAAR is the net acid addition rate. We estimated that this soil is well buffered with a pHBC of 22 mmol H$^+$ kg^{-1} pH^{-1} and it will take 17 years for the pH to drop by 1 unit.

We can predict many soil physical and chemical properties. So far we did not have any functions for biological properties but these should be developed. With respect to the sensitivity of the properties we are predicting, we imagine the predictions could be improved by some input information on soil structural form and stability, electrolyte concentration and clay mineral type or charge of the soil. Pedotransfer functions that use such characteristics we believe would have improved predictability. So there is much to be done on the development of pedotransfer functions.

5. GENERAL DISCUSSION AND CONCLUSIONS

We have illustrated the concept of soil inference system. Using pedotransfer functions as the knowledge rules for an inference engine, we can predict various important physical and chemical properties from the limited information we have. Although it is not a full inference engine, we have built the essential frame. An ideal inference system would have a user interface, be populated with initial values and would return the minimum variance (or some other optimality criterion) prediction of desired quantities (Figure 8). In this case, the input interface would prompt for:
1. The contextual (general) information. Basically, the region, soil class or texture class of interest. Depending on what is known, the mean values for up to three would be generated and the minimum variance estimate filled in.
2. Any better estimates of any of the variables would be filled in along with the uncertainty, e.g., field pH or laboratory-measured clay. The system should have standardized uncertainty estimates for routine laboratory methods.
3. The target variables are defined.

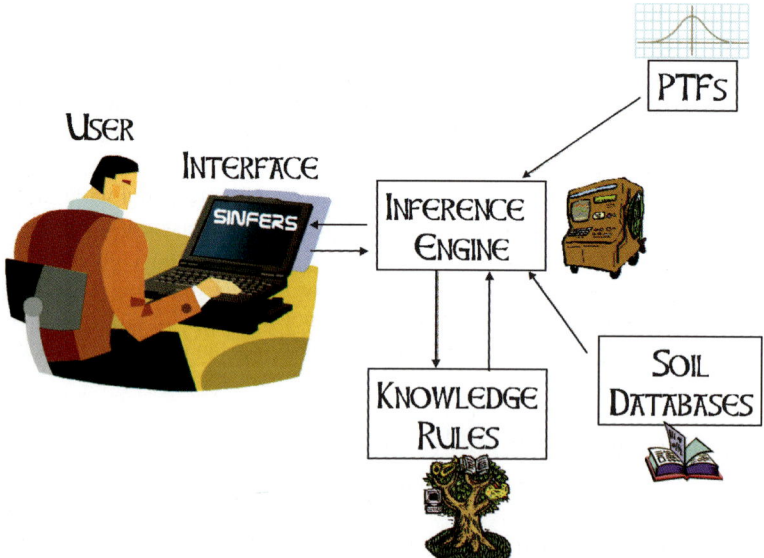

Figure 8. The future: soil inference systems.

We envision that a complete soil inference system will be built and can be adapted in a GIS context. The system should have a database of mean soil properties, such as particle-size distribution and organic matter content, for different soil types. The inference engine has knowledge rules that will determine what functions to use realizing their uncertainties. The output will be the predicted physical and chemical properties along with their uncertainties. This can be incorporated into a spatial framework, where a point in space can be predicted from the neighborhood basic soil properties or soil class.

Bouma (1989) defined pedotransfer functions in terms of *data* translation. We can describe this translation function as *information*. This information, when properly and logically conjoined, constitutes *knowledge*. Knowledge can generate various data. If we take pedotransfer functions as a data mining process, soil inference system is data generating. Soil inference systems take measurements we know with a certain precision and infer properties we don't know with given precision, by means of properly and logically linked pedotransfer functions.

REFERENCES

Acutis, M., Donatelli, M., 2003. SOILPAR 2.00: software to estimate soil hydrological parameters and functions. Eur. J. Agron. 18, 373-377.

Anderson, A.N., McBratney, A.B., 1995. Soil aggregates as mass fractals. Aust. J. Soil Res. 33, 757-772.

Arya, L.M., Paris, J.F., 1981. A physicoempirical model to predict soil moisture characteristics from particle-size distribution and bulk density data. Soil Sci. Soc. Am. J. 45, 1023-1030.

Bouma, J., 1989. Using soil survey data for quantitative land evaluation. Adv. Soil Sci. 9, 177-213.

Campbell, G.S., 1974. A simple method for determining unsaturated conductivity from moisture retention data. Soil Sci. 117, 311-314.

Campbell, G.S., 1985. Soil physics with BASIC. Elsevier, New York.

Collis-George, N., 1959. The physical environment of soil animals. Ecology 40, 550-557.

Cresswell, H.P., Pierret, C., Brebner, P., Paydar, Z., 2000. The SH-Pro V1.03 software for predicting and analysing soil hydraulic properties. CSIRO Land and Water, Canberra, Australia.

Da Silva, A., Kay, B.D., 1997. Estimating the least limiting water range of soils from properties and management. Soil Sci. Soc. Am. J. 61, 877-883.

Dale, M.B., McBratney, A.B., Russell, J.S., 1989. On the role of expert systems and numerical taxonomy in soil classification. J. Soil Sci. 40, 223-234.

De Gruijter, J.J., McBratney, A.B., 1988. A modified fuzzy k-means method for predictive classification. *In*: Bock, H.H. (Ed.), Classification and Related Methods of Data Analysis. Elsevier Science, North-Holland, pp. 97-104.

De Vries, D.A., 1966. Thermal properties of soils. *In*: Van Wijjk, W.R. (Ed.), Physics of Plant Environment. North Holland Publishing Company, Amsterdam, pp. 210-235.

Dexter, A.R., 2004. Soil physical quality. Part I; theory, effects of soil texture, density, and organic matter, and effects on root growth. Geoderma, 120, 201-214.

Dexter, A.R., Bird, N.R.A., 2001. Methods for predicting the optimum and the range of soil water contents for tillage based on the water retention curve. Soil Till. Res. 57, 203-212.

Dobson, M.C., Ulaby, F.T., Hallikainen, M.T., El-Rayes, M.A., 1985. Microwave dielectric behavior of wet soil – Part II: Dielectric mixing models. IEEE Trans. Geosci. Remote Sens. 23, 35-46.

Environmental System and Technologies, 1990. SOILPROP, a program to estimate unsaturated soil hydraulic properties from particle size distribution. User's guide. Blacksburg, VA.

Farouki, O.T., 1986. Thermal Properties of Soils. Trans. Tech. Publications, Clausthal-Zellerfeld, Federal Republic of Germany.

Fredlund, M.D., Wilson, G.W., Fredlund, D.G., 1999. Estimation of hydraulic properties of an unsaturated soil using a knowledge-based system. *In*: van Genuchten, M.Th., Leij, F.J., Wu, L. (Eds.), Proceedings of the International Workshop on Characterization and Measurement of the Hydraulic Properties of Unsaturated Porous Media. University of California, Riverside, pp. 1295-1305.

Hamaker, J.W., Thompson, J.M., 1972. Adsorption. *In*: Hamaker, J.W., Goring, C.A.I. (Eds.), Organic Chemicals in the Soil Environment, Vol. 1. Marcel Dekker Inc, New York, pp. 49-144.

Helyar, K.R., Cregan, P.D., Godyn, D.L., 1990. Soil acidity in New South Wales – current pH values and estimates of acidification rate. Aust. J. Soil Res. 28, 523-527.

Henderson, B.L., Bui, E.N., 2002. An improved calibration curve between soil pH measured in water and $CaCl_2$. Aust. J. Soil Res. 40, 1399-1405.

Hendrickx, J.M.H., van Dam, R.L., Borchers, B., Curtis, J., Lensen, H.A., Harmon, R., 2003. Worldwide distribution of soil dielectric and thermal properties, *In*: Detection and remediation technologies for mines and minelike targets VIII, Proceedings of the SPIE, Vol. 5089.

Hubrechts, L., Feyen, J., 1996. Pedotransfer functions for thermal soil properties. Institute for Land and Water Management, Katholieke University, Leuven.

Letey, J., 1985. Relationship between soil physical properties and crop production. Adv. Soil Sci. 1, 277-294.

McBratney, A.B., Minasny, B., Cattle, S.R., Vervoort, R.W., 2002. From pedotransfer functions to soil inference systems. Geoderma 109, 41-73.

Minasny, B., McBratney, A.B., 2000. Hydraulic conductivity pedotransfer functions for Australian soil. Aust. J. Soil Res. 38, 905-926.

Minasny, B., McBratney, A.B., 2002. The neuro-m method for fitting neural network parametric pedotransfer functions. Soil Sci. Soc. Am. J. 66, 352-361.

Minasny, B., McBratney, A.B., 2003a. Neuro θ. Neural networks pedotransfer functions for predicting soil hydraulic properties for Australian soil. Australian Centre for Precision Agriculture, The University of Sydney, http://www.usyd.edu.au/su/agric/acpa

Minasny, B., McBratney, A.B., 2003b. Integral energy as a measure of soil-water availability. Plant Soil 249, 253-262.

Minasny, B., Hopmans, J.W., Harter, T., Eching, S.O., Tuli, A., Denton, M., 2004. Neural networks prediction of soil water retention and unsaturated hydraulic conductivity functions of alluvial soils, as estimated from multi-step outflow experiments. Soil Sci. Soc. Am. J. 68, 417-429.

Mishra, S., Parker, J.C., Singhal, N., 1989. Estimation of soil hydraulic-properties and their uncertainty from particle-size distribution data. J. Hydrol. 108, 1-18.

Moldrup, P., Olesen, T., Schjønning, P., Yamaguchi, T., Rolston, D.E., 2000. Predicting the gas diffusion coefficient in undisturbed soil from soil water characteristics. Soil Sci. Soc. Am. J. 64, 94-100.

Noble, A.D., Cannon, M., Muller, D., 1997. Evidence of accelerated soil acidification under Stylosanthes-dominated pastures. Aust. J. Soil Res. 35, 1309-1322.

Palisade, 2000. @RISK Version 4. Palisade Corporation, New York, USA.

Parlange, J.Y., 1975. On solving the flow equation in unsaturated soils by optimization: horizontal infiltration. Soil Sci. Soc. Am. Proc. 39, 415-418.

Peplinski, N.R., Ulaby, F.T., Dobson, M.C., 1995. Dielectric properties of soils in the 0.3–1.3 GHz range. IEEE Trans. Geosci. Remote Sens. 33, 803-807.

Perfect, E., Sukop, M.C., Haszler, G.R., 2002. Prediction of dispersivity for undisturbed soil columns from water retention parameters. Soil Sci. Soc. Am. J. 66, 696-701.

Ritchie, J.T., Gerakis, A., Suleiman, A., 1999. Simple model to estimate field-measured soil water limits. Trans. ASAE 42, 1609-1614.

Romano, N., Santini, A., 2003. Water retention and storage. Field. *In*: Dane, J.H., Topp, G.C. (Eds.), Methods of Soil Analysis Part 4. Physical Methods, SSSA Book Series No. 5. Soil Science Society of America, Madison, WI, pp. 721-738.

Saarenketo, T., 1998. Electrical properties of water in clay and silty soils. J. Appl. Geophys. 40, 73-88.

Saxton, K.E., Rawls, W.J., Romberger, J.S., Papendick, R.I., 1986. Estimating generalized soil-water characteristics from texture. Soil Sci. Soc. Am. J. 50, 1031-1036.

Schaap, M.G., Leij, F.L., van Genuchten, M.Th., 1998. Neural network analysis for hierarchical prediction of soil hydraulic properties. Soil Sci. Soc. Am. J. 62, 847-855.

Schaap, M.G., Leij, F.L., van Genuchten, M.Th., 2001. Rosetta: a computer program for estimating soil hydraulic parameters with hierarchical pedotransfer functions. J. Hydrol. 251, 163-176.

Šimunek, J., Šejna, M., van Genuchten, M.Th., 1999. The HYDRUS-2D software package for simulating two-dimensional movement of water, heat, and multiple solutes in variable saturated media, Version 2.0, IGWMC-TPS-53. International Ground Water Modeling Center, Colorado School of Mines, Golden, Colorado.

Suleiman, A.A., Ritchie, J.T., 2001. Estimating saturated hydraulic conductivity from soil porosity. Trans. ASAE. 44, 235-239.

Ungaro, F., Calzolari, C., 2001. Using existing soil databases for estimating retention properties for soils of the Pianura Padano-Veneta region of North Italy. Geoderma 99, 99-121.

Ungaro, F., Calzolari, C., Borselli, L., Torri, D., 2001. PEDON-E-Pedotransfer function for estimating soil hydraulical parameters. Java-applet, beta version, http://www.area.fi.cnr.it/irpi/pedone/Pedon_introd.htm

van Dam, J.C., Huygen, J., Wesseling, J.G., Feddes, R.A., Kabat, P., van Walsum, R.E.V., Groenendijk, P., van Diepen, C.A., 1997. Theory of SWAP version 2.0. SC-DLO, Report 71. Department of Water Resources, Wageningen Agricultural University, The Netherlands.

van Genuchten, M.Th., 1980. A closed-form equation for predicting the hydraulic conductivity of unsaturated soils. Soil Sci. Soc. Am. J. 44, 892-898.

White, I., Sully, M.J., 1987. Macroscopic and microscopic capillary length and time scales from field infiltration. Water Resour. Res. 23, 1514-1522.

White, I., Smiles, D.E., Melville, M.D., 1989. Use and hydrological rubusteness of time to incipient ponding. Soil Sci. Soc. Am. J. 53, 1343-1346.

Williams, J., Ross, P.J., Bristow, K.L., 1992. Prediction of the Campbell water retention function from texture, structure and organic matter. *In*: van Genuchten, M.Th., Leij, F.J., Lund, L.J. (Eds.), Proceedings of the International Workshop on Indirect Methods for Estimating the Hydraulic Properties of Unsaturated Soils. University of California, Riverside, CA, pp. 427-441.

Wösten, J.H.M., Lilly, A., Nemes, A., Le Bas, C., 1999. Development and use of a database of hydraulic properties of European soils. Geoderma 90, 169-185.

Wu, L., Feng, G., Letey, J., Ferguson, L., Mitchell, J., McCullough-Sanden, B., Markegard, G., 2003. Soil management effects on the nonlimiting water range. Geoderma 114, 401-414.

Chapter 19

GRAPHIC USER INTERFACES FOR PEDOTRANSFER FUNCTIONS

M.G. Schaap

George E. Brown, Jr. Salinity Laboratory (USDA/ARS), 450 Big Springs Road, Riverside, CA 92507, USA
Tel.: + 1-909-369-4844; E-mail: mschaap@ussl.ars.usda.gov

Pedotransfer functions (PTFs) can be established with a number of methods, yielding models that can range from very simple, such as univariate expressions or lookup tables to more complex multivariate expression (e.g., Cosby et al., 1984; Carsel and Parrish, 1988; Vereecken et al., 1989; Wösten et al., 1995). Until the mid-nineties even the more complex expressions for PTFs were simple enough to be published in printed journals. To get quick access to PTF estimates, all a user of such a PTFs had to do is to copy the expressions into a spreadsheet or to implement these into a computer code. More recently, increasingly sophisticated techniques have been used for PTF development, leading to models that cannot be easily published and/or implemented. Especially artificial neural network or related models come to mind here (e.g., Tamari et al., 1996; Pachepsky et al., 1996). These models generally have a "black-box" nature and often contain a large number of coefficients, especially when combined with the Bootstrap Method (e.g., Schaap et al., 1998). Direct implementation of such PTFs in software with some kind of user-interface is often the only way to make them useful for a wider audience. In addition, an attractive and useful user-interface can also open up the PTF to a wider audience. Several PTF implementations have emerged in recent years and in the following we will dicuss four codes that can be easily obtained through the world-wide-web.

1. SOIL WATER CHARACTERISTICS FROM TEXTURE

Soil Water Characteristics from Texture (SWCT) is part of the Soil Plant Atmosphere Water Field and Pond Hydrology (SPAW) package developed by K. E. Saxton (USDA/ARS) and is based on Saxton et al. (1986). The windows-based SPAW package is targeted at farmers and resource managers interested in water and nutrient budgeting in soil and ponds and is available at http://www.bsyse.wsu.edu/saxton/spaw/ SPAW uses SWCT to estimate soil hydraulic data such as wilting point, field capacity and available water content. The main window (Figure 1) allows the user to click on textural classes in the textural triangle. Estimated quantities at the top right hand side of the window ("Soil Characteristics") and plotted hydraulic characteristics at the bottom are updated immediately. A more fine-grained control of soil texture is possible through a horizontal and vertical slider bars, for clay and sand percentages, respectively. In addition, it is

DEVELOPMENTS IN SOIL SCIENCE
VOLUME 30 ISSN 0166–2481/DOI 10.1016/S0166-2481(04)30019-X

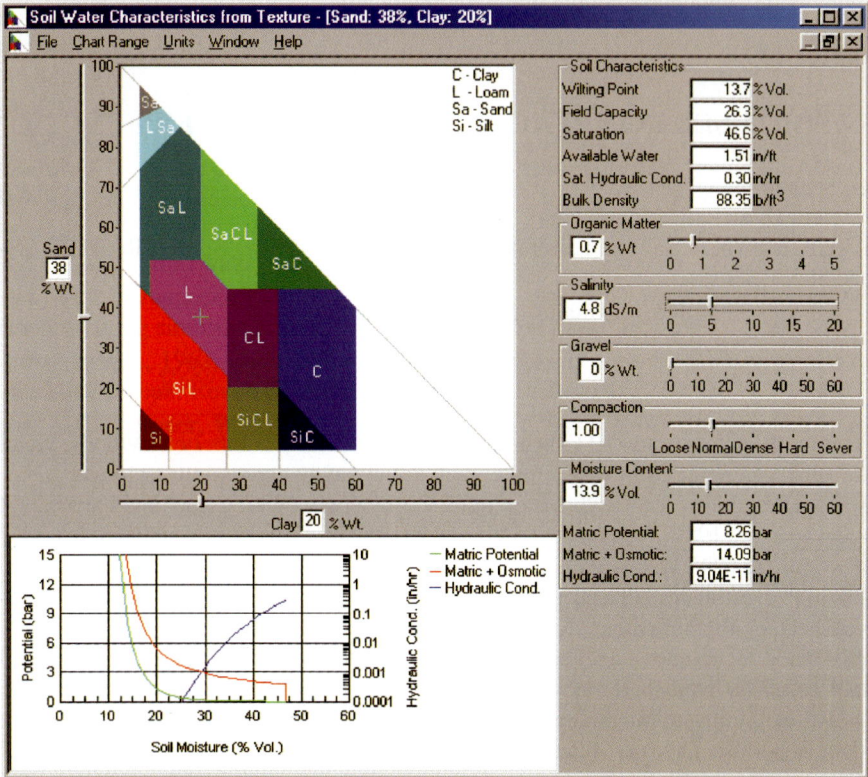

Figure 1. The main window of SWCT, see text for explanation.

possible to input, organic matter content, gravel content, salinity and soil compaction through slider bars at the right hand side of the screen. A separate retention curve is plotted for non-zero salinities to account for salinity-induced osmotic potentials. The resulting graph provides an effective retention characteristic as experienced by vegetation. The program allows for the export of graphical and numerical results and includes a help-system that explains the background of the program.

2. SOILPAR

SOILPAR was developed by M Donatelli and M. Acutis at the Research Institute for Industrial Crops (ISCI), Bolongna, Italy. The program, and its update, can be downloaded from http://www.sipeaa.it/ASP/ASP2/SOILPAR.asp. The program implements ten point PTFs and four parametric PTFs and provides a wide range of output data (Figure 2). Required input data and estimated output data depend on the model used and include soil texture, organic carbon, soil pH, and cation exchange capacity. The window depicted in Figure 2 consists of an input area (top) and output area. The "tabs" at the bottom of the

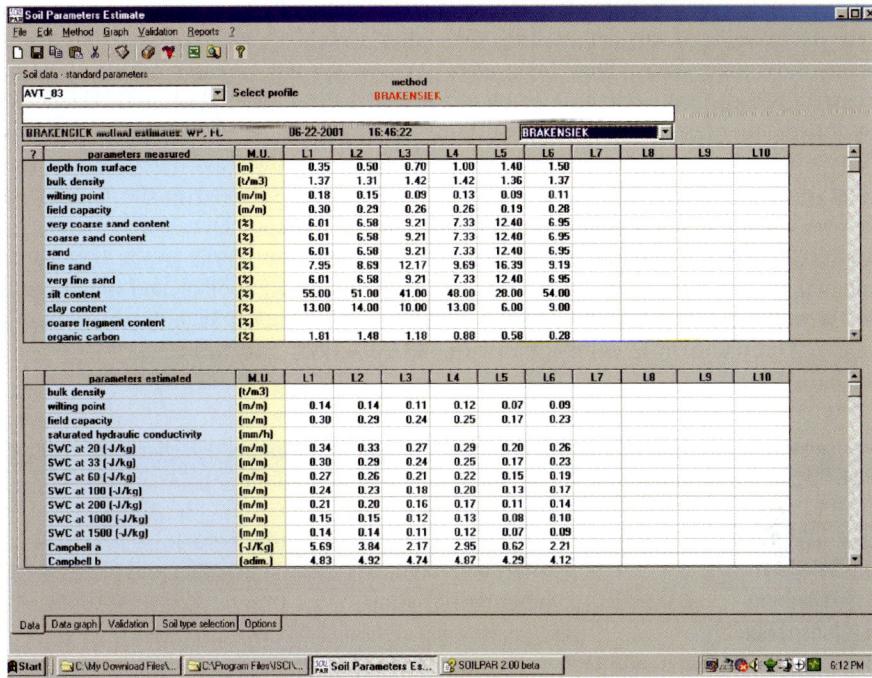

Figure 2. The main window of SOILPAR.

window provide access to various other program functions including plotting and validation routines, and a utility to convert various systems textural data. The program also allows fitting retention data to four types of retention equations. Soil data are stored in a geo-referenced soil database and various input and output data formats are supported (such as EXCEL and ArcView/ArcInfo). The program is well documented through a "help" system that describes the program background, data organization and possible operations on the data.

3. ROSETTA

Rosetta is a windows-based program that implements artificial neural network results published by Schaap et al. (1998), Schaap and Leij (1998), and Schaap and Leij (2000) and is available from http://www.ussl.ars.usda.gov/models/rosetta/rosetta.HTM. The program implements five PTFs in a so-called hierarchical approach (Schaap et al., 2001). This approach was chosen to maximize the accuracy of the PTF estimates given a particular data availability. All models in Rosetta estimate saturated hydraulic conductivity, and van Genuchten (1980) retention and unsaturated hydraulic conductivity parameters. Through use of the Bootstrap Method, Rosetta is also able to estimate the uncertainty of the estimated hydraulic parameters.

Rosetta uses a database to store its input and output data. In general the user creates a new database or opens an existing one, after which a main window appears that contains three main areas. The "Input Data" box on the left contains information about the current record in the database (shown is record 1 out of a total of 564). Under these entry boxes the textural distribution, bulk density and water retention at 33 and 1500 kPa are shown (data that are not available are shown as -9.9). Entry boxes become gray or white, depending on the type of model selected in the bottom area. Estimates and associated uncertainties appear in the "Output Data" area on the right and are made by clicking the single or double exclamation marks in the tool bar, for estimation for the current or all records, respectively. Other toolbar and menu options serve to edit or navigate through the database, or to import data from or into the database. The database is compatible with MS-ACCESS™ software.

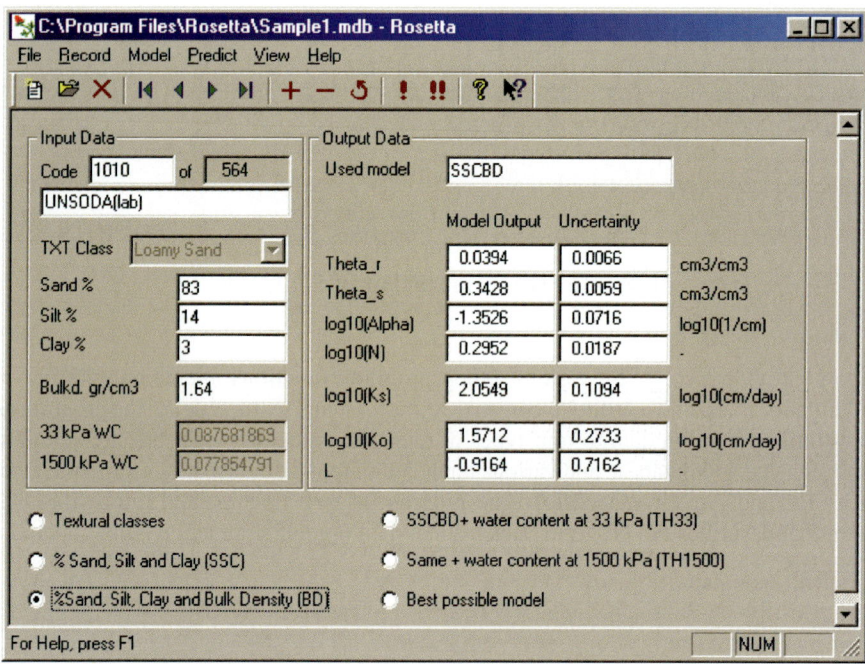

Figure 3. The main window of Rosetta.

A simplified version of Rosetta (Rosetta-Lite) is available as plugin for external software for the computation of water retention and saturated hydraulic conductivity parameters. This plugin is currently included in the Hydrus-1D and Hydrus 2-D software, but can be easily implemented in other types of software. Rosetta-Lite comes with a user-interface that is similar to the one depicted in Figure 4 but without database support.

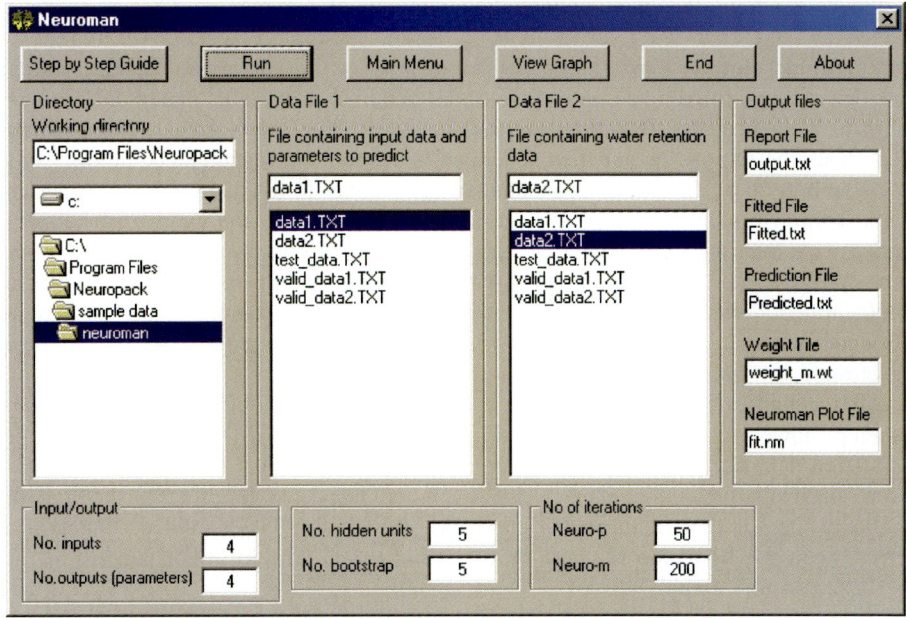

Figure 4. The main window of NeuroMan.

4. NEUROPACK

Minasny and McBratney (2002) developed the Neuropack software package (http://www.usyd.edu.au/su/agric/acpa/neuropack/neuropack.htm). This package differs from the previously discussed software in that it is primarily intended to develop PTFs using neural network-based techniques using data that is supplied by the user. Except from some sample data, no previously calibrated PTF is shipped with Neuropack. The Neuropack package comes with a complete technical guide and users manual. The technical guide describes the scientific background of both programs and includes a ANN primer, an explanation of the Bootstrap Method, and a description of optimization and error-criteria. The users, manual provides a detailed description of the various program options. A small drawback of Neuropack is that it can only work with 100 data points, a full version is available from Dr. B. Minasny upon request.

The package consists of two separate programs, NeuroPath and NeuroMan. NeuroPath is somewhat simpler than Neuroman and can be used to develop ANN and Bootstrap Method-based PTFs that estimate water retention points in a cycle of calibration, validation and prediction. A similar but more elaborate structure is used for NeuroMan. This program can be used to develop parametric-PTFs, again based on a combination of ANNs and the Bootstrap Method. Neuroman uses a two-step approach outlined in Minasny and McBratney (2002) for establishing parametric PTFs by defining the objective function in terms of water contents (see also Section 1 of Chapter 3).

Figure 4 shows the main optimization window of NeuroMan; other windows in NeuroMan and NeuroPath are similar in setup. Essentially, NeuroMan requires a working folder (leftmost box "Directory"), a data file with basic input data (e.g., texture, bulk density) and fitted retention parameters (second box from left, "Data File 1"), and a data file with measured retention points ("Data File 2"). The rightmost box ("Output files") lists the various output files generated by Neuroman (coefficients, plotting files, etc). The bottom of the window contains six entry boxes to list the number of input and output parameters, or to control the neural network topology, the number of bootstrap replicas (see Section 3 of Chapter 3), and the number of optimization steps in the two-step optimization. The top of the window contains buttons for a step-by-step specification of the necessary data files, running the optimization, to get back to an introductory menu, to view graphs, as well as to end the program or to get more background information.

After running the optimization, a graph like Figure 5 (left hand side) appears, showing how well measured and estimated water contents match for the entire calibration data set; the root mean square error (RMSE, Section 2 of Chapter 3) is also given. Individual retention characteristics can be inspected (Figure 6), showing the originally measured data, the mean curve, as well as the 95% confidence interval as derived with the Bootstrap Method. Pertinent numerical data for this characteristic are also listed in this graph.

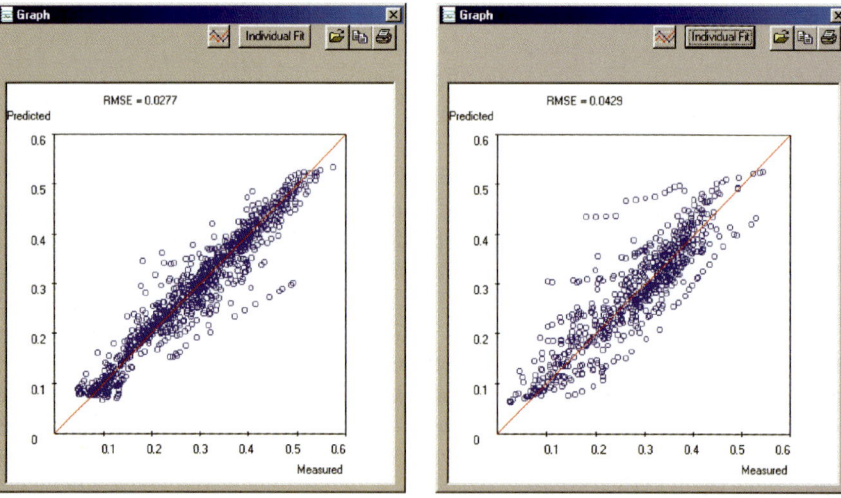

Figure 5. Calibration (left) and validation results (right) for the sample data set in NeuroMan.

Calibrated models can be tested with a validation step, using a window that is similar to that in Figure 5 (left hand side). Figure 5 (right hand side) shows a scattergram that indicates that the model does not perform as well on the validation dataset as for the calibration dataset (Figure 5, left hand side). Other program options include making estimates and the printing and saving of results.

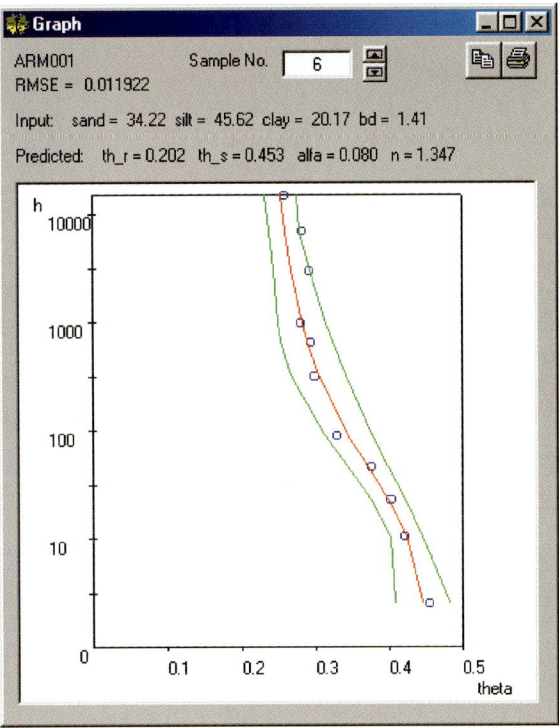

Figure 6. NeuroMan results for one particular soil sample. Circles represent measured data points, the estimate with the 95% interval is shown as lines.

REFERENCES

Carsel, R.F., Parrish, R.S., 1988. Developing joint probability distributions of soil water retention characteristics. Water Resour. Res. 24, 755-769.

Cosby, B.J., Hornberger, G.M., Clapp, R.B., Ginn, T.R., 1984. A statistical exploration of the relationships of soil moisture characteristics to the physical properties of soils. Water Resour. Res. 20, 682-690.

Minasny, B., McBratney, A.B., 2002. The neuro-m method for fitting neural network parametric pedotransfer functions. Soil Sci. Soc. Am. J. 66, 352-362.

Pachepsky, Ya.A., Timlin, D., Varallyay, G., 1996. Artificial neural networks to estimate soil water retention from easily measurable data. Soil Sci. Soc. Am. J. 60, 727-733.

Saxton, K.E., Rawls, W.J., Romberger, J.S., Papendick, R.I., 1986. Estimating generalized soil–water characteristics from texture. Soil Sci. Soc. Am. J. 50, 1031-1036.

Schaap, M.G., Leij, F.J., 1998. Database related accuracy and uncertainty of pedotransfer functions. Soil Sci. 163, 765-779.

Schaap, M.G., Leij, F.J., 2000. Improved prediction of unsaturated hydraulic conductivity with the Mualem-van Genuchten model. Soil Sci. Soc. Am. J. 64, 843-851.

Schaap, M.G., Leij, F.J., van Genuchten, M.Th., 1998. Neural network analysis for hierarchical prediction of soil water retention and saturated hydraulic conductivity. Soil Sci. Soc. Am. J. 62, 847-855.

Schaap, M.G., Leij, F.J., van Genuchten, M.Th., 2001. Rosetta: a computer program for estimating soil hydraulic parameters with hierarchical pedotransfer functions. J. Hydrol. 251, 163-176.

Tamari, S., Wösten, J.H.M., Ruiz-Suárez, J.C., 1996. Testing an artificial neural network for predicting soil hydraulic conductivity. Soil Sci. Soc. Am. J. 60, 1732-1741.

van Genuchten, M.Th., 1980. A closed form equation for predicting the hydraulic conductivity of unsaturated soils. Soil Sci. Soc. Am. J. 44, 892-989.

Vereecken, H., Maes, J., Feyen, J., Darius, P., 1989. Estimating the soil moisture retention characteristic from texture, bulk density, and carbon content. Soil Sci. 148, 389-403.

Wösten, J.H.M., Finke, P.A., Jansen, M.J.W., 1995. Comparison of class and continuous pedotransfer functions to generate soil hydraulic characteristics. Geoderma 66, 227-237.

Chapter 20

METHODS TO EVALUATE PEDOTRANSFER FUNCTIONS

M. Donatelli[1,*], J.H.M. Wösten[2] and G. Belocchi[1]

[1]ISCI (Research Institute for Industrial Crops), Via di Corticella, 133, 40128 Bologna, Italy

[2]Alterra, Droevendaalsesteeg 3, 6700 AA Wageningen, The Netherlands

*Corresponding author: Tel.: +39-051-6316843; fax: +39-051-7456931.
E-mail: m.donatelli@isci.it

In this chapter, procedures are presented which can be used for pedotransfer functions evaluation. Section 1 presents an overview of problems and methodologies, which have an impact on the evaluation; Appendix A contains a compendium of formulas for the evaluation. In Section 2 two integrated indices to evaluate pedotransfer functions estimating soil hydrological parameters are introduced and illustrated. The Appendix B shows an example calculation. In Section 3 the functional evaluation of pedotransfer functions is discussed.

1. EVALUATION OF PEDOTRANSFER FUNCTIONS
M. Donatelli, H. Wösten and G. Belocchi

One of the principles of validating models dictates that complete testing is not possible (Balci, 1997). Consequently, in model validation, the purpose is to increase confidence in model accuracy with reference to the specific goal of model use, rather than trying to test the model application completely. The same principles are relevant to the pedotransfer functions validation. The "confidence building" activity must be performed by considering not only model accuracy, but also other quality characteristics that affect acceptability. For example, if a pedotransfer function requires available and reliable inputs, users may have higher confidence in its accuracy as opposed to other PTFs which require inputs that are unlikely to be determined.

Model testing requires that both model inputs and measured data vs. model estimates must be carefully assessed to avoid a "blind" evaluation, which would produce unreliable results. In other words, preliminary data quality evaluation is key to provide the basis for model testing, with reference both to predictive capabilities and to correct implementation. The following discussion highlights the major problems in PTF validation.

1.1. Evaluating uncertainty in equations and data sets

The answer to the question of how adequate a PTF is for a particular purpose depends on a combination of field/laboratory studies (i.e., input/output data collection), parameter estimation, and evaluation of results, and very much relies on the skills and knowledge of the modeling teams. Currently, there are many PTFs, but PTFs development still seems to be a developing process. Increasing accessibility of PTFs may increase improper use of them, because as modeling studies become more complex pedotransfer functions are used by parties less familiar with the underlying assumptions of the method used. In general, variability of soil hydraulic characteristics depends on the extent of the area studied, on the spatial variability of soils within the area, and on the methods used for sampling and measurement. Demonstration that PTF output more or less fits a set of data is a necessary but not sufficient indication of validity. When comparing model results with experimental data, deviations should not only lead to a critical evaluation of the equation, but data should also be critically considered. A quality indicator can be assigned to experimental data, indicating how reliable a particular set of observations is. Such a quality indicator can be assigned subjectively based on visual data inspection, or it can be assigned automatically according to certain criteria. Such criteria may include ranges for single parameters, or a cross-validation across subsets of data which tests the robustness of the equations and the heterogeneity of the data. Caution must be paid to both the subjective nature of the data inspection and the inability of the automated procedure to reveal peculiar errors possibly associated to the data. Any particular data handling for a given data set should be based on an expert consensus to be built up within the scientific and operational community which produces and uses these data.

General principles for pedotransfer functions are (McBratney et al., 2002): (1) nothing has to be estimated that is easier to measure than its estimator, (2) pedotransfer functions should not be used unless their uncertainty can be evaluated and, for a given problem, if a set of alternative pedotransfer functions is available, the one with minimum variance must be used. For example, to estimate the saturated hydraulic conductivity of a soil from its structural features inferred from the image analysis would not constitute an efficient pedotransfer function, as it takes more effort to use the image analysis technique than making measurements on the real system. The uncertainty associated with the model can be calculated from bootstrap methods (Efron and Tibshirani, 1993; Schaap and Leij, 1998) or the standard first-order Taylor analysis if pedotransfer parameters are generated using the least-squares method (Heuvelink, 1998). The uncertainty in input data can be evaluated using Monte Carlo simulation, by repeatedly sampling the assumed distribution of the input data and evaluating the output of the pedotransfer (Minasny and McBratney, 2002). When the uncertainty of the PTF and the statistics of the training data are not defined, procedures based on fuzzy rules can be implemented for the uncertainty assessment (McBratney et al., 2002).

1.2. Comparing estimates and measurements

As predictive equations, PTFs are routinely evaluated in terms of correspondence between measured and predicted values. When measured values are those used to develop the equation, the accuracy of the equation is evaluated. The reliability of pedotransfer functions needs to be assessed by examining the correspondence between measured and estimated data for data set(s) other than the one used to develop a PTF. A multitude of statistics is available to evaluate both accuracy and reliability (Pachepsky et al., 1999).

Although many accuracy evaluation techniques are available, only a limited number of such techniques are used in modeling projects due to time and resource constraints. This is also because different users may have different thresholds for confidence: some users may derive their confidence simply from the model reports display, others may require more in-depth validation before being willing to accept the results. Limited testing hinders modelers' ability to substantiate sufficient accuracy of PTFs. As a general rule, the more tests are performed in which it cannot be proved that the function is incorrect, the more increase in confidence in the function. The indices most commonly used for PTF evaluation are the mean error (ME), mean absolute error (MAE), the coefficient of determination (R^2), the root mean square error (RMSE), also called root mean squared deviation (RMSD) or root mean square residual (RSMR). The indices can be calculated directly on the residuals (e.g., Pachepsky et al., 1998; Schaap et al., 2001; van Alphen et al., 2001; Romano and Palladino, 2002) but often statistics are computed relative to observed average or observed variance as proposed by Janssen and Heuberger (1995) for the cases with a wide range in measured and estimated values (e.g., Tietje and Hennings, 1996; Wagner et al., 2001). The estimated retention curves can be also compared with experimental ones using the ME or the root of the mean square error between estimated and measured water content, calculated for any given matric potential in an integral form (Tietje and Tapkenhinrichs, 1993).

The accuracy of predictions may serve as benchmarks. However, it should be compared to variability in measured input data. In general, models should not be more accurate than data used in model development (Pachepsky et al., 1999). Therefore, PTFs can be considered to be sufficiently accurate if the variability of the PTF errors does not differ significantly from variability in other data, and if the average error does not significantly differ from zero.

Using this approach, El-Kadi (1985) compared several equations using mean square error of water content at several pressure heads and found Brutsaert's equation to be most accurate. In the paper of Tietje and Tapkenhinrichs (1993), accuracy of the retention prediction was quantified by the RMSE, using test data sets with a broad range of soils.

Pachepsky and Rawls (1999) tested whether grouping according to taxonomic unit, soil moisture regime, soil temperature regime, and soil textural class would improve both pedotransfer function accuracy and reliability, using RMSE as a evaluation criterion. In the evaluation of PTFs giving volumetric water content at given values of matric potential, Calzolari et al. (2000) used the statistics ME and RMSE. ME indicates weather a PTF over-estimates (ME > 0) or under-estimates (ME < 0) the water content, whereas RMSE, always positive, can be viewed as the continuous analogue of the standard deviation over the whole retention curve. In order to define ranges for optimal usage of different PTFs, the two error indices were computed not only for the whole data set, but also for different subsets formed according to USDA textural classes, organic carbon content, bulk density and matric potential (Ungano and Calzolari, 1998).

Pedotransfer functions for soil bulk and particle densities have been subject to statistical analysis and to validation by Leonavičiutę (2000) who used three error criteria: correlation coefficient, relative RMSE (RRMSE), and modeling efficiency (EF). EF gives a comparison of predictive errors around the mean measured value. It indicates whether the method describes the data better than simply the arithmetic average of the observations. In Cornelis et al. (2001), absolute and square measures of error were used together with the correlation coefficient to evaluate some pedotransfer functions with respect to their accuracy in predicting the soil moisture retention curve.

Table 1
Typical examples of the water retention PTF accuracy (afterWöstern et al., 2001)

Source	Pressure head (kPa)	RMSE, (m^3 m^{-3})	PTF input variables
Ahuja et al., 1985	−33	0.05	Clay, silt, organic matter, bulk density
	−1500	0.05	
Beke and MacCormick, 1985	−33	0.05	Clay, silt, organic matter, bulk density
	−1500	0.04	
Bell, 1993	−1500	0.04	Organic matter, sand
Boix Fayos, 1997	−10	0.03	Organic matter, aggregates 0.1–1 mm
	−1500	0.05	Organic matter, aggregates 1–2 mm, fine silt
Bruand et al., 1996	−33	0.03	Bulk density
	−1500	0.03	
Calhoun et al., 1972	−1500	0.05	Clay
Gupta and Larson, 1979	−1500	0.05	Clay, silt, organic matter content, bulk density
Koekkoek and Bootlink, 1999	−10	0.05	Sand, clay, silt, organic matter content, bulk density
	−1500	0.05	
Lenhard, 1984	−33	0.07	Clay
	−1500	0.05	
Mayr and Jarvis, 1999	A[a]	0.03	
		0.06[b]	

Reference	Suction	RMSE[a]	Inputs
Minasny et al., 1999	−33	0.07	Clay, silt, sand, bulk density, porosity, mean particle diameter, geometric standard deviation
	−1500	0.07	
Pachepsky et al., 1996	−33	0.02	Sand, silt, clay, bulk density
	−1500	0.02	
Paydar and Cresswell, 1996	A	0.02	Slope of the particle size distribution curve + one measured point on the WRC
		0.04[c]	
Paydar and Cresswell, 1996	A	0.03	Clay, silt, coarse sand, fine sand, organic matter content
		0.05[c]	
Schaap et al., 1998	A	0.11	Textural class only
Schaap et al., 1998	A	0.09	Sand, silt, clay, bulk density
Schaap and Leij, 1998	A	0.10	Sand, silt, clay
Sinowski et al., 1997	−30	0.04	Clay, silt, sand, bulk density, porosity, median particle diameter and standard deviation
	−1500	0.04	
Tomasella and Hodnett, 1998	A	0.06	Clay, silt, sand

[a] Average RSME along the measured water retention curves obtained after estimating parameters of a water retention equation and using this equation to compute water contents at all suction where the water retention was measured.
[b] Various genetic groups.
[c] Various textural classes.

The above discussion was meant not only to review the performance measures that are used in pedotransfer function validation, but also to emphasize that no single statistic can adequately describe PTF performance. Accuracy of existing pedotransfer functions varies appreciably. Table 1 presents a sample from literature that gives typical values of RMSE achieved with pedotransfer functions to predict soil water retention. The RMSE of volumetric water contents ranges from 0.02 to 0.11 $m^3\ m^{-3}$. The smallest RMSE values of 0.02 $m^3\ m^{-3}$, are obtained in studies where either a preliminary grouping has been applied or one or more measured points were used. The largest RMSE value of 0.11 $m^3\ m^{-3}$ was obtained in a study where the textural class was the sole predictor. Accuracy of predicting a complete characteristic is lower than accuracy of the prediction for a specific pressure head. A clear trend in dependence of accuracy on pressure head cannot be established with only two pressure heads being evaluated. Some authors who tried to observe such a trend reported the lowest accuracy somewhere between -10 and -100 kPa (Rajkai and Várallyay, 1992). Nemes et al. (2003) found lowest prediction accuracy for the range -50 to -250 kPa when soil texture, bulk density and organic matter (OM) content were used as predictors. Errors in that pressure range were greatly reduced when one or two measured water retention points were additionally used as input.

The RMSE of log(Ksat) predictions is not better than 0.5 (Jaynes and Tyler, 1984; Ahuja et al., 1989; Tietje and Hennings, 1996; Schaap et al., 1998). The highest accuracy was attained using the Brooks and Corey water retention parameters along with basic soil properties (Timlin et al., 1999; Pachepsky et al., 1999). Schaap and Leij (1998) reported an RMSE of 0.10 $m^3\ m^{-3}$ in approximating water retention curves in their database with the Van Genuchten equation.

When evaluating PTFs, the one statistic that normally takes precedence over the others is the mean square error (MSE) within the range of estimated values, or its square root, the RMSE, or derived statistics such as the RRMSE. MSE is also the statistic that is usually minimized during the parameter calibration process. Although widely used, MSE, RMSE, and other common indices are not exhaustive in analyzing model capability to fit measured data. A review on performance measures for possible use in pedotransfer function evaluation has been given by Imam et al. (1999). Many authors believe that several methods need to be used together to give a comprehensive check, as emphasized by Donatelli et al. (2002). There is little attention, however, by many other users for aspects of PTF evaluation related to the patterns of residuals, whereas most of the indices used are based on the difference or the statistical association between calculated and measured data. Biases as a function of some soil characteristic (e.g., clay content, or organic matter content) are typically found with pedotransfer functions (Gijsman et al., 2002). Examples of pattern analysis are in Belocchi et al. (2002) and Donatelli et al. (2004).

1.3. Pedotransfer as inputs for simulation models: sensitivity analysis

Modeling water movement and storage in soils involves a number of assumptions and approximations. The soil water redistribution models incorporated into agro-ecological models offer choices of such approximations. Among these choices are the methods of deriving soil hydraulic properties from simpler, more readily available data. The simulated soil water regimes are sensitive to such choices because pedotransfer functions may differ remarkably in their estimation from the same inputs. Although a moderate level of accuracy can be acceptable for some simulation projects, it can be the source of large errors in simulation outputs for models developed to estimate soil water balance with greater detail. Hence, PTF selection may significantly affect the performance of the model at different levels (e.g., Sonneveld et al., 2003). Caution is therefore advised when applying models to environments for which they have not been calibrated and for which inputs have to be obtained indirectly.

When using PTFs to estimate model inputs, a sensitivity analysis is required (e.g., Arah and Hodnett, 1997), covering a range of soils, environment inputs and models of root distributions.

2. INTEGRATED INDICES FOR PEDOTRANSFER FUNCTION EVALUATION
M. Donatelli, M. Acutis, A. Nemes and H. Wösten

Several statistical indices are available for quantifying how well model fits the measurements (Appendix A). Many authors (e.g., Smith et al., 1997; Yang et al., 2000) advocate that there is no robust statistic that can be used to draw conclusions in model evaluations and, therefore, several methods need to be used together to give a comprehensive check. According to Belocchi et al. (2002), three elements are relevant in a model quality judgment: (i) the ability of the model to produce small residuals that is similar to the minimum variance principle; (ii) the correlation between estimates and measurements; (iii) the absence of systematic pattern in residuals. As a consequence, the evaluation of PTFs should include simultaneous evaluation of different indices. The combination of multiple statistics into a single, integrated index is not an obvious procedure, because weights are needed to capture subjective judgment and describe it in mathematical terms. This, in principle, may be achieved via aggregating indices by summation, multiplication, or a combination of both. Such aggregation poses mathematical and conceptual problems (Keeney and Raiffa, 1993), since evaluation statistics differ in their nature, dimensions and range of possible values. A different approach to evaluate model performance can rely in setting up a fuzzy expert system (Hall and Kandel, 1991) using decision rules. This technique is applicable to uncertain and imprecise data such as subjective judgments, and allows the aggregation of dissimilar measures in a consistent and reproducible way (Bouchon-Meunier, 1993). This chapter describes the development and application of integrated indices for PTF evaluation.

2.1. Integrated indices to evaluate PTFs for soil water retention

In this section, we define two integrated indices to estimate soil water point PTFs (I_{PTFSW}) and soil water retention curve PTFs (I_{PTFRC}). We define three indicator modules, named "Accuracy", "Correlation", and "Pattern". The value of each module depends on one or more accuracy indices (Table 1) and a set of decision rules. For each module, a dimensionless value between 0 (best model response) and 1 (worst model response) is calculated. The procedure, based on the multivalued fuzzy set theory introduced by Zadeh (1965), follows the so-called Sugeno or Takagi-Sugeno-Kang method of fuzzy inference (Sugeno, 1985). This approach is computationally efficient and well suited for mathematical analysis. It has been applied to a wide variety of problems, such as the design of an indicator for assessing environmental impact of pesticides (van der Werf and Zimmer, 1998), and the development of novel approaches to support decisions regarding sustainable development (Cornelissen et al., 2001). Three membership classes (or subsets) are defined for all indices given in Table 2, according to an expert judgment, namely

Table 2
The modules "Accuracy", "Correlation" and "Pattern", and their inputs (see text for details)

Inputs	Indicator modules		
	"Accuracy"	"Correlation"	"Pattern"
Integrated index "Pedotransfer Soil Water", I_{PTFSW}			
RRMSE	x		
EF	x		
r		x	
PI_{D50}			x
PI_{OC}			x
Integrated index "Pedotransfer Retention Curves", I_{PTFRC}			
RRMSE	x		
EF	x		
r		x	
PI_{D50}			x
PI_{OC}			x
PI_{ψ}			x

favorable (F), unfavorable (U), and partial (or fuzzy) membership. Several indices are aggregated into modules, and the modules in the final integrated index, using fuzzy-based logic rules. The procedure is explained in detail in Appendix B.

2.1.1. The "Accuracy" module

The composition of the module "Accuracy" was based essentially on the suggestions of Yang et al. (2000). These authors found that a sound conclusion on model accuracy may be drawn using an index of the amount of residuals, (e.g., RMSE) (Fox, 1981), and a measure of modeling efficiency (e.g., EF: efficiency) (Loague and Green, 1991), that are defined as follows

$$\text{RMSE} = \left[\frac{\sum_{i=1}^{n} (R_i)^2}{n} \right]^{0.5} \tag{1}$$

$$\text{EF} = 1 - \frac{\sum_{i=1}^{n} (R_i)^2}{\sum_{i=1}^{n} (M_i - \bar{M})^2} \tag{2}$$

where R_i is the difference $E_i - M_i$, E_i is the ith estimated value, M_i is the ith measured value n is the number of pairs E_i/M_i, $\bar{M}$ is average value of measured values of volumetric soil water content. We did not include the two-tailed paired t-test suggested by Yang et al. (2000) as it did not provide, in our experience, any significant contribution in differentiating methods in the comparison process of pedotransfer functions.

Instead of RMSE, we found it more appropriate to use a relative measure, the RRMSE, where:

$$\text{RRMSE} = 100 \frac{\text{RMSE}}{\bar{M}} \tag{3}$$

The index RRMSE may vary from 0 to positive infinity. The smaller RRMSE, the better the model performance. The RRMSE is dimensionless allowing comparison among different model responses, regardless of units.

The limits for favorable and unfavorable subsets can be set according to the available information or based on the data set used, and this applies to all the indices considered here and in two following paragraphs. Assuming that the data set is adequate in terms of representativeness (size and ranges for the input variables under evaluation), a conservative way to set the limits is to compute, for each index, the median of the pedotransfer function performance, and select an interval which includes the median. Such interval allows discriminating performance of the methods used if the limits are chosen in order to classify them as F, transition, or U. The median, rather than the average, is chosen to minimize the effect of possible outliers. The interval does not necessarily need to be median-centered, given that some indices may have known properties (e.g., EF cannot have a U limit <0). Using the procedure above, the limit to the favorable subset F for RRMSE was set equal to 30 (RRMSE ≤ 30 is F), while the limit to the unfavorable subset U was established equal to 60 (RRMSE ≥ 60 is U).

The index EF is very informative because it allows the immediate identification of inefficient models. EF is bounded from above by 1, and it can assume negative values (bounded from below at negative infinity). Negative values of EF indicate that the average value of all measured values is a better predictor than the model used. When estimating soil water content at given pressure heads, the limit for the subset U, EF = 0.00 and the one for the subset F, EF = 0.50, were chosen (EF ≤ 0.00 is U, EF ≥ 0.50 is F). Again, both limits are derived from the experience made evaluating pedotransfer functions as described in the previous paragraph.

The value of the module "Accuracy" was calculated from the basic indices according to four decision rules, as summarized in Table 3. The expert reasoning runs as follows: if all indices are F, the value of the module is 0 (identity of estimates and measurements); if all indices are U, the value of the module is 1. In setting up the decision rules for the other combinations we had to decide on the relative importance of each index. In our experience,

Table 3
Summary of decision rules describing the effect of the input indices RRMSE, and EF on the module "Accuracy"

Membership class		Expert weight
RRMSE	EF	
F	F	0.00
F	U	0.50
U	F	0.50
U	U	1.00

F, favorable; U, unfavorable (see text for details).

RRMSE and EF assume an equal importance in the evaluation process, thus the weight of 0.50 was attributed to both RRMSE and EF. If all indices are F, the value of the module is 0; if all indices are U, the expert weight is 1; if one index is F and the other is U, the weight is 0.50.

2.1.2. The "Correlation" module

The value of the module "Correlation" depends on one basic index, which is the correlation coefficient r (Addiscott and Whitmore, 1987).

The coefficient r is derived from the Pearson's linear correlation coefficient:

$$r = \frac{\sum_{i=1}^{n}(E_i - \bar{E})(M_i - \bar{M})}{\left[\sum_{i=1}^{n}(E_i - \bar{E})^2 \sum_{i=1}^{n}(M_i - \bar{M})^2\right]^{0.5}} \quad (5)$$

where $\bar{E}$ is the average of estimates. The value of r may vary from -1 (full negative correlation) to 1 (full positive correlation). The closer r is to 1, the better the model. Besides the indices based on differences, the coefficient of correlation r between estimates and measurements is commonly computed. The use of this index is questioned (e.g., Willmott, 1982), since its value is not related to the accuracy of the estimate. However, the index r is a universal measure with multiple interpretations. For instance, Cahan (1987) looks at r as a measure of similarity between standardized values. Moreover, the value of r may help recognize the fluctuation of the estimates among the measurements (Kobayashi and Salam, 2000). For these reasons, the index r is generally still regarded as a useful measure of model performance. The membership limits attributed here to the correlation coefficient basically come from the general categorization made by Hinkle et al. (1994). We consider correlation coefficients equal to or greater than 0.90 as very high correlations (limit for F: $r = 0.90$), and coefficients equal to or lower than 0.70 as moderate and little correlations (limit for $U : r = 0.70$). It must be pointed out that statistical significance for correlation coefficients does not always imply practical significance. The limits attributed here are mere descriptors for the practical interpretation of correlation coefficients, and do not take into account statistical significance. The latter depends on the number of data points and can be verified by a t-test, provided that both estimated and measured series conform the assumptions required for the appropriate application of the test.

Given that there is only one index in the module, the computation of "Correlation" is simplified to two decision rules: if r is F then 0, if r is U then 1.

2.1.3. The "Pattern" module

The module "Pattern" accounts for two relevant independent variables in pedotransfer functions, the particles median diameter (D50) and the organic carbon content (OC). This does not mean that a PTF, to be evaluated using this procedure, must explicitly include either D50 or OC, or both. Other variables, such as bulk density, could have been used as an independent variable; however, the scarcely available (and reliable) bulk density data would have likely made the index of little use. For the computation of pattern indices, the range of D50 and OC is divided into three sub-ranges of equal length, thus producing three groups of residuals. The pattern index chosen (PI) is based on the pair-wise differences between average residuals of each group (Donatelli et al., 2004):

$$\text{PI} = \max_{l,m=1,\ldots,3;\ l\neq m} \left| \frac{1}{q_l} \sum_{i_l=1}^{q_l} R_{i_l} - \frac{1}{q_m} \sum_{i_m=1}^{q_m} R_{i_m} \right| \tag{6}$$

where R is the model residual, l and m indicate two groups being compared, q_l and q_m represent the number of residuals in the groups, i_l and i_m identify each value of residuals in the groups. Three groups were chosen by visual inspection of the residuals produced by PTF estimates plotted vs. the independent variables; two, four, or five groups could have also been used according to the definition of PI (Donatelli et al., 2004), but according to the preliminary analysis they did not show the same discriminating power among PTFs given by the three groups option. PI values have the same units as the output variable under study (in this case $m^3\,m^{-3}$).

The pattern indices are targeted at pointing out macro-patterns. The presence of patterns usually means that the residuals contain structure that is not accounted for in the model. When applied to different types of plots of the residuals, pattern indices may provide meaningful information on the adequacy of different aspects of the model. In the PTF evaluation PI allows to find out whether the goodness of the method changes according to the input value. We refer here to pattern indices computed against D50 (PI_{D50}) and OC (PI_{OC}) because of estimate residuals from pedotransfer functions often show non-random distribution of residuals over the range of such variables. In principle, if residuals show a pattern with respect to other than D50 and OC independent variables available to characterize each measurement, such variables can also be used to define a basic PI. Hence, quantitative evaluation of pattern allows both setting limits of method application within the ranges of the input variables and comparing methods. If water retention curves are evaluated, a third PI should be computed against the soil water potential (PI_ψ), as methods for estimating the soil water retention curve may show a different fit at different potentials. To account for the non-linearity of soil water retention with respect to water potential, the logarithm of the water potential is used to compute PI_ψ, thus the sub-ranges become 5–50, 50–280, 280–1500 kPa.

Whereas modules "Accuracy" and "Correlation" are the same for all the integrated indices presented, the module "Pattern" is different according to the parameter estimated by pedotransfer functions. In fact, the module "Pattern" is made of two basic indices (PI_{D50} and PI_{OC}) for methods which estimate single values (e.g., soil water content at -33 and -1500 kP), and of three basic indices (PI_{D50}, PI_{OC} and PI_ψ) for methods which estimate the soil water retention curve considering measurements at different potentials.

The limits for F and U were set with the same procedure of the paragraph above. The limit to the fuzzy subset F for PI_{D50} was set to 0.030 ($\text{PI}_{\text{D50}} \leq 0.030$ corresponds to F, units are volumetric water content), while the limit to the subset U was set to 0.060 ($\text{PI}_{\text{D50}} \geq 0.060$ corresponds to U). The limits for PI_{OC} are 0.050 ($\text{PI}_{\text{D50}} \leq 0.050$ corresponds to F) and 0.080 ($\text{PI}_{\text{D50}} \leq 0.080$ corresponds to F). The difference is due to the pedotransfer function performance with respect to soil organic carbon content, which appeared to be worse than with respect to particle median diameter. The performance was evaluated by comparing the values of the two PIs, for each method, in a first explorative PI computation. The limits for PI_ψ are 0.030 ($PI_{D50} \leq 0.030$ corresponds to F) and 0.060 ($PI_{D50} \leq 0.060$ corresponds to F). F and U limits can be changed easily by using a dedicated free software (http://www.sipeaa.it/tools, page "IRENE_dll"); this might be advisable when working with a much larger dataset.

The value of the module "Pattern" for methods that estimate point values depends on the input indices according to decision rules summarized in Table 4. The same weight was attributed to PI_{D50} and PI_{OC}. If all indices are F, the value of the module is 0 (i.e., no pattern); if all indices are U, the expert weight is 1; if one index is F and the other is U, the weight is 0.50.

Table 4
Summary of decision rules describing the effect of the input indices PI_{D50}, PI_{OC}, and PI_ψ on the module "Pattern"

	Membership class		
PI_{D50}	PI_{OC}	PI_ψ	Expert weight
Module pattern in I_{PTFSW}			
F	F		0.00
F	U		0.50
U	F		0.50
U	U		1.00
Module pattern in I_{PTFRC}			
F	F	F	0.00
F	F	U	0.40
F	U	F	0.30
F	U	U	0.70
U	F	F	0.30
U	F	U	0.70
U	U	F	0.60
U	U	U	1.00

F, favorable, U, unfavorable (see text for details).

The value of the module "Pattern" for methods that estimate soil water retention curves depends on the input indices according to decision rules summarized in Table 4. The same weight was attributed to PI_{D50} and PI_{OC} (0.3); the value attributed to PI_ψ is consequently 0.4. If all indices are F, the value of the module is 0 (i.e., no pattern); if all indices are U, the expert weight is 1. If either PI_{D50} or PI_{OC} is U and PI_ψ is F the weight is 0.3. If both PI_{D50} and PI_{OC} are U, and PI_ψ is F, the weight is 0.6. If either PI_{D50} or PI_{OC} is U and PI_ψ is U the weight is 0.7. Finally, if both PI_{D50} and PI_{OC} are F, and PI_ψ is U, the weight is 0.4.

2.1.4. Aggregation of modules

The three modules described above can be used to compare different pedotransfer functions by aggregating the modules (second-level aggregation) into overall indicators, I_{PTFSW} and I_{PTFRC}. Such indicators reflect a global judgment about method performance, again on a 0 to 1 scale. The aggregation of the three modules, which uses decision rules, is done as described above for the aggregation of indices into the modules.

The value of the indicator I_{PTFSW} and I_{PTFRC} depend on the modules "Accuracy", "Correlation", and "Pattern" according to a set of eight decision rules (Table 5). The definition of the limits of the transition interval is the same for the three modules: we assigned complete membership to F if the value of the module is 0 (i.e., identity of

Table 5
Summary of decision rules describing the effect of the three modules on the value of the indicators I_{PTFSW} and I_{PTFRC}

	Membership class		
"Accuracy"	"Correlation"	"Pattern"	Expert weight
F	F	F	0.00
F	F	U	0.35
F	U	F	0.15
F	U	U	0.50
U	F	F	0.50
U	F	U	0.85
U	U	F	0.65
U	U	U	1.00

estimates and measurements, unit correlation, no pattern of residuals vs. independent variables), and complete membership to U if the value of the module is 1. In setting up the other decision rules we had to establish the relative importance of each module. As a general rule in model evaluation more emphasis is given to the residuals, whereas the correlation of estimates vs. measurements carries less weight. Because of their recent development, pattern indices are rarely used in model evaluation. Based on previous experience, the importance of modules decreases in the sequence "Accuracy", "Pattern", and "Correlation". Thus, for instance, if "Accuracy" is U, then the weight is 0.50. If both "Accuracy" and "Pattern" are U, then the expert conclusion is 0.85. If both "Accuracy" and "Correlation" are U, then the conclusion is 0.65. The relative effect of each index on the indicator can be deduced by combining the weights of the indices into their own module with the ones of the modules into the indicator (Table 6).

2.1.5. The soil data set

Soils to test the methodology were selected from the HYPRES database (Wösten et al., 1999). The database was searched for soils for which data on soil texture, organic matter content, bulk density, saturated hydraulic conductivity (K_s) and soil water retention measured at a minimum of 6 matric potentials were available. From the selected soils only those that had water retention measured at saturation, at -33 kPa and at (or close to) -1500 kPa were retained. Data were filtered to exclude samples with obvious inconsistency in physical data. Basic statistics of selected properties of the remaining data set ($N = 53$) are provided in Table 7. The selected soils represent a wide range of soils in all selected physical and hydraulic properties. Figure 1 shows the textural distribution of the soils in the texture triangle. According to the USDA classification, 9 of the 12 texture groups are represented in this selection. The sand, loamy sand and clay loam classes are better represented than the other six (loam, sandy loam, sandy clay loam, silty

Table 6
Relative incidence of each basic evaluation index on the value of the indicators I_{PTFSW} and I_{PTFRC}

Index	Relative incidence on I_{PTFSWC}
Index I_{PTFSW}	
RRMSE	$0.5 \times 0.50 = 0.250$
EF	$0.5 \times 0.50 = 0.250$
r	$1.0 \times 0.15 = 0.150$
PI_{D50}	$0.5 \times 0.35 = 0.175$
PI_{OC}	$0.5 \times 0.35 = 0.175$
Index I_{PTFRC}	
RRMSE	$0.5 \times 0.50 = 0.250$
EF	$0.5 \times 0.50 = 0.250$
r	$1.0 \times 0.15 = 0.150$
PI_{D50}	$0.3 \times 0.35 = 0.105$
PI_{OC}	$0.3 \times 0.35 = 0.105$
PI_ψ	$0.4 \times 0.35 = 0.140$

Table 7
The soil data set

Variable	Unit	Min	Max	Mean	Std. dev.	Median
Sand$_{(0.05-2 \text{ mm})}$	(g g^{-1})	0.041	0.954	0.588	0.287	0.623
Silt$_{(0.002-0.05 \text{ mm})}$	(g g^{-1})	0.019	0.679	0.234	0.154	0.222
Clay$_{(<0.002 \text{ mm})}$	(g g^{-1})	0.011	0.495	0.178	0.154	0.159
ρ_b	(g cm^{-3})	1.120	1.780	1.511	0.140	1.525
Org. carbon	(%)	0.104	1.676	0.445	0.336	0.354
K_s	(ln(cm^{-1}))	-1.897	6.250	2.876	2.823	4.203
θ_s	(m^3 m^{-3})	0.2305	0.5300	0.3599	0.0711	0.3415
$\theta_{(-33 \text{ kPa})}$	(m^3 m^{-3})	0.0535	0.3675	0.1919	0.0957	0.1958
$\theta_{(-1500 \text{ kPa})}$	(m^3 m^{-3})	0.0000	0.2400	0.0749	0.0461	0.0701

clay loam, silty clay and clay) classes. Soil hydraulic data were obtained using various techniques. Water retention in the wet range was characterized using tension tables, ceramic plates or the hanging water column technique. In the dry range pressure chambers were used along with vapor equilibrium techniques. Saturated hydraulic conductivity was determined in the field using Guelph permeameter, crust infiltrometer or ring infiltrometer. The data sets were obtained from different sources and had different numbers and positions of points at the water retention curve. To obtain uniform description of all the water retention curves, the volumetric soil water content, θ, as a function of matric potential, h, the equations of van Genuchten (1980) were fitted to the data.

2.1.6. The pedotransfer functions evaluated

Several pedotransfer functions are used to illustrate the development and use of integrated indices for pedotransfer evaluation. They are a subset of PTFs implemented in

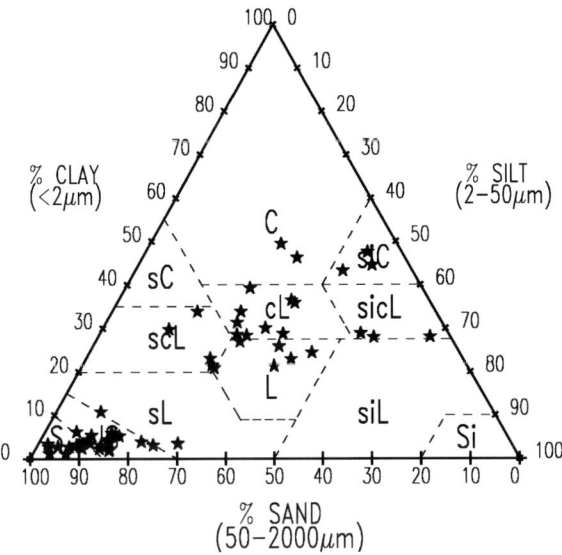

Figure 1. Particle size of soil samples in this study.

the software *SOIL*PAR 2.00 (Acutis and Donatelli, 2003). During the development of *SOIL*PAR, the estimates made using each method were compared with implementations made in already available software (if available), in order to avoid errors found in the original papers describing the methods (Gijsman et al., 2002). The equations of the pedotransfer functions used are reported below. The pedotransfer inputs and outputs are reported in Table 8. Refer to Table 8 for the acronym explanations.

The following methods reproduce the results of the LeachM model implementation (Hutson and Wagenet, 1992):

(1) Brakensiek–Rawls
   The water content at −33 kPa is calculated as:

$$SWC_{-33} = 0.3486 - 0.0018\text{sand} + 0.0039\text{clay} + 0.0228\text{OM} - 0.0738\text{BD}$$

The water content at −1500 kPa is calculated as:

$$SWC_{-1500} = 0.0854 - 0.0004\text{sand} + 0.0044\text{clay} + 0.0122\text{OM} - 0.0182\text{BD}$$

(2) Hutson
   The water content at −33 kPa is calculated as:

$$SWC_{-33} = \text{Exp}(-3.43 + 0.419(\text{clay} + \text{silt})^{0.5} - 1.83 \times 0.001(\text{clay} + \text{silt})^{1.5})$$

The water content at −1500 kPa is calculated as:

$$SWC_{-1500} = \text{Exp}(-4.384 + 0.404(\text{clay} + \text{silt})^{0.5} - 9.85 \times 0.0000001(\text{clay} + \text{silt})^{3})$$

Table 8
Compared PTFs

Method	Variables requested	Parameter estimated	Source
Brakensiek/Rawls	PSD, OC, BD	SWC	LeachM (Hutson and Wagenet, 1992)
Hutson British Soil Survey topsoil	PSD, OC, BD	SWC	LeachM (Hutson and Wagenet, 1992)
British Soil Survey topsoil	PSD, OC, BD	SWC	LeachM (Hutson and Wagenet, 1992)
Baumer	PSD, OC	BD, FC WP	ASW/EPIC
Rawls	PSD	BD, FC WP	ASW/EPIC
Manrique	PSD, BD	FC WP	ASW/EPIC
Campbell	PSD, BD	Campbell retention function parameters	Campbell, 1985
Mayr-Jarvis	PSD, OC, BD	Hutson-Cass retention function parameters	Mayr and Jarvis, 1999
Rawls/Brakensiek	PSD, Porosity	Brooks–Corey retention function parameters	Rawls and Brakensiek, 1989
Vereecken	PSD, OC, BD	Van Genuchten retention function parameters	Vereecken et al., 1989
HYPRES	PSD, OM, BD	Van Genuchten retention function parameters	Wösten et al., 1999
EUR-M3	PSD, OM, BD	Van Genuchten retention function parameters	Nemes et al., 2003

PSD, particle size distribution; OC, organic carbon (%); OM, organic matter (%); BD, bulk density (t m^{-3}); FC, soil water content at field capacity (m^3 m^{-3}); WP, soil water content at wilting point (m^3 m^{-3}); SWC, soil water content at several pressures; Porosity: the authors propose to obtain porosity from (2.65-BD)/2.65.

(3) British soil service, topsoil

The method does not estimate directly the SWC at -33 kPa, but at -20, -40, -200, and -1500 kPa. Therefore, in *SOIL*PAR, the -33 kPa value is obtained via an interpolation done using the 2nd degree polynomial passing for the three known points closest to the unknown one. The water content at -33 kPa is interpolated from the water content at -20, -40 and -200 kPa:

$$SWC_{-20} = 0.403 + 0.0034\text{clay} + 0.0013\text{silt} + 0.004\text{OC} - 0.125\text{BD}$$

$$SWC_{-40} = 0.2668 + 0.0039\text{clay} + 0.0013\text{silt} + 0.0046\text{OC} - 0.0764\text{BD}$$

$$SWC_{-200} = 0.0938 + 0.0047\text{clay} + 0.0011\text{silt} + 0.0069\text{OC}$$

The water content at -1500 kPa is calculated as:

$$SWC_{-1500} = 0.0611 + 0.004\text{clay} + 0.0005\text{silt} + 0.005\text{OC}$$

These methods reproduce the implementation of the EPIC-ASW Utilities:

(4) Baumer

A detail of the procedure is not given here, because they are based on large tables. This procedure considers the PSD, and the CEC of a soil. If CEC was not available, it was estimated from clay content.

(5) Rawls

$$SWC_{-33} = 0.2756 + 0.64 SWC_{-1500} - 0.0013\text{sand} + 0.00217\text{BD} \times \text{clay} - 0.00196\text{BD}^2 \times \text{clay}$$

$$SWC_{-1500} = -0.0208 + 0.007\text{clay} - 0.00003\text{clay}^2 + 0.0224\text{BD}^2 \times \text{CEC}/\text{clay}$$

(6) Manrique

$$\begin{cases} SWC_{-33} = 0.73426 - \text{sand}0.00145 - \text{BD}0.29176 & |\text{sand} \geq 75 \\ SWC_{-33} = 0.5784 + \text{clay}0.00227 - \text{BD}0.28438 & |\text{sand} < 75 \end{cases}$$

$$SWC_{-1500} = 0.02413 + \text{clay}0.00373$$

The following methods were not implemented in free software, and are originally implemented in *SOIL*PAR:

(7) Campbell method (Campbell, 1985)

This method estimates the coefficient of the Campbell retention function on the basis of geometric mean and standard deviation of the particle size and bulk density, using the following relation:

$$\psi_m = \psi_e \left(\frac{\theta_{act}}{\theta_s} \right)^{-b}$$

where ψ_m is the pressure corresponding to θ_{act}; ψ_e the pressure at air entry point; b the empirical coefficient; θ_e the soil water content at saturation; θ_{act} actual soil water content.

Their parameters are estimated according to the following equation:

$$b = 2 \times 0.5_{d_{50}}^{-1/2} + 0.2\sigma_g$$

$$\psi_e = -0.5_{d_{50}}^{-1/2}(BD/1.3)^{0.67b}$$

where d_{50} is the median diameter and σ_g the geometric standard deviation of particles diameter.

The SWC values at -33 and -1500 kPa are obtained by the Campbell function evaluation.

(8) Mayr–Jarvis (1999) method

This method estimates the parameters of the Hutson–Cass retention function (a modification of the Brooks – Corey retention function) using sand, clay, organic matter, and bulk density. This function is calibrated on the basis of sand in the interval 2–0.063 mm, silt from 0.063 to 0.002 mm and clay < 0.002. In our example, because the data are expressed using the three class USDA classification, we have used an option of *SOIL*PAR 2.00 that converts from the USDA classification scheme to the required classification scheme under the hypothesis of log normal distribution of particle size.

The Huston–Cass (Hutson and Cass, 1987) retention function is:

$$\begin{cases} \psi_m = \dfrac{a(1 - \theta_{act}/\theta_s)^{1/2}(\theta_c/\theta_s)^{-b}}{(1 - \theta_c/\theta_s)^{1/2}} & \text{if } 0 \leq \psi_m \leq \psi_c \\ \psi_m = \psi_e\left(\dfrac{\theta_{act}}{\theta_s}\right)^{-b} & \text{if } \psi_m > \psi_c \end{cases}$$

where

$$\psi_m = \psi_e[2b/(1+2b)]^{-b}$$

$$\theta_c = 2b\theta_s/(1+2b)$$

and the symbols have the same meaning of those used in the Campbell's retention function

The Huston–Cass parameters are estimated with the following relation:

$$\psi_e = \exp(-4.9840297533 + 0.0509226283\text{sand} + 0.1575152771\text{silt}$$
$$+ 0.1240901644\text{BD} - 0.1640033143\text{OC} - 0.0021767278\text{silt}^2$$
$$+ 0.0000143822\text{silt}^3 + 0.0008040715\text{clay}^2 + 0.0044067117\text{OC}^2)$$

$$b = 1/\exp(-0.8466880654 - 0.0046806123\text{sand} + 0.0092463819\text{silt}$$
$$- 0.4542769707\text{BD} - 0.0497915563\text{OC} + 0.0003294687\text{sand}^2$$
$$- 0.000001689056\text{sand}^3 + 0.0011225373\text{OC}^2)$$

The SWC at -33 and -1500 kPa are obtained by the Huston–Cass function evaluation.

(9) Rawls–Brakensiek (Rawls and Brakensiek, 1989)

This method estimates the parameters of the Brooks–Corey retention function using sand, clay, and bulk density. The Brooks–Corey retention function is the following:

$$\theta(h) = \theta_r + \frac{\theta_s - \theta_r}{(h_b/h)^\lambda}$$

where h_b is the air entry value; λ is the pore size index; θ_r is the residual water content. And their parameters are estimated using the following equations:

$$H_b = \text{Exp}(5.34 + 0.184\text{clay}/100 - 2.484\text{porosity} - 0.00214(\text{clay}/100)^2$$
$$- 0.0436\text{sand}/100\text{porosity} - 0.617\text{clay}/100\text{porosity}$$
$$+ 0.00144(\text{sand}/100)^2\text{porosity}^2 - 0.00855(\text{clay}/100)^2\text{porosity}^2$$
$$- 0.0000128(\text{sand}/100)^2\text{clay}/100 + 0.00895(\text{clay}/100)^2\text{porosity}$$
$$- 0.000724(\text{sand}/100)2\text{porosity} + 0.0000054(\text{clay}/100)^2\text{sand}/100$$
$$+ 0.5\text{porosity}^2\text{clay}/100)$$

$$\lambda = \text{Exp}(-0.784 + 0.0177\text{sand}/100 - 1.062\text{porosity} - 0.000053(\text{sand}/100)^2$$
$$- 0.00273(\text{clay}/100)^2 + 1.111\text{porosity}^2 - 0.0309\text{sand}/100\text{porosity}$$
$$+ 0.000266(\text{sand}/100)^2\text{porosity}^2 - 0.0061(\text{clay}/100)^2\text{porosity}^2$$
$$- 0.00000235(\text{sand}/100)^2\text{clay}/100 + 0.00799(\text{clay}/100)^2\text{porosity}$$
$$- 0.00674\text{porosity}^2\text{clay}/100)$$

$$\theta_r = -0.0182 + 0.000873\text{sand}/100 + 0.00513\text{clay}/100 + 0.0294\text{porosity}$$
$$- 0.000154(\text{clay}/100)^2 - 0.00108\text{sand}/100\text{porosity} - 0.000182(\text{clay}/100)^2\text{porosity}^2$$
$$+ 0.000307(\text{clay}/100)^2\text{porosity} - 0.00236\text{porosity}^2\text{clay}/100$$

The following PTFs evaluate the parameters of the van Genuchten retention function:

$$\theta(h) = \theta_r + \frac{\theta_s - \theta_r}{[1 + (\alpha h)^n]^m}$$

(10) Vereecken (Vereecken et al., 1989)

$\theta_s = 0.803 - 0.283\text{BD} + 0.0013\text{clay}$

$\theta_r = 0.015 + 0.005\text{clay} + 0.014\text{OC}$

$\ln(\alpha) = -2.486 + 0.025\text{sand} - 0.351\text{clay}$

$\ln(n) = -0.035 - 0.009\text{sand} - 0.013\text{clay} + 0.015\text{sand}^2$

$m = 1$

(11) HYPRES (Wösten et al., 1999)

$\theta_s = 0.7919 + 0.001691\text{clay} - 0.29619\text{BD} - 0.000001491\text{silt}^2 + 0.0000821\text{OM}^2$
$+ 0.02427\text{clay}^{-1} + 0.01113\text{silt}^{-1} + 0.01472\ln(\text{silt}) - 0.0000733\text{OM} \times \text{clay}$
$- 0.000619\text{BD} \times \text{clay} - 0.001183\text{BD} \times \text{OM} - 0.0001664\text{topsoil} \times \text{silt}$

$\ln(\alpha) = -14.96 + 0.03135\text{clay} + 0.0351\text{silt} + 0.646\text{OM} + 117.29\text{BD} - 0.192\text{topsoil}$
$- 4.671\text{BD}^2 - 0.000781\text{clay}^2 - 0.00687\text{OM}^2 + 0.0449\text{OM}^{-1} + 0.0663\ln(\text{silt})$
$+ 0.1482\ln(\text{OM}) - 0.04546\text{BD} \times \text{silt} - 0.4852\text{BD} \times \text{OM} + 0.00673\text{topsoil} \times \text{clay}$

$\ln(n - 1) = -217.23 - 0.02195\text{clay} + 0.0074\text{silt} - 0.1940\text{OM} + 417.5\text{BD} - 7.24\text{BD}^2$
$+ 0.0003658\text{clay}^2 + 0.002885\text{OM}^2 - 12.81\text{BD}^{-1} - 0.1524\text{silt}^{-1}$
$- 0.01958\text{OM}^{-1} - 0.2876\ln(\text{silt}) - 0.0709\ln(\text{OM}) - 44.6\ln(\text{BD})$
$- 0.02264\text{BD} \times \text{clay} + 0.0896\text{BD} \times \text{OM} + 0.00718\text{topsoil} \times \text{clay}$

$\theta_r = 0.1$

$m = 1 - 1/n$

(12) EUR-M3 (Nemes et al., 2003)

EUR-M3 is a pedotransfer model that was developed from a subset ($N = 2464$) of the European HYPRES database (Wösten et al., 1999) using neural networks (NN). A NN model consists of many simple computing elements (neurons or nodes) that are organized into subgroups (layers) and are interconnected as a network by weights. The weight matrices are obtained through a calibration (training) procedure, which can then be used to make estimations on independent data. For a more thorough description on NNs, we refer the reader to Hecht-Nielsen (1990). Nemes et al. (2003) used a three-layer back-propagation NN model. The number of nodes in the hidden layer was set to six. Model EUR-M3 estimates water retention – through the parameters of the van Genuchten equation (van Genuchten, 1980) – from sand, silt and clay content, bulk density and the organic matter content of the soil.

NNs can be combined with the data selection procedure of the bootstrap method (Efron and Tibshirani, 1993) to generate internal calibration–validation data set pairs for an early stopping procedure. The bootstrap method is a non-parametric technique that simulates alternative (replica) data sets out of a single data set by random selection with replacement. Multiple realizations of subsets can help to avoid bias towards any particular calibration–validation data set pairs. In this case, 50 replica data sets were generated, each of which was used to calibrate the NN models. The final estimate of the PTF for each value was then calculated by averaging the 50 substimates of the value.

2.2. Evaluating pedotransfer functions using the integrated indices

Aggregated indices are not meant to replace single indices; instead they must be used as summary indices for preliminary evaluations, in method comparisons, and possibly as cost functions in iterative procedures. Different modules allow creating information, which can be extremely useful in understanding why a given pedotransfer function does not perform well. As an example, a medium value for an integrated index that is the result of a large value for the module accuracy, and close to 0 for the module pattern, provides indication that the method performs consistently across the range of the independent variables evaluated via the PIs, but with a large variance in estimates. Another example, a medium value for an integrated index which is the result of a very small value for the module accuracy, a noticeable overall value of the module PI, a close to 0 value for the PI_{D50}, and finally a large value for PI_{OC} informs that the method performs well for soils within a specific sub-range of soil organic carbon content. A good performance with respect to a PI (values close to 0) does not necessarily mean that the independent variable (D50 or OC in our study) is specifically accounted for by the method, i.e., that the method has such variables as inputs. Other independent variables included in the method may surrogate D50 or OC, but in this case the use of a different dataset with a larger range of values for D50 and OC may lead to less favorable values of the PIs.

The indices computed for the pedotransfer functions described in Section 2.1.6 are discussed to illustrate the integrated indices applied to pedotransfer functions, and should not be viewed as a comparison of the pedotransfer functions used in this analysis. In this respect, none of the basic indexes calculated in this study should be seen in terms of their absolute values with respect to the specific PTF; instead, they should be seen as example results to illustrate the use of integrated indicators. The results presented below are related to the data set used, and they are presented for illustrative purposes only. A larger data set, screened prior to using PTFs according to the rules reported in Section 1, may produce results that are different with respect to methods performance.

2.2.1. The index I_{PTFSW}

This index can be used for the evaluation of pedotransfer functions which estimate either a single value of soil volumetric water content (SWC_θ) such as $SWC_\theta FC$ or $SWC_\theta PWP$.

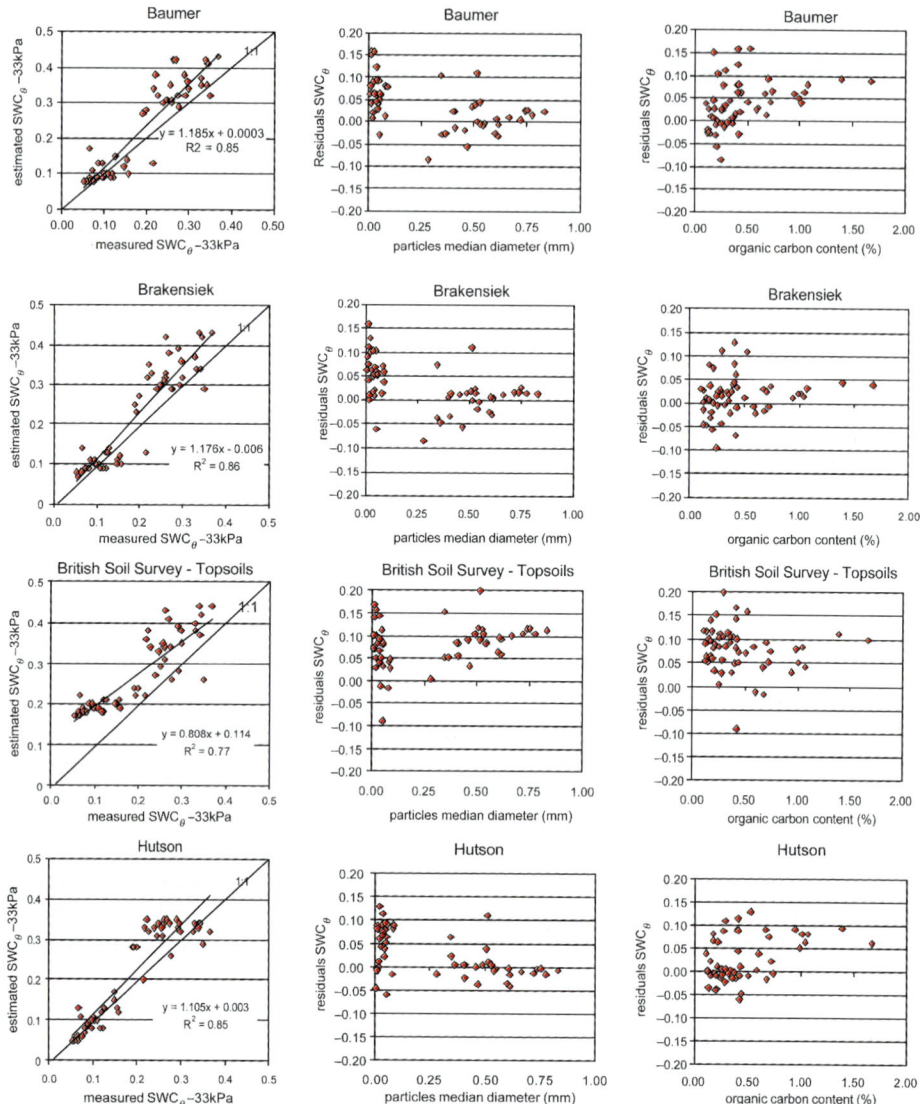

Figure 2. Pedotransfer estimates of volumetric soil water content at −33 kPa vs. measured values (on the left), residuals vs. soil particles median diameter (center), and residuals vs. soil organic carbon (right).

The graphs of Figure 2 show the pedotransfer performance for SWC_θ at −33 kPa as estimated vs. measured values (graphs on the left column). The graphs of residuals vs. D50 (center column) and vs. OC (right column) are also shown. Figure 3 shows the same graphs for SWC_θ at −1500 kPa. With reference to the basic indices used to build the integrated index, graphs on the left contain an indication about the RMSE (dispersion of data around

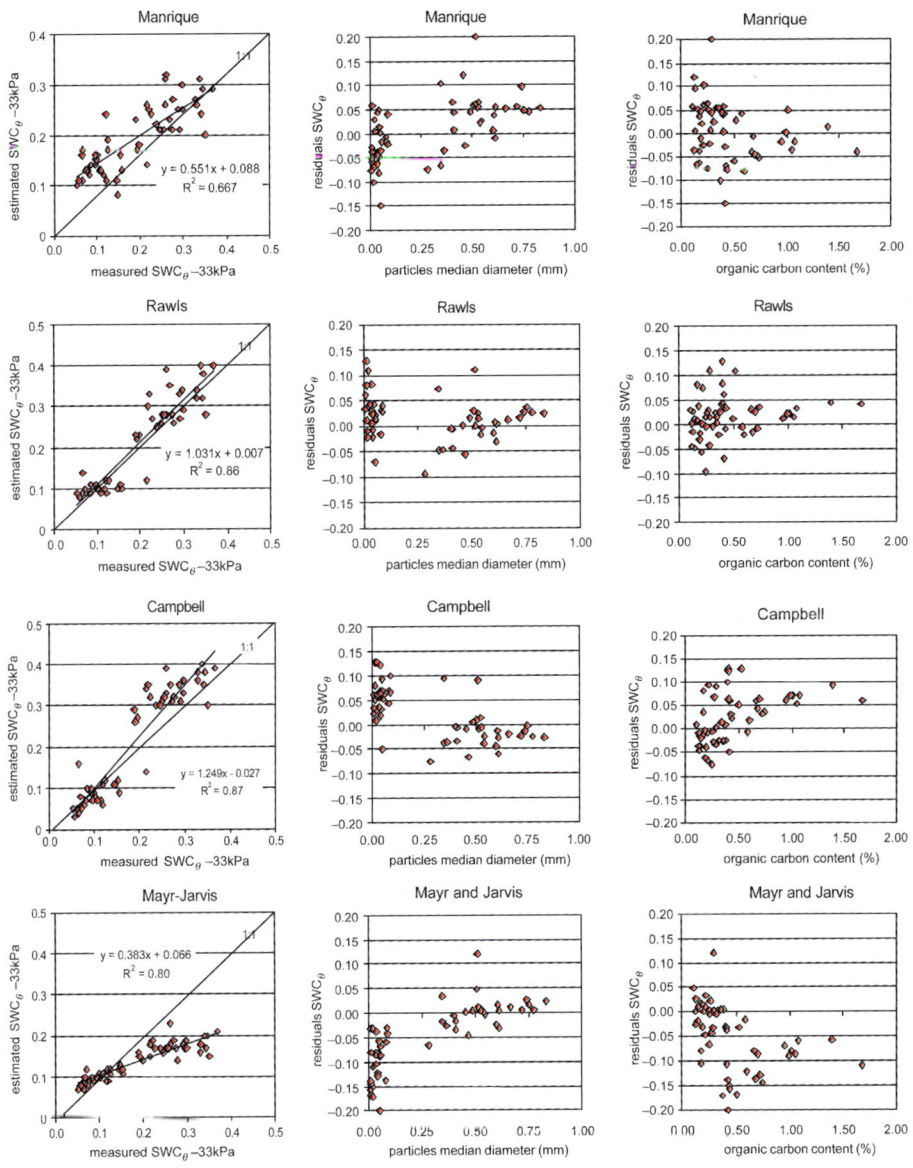

Figure 2 (*continued*)

the linear fitting), and the slope (correlation estimated vs. measured), whereas no clear indication about the value of EF is given. The data on graphs on the center and right columns allow a rough qualitative assessment of PIs (that is the difference of mean values of residuals grouped in three subgroups according to D50 and OC). The basic indices are computed for each pedotransfer function, and, finally, the integrated index is computed. The data of basic indices, modules, and of the integrated index are reported in Table 9.

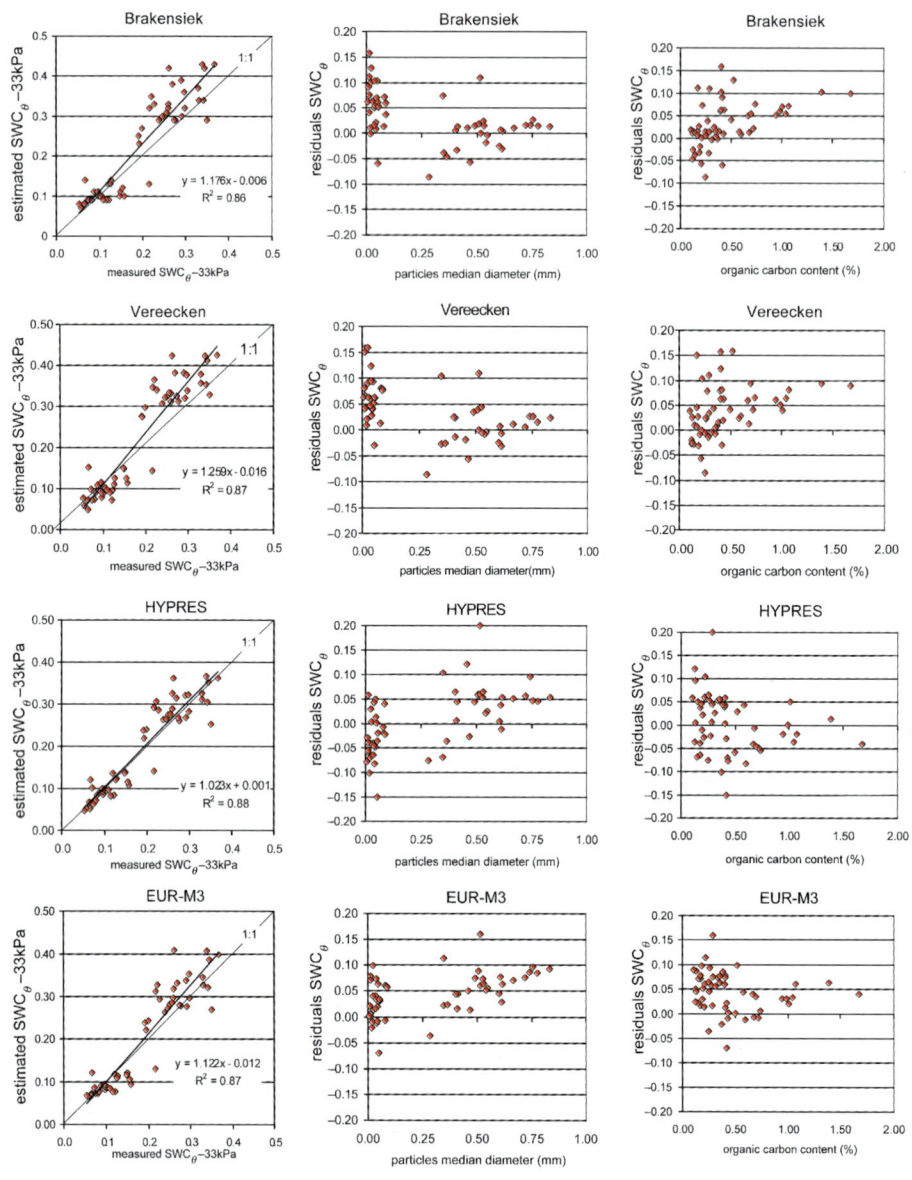

Figure 2 (*continued*)

The performance of different methods in estimating SWC at −1500 kPa was remarkably worse than the one at −33 kPa. Only one method showed a modeling efficiency greater than 0, meaning that for all other methods the mean of the measured values is a better predictor than the pedotransfer function estimate. The negative impact of the basic indices of the module "Accuracy" on the integrated index is evident. It must be pointed out that values for

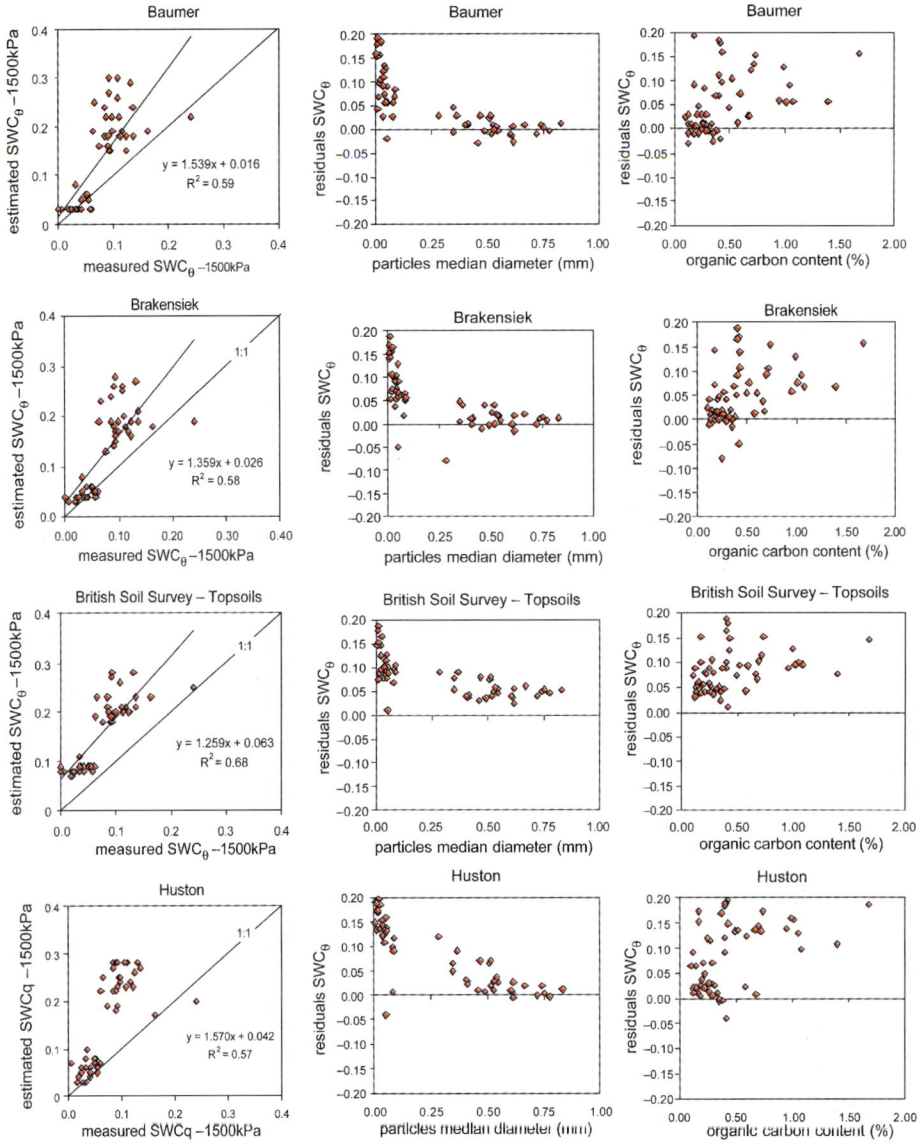

Figure 3. Pedotransfer estimates of volumetric soil water content at −1500 kPa vs. measured values (on the left), residuals vs. soil particles median diameter (center), and residuals vs. soil organic carbon (right).

the integrated index should be smaller than 0.5 to be acceptable, meaning that in this case no pedotransfer function for estimating soil water content at −1500 kPa is acceptable according to the fuzzy limits chosen. This is qualitatively confirmed by the graphs of Figure 3, even ignoring the component due to the pattern of residuals.

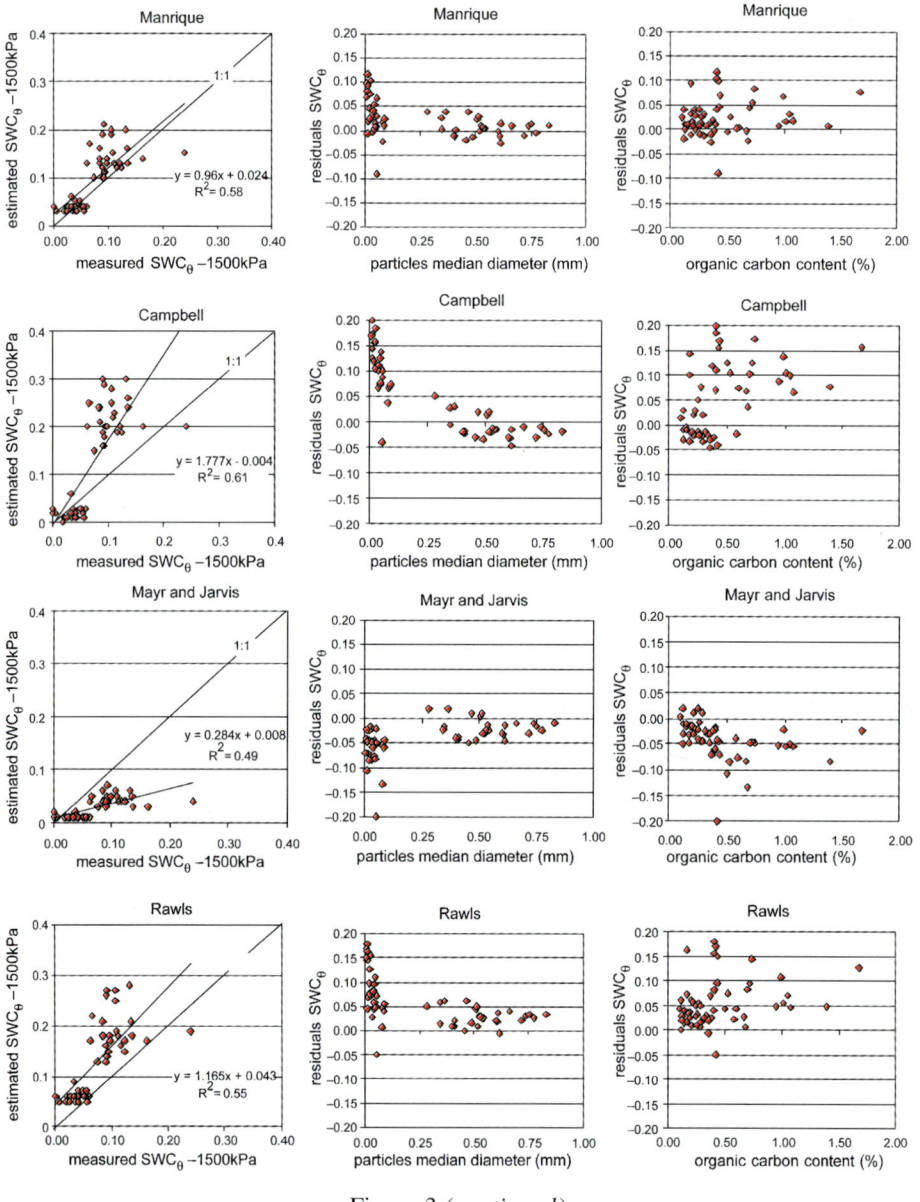

Figure 3 (*continued*)

In general, most pedotransfer functions showed a pattern in the residuals with respect to both soil particle median diameter and soil organic carbon. The values of Table 9 can be evaluated with a top-down approach, meaning considering the integrated index first, then the values of the modules, and finally the basic indices. This methodology allows understanding both strong and weak points for each pedotransfer function, suggesting

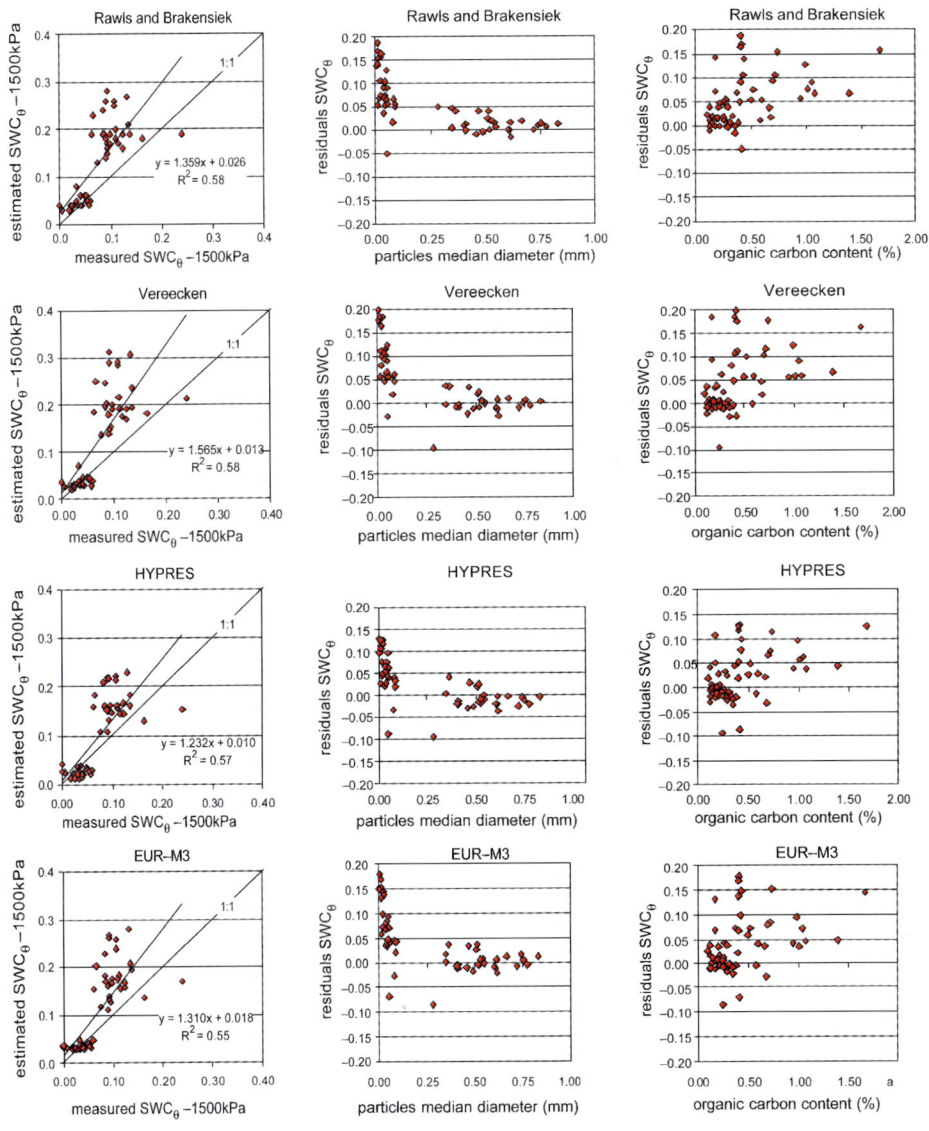

Figure 3 (*continued*)

further testing and/or areas which require an improvement. The values in bold, for group of 12 rows (-33 and -1500 kPa), are the most favorable, the lowest values being the best for RRMSE, PI_{D50} and PI_{OC}, all module values, and I_{PTFSWC}. The best value for EF and r is the highest (optimum value is 1). The procedure shown can also be used to develop other integrated indices related to pedotransfer functions, such us I_{PTFKs} (for saturated conductivity) and I_{PTFBD} (for bulk density).

Table 9

Values of basic indices (RRMSE, EF, r, PI_{D50}, PI_{OC}, PI_ψ), modules (Accuracy, Correlation, Pattern), and integrated index (I_{PTFRC}) to evaluate pedotransfer functions. Values in bold refer to the best functions

Pedotransfer Function	PRMSE	EF	r	PI_{OC}	PI_{D50}	Mod Acc	Mod Corr	Mod PI_{SWC}	I_{PTFSWC}
				−33 kPa					
Baumer	32.7		0.920	0.0637	0.0661	0.008	**0.000**	0.709	0.291
Brakensiek	29.5		0.926	0.0827	0.0642	**0.000**	**0.000**	1.000	0.350
British Soil Survey Topsoils	46.8		0.879	0.0535	**0.0304**	0.698	0.022	0.014	0.409
Hutson	26.8		0.921	0.0633	0.0630	**0.000**	**0.000**	0.698	0.286
Manrique	29.7		0.818	0.0310	0.0716	**0.000**	0.340	0.500	0.222
Rawls	29.2		0.935	**0.0275**	0.0454	**0.000**	**0.000**	0.264	0.049
Campbell	29.1		0.933	0.0672	0.0844	**0.000**	**0.000**	0.819	0.327
Mayr–Jarvis	42.2		0.896	0.0562	0.1051	0.407	0.001	0.543	0.392
Rawls–Brakensiek	22.3		0.925	0.0310	0.0380	**0.000**	**0.000**	0.070	0.003
Vereecken	32.5		0.935	0.0800	0.0783	0.007	**0.000**	1.000	0.350
HYPRES	**19.2**		**0.938**	0.0443	0.0309	**0.000**	**0.000**	**0.001**	**0.000**
EUR-M3	23.4		0.932	0.0518	0.0513	**0.000**	**0.000**	0.420	0.124
				−1500 kPa					
Baumer	116.0	−2.334	0.724	0.0667	0.1027	1.000	0.972	0.803	0.972

Brakensiek	104.8	−1.725	0.724	0.0734	0.0840	1.000	0.970	0.952	0.997
British Soil Survey Topsoils	125.2	−2.883	**0.783**	0.0400	0.0640	1.000	**0.656**	0.500	0.777
Hutson	146.1	−.4289	0.732	0.0808	0.1261	1.000	0.950	1.000	0.999
Manrique	**60.6**	**0.088**	0.725	0.0268	**0.0349**	**0.969**	0.970	**0.027**	**0.648**
Rawls	38.4	−0.938	0.728	0.0656	0.0949	1.000	0.960	0.770	0.961
Campbell	125	−2.871	0.747	0.0835	0.1356	1.000	0.891	1.000	0.996
Mayr–Jarvis	78.9	−0.542	0.676	**0.0181**	0.0403	1.000	1.000	0.117	0.660
Rawls–Brakensiek	101.2	−1.538	0.701	0.0401	0.0607	1.000	1.000	0.500	0.825
Vereecken	118.0	−2.452	0.719	0.0768	0.1055	1.000	0.982	0.988	1.000
HYPRES	78.9	−0.542	0.722	0.0708	0.0733	1.000	0.975	0.905	0.993
EUR-M3	95.1	−1.242	0.709	0.0660	0.0766	1.000	0.996	0.783	0.967
Maximum	146.1	0.853	0.938	0.0835	0.1356				
Median	53.7	0.106	0.800	0.0635	0.0689				
Minimum	19.2	−4.289	0.676	0.0181	0.0304				
Limit U	50.0	0.000	0.700	0.0800	0.0600				
Limit F	30.0	0.500	0.900	0.0500	0.0300				

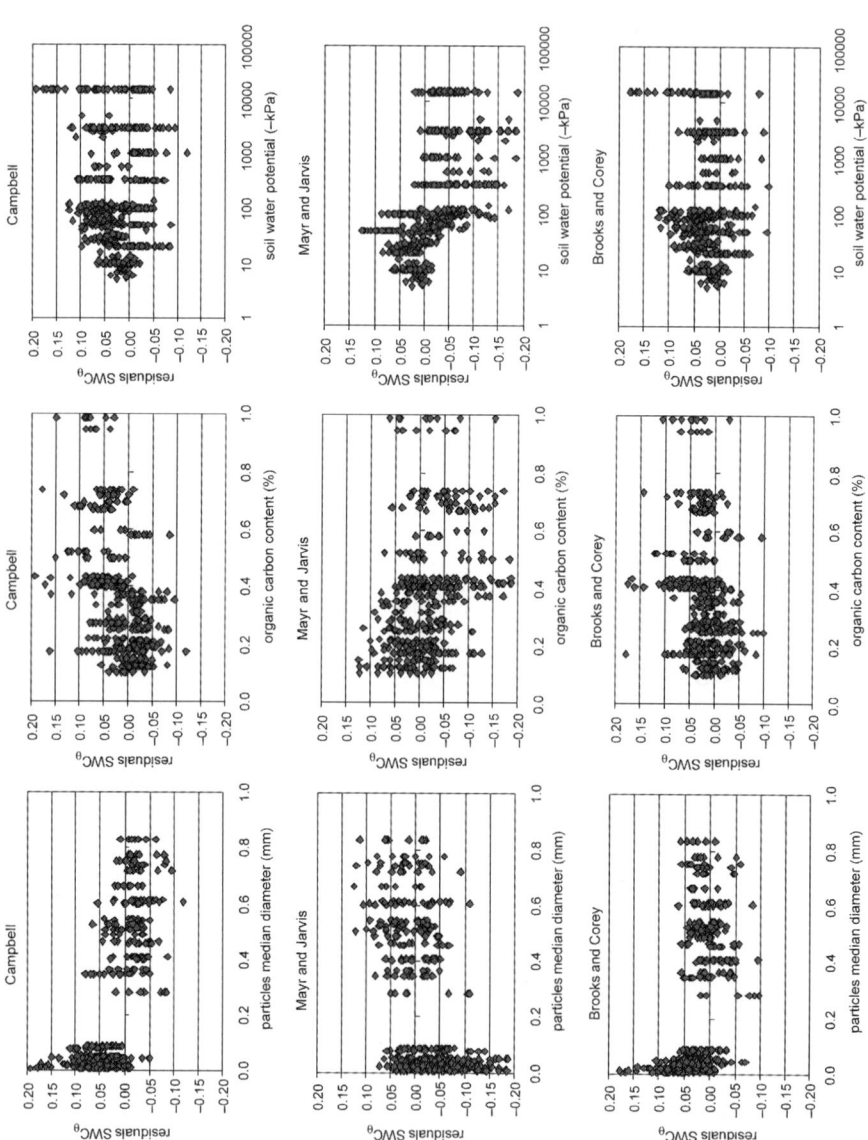

Figure 4. Residuals of pedotransfer estimates vs. soil particles median diameter (left), vs. soil organic carbon (center), and vs. soil water tension (right).

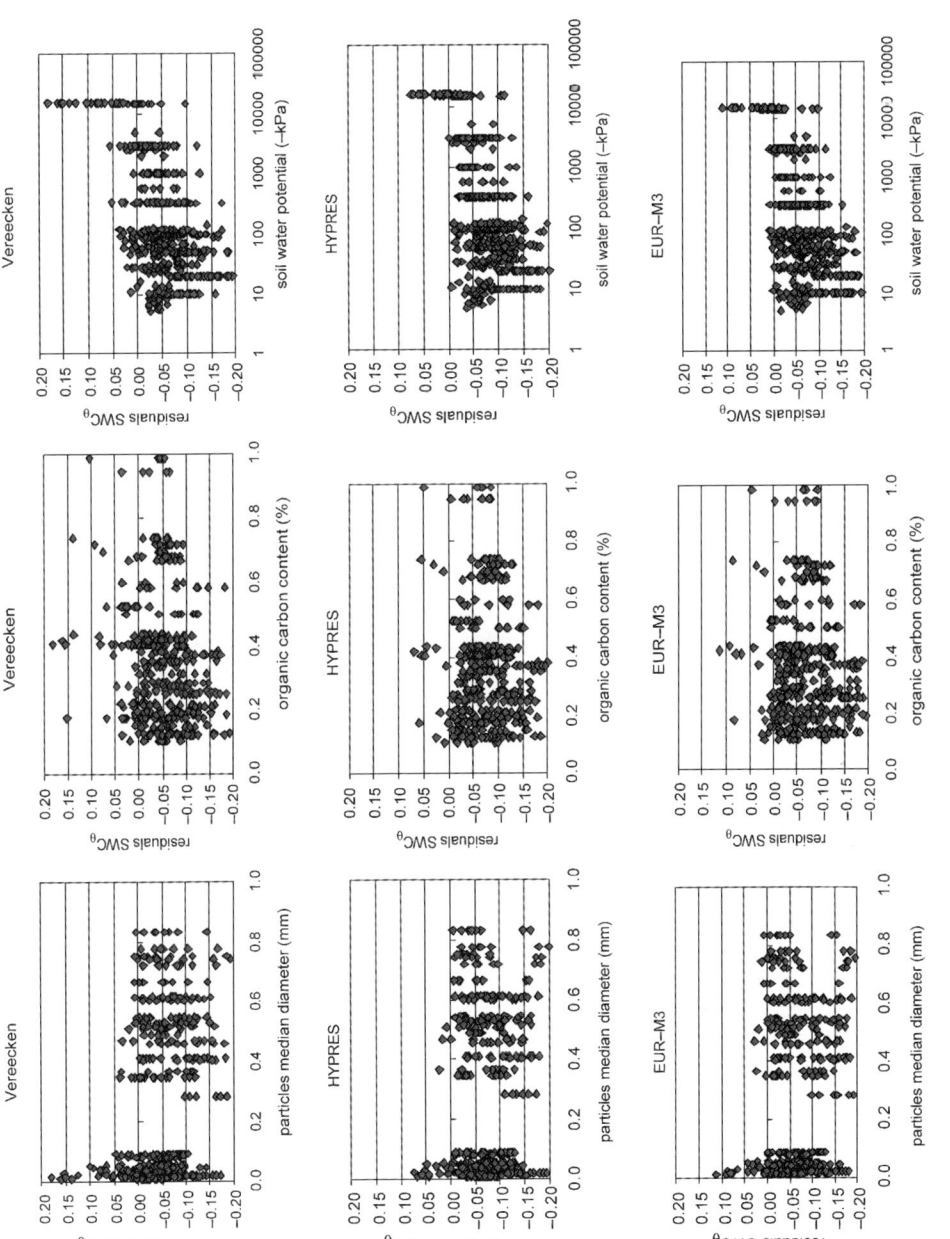

Figure 4. (*continued*).

2.2.2. The index I_{PTFRC}

This index can be used for the evaluation of pedotransfer functions which provide estimates over a range of soil matric potentials. It differs from the index I_{PTFSW} in the structure of the module "Pattern" (Section 2.1.3), where the index PI_ψ is included to account for patterns in the residuals with respect to soil water potential. The data used come from the same soil profiles of the section above, but soil water content was estimated via the soil water retention curves of Campbell (1985), Mayr and Jarvis (1999), Brooks and Corey (1964), Vereecken et al. (1992), HYPRES (Wösten et al., 1999), and EUR-M3 (Nemes et al., 2003) corresponding to soil water tensions from -10 to -1500 kPa. The estimates made were 480 for each method.

The graphs in Figure 4 show, for each method, the residuals vs. D50, OC, and $-\ln(\psi)$. The graphs in which the residuals are plotted against D50 and OC resemble the ones of Figures 2 and 3, whereas the "funnel" effect visible in the graphs of residuals vs. $-\ln(\psi)$ confirms the worse accuracy of water content estimates at low soil water potentials seen when evaluating pedotransfer functions at -1500 kPa. The data of basic indices, modules, and of the integrated index are reported in Table 10.

The data in Table 10 show that different methods differ mostly with respect to the PI, and the Brooks and Corey' method resulted as the best overall performer, even if not the best for each basic index. An evaluation with more methods could suggest a change in the fuzzy limits for each basic index, targeting at a sharper discrimination among methods with the contribution of the module "Accuracy". With reference to the specific data set used, the inclusion of PI has allowed quantifying differences in method performance which reproduce the qualitative evaluation made with the graphs of residuals of Figure 4. Also, the various components of I_{PTFRC} allow quantifying components of the performance, which highlight both strengths and weaknesses of the methods.

2.2.3. Final remarks about integrated indices for pedotransfer function evaluation

Integrated indices summarize information on method performance, and are not meant to replace basic indices. The integrated indices proposed here refer to as some of the pedotransfer functions available, and could be easily extended and customized to evaluate other variables, such as saturated hydraulic conductivity, bulk density, etc. Two major points constitute the importance of integrated indices: summarizing various aspects of the method performance into a single indicator, and quantifying method performance with respect to independent variables, that is either inputs of the methods or variables not explicitly used as inputs. The former point has its value in allowing a consistent method comparison, and it also allows defining cost functions which can be used in parameters optimization (when developing a new method). The latter point allows getting an insight about model performance, and it allows highlighting ranges of independent values in which the method can be reliably used. The modular structure also presents advantages, providing transparency of each step and a control opportunity for anybody concerned by the process itself. The mode of module aggregation can be changed at any time, and new modules can possibly be added. The multivariated nature of the issue is explicitly stated, the rules are easy to read, and the numerical scores easy to tune to match expert opinions.

The weakness of integrated indices lie in the subjective setting of both limits and weights applied to indices and aggregated modules. The limits for U and F, which

Table 10
Values of basic indices (RRMSE, EF, r, PI_{D50}, PI_{OC}, PI_ψ), modules (Accuracy, Correlation, Pattern), and integrated index (I_{PTFRC}) to evaluate pedotransfer functions. Values in bold refer to the best functions

Pedotransfer Function	PRMSE	EF	r	PID_{50}	PI_{OC}	PI_ψ	Mod Acc	Mod Corr	Mod PI_{RC}	I_{PTRRC}
Campbell	24.0	0.804	0.945	0.0554	0.0820	0.0328	**0.000**	**0.000**	0.337	0.0794
Mayr–Jarvis	26.5	0.759	0.903	0.0345	0.0587	0.0501	**0.000**	**0.000**	0.609	0.2430
Brooks–Corey	**19.7**	**0.868**	**0.958**	0.0348	0.0401	**0.0315**	**0.000**	**0.000**	**0.072**	**0.0037**
Vereecken	31.4	0.665	0.898	0.0382	0.0392	0.0969	0.002	**0.000**	0.456	0.1455
HYPRES	37.3	0.527	0.931	0.0347	**0.0135**	0.0750	0.059	**0.000**	0.400	0.1197
EUR-M3	35.6	0.568	0.911	**0.0295**	0.0227	0.0813	0.035	**0.000**	0.400	0.1147
Maximum	37.3	0.868	0.958	0.0554	0.0820	0.0969				
Median	29.0	0.712	0.921	0.0347	0.0396	0.0625				
Minimum	19.7	0.527	0.898	0.0295	0.0135	0.0315				
Limit U	60.0	0.000	0.700	0.0600	0.0500	0.0600				
Limit F	30.0	0.500	0.900	0.0300	0.0800	0.0300				

discriminate between fuzzy and crisp areas according to the approach presented, can be based on information available for a given statistic (e.g., the known range of variation for RRMSE), or they can be tuned based on the data being analyzed. The former option is preferable, although not often available. If the latter must be chosen, the reliability of the data set is critical, as it becomes the only source of information to build expert judgment. More specifically, choosing U and F limits with respect to pattern indices requires that the data set used is large enough to explore the range of the independent variable under evaluation. The subjective judgment in defining index and module weights, rather than being a weakness, may as well be the procedure's strength. In fact, it can allow targeting the development of a new index to specific goals (e.g., having accuracy prevailing on patterns when the range of independent variables is not wide; increasing the relative incidence of patterns when the application of the model must be over a wide range of soils).

Finally, although integrated indices can be built using different basic indices and aggregating them in different modules, some general rules must or should be followed. In particular, it is needed to use indices, which quantify different components of the error; avoiding tight correlations, which would lead to "falsely" integrated indices. The use of pattern indices, although not mandatory, boosts the power of the indices as explained above.

3. FUNCTIONAL EVALUATION OF PEDOTRANSFER FUNCTIONS
H. Wösten, A. Nemes and M. Acutis

Hydraulic characteristics are not an aim in itself but they are essential input data for the calculation of water and solute transport in the unsaturated zone. As a consequence, characterization of PTF accuracy does not provide information about the performance of PTFs in specific applications. Therefore, a functional evaluation was first proposed by Wösten et al. (1986) who used criteria directly related to applications rather than statistics to characterize the accuracy.

3.1. Example of functional evaluation of PTF uncertainty

An example of how to evaluate in a functional sense the uncertainty in PTFs is presented by Finke et al. (1996). For a soil mapping unit in the Netherlands, they quantified and statistically evaluated the contribution to the explanation of variability in modeling results, of two major sources of uncertainty in model input parameters. These two sources of uncertainty were: (i) spatial variability of basic soil properties such as profile composition, soil texture and water-table depths; and (ii) uncertainty associated with the use of pedotransfer functions to predict soil hydraulic characteristics. When pedotransfer functions are used, the spatial variability of the basic soil properties is directly translated to variations in hydraulic characteristics and, subsequently, to variations in simulated functional soil behavior. In addition, the prediction error of the pedotransfer function itself results in variations in the predicted soil hydraulic characteristics. This also contributes to uncertainties in simulated functional soil behavior.

Spatial variability of soil properties within the soil mapping unit was expressed by selecting a statistically representative set of 88 profiles within the mapping unit.

Uncertainty in pedotransfer function predictions was quantified by drawing 20 error terms from the covariance matrix of differences between measured and predicted soil hydraulic characteristics using a Cholesky decomposition. Adding the 20 error terms to the hydraulic characteristic directly predicted with the PTF, resulted in a set of 20 new hydraulic characteristics. As such this set quantifies the uncertainty involved in applying the PTF. As a result, $88 \times 20 = 1760$ possible combinations of soil profiles and hydraulic characteristics were generated. These combinations were used in a Monte Carlo procedure to calculate the following functional aspects of soil behavior 'days with good workability', 'days with sufficient aeration', 'chloride breakthrough', 'cadmium breakthrough' and 'isoproturon breakthrough'. The contribution of the two sources of uncertainty to the explanation of variability in modeling results of the five functional aspects of soil behavior were statistically evaluated by an analysis of variance.

Figure 5 shows that uncertainty in pedotransfer functions plays an important role when functional aspects of soil behavior with a physical nature are calculated, such as 'days with good workability' and 'sufficient aeration'. If adsorbing chemicals, such as 'cadmium breakthrough' and 'isoproturon breakthrough', are considered, the uncertainty caused by variability in basic soil properties largely explains the uncertainty in the modeling results. In case of the inert tracer, 'chloride breakthrough' modeling results are far less sensitive to variability in basic soil properties when compared to the adsorbing chemicals cadmium and isoproturon. This is understandable because variability in basic soil properties means essentially variability in percentages clay, silt and organic matter. In turn, these variabilities have a large influence on the adsorption of cadmium and isoproturon.

This example illustrates that there is not a single source of variability, either PTF-related or soil-related, that can explain the uncertainty in every calculated functional aspect of soil behavior. Contrarily, depending on the type of functional aspect of soil behavior one is interested in, variability in different types of input parameters will dominate uncertainty in results. As a consequence, it is rewarding to first identify which type of input parameter dominates the explanation of uncertainty in results. Next, the variability of this type of input parameters can be assessed and possibly be reduced. An analysis along these lines will also help to indicate whether the use of an existing PTF and its associated uncertainty is sufficiently accurate to generate the soil hydraulic characteristics for a specific application or if additional, costly soil physical measurements are justified.

3.2. Application of PTFs in a functional context

Vereecken et al. (1992) defined functional evaluation as the statistical examination of the variability in the outcome of a simulation model for a specific application when the variability arises solely from uncertainty in the PTFs. Vereecken et al. (1992) found that prediction errors in hydraulic characteristics account for about 90% of variations in moisture supply capacity of soil. The remaining 10% are due to the within map variability. These authors presented the seemingly paradoxical case study in which errors of PTFs with better accuracy result in higher variability in simulated moisture supply capacity and in simulated downward fluxes beyond the root zone.

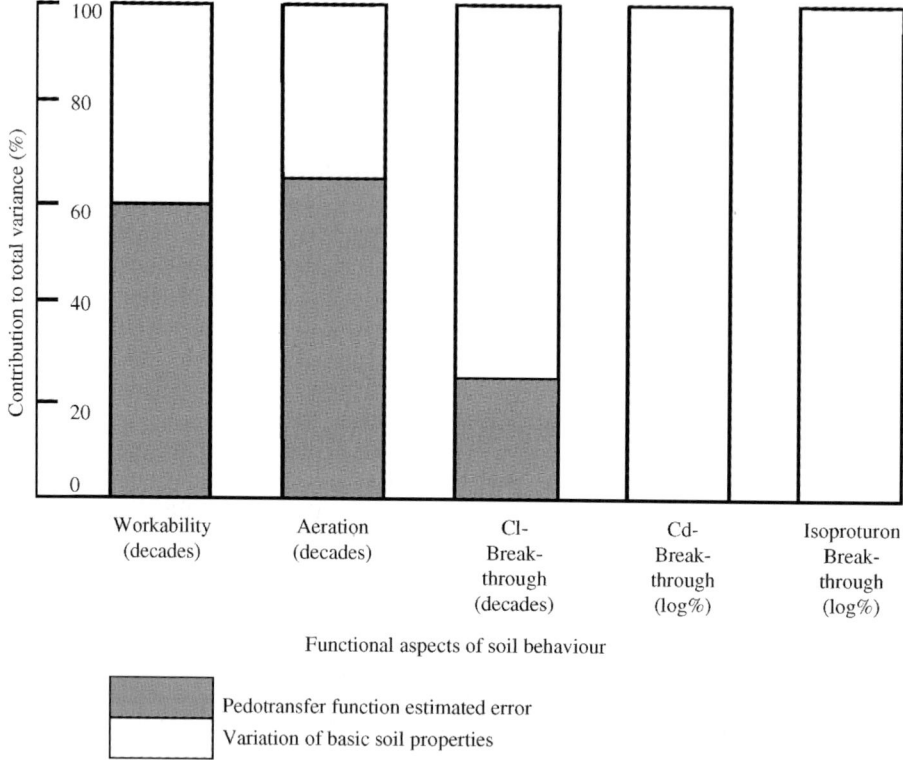

Figure 5. Relative importance of different sources of variability in explaining the uncertainty in five simulated functional aspects of soil behavior (after Finke et al., 1996).

At least four factors affect the performance of a PTF in simulations. These are the accuracy of basic soil data used as inputs in PTFs, the accuracy of PTF itself, specific features of the simulation model, and the output used as a functional criteria. Workman and Skaggs (1992) compared performance of PTFs in two models, PREFLO and DRAINMOD, in simulations of evapotranspiration, infiltration, runoff and drainage. PREFLO was more sensitive than DRAINMOD to predictions of the saturated water content. Differences between measured and PTF predicted hydraulic characteristics created larger differences in model output than differences in the two models.

An important limitation of many studies that evaluate PTFs is that it mostly remains unclear what the main sources of the estimation errors are. In most cases it is not clear whether differences between data sets used to derive PTFs (size, origin, reliability), differences between the algorithms of PTF development (e.g., different regression types vs. NN models) or differences among the predictors cause a particular PTF to perform better than others. Nemes et al. (2003) used national, continental and intercontinental scale data collections to derive PTFs to estimate soil water retention. The same methodology and the same sets of predictors were used to allow the source database to be the only variable that is changed. Eleven different PTFs were developed from each of the data sets, in order to study the influence of different combinations of predictors. Water retention estimations showed an improving trend as the list of input variables increased. Using a small set of relevant (local) data was better than using a large but more general data set. Estimated water retention curves were then used to simulate soil moisture time series of seven soils. Sum of square residuals were evaluated for simulated water contents and water contents observed in the field. Small differences were found among the PTFs derived from different scale data collections. Moreover, PTF estimates were only marginally worse than estimates using laboratory measured WRCs. Differences between estimations using different scale data sets (or measured WRCs) were not the main source of error in the simulation model.

When the sensitivity of a model to hydraulic characteristics is not substantial, the accuracy of PTF may not be an issue. Wopereis et al. (1993) presented a functional evaluation of PTF predicted hydraulic conductivity for simulating water content in the upper 40 cm soil layer under dryland rice cultivation. Although PTF predicted hydraulic conductivity differed substantially from measured values, good agreement was obtained between observed and simulated soil water contents, just because simulation results were relatively insensitive to this parameter. A similar conclusion was reached by Wösten et al. (1995) when they compared the functional behavior of hydraulic characteristics predicted with class and continuous PTFs. The simulated number of workable days did not depend on the type of pedotransfer function applied whereas the number of days with good aeration did.

Functional evaluation of PTFs can be done without taking into account PTF accuracy *per se*. Wösten et al. (1990a) compared the performance of four different PTFs in the same model and found in an analysis of variance no significant differences in terms of simulating soil water content in the upper 50 cm soil layer. In this case, PTF cost rather than its accuracy was used as a criterion. Verburg et al. (1996) applied four different PTFs to predict parameters for water and bromide transport in soil profiles, and found that differences in PTF predictions did not affect bromide transport simulations with their model. Functional evaluation showed that differences in PTF performance in modeling studies may depend on other environmental variables. No difference was found between performance of several PTFs when only the type of PTF was considered as input for the analysis of variance (Wösten et al., 1990b). However, when the same data were re-analyzed using rainfall deficit as a co-variable, the analysis of variance showed a clear effect of the type of PTF.

Data collected at different scales may be available for the functional evaluation of PTFs. Including scale information in the PTF evaluation process involves the following three operations: interpolation, running the model and aggregation of results (Heuvelink and Pebesma, 1999). In principle, it is possible to aggregate first and to simulate later, or alternatively to simulate first and to aggregate later. Heuvelink (1998) recommends to run the model in many points at small scale within a large scale resolution unit and then to aggregate simulation results.

The majority of PTFs are developed from measurements on standard small samples. Soil sample size may affect both average values of soil hydraulic characteristics as well as their spatial distributions (Shein et al., 1995; Giménez et al., 1999; Pachepsky et al., 2001b). Such scale effects, albeit documented, are still not understood to the extent that scale can be incorporated in PTFs. Therefore, local application of PTFs developed at the sample scale remains a typical feature of its use.

APPENDIX A. NUMERICAL INDICES AND TEST STATISTICS FOR MODEL EVALUATION
G. Belocchi

A.1. LIST OF ABBREVIATIONS

i	ith data point
n	number of data pairs
E_i	estimated value
M_i	measured value
$\bar{E}$	mean of estimates
$\bar{M}$	mean of measurements
s_E^2	variance of estimates
s_M^2	variance of measurements
s_E	standard deviation of estimates
s_M	standard deviation of measurements
$F_E(x)$	cumulative empirical distribution function for estimated values
$F_M(x)$	cumulative empirical distribution function for measured values
OLS	ordinary least squares method
RMA	reduced major axis method
s.e.(b_1)	standard error of b_1
s.e.(b_0)	standard error of b_0
$\hat{E}_i$	estimated value from the regression E_i vs. M_i
s.e.$_{\hat{E}}$	standard error of the estimates from the regression E_i vs. M_i
l, m	group index
p	number of groups (2–5)
R	model residual $(E_i - M_i)$
q_l, q_m	numerosity of groups
i_l, i_m	residual value inside group
s_p^2	between-groups variance
s_e^2	error variance

A.2. DIFFERENCE-BASED STATISTICS

Various difference-based performance measures follow, grouped into simple, square and absolute indices, and test statistics (Mayer and Butler, 1993; Janssen and Heuberger, 1995; Martorana and Belocchi, 1999; Metselaar, 1999).

Statistic	Symbol	Formulation
Simple differences		
Mean bias error	MBE	$\text{MBE} = \bar{E} - \bar{M}$
Relative error	E	$E = \dfrac{100}{n} \sum_{i=1}^{n} \dfrac{E_i - M_i}{M_i}$
Maximum error	MaxE	$\text{MaxE} = \text{Max}(E_i - M_i)$
Maximum percent error	MaxE%	$\text{MaxE\%} = \dfrac{100}{M_{\max}} \text{Max}(E_i - M_i)$
Normalized average error	NAE	$\text{NAE} = \dfrac{\bar{E} - \bar{M}}{\bar{M}}$
Fractional bias	FB	$\text{FB} = 2\dfrac{\bar{E} - \bar{M}}{\bar{E} + \bar{M}}$
Relative mean bias	rB	$rB = \dfrac{\bar{E} - \bar{M}}{s_M}$
Coefficient of residual mass	CRM	$\text{CRM} = \dfrac{\sum_{i=1}^{n} M_i - \sum_{i=1}^{n} E_i}{\sum_{i=1}^{n} M_i}$
Square differences		
Simulation bias	SB	$\text{SB} = (\bar{E} - \bar{M})^2$
Root mean square variation	RMSV	$\text{RMSV} = \sqrt{\dfrac{\sum_{i=1}^{n} [(E_i - \bar{E}) - (M_i - \bar{M})]^2}{n}}$
Sum of square error	SSE	$\text{SSE} = \sum_{i=1}^{n} (E_i - M_i)^2$
Mean square error	MSE	$\dfrac{\sum_{i=1}^{n} (E_i - M_i)^2}{n}$
Root mean square error	RMSE	$\text{RMSE} = \sqrt{\dfrac{\sum_{i=1}^{n} (E_i - M_i)^2}{n}}$
Relative root mean square error	RRMSE	$\text{RRMSE} = \text{RMSE} \dfrac{100}{\bar{M}}$
Normalized mean square error	NMSE	$\text{NMSE} = \dfrac{\sum_{i=1}^{n} (E_i - M_i)^2}{\sum_{i=1}^{n} E_i M_i}$
Modeling efficiency	EF	$\text{EF} = 1 - \dfrac{\sum_{i=1}^{n} (E_i - M_i)^2}{\sum_{i=1}^{n} (M_i - \bar{M})^2}$
Modeling percent efficiency	EF%	$\text{EF\%} = 100(1 - \text{EF})$
Index of agreement	d	$d = 1 - \dfrac{\sum_{i=1}^{n} (E_i - M_i)^2}{\sum_{i=1}^{n} (|E_i - \bar{M}| + (M_i - \bar{M}))}$

Appendix A.2. Continued

Ratio of scatter	RS	$\mathrm{RS} = \dfrac{\sum_{i=1}^{n}(M_i - \bar{M})^2}{\sum_{i=1}^{n}(E_i - \bar{M})^2}$
Fractional variance	FV	$FV = \dfrac{s_E^2 - s_M^2}{0.5(s_E^2 + s_M^2)}$
Variance ratio	VR	$\mathrm{VR} = \dfrac{s_E^2}{s_M^2}$
Absolute differences		
Absolute fractional bias	AFB	$\mathrm{AFB} = 2\dfrac{|\bar{E} - \bar{M}|}{\bar{E} + \bar{M}}$
Maximum absolute error	MaxAE	$\mathrm{MaxAE} = \max_{i,\ldots,n}|E_i - M_i|$
Mean absolute error	MAE	$\mathrm{MAE} = \sum_{i=1}^{n}\dfrac{|E_i - M_i|}{n}$
Mean absolute percent error	MA%E	$\mathrm{MA\%E} = 100\sum_{i=1}^{n}\dfrac{|E_i - M_i|}{M_i}\dfrac{1}{n}$
Relative mean absolute error	RMAE	$\mathrm{RMAE} = \mathrm{MAE}\dfrac{100}{\bar{M}}$
Modified modeling efficiency	EF_1	$EF_1 = 1 - \dfrac{\sum_{i=1}^{n}|E_i - M_i|}{\sum_{i=1}^{n}|M_i - \bar{M}|}$
Alternative index of agreement	Ad	$\mathrm{Ad} = 1 - \dfrac{\sum_{i=1}^{n}|E_i - M_i|}{\sum_{i=1}^{n}(|E_i - \bar{M}| + (M_i - \bar{M}))}$
Non-parametric differences		
Median absolute error	MdAE	$\mathrm{MdAE} = \mathrm{median}_{i=1,\ldots,n}|E_i - M_i|$
Relative median absolute error	RMdAE	$\mathrm{RMdAE} = \mathrm{median}_{i=1,\ldots,n}|E_i - M_i|\dfrac{100}{\bar{M}}$
Relative modeling efficiency	REF	$\mathrm{REF} = \mathrm{median}_{i=1,\ldots,n}\left(\dfrac{\mathrm{median}_{i=1,\ldots,n}|M_i - \bar{M}| - \mathrm{median}_{i=1,\ldots,n}|M_i - E_i|}{\mathrm{median}_{i=1,\ldots,n}|M_i - \bar{M}|}\right)$
Upper quartile absolute error	UppAE	$\mathrm{UppAE} = 75\text{th percentile }|E_i - M_i|$

Test statistics
t-test (MBE = 0) t

$$t = \frac{\text{MBE}}{s_d^2} = \sqrt{\frac{(n-1)\text{MBE}^2}{\text{MSE} - \text{MBE}^2}}$$

chi-square χ^2
(MSE = 0)

$$\chi^2 = \frac{\sum_{i=1}^{n}(E_i - M_i)}{\frac{1}{n-1}\sum_{i=1}^{n}[(E_i - M_i)^2 - \text{MSE}]^2}$$

Kolomogorov-   KS
Smirnov

$$\text{KS} = \max_x |F_E(x) - F_M(x)|$$

Indices such as MBE (simple) or SB (square) compare estimates and measurements on an average level. Different ways of normalizing the average error result in the dimensionless measures NAE, FB and rB. Averaging smooths out the comparison features. The signs of the simple differences denote over- and under-estimation. Statistics such as FV and VR are two ways of comparing the variances. As well as the tests statistics t(MBE = 0), χ^2(MSE = 0) and KS, they refer to a comparison between estimates and measurements on a population level. Parametric test statistics t(MBE = 0) and χ^2(MSE = 0) are only appropriate if the involved quantities can be considered as random samples from normal distributions, otherwise KS can be applied (Shea, 1989; Jacovides and Kontoyiannis, 1995).

The most part of difference-based indices compare the estimates and measurements on the individual level, and try to express the spread $E_i - M_i$. The common RMSE is an example of difference-based performance measure in a quadratic form, therefore it is rather very sensitive to outliers. The index of agreement (d) by Willmott (1981) can be viewed as a standardized measure of the mean square error, and express the matching between estimates and measurements more directly (ranging from 0 with no match to 1 with perfect match). An alternative way of standardizing the RMSE is dividing it by the mean of measurements (e.g., Jørgensen et al., 1986), thus rendering a kind of coefficient of variation of the discrepancies $E_i - M_i$, around the mean (RRMSE). The normalization by the product $E_i M_i$ prevents NMSE from the effect of large discrepancies (Kukkonen et al., 2001). Difference based indices, such as MSE or its squared root RMSE, have also been extensively used as cost functions in model parameter estimation (i.e., Sorooshian et al., 1983, 1993; Wallach, 1999), and in optimization algorithms (Wallach et al., 2001).

Accounting for absolute deviations $E_i - M_i$ renders the MAE. This measure is less sensitive to outliers than RMSE, and unlike MBE, does not account for compensations of positive and negative discrepancies. Dividing the MAE by the observed mean, the dimensionless RMAE is obtained. Other simple (FB) and square indices (d, EF) have their absolute counterpart (AFB, Ad, EF_1), which are less sensitive to outliers. The maximum absolute error is most sensitive to outliers. The least sensitivity to outliers is obtained with non-parametric indices.

In situations where the compared quantities vary substantially over the considered range of values, the abovementioned performance measures tend to over-emphasize points where the higher value occur. To obtain a more even analysis of the discrepancy over the complete span of values, it is recommended to apply measures based on the ratios $(E_i - M_i)/M_i$.

Measures such as RS (or its inverse, designated as coefficient of determination) and EF (Loague and Green, 1991) can be used to interpret the variability of errors when replicates are unavailable (Smith et al., 1997). Their best value is 1. They relate the model estimates and measurements to a "nominal" or "benchmark" situation, namely the mean of the observations.

A.3. CORRELATION-BASED STATISTICS

Statistic	Symbol	Formulation
Indices		
Pearson correlation coefficient (E_i vs. M_i)	r	$r = \dfrac{\sum_{i=1}^{n} (E_i - \bar{E})(M_i - \bar{M})}{\sqrt{\sum_{i=1}^{n} (E_i - \bar{E})^2 \sum_{i=1}^{n} (M_i - \bar{M})^2}}$
Pearson correlation coefficient ($E_i - M_i$ vs. $E_i + M_i$)	$r_{d,s}$	$r_{d,s} = \dfrac{\sum_{i=1}^{n} (E_i - M_i)(E_i + M_i)}{\sqrt{\sum_{i=1}^{n} (E_i - M_i)^2 \sum_{i=1}^{n} (E_i + M_i)^2}}$
Spearman correlation coefficient (E_i vs. M_i)	r_s	$r_s = \dfrac{6 \sum_{i=1}^{n} (E_i - M_i)^2}{n^3 - n}$
Slope of the regression E_i vs. M_i	b_1	OLS: $b_1 = \dfrac{\sum_{i=1}^{n} (E_i - \bar{E})(M_i - \bar{M})}{\sum_{i=1}^{n} (E_i - \bar{E})^2}$ RMA: $b_1 = \pm \dfrac{s_E}{s_M}$
Intercept of the regression E_i vs. M_i	b_0	$b_0 = \bar{E} - b_1 \bar{M}$
Systematic root mean square error[a]	RMSE$_s$	$\text{RMSE}_s = \sqrt{\dfrac{\sum_{i=1}^{n} (\hat{E}_i - M_i)^2}{n}}$
Erratic root mean square error[a]	RMSE$_e$	$\text{RMSE}_e = \sqrt{\dfrac{\sum_{i=1}^{n} (E_i - \hat{E}_i)^2}{n}}$
Test statistics[b]		
t-test ($\rho = 0$)	t	$t = \sqrt{\dfrac{r^2(n-2)}{1 - r^2}}$
t-test ($\rho_{d,s} = 0$)	t	$r_{d,s}^2 = \sqrt{\dfrac{r_{d,s}^2(n-2)}{1 - r_{d,s}^2}}$
t-test ($\beta_1 = 0$)	t	$t = \dfrac{b_1}{\text{s.e.}(b_1)}$
t-test ($\beta_1 = 1$)	t	$t = \dfrac{1 - b_1}{\text{s.e.}(b_1)}$

Statistic	Symbol	Formulation
t-test ($\beta_0 = 0$)	t	$t = \dfrac{b_0}{\text{s.e.}(b_0)}$
F-test ($\beta_0 = 0$ and $\beta_1 = 1$)	F	$F = \dfrac{(n-2)[nb_0^2 + 2n\bar{E}b_0(b_1-1) + \sum_{i=1}^{n} E_i^2(b_1-1)^2]}{2n\,\text{s.e.}_{\hat{E}}^2}$

[a]Disaggregating into systematic and erratic portions can be extended to any statistic based on the difference $E_i - M_i$.
[b]Greek letters indicate population parameters (ρ: Pearson correlation coefficient; β_0, β_1: regression parameters).

Linear regression can be used to compare the model-estimates and measurements. The deviation of the intercept term b_0, and the slope b_1, in the regression relationship $E_i = b_0 + b_1 \cdot M_i + \epsilon_i$ from 0 and 1, respectively, and the deviation of r from 1 are useful indicators for the model-data agreement (Dent and Blackie, 1979; Addiscott and Whitmore, 1987; Mayer and Butler, 1993). The regression results should, however, be carefully used and interpreted, especially if the underlying conditions are not completely fulfilled or if there is much variability in measurements and estimates. Compared to ordinary least squares method, reduced major axis method provides a less biased estimate of the underlying functional relation as described by the regression estimates vs. measurements (Ricker, 1984).

According to Willmott (1981), regression values can be used to disaggregate the differences $E_i - M_i$ into a systematic (difference between regression estimates and measurements) and an erratic portion (difference between model and regression estimates). See also Kobayashi and Salam (2000).

The Pearson correlation coefficient may be inappropriate when one or both the series of data come from non-normally distributed populations and the departure from normality is serious. In such cases, rank transformation can be performed and the non-parametric Spearman coefficient can be applied by computing the differences among all the ranks assigned to each pair of data. The significance of r_s may be tested comparing it with the specific tables. Significance of the correlation coefficient $r_{d,s}$ can be tested to verify the equality of variances of estimates and measurements (Pitman-Morgan procedure: correlation means heterogeneous variances, Kleijnen, 1987).

Statistic	Symbol	Formulation
Pattern indices		
Range-based pattern index	PI	$\text{PI} = \max_{l,m=2,\ldots,p; l \neq m} \left\lvert \dfrac{1}{q_l}\sum_{i_l=1}^{q_l} R_{i_l} - \dfrac{1}{q_m}\sum_{i_m=1}^{q_m} R_{i_m} \right\rvert$
F-based pattern index	PI-F	$\text{PI-}F = \dfrac{s_p^2}{s_e^2}$

Graphical analysis of residuals against various relevant quantities is recommended for virtually any regression analysis (e.g., Miklós et al., 1995; Cook and Weisberg, 1999;

Wisniak and Polishuk, 1999), and modeling applications (e.g., Dent and Blackie, 1979; Mayer and Butler, 1993). When the plot of residuals shows a structure in terms of 2–5 groups with somewhat different mean values, such a structure is defined a "macro-pattern" (Donatelli et al., 2004).

Pattern indices are used to investigate the relationships between the differences $E_i - M_i$ and an external variable. They are computed by dividing the range of the external variable in 2, 3, 4, or 5 sub-ranges. Both range- and F-based indices may be computed given either fixed or varying sub-ranges, the latter according to an optimization exhaustive algorithm with the constraint of a minimum number of data (5, 10, 15, 20%) in each sub-range. Specific tables, developed from Monte Carlo simulations, are available to test the significance of pattern indices at specific conditions (Donatelli et al., 2004).

APPENDIX B. FUZZY EXPERT SYSTEMS
G. Fila and G. Belocchi

Zadeh (1965) proposed the use of fuzzy set theory to describe relationships that are best characterized by compliance to a collection of attributes. In setting up the set of decision rules in model evaluation, the attributes are the basic evaluation indices, and the user must decide on the relative importance of each one. The threshold values, beyond which the index is "certainly" favorable or unfavorable, must be given by the user. With this procedure three membership classes are created for the index values: favorable (F), unfavorable (U), partial (or fuzzy) membership.

The fuzzy set theory addresses this type of problem by allowing one to define the "degree of membership" of an element in a set by means of membership functions (transition functions), that can take any value from the interval [0, 1]. The value 0 represents complete non-membership, the value 1 represents complete membership, and values in between are used to represent partial membership. For classical or "crisp" sets, the membership function only takes two values: 0 (non-membership) and 1 (membership).

The hierarchical structure of this technique is used to aggregate indices into first-level fuzzy indicators (modules) and, next, into a second-level fuzzy indicator. This process of aggregation may continue, hypothetically, until a final-level fuzzy indicator is achieved. The indicator developed here is a second-level indicator. For simplicity, only first-level aggregation is developed in the example below.

The aggregation process is accomplished by combining weighted fuzzy values. According to this approach we can characterize the shape of the membership function of each input index by the two limits of the "transition interval". Fuzzy membership functions may have different shapes, depending on experience or preference. We used membership functions that are S-shaped in the transition interval, since they provide smoother variations of the values of the inputs than functions that are linear in the transition interval. If x is the value of the index, α and γ lower and upper bounds, respectively, the S-function is flat at a value of 0 and 1 for $x \leq \alpha$ and for $x \geq \gamma$

respectively. Between α and γ the S-function is a quadratic function of x (Liao, 2002):

$$S(x;\alpha;\gamma) = \begin{cases} 0 & x \leq \alpha \\ 2\left(\dfrac{x-\alpha}{\gamma-\alpha}\right)^2 & \alpha \leq x \leq \beta \\ 1 - 2\left(\dfrac{x-\gamma}{\gamma-\alpha}\right)^2 & \beta \leq x \leq \gamma \\ 1 & x \geq \gamma \end{cases} \quad (B1)$$

where $\beta = (\alpha + \gamma)/2$. Two adjacent fuzzy terms with S-shaped membership functions have 0.5 overlap at the midpoint between the two extremes. Equation (B1) can only represent the left-hand side. For the right-hand side, the complement of Equation (B1) is needed.

For each module we formulated a set of decision rules attributing values between 0 and 1 to an output variable according to the membership of its input indices to the fuzzy subsets F and U.

The linguistic description of these components is accomplished in the form of fuzzy rules with a relatively simple syntax. In fact, fuzzy rules inference involves computation of fuzzy rules which are mostly the "if...then..." statements. When two indices are aggregated, the method makes use of the conjunctive operator AND, as formalized by four rules:

if ($x1$ is A11) AND ($x2$ is A12) then ($y1$ is B1) [rule 1]
if ($x1$ is A21) AND ($x2$ is A22) then ($y2$ is B2) [rule 2]
if ($x1$ is A31) AND ($x2$ is A32) then ($y3$ is B3) [rule 3]
if ($x1$ is A41) AND ($x2$ is A42) then ($y4$ is B4) [rule 4]

where $xj(j = 1,2)$ is an input index, Aij is a fuzzy subset, yi is an output variable, Bi(i = 1,2,3,4) is a number. "xj is Aij" is called a premise of the ith rule; "yi is Bi" is called conclusion (or expert weight) of the ith rule. In this case the decision rules consist of two premises (if... AND if...) linked by and followed by a conclusion (then...).

The process continues, thus quantifying the degree of truth in the premise portion of each rule and, then, aggregating the truth of linked conditions. Let '$x1$ and '$x2$ be the values taken by $x1$ and $x2$, and Aij('xj) the membership value of 'xj to the fuzzy set Aij (given by the membership function that defines Aij). According to Sugeno's inference method (Sugeno, 1985), when the premises are linked by a conclusion, the true value of a decision rule is defined as the smallest of the true values of its premises. Therefore, a fuzzy subset is assigned to each output variable for each rule using "min" aggregator, where min means "minimum value of". One can define $w1$, $w2$, $w3$ and $w4$ the true values of the rules, as follows:

$w1 = \min(A11('x1), A12('x2))$ [truth value of the rule 1]
$w2 = \min(A21('x1), A22('x2))$ [truth value of the rule 2]
$w3 = \min(A31('x1), A32('x2))$ [truth value of the rule 3]
$w4 = \min(A41('x1), A42('x2))$ [truth value of the rule 4]

The first rule infers $w1B1$, the second one $w2B2$, and so on. The fuzzy sets that represent the outputs of each rule are combined by summation into a single fuzzy solution set ('$y0$):

'$y0 = w1B1 + w2B2 + w3B3 + w4B4$

A solution of this type is sometimes known as "singleton" output membership function, and it can be thought of as a pre-defuzzified fuzzy set (Jones and Barnes, 2000).

The global output 'y is

'$y = (w1B1 + w2B2 + w3B3 + w4B4)/(w1 + w2 + w3 + w4)$

This last operation (the weighted average of several data points) is a common method (centroid calculation) adopted to reduce the final fuzzy set to a crisp value (defuzzification) in the Sugeno-type systems (Sugeno, 1985).

The computation of the aggregated indicator is primarily influenced by the "truth value" of each rule (wi), for which the expert weight (Bi) is a multiplication factor. As a result of this insight, it should be clear that weights on rules are not simply measures of the "relative importance" of each rule. They are essentially measures of the importance of the increase from the worst to the best level of performance on one rule. The terms Bi are subjective elements introduced in the process of aggregation, which the global output may be sensitive to. This means that the quality of outputs is influenced by the weights of the individual rules. This crucial issue should be investigated within the context of interest, which the true values depend on.

To illustrate Sugeno's inference method, a numerical example will be given for the composition of the module "Pattern", obtained by aggregating two pattern indices, PI_{D50} and PI_{OC} (see details in Section 2.1.3). We gave three classes of model response with respect to the presence of patterns in the residuals against one independent variable. See for instance Figure B1 for PI_{D50}.

We assigned the membership value of 1 for the fuzzy subset U and the membership value of 0 for the fuzzy subset F to the response classified as "pattern" ($PI_{D50} \geq 0.060$ mm, $PI_{OC} \geq 0.080\%$). The Model response classified as "no pattern" ($PI_{D50} \leq 0.030$ mm, $PI_{OC} \leq 0.050\%$) receives the membership value of 0 for the fuzzy subset U and the membership value of 1 for the fuzzy subset F (Figure B1 for PI_{D50}).

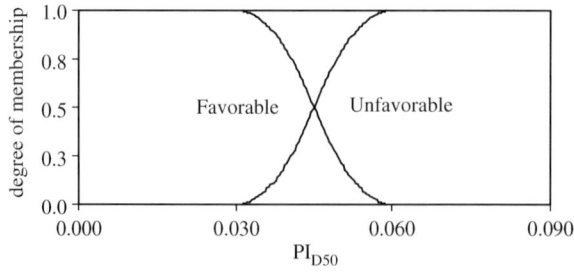

Figure B1. Graphical presentation of fuzzy sets for PI_{D50}.

The "borderline values" (0.030 mm < PI_{D50} < 0.060 mm, 0.050% < PI_{OC} < 0.080%) fall within the "transition interval" in which the membership value for U increases from 0 (at PI_{D50} = 0.03 mm, PI_{OC} = 0.05%) to 1 (at PI_{D50} = 0.06 mm, PI_{OC} = 0.08%), and the membership value for F decreases from 1 to 0 (thus the functions characterizing F and U are complementary). In our example the reasoning for the four rules is:

if (PI_{D50} is F) AND if (PI_{OC} is F) then (B1 is 0.0) [rule 1]
if (PI_{D50} is F) AND if (PI_{OC} is U) then (B2 is 0.5) [rule 2]
if (PI_{D50} is U) AND if (PI_{OC} is F) then (B3 is 0.5) [rule 3]
if (PI_{D50} is U) AND if (PI_{OC} is U) then (B4 is 1.0) [rule 4]

Notice the same weight is attributed here to each basic index.

Suppose that PI_{D50} was equal to 0.050 mm and PI_{OC} equal to 0.075%. For both indices membership to fuzzy subsets F and U has to be defined. The membership functions of Figure B2 show the calculation of the truth value (see above) of the premises, i.e., the degree of membership to the fuzzy subset concerned for PI_{D50} (Figure B2, top) and PI_{OC} (Figure B2, bottom).

Results are shown in Table B1

The value of "Pattern" is calculated as the sum of the conclusions of the decision rules, weighted by the sum of their truth values:

$$\text{"Pattern"} = \frac{0.00 \; 0.0556 + 0.50 \; 0.2222 + 0.50 \; 0.0556 + 1.00 \; 0.7778}{0.0556 + 0.2222 + 0.0556 + 0.7778} = 0.8250$$

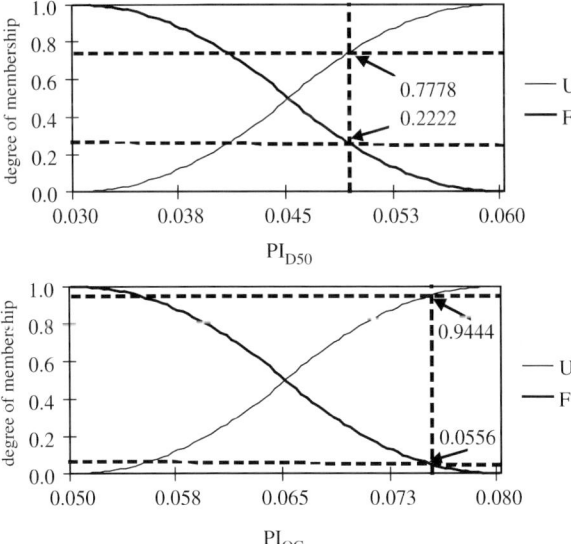

Figure B2. Membership to the fuzzy sets favorable (F) and unfavorable (U) for a hypothetical model response in terms of PI_{D50} (top) and PI_{OC} (bottom). PI_{D50} = 0.050 mm and PI_{OC} = 0.075%.

Table B1
Summary of decision rules describing the effect of the input indices PI_{D50} and PI_{OC} on the module "Pattern". Truth values of premises (w_i) and conclusions (B_i) for $PI_{D50} = 0.050$ and $PI_{OC} = 0.750$ are shown

PI_{D50}		PI_{OC}		Expert weight (B_i)	Truth value (w_i)	$w_i B_i$
Membership class	Membership value	Membership class	Membership value			
F	0.2222	F	0.0556	0.00	0.0556	0.0000
F	0.2222	U	0.9444	0.50	0.2222	0.1111
U	0.7778	F	0.0556	0.50	0.0556	0.0278
U	0.7778	U	0.9444	1.00	0.7778	0.7778
Fuzzy solution set ('y_0')						0.9167
Sum of truth values					1.1112	
Global output ('y')					0.8250	

F, favorable; U, unfavorable (for details, see the text).

Note that the first term of the numerator (0.00·0.0556) refers to the case when the value of membership class is F for all the indices considered, therefore the resulting value is always 0; the term is reported for completeness.

REFERENCES

Acutis, M., Donatelli, M., 2003. Soilpar 2.00: software to estimate soil hydrological parameters and functions. Eur. J. Agron. 18, 373-377.

Addiscott, T.M., Whitmore, A.P., 1987. Computer simulation of changes in soil mineral nitrogen and crop nitrogen during autumn, winter and spring. J. Agric. Sci. (Cambridge) 109, 141-157.

Ahuja, L.R., Naney, J.W., Williams, R.D., 1985. Estimating soil water characteristics from simpler properties or limited data. Soil Sci. Soc. Am. J. 49, 1100-1105.

Ahuja, L.R., Nofziger, D.L., Swartzendruber, D., Ross, J.D., 1989. Relationship between Green and Ampt parameters based on scaling concepts and field-measured hydraulic data. Water Resour. Res. 25, 1766-1770.

Arah, J., Hodnett, M.J., 1997. Handling soil hydrology in agroforestry models. In: Lawson, G.J., et al. (Eds.), Agroforestry modeling and research co-ordination, Annual Report July 1996, June 1997. Institute of Terrestrial Ecology, Natural Environment Research Council, Edinburgh, UK, pp. 35-42.

Balci, O., 1997. Principles of simulation model validation, verification, and testing. Trans. S.C.S.I. 14, 3-12.

Beke, G.L., McCormic, M.J., 1985. Predicting volumetric water retention for subsoil materials form Colchester County. Nova Scotia. Can. J. Soil Sci. 65, 233-236.

Bell, M.A., 1993. Organic matter, soil properties and wheat production in the high Valley of Mexico. Soil Sci. 156, 86-93.

Belocchi, G., Acutis, M., Fila, G., Donatelli, M., 2002. An indicator of solar radiation model performance based on a fuzzy expert system. Agron. J. 94, 1222-1233.

Bouchon-Meunier, B., 1993. La Logique Floue. Presses Universitaires de France, Paris, France.

Brooks, R.H., Corey, A.T., 1964. Hydraulic properties of porous media, Hydrology Paper No. 3. Colorado State University, Fort Collins, Colorado.

Bruand, A., Duval, O., Gaillard, H., Darthout, R., Jamagne, M., 1996. Variabilité des propriétés de rétention en eau des sols, importance de la densité apparante. Etude Gestion Sols 3, 27-40.

Cahan, S., 1987. Stability and change in human characteristics. Wiley, New York.

Calhoun, F.G., Hammond, L.C., Caldwell, R.E., 1972. Influence of particle size and organic matter on water retention in selected Florida soils. Soil Crop Sci. Fla. Proc. 32, 111-113.

Calzolari, C., Ungano, F., Busoni, E., 2000. The SINA project in the Padano-Veneto basin. In: Proceedings of the International Congress Soil Vulnerability and Sensitivity, pp. 287-307. Florence, Italy.

Campbell, G.S., 1985. Soil physics with basic: transport models for soil–plant system. Elsevier, Amsterdam.

Cook, R.D., Weisberg, S., 1999. Graphs in statistical analysis: is the medium the message? Am. Stat. 53, 29-37.

Cornelis, M., Ronsyn, J., Van Meirvenne, M., Hartmann, R., 2001. Evaluation of Pedotransfer Functions for Predicting the Soil Moisture Retention Curve. Soil Sci. Soc. Am J. 65, 638-648.

Cornelissen, A.M.G., van den Berg, J., Koops, W.J., Grossman, M., Udo, H.M.J., 2001. Assessment of the contribution of sustainability indicators to sustainable development: a novel approach using fuzzy set theory. Agric. Ecosyst. Environ. 86, 173-185.

Dent, J.B., Blackie, M.J., 1979. Systems simulation in agriculture. Applied Science Publisher Ltd, London, United Kingdom.

Donatelli, M., van Ittersum, M.K., Bindi, M., Porter, J.R., 2002. Modelling cropping systems – highlights of the symposium and preface to the special issues. Eur. J. Agron. 18, 1-11.

Donatelli, M., Acutis, M., Belocchi, G., Fila, G., 2004. New indices to quantify patterns of residuals produced by model estimates. Agron. J. 96, 631-645.

Efron, B., Tibshirani, R.J., 1993. An Introduction to the Bootstrap, Monographs on Statistics and Applied Probability, Vol. 57. Chapman & Hall, London, UK.

El-Kadi, A.I., 1985. On estimating the hydraulic properties of soil. 1. Comparison between forms to estimate the soil-water characteristic function. Adv. Water Resour. 8, 136-147.

Finke, P.A., Wösten, J.H.M., Jansen, M.J.W., 1996. Effects of uncertainty in major input variables on simulated functional soil behaviour. Hydrol. Process. 10, 661-669.

Fox, D.G., 1981. Judging air quality model performance: a summary of the AMS workshop on dispersion models performance. Bull. Am. Meteorol. Soc. 62, 599-609.

Gijsman, A.J., Jagtap, S.S., Jones, J.W., 2002. Wading through a swamp of complete confusion: how to choose a method for estimating soil water retention parameters for crop models. Eur. J. Agron. 18, 75-105.

Giménez, D., Rawls, W.J., Lauren, J., 1999. Scaling properties of saturated hydraulic conductivity in soil. Geoderma 88, 205-220.

Gupta, S.C., Larson, W.E., 1979. Estimating soil water characteristic from particle size distribution, organic matter percent, and bulk density. Water Resour. Res. 15, 1633-1635.

Hall, L.O., Kandel, A., 1991. The evolution from expert systems to fuzzy expert systems. In: Kandel, A. (Ed.), Fuzzy Expert Systems Theory. CRC Press, Boca Raton, USA, pp. 3-21.

Hecht-Nielsen, R., 1990. Neurocomputing. Addison-Wesley, Reading, USA.

Heuvelink, G.B.M., 1998. Error propagation in environmental modelling with GIS. Taylor and Francis, London.

Heuvelink, G.B.M., Pebesma, E.J., 1999. Spatial aggregation and soil process modelling. Geoderma 89, 47-65.

Hinkle, D., Wiersma, W., Jurs, S., 1994. Applied statistics for the behavioural sciences, Third edition Houghton Mifflin Company, Boston, USA.

Hutson, J.L., Cass, A., 1987. A retentivity function for use in soil-water simulation model. J. Soil Sci. 38, 105-113.

Hutson, J.L., Wagenet, R.J., 1992. LEACHM, Leaching estimation and chemistry model, Department of Soil, Crop and Atmospheric Sciences Research series no. 92.3. Cornell University, New York.

Imam, B., Sorooshian, S., Mayr, T., Schaap, M., Wösten, J.H.M., Scholes, B., 1999. Comparison of pedotransfer functions to compute water holding capacity using the van Genuchten model in inorganic soils. IGBP-DIS Working Paper #22, The International Geosphere-Biosphere Programme, Stockholm, Sweden.

Jacovides, C.P., Kontoyiannis, H., 1995. Statistical procedures for the evaluation of evapotranspiration computing models. Agric. Water Manag. 27, 365-371.

Janssen, P.H.M., Heuberger, P.S.C., 1995. Calibration of process-oriented models. Ecol. Model. 83, 55-66.

Jaynes, D.B., Tyler, E.J., 1984. Using soil physical properties to estimate hydraulic conductivity. Soil Sci. 138, 298-305.

Jones, D., Barnes, E.M., 2000. Fuzzy composite programming to combine remote sensing and crop models for decision support in precision crop management. Agric. Syst. 65, 137-158.

Jørgensen, S.E., Kamp-Nielsen, L., Christensen, T., Windolf-Nielsen, J., Westergaard, B., 1986. Validation of a prognosis based upon a eutrophication model. Ecol. Model. 35, 165-182.

Keeney, R.L., Raiffa, H., 1993. Decisions with Multiple Objectives: Preferences and Value Tradeoffs. Cambridge University Press, New York.

Kleijnen, J.P.C., 1987. Statistical Tools for Simulation Practitioners. Dekker, New York.

Kobayashi, K., Salam, M.U., 2000. Comparing simulated and measured values using mean squared deviation and its components. Agron. J. 92, 345-352.

Koekkoek, E., Bootlink, H., 1999. Neural network models to predict soil water retention. Eur. J. Soil Sci. 50, 489-495.

Kukkonen, J., Härkönen, J., Walden, J., Karppinen, A., Lusa, K., 2001. Evaluation of the dispersion model CAR-FMI against data from a measurement campaign near a major road. Atmos. Environ. 35, 949-960.

Lenhard, R.J., 1984. Effects of clay–water interactions on water retention in porous media. Ph.D. Thesis, 145 pp. Oregon State University, Corvallis, OR, USA.

Leonavičiutę, N., 2000. Predicting soil bulk and particle densities by pedotransfer functions from existing soil data in Lithuania. Geografijos metraðtis 33, 317-330.

Liao, T.W., 2002. A fuzzy C-medians variant for the generation of fuzzy term sets. Int. J. Intelligent Systems 17, 21-43.

Loague, K., Green, R.E., 1991. Statistical and graphical methods for evaluating solute transport models: overview and application. J. Contam. Hydrol. 7, 51-73.

Martorana, F., Belocchi, G., 1999. A review of methodologies to evaluate agroecosystem simulation models. Ital. J. Agron. 3, 19-39.

Mayer, D.G., Butler, D.G., 1993. Statistical validation. Ecol. Model. 68, 21-32.

Mayr, T., Jarvis, N.J., 1999. Pedotransfer function to estimate soil water retention parameter for a modified Brooks–Corey type model. Geoderma 91, 1-9.

McBratney, A.B., Minasmy, B., Cattle, S.R., Vervoort, R.W., 2002. From pedotransfer functions to soil inference systems. Geoderma 109, 41-73.

Metselaar, K., 1999. Auditing predictive models: a case study in crop growth. Thesis, Wageningen Agricultural University, Wageningen, The Netherlands.

Miklós, D., Kemény, S., Almásy, G., Kollár-Hunek, K., 1995. Thermodynamic consistency of data banks. Fluid Phase Equilibr. 110, 89-113.
Minasny, B., McBratney, A.B., 2002. Uncertainty analysis for pedotransfer functions. Eur. J. Soil Sci. 53, 417-430.
Minasny, B., McBratney, A.B., Bristow, K.L., 1999. Comparison of different approaches to the development of pedotransfer functions for water-retention curves. Geoderma 93, 225-253.
Nemes, A., Schaap, M.G., Wösten, J.H.M., 2003. Functional evaluation of pedotransfer functions derived from different scales of data collection. Soil Sci. Soc. Am. J. 67, 1093-1102.
Pachepsky, Ya.A., Rawls, W.J., 1999. Accuracy and reliability of pedotransfer functions as affected by grouping soils. Soil Sci. Soc. Am. J. 63, 1748-1757.
Pachepsky, Ya.A., Timlin, D., Várallyay, G., 1996. Artificial neural networks to estimate soil water retention from easily measurable data. Soil Sci. Soc. Am. J. 60, 727-773.
Pachepsky, Ya.A., Rawls, W.J., Gimenez, D., Watt, J.P.C., 1998. Use of soil penetration resistance and group method of data handling to improve soil water retention estimates. Soil Till. Res. 49, 117-128.
Pachepsky, Ya.A., Rawls, W.J., Timlin, D.J., 1999. The current status of pedotransfer functions, their accuracy, reliability, and utility in field- and regional-scale modelling. *In*: Corwin, D.L., Loague, K., Ellsworth, T.R. (Eds.), Assessment of non-point source pollution in the vadose zone, Geophysical monograph 108. American Geophysical Union, Washington, DC, USA, pp. 223-234.
Pachepsky, Ya.A., Rawls, W.J., Giménez, D., 2001b. Comparison of soil water retention at field and laboratory scales. Soil Sci. Soc. Am. J. 65, 460-462.
Paydar, Z., Cresswell, H.P., 1996. Water retention in Australian soils. II. Prediction using particle size, bulk density and other properties. Aust. J. Soil Res. 34, 679-693.
Rajkai, K., Várallyay, G., 1992. Estimating soil water retention from simpler properties by regression techniques. *In*: van Genuchten, M.Th., Leij, F.J., Lund, L.J. (Eds.), Indirect methods for estimating the hydraulic properties of unsaturated soils, Proceedings of the International Workshop on Indirect Methods for Estimating the Hydraulic Properties of Unsaturated Soils. Riverside, California, USA, pp. 417-426.
Rawls, W.J., Brakensiek, D.L., 1989. Estimation of soil water retention and hydraulic properties. *In*: Morel, S. (Ed.), Unsaturated flow in hydrologic modeling. Theory and practice. Kluwer academic publishers, Beltsville, USA, pp. 275-300.
Ricker, W.E., 1984. Computation and uses of central trend lines. Can. J. Zool. 62, 1897-1905.
Romano, N., Palladino, M., 2002. Prediction of soil water retention using soil physical data and terrain attributes. J. Hydrol. 265, 65-75.
Schaap, M.G., Leij, F.J., 1998. Database-related accuracy and uncertainty of pedotransfer functions. Soil Sci. 163, 765-779.
Schaap, M.G., Leij, F.J., van Genuchten, M.Th., 1998. Neural network analysis for hierarchical prediction of soil hydraulic properties. Soil Sci. Soc. Am. J. 62, 847-855.
Schaap, M.G., Leij, F.J., van Genuchten, M.T., 2001. ROSETTA: a computer program for estimating soil hydraulic parameters with hierarchical pedotransfer functions. J. Hydrol. 251, 163-176.

Shea, D., 1989. Spectral analysis of performance evaluation for unsaturated flow modeling. MS thesis, Department of Civil Engineering, MIT, Cambridge, USA.

Shein, E.V., Pachepsky, Ya.A., Guber, A.K., Checkhova, T.I., 1995. Experimental determination of hydrophysical and hydrochemical parameters in mathematical models for moisture- and salt transfer in soils. Pochvovedenie 11, 1479-1486.

Sinowski, W., Scheinost, A.C., Auerswald, K., 1997. Reginalization of soil water retention curves in highly variable soilscape. II. Comparison of regionalization procedures using a pedotransfer function. Geoderma 78, 145-159.

Smith, P., Smith, J.U., Powlson, D.S., McGill, W.B., Arah, J.R.M., Chertov, O.G., Coleman, K., Franko, U., Frolking, S., Jenkinson, D.S., Jensen, L.S., Kelly, R.H., Klein-Gunnewiek, H., Komarov, A.S., Li, C., Molina, J.A.E., Mueller, T., Parton, W.J., Thornley, J.H.M., Whitmore, A.P., 1997. A comparison of the performance of nine soil organic matter models using datasets from seven long-term experiments. Geoderma 81, 153-225.

Sonneveld, M.P.W., Backx, M.A.H.M., Bouma, J., 2003. Simulation of soil water regimes including pedotransfer functions and land-use related preferential flow. Geoderma 112, 97-110.

Sorooshian, S., Gupta, V.K., Fulton, J.L., 1983. Evaluation of maximum likelihood parameter estimation techniques for conceptual runoff models: influence of calibration data variability and length on model credibility. Water Resour. Res. 1, 251-259.

Sorooshian, S., Duan, Q., Gupta, V.K., 1993. Calibration of rainfall-runoff models: application of global optimization to the Sacramento Soil Moisture. Model. Water Resour. Res. 29, 1185-1194.

Sugeno, M., 1985. An introductory survey of fuzzy control. Inform. Sci. 36, 59-83.

Tietje, O., Hennings, V., 1996. Accuracy of the saturated hydraulic conductivity prediction by pedo-transfer functions compared to the variability within FAO textural classes. Geoderma 69, 71-84.

Tietje, O., Tapkenhinrichs, M., 1993. Evaluation of pedo-transfer functions. Soil Sci. Soc. Am. J. 57, 1088-1095.

Timlin, D.J., Ahuja, L.R., Pachepsky, Ya.A., Williams, R.D., Gimenez, D., Rawls, W.J., 1999. Use of Brooks Corey parameters to improve estimates of saturated conductivity from effective porosity. Soil Sci. Soc. Am. J. 63, 1086-1092.

Tomasella, J., Hodnett, M.G., 1998. Estimating soil water retention characteristics from limited data in Brazilian Amazonia. Soil Sci. 163, 190-202.

Ungaro, F., Calzolari, C., 1998. Caratterizzazione delle proprietà idrologiche dei suoli: modellizazione delle curve di ritenzione, applicazione e validazione delle pedofunzioni di trasferimento. Ministero per l'Ambiente- Progetto SINA Carta Pedologica, Sottoprogetto 2, Rapporto 4.1. CNR-IGES Istituto per la Genesi e l'Ecologia del Suolo, Florence, Italy.

van Alphen, B.J., Booltink, H.W.G., Bouma, J., 2001. Combining pedotransfer functions with physical measurements to improve the estimation of soil hydraulic properties. Geoderma 103, 133-147.

van der Werf, H.M.G., Zimmer, C., 1998. An indicator of pesticide environmental impact based on a fuzzy expert system. Chemosphere 36, 2225-2249.

van Genuchten, M.Th., 1980. A closed-form equation for predicting the hydraulic conductivity of unsaturated soil. Soil Sci. Soc. Am. J. 44, 892-899.

Verburg, K., Bond, W.J., Bristow, K.L., Cresswell, H.P., Ross, P.J., 1996. Functional sensitivity and uncertainty analysis of water and solute transport predictions. In: Proceedings of the Australian and New Zealand National Soil Conference, pp. 299–300. Melbourne, Australia.

Vereecken, H., Feyen, J., Maes, J., Darius, P., 1989. Estimating the soil moisture retention characteristic from texture, bulk density, and carbon content. Soil Sci. 148, 389-403.

Vereecken, H., Diels, J., van Orshoven, J., Feyen, J., Bouma, J., 1992. Functional evaluation of pedotransfer functions for the estimation of soil hydraulic properties. Soil Sci. Soc. Am. J. 56, 1371-1378.

Wagner, B., Tarnawski, V.R., Hennings, V., Muller, U., Wessolek, G., Plagge, R., 2001. Evaluation of pedo-transfer functions for unsaturated soil hydraulic conductivity using an independent data set. Geoderma 102, 275-297.

Wallach, D., 1999. Linking model validation, uncertainty analysis and parameter adjustment. In: Proceedings of the 1st International Symposium on Modelling Cropping Systems, Lleida, Spain, pp. 271–272.

Wallach, D., Goffinet, B., Bergez, J.-E., Debaeke, P., Leenhardt, D., Aubertot, J.N., 2001. Parameter estimation of crop models: a new approach and application to a corn model. Agron. J. 93, 757-766.

Willmott, C.J., 1981. On the validation of models. Phys. Geogr. 2, 184-194.

Willmott, C.J., 1982. Some comments on the evaluation of model performance. Bull. Am. Meteorol. Soc. 63, 1309-1313.

Wisniak, J., Polishuk, A., 1999. Analysis of residuals – a useful tool for phase equilibrium data analysis. Fluid Phase Equilibr. 164, 61-82.

Wopereis, M.C.S., Wösten, J.H.M., ten Berge, H.F.M., Woodhead, T., de San Augustin, E.M., 1993. Comparing the performance of a soil–water balance model using measured and calibrated hydraulic conductivity data, a case study for the dryland rice. Soil Sci. 156, 133-140.

Workman, S.R., Skaggs, R.W., 1992. Sensitivity of water management models to errors in soil hydraulic properties, Paper presented at the 1992 International Winter Meeting of ASAE, Nashville, USA, Paper No. 922567.

Wösten, J.H.M., Bannink, M.H., de Gruijter, J., Bouma, J., 1986. A procedure to identify different groups of hydraulic conductivity and moisture retention curves for soil horizons. J. Hydrol. 86, 133-145.

Wösten, J.H.M., Schuren, C.H.J.E., Bouma, J., Stein, A., 1990a. Functional sensitivity analysis of four methods to generate soil hydraulic functions. Soil Sci. Soc. Am. J. 54, 832-836.

Wösten, J.H.M., Schuren, C.H.J.E., Bouma, J., Stein, A., 1990b. Use of practical aspects of soil behavior to evaluate different methods to generate soil hydraulic functions. Hydrol. Process. 4, 299-310.

Wösten, J.H.M., Finke, P.A., Jansen, M.J.W., 1995. Comparison of class and continuous pedotransfer functions to generate soil hydraulic characteristics. Geoderma 66, 227-237.

Wösten, J.H.M., Lilly, A., Nemes, A., Le Bas, C., 1999. Development and use of a database of hydraulic properties of European soils. Geoderma 90, 169-185.

Wösten, J.H.M., Pachepsky, Ya.A., Rawls, W.J., 2001. Pedotransfer functions: bridging the gap between available basic soil data and missing soil hydraulic characteristics. J. Hydrol. 251, 123-150.

Yang, J., Greenwood, D.J., Rowell, D.L., Wadsworth, G.A., Burns, I.G., 2000. Statistical methods for evaluating a crop nitrogen simulation model, N_ABLE. Agric. Syst. 64, 37-53.

Zadeh, L.A., 1965. Fuzzy sets. Information Control 8, 338-353.

PART VI

PEDOTRANSFER FUNCTIONS DEVELOPED FOR DIFFERENT REGIONS OF THE WORLD

Chapter 21

PEDOTRANSFER FUNCTIONS FOR TROPICAL SOILS

J. Tomasella[1,*] and M. Hodnett[2]

[1]INPE/CPTEC, Rod. Presidente Dutra km. 39, 12630-000 Cachoeira Paulista/SP, Brazil

[2]Centre for Ecology and Hydrology, Wallingford, OX10 8BB, United Kingdom

*Corresponding author: Tel.: +55-12-3186-8461; fax: + 55-12-3101-2835

1. INTRODUCTION

Most of the pedotransfer functions (PTFs) available in the literature have been derived and tested using extensive databases of soils of temperate regions. The lack of data for tropical soils has been pointed out as a major constraint for the development of PTFs for tropical soils (Hodnett and Tomasella, 2002). More recently, the increased interest of understanding the effect of land use and land cover change in the tropics on global climate has raised the need for an improved knowledge of the hydrological functioning of tropical soils. The impacts of such changes are usually assessed using general circulation models, or GCMs, which require detailed soil information on a global scale. Since tropical soils have been surveyed from a pedological perspective, with very little information of the hydraulic characteristics, PTFs are the only tool that can provide the necessary hydraulic information on the spatial scale needed by GCMs.

Kaolinitic tropical soils have usually clay contents ranging from 60 to 90%. In temperate climates, soils with more than 60% of clay are considered as low permeability heavy clays and are regarded as "non-agricultural soils" (Carsel and Parrish, 1988). Measurements from Correa (1984) and Tomasella and Hodnett (1996), among others, clearly showed that kaolinitic tropical soils show "unusual" properties when compared with typical temperate clayey soils: low bulk density (0.7–1.2 Mg m^{-3}), high permeability (K_{sat} usually 10–1000 mm h^{-1}), have low available water capacity (AWC) (70 mm m^{-1}), and almost 80% of the plant available water between −10 and −100 kPa (Demattê, 1988).

The pronounced differences between temperate and tropical clayey soils are usually explained by the micro-aggregated structure of oxisols. In kaolinitic tropical soils, major cations such as Ca^{2+}, Mg^{2+}, K^+ and silica are eliminated from the soil profile as a result of the high rainfall and continuous leaching (Vieira and Santos, 1987). The increasing concentration of hydrogen relative to basic cations results in low pH. The removal of alkali elements, added to the transport of oxides from the upper horizon, increase the concentration of sesquioxides (compounds of Fe^{3+} and Al^{3+}) in the B horizon (Sanchez, 1976; Vieira and Santos, 1987). In oxisols and nitosols, Fe and Al oxides play an important role as binding agents of negatively charged clay minerals, creating stable micro-aggregates within the size range of silt to fine sand. This explains why their field texture is loamy rather than clayey as determined by laboratory analysis (Cassel and Lal, 1992). Since oxisols are very permeable, most of the soil water is released between saturation and water potentials above -10 kPa. This behavior resembles that of sandy soils, although the water contents are comparatively higher because of the clayey character of oxisols. (Sharma and Uehara, 1986). At lower potentials a significant proportion of water is held within the micro-aggregates. Therefore, the water retention curve is almost flat from -100 kPa up to water potentials as low as -4000 kPa (Chauvel et al., 1991), where a sudden drop of soil water content occurs, which probably coincides with the air-entry value of the aggregates. Since water below 1500 kPa is no longer available for the majority of plants, the plant available water in oxisols is lower compared to "temperate" clays.

Although the mineralogical composition and the characteristic chemical processes cause oxisols and related soils to be characterized by uniform texture, high friability, and the presence of extremely stable micro-aggregates (Demattê, 1988), this observation cannot be generalized to other soils of the tropics, particularly those under intensive land use. Surface horizons of many soils of the tropics, with relatively less clay and organic matter, are less aggregated than soils of temperate zones (Cassel and Lal, 1992).

Physical and chemical differences between temperate and tropical soils might explain why the PTFs derived for soils of temperate climate appeared to be inadequate for oxisols and related soils (van den Berg et al., 1997). It may also be argued that the poor performance of "temperate soil" PTFs arises because the clay content of oxisols frequently exceeds 60%, while PTFs developed for temperate soils often do not cover that range: as an example, the PTF by Rawls and Brakensiek (1985) is only valid for soils with clay contents between 5 and 60%. Tomasella et al. (2000) compared the performance of a PTF derived for Brazilian soils with a range of temperate soil PTFs and concluded that the former performed substantially better, even within the range of validity of latter. This result implies that there must be a marked difference in the hydraulic properties of tropical and temperate soils (Hodnett and Tomasella, 2002). It is not surprising then that the PTF proposed by Tomasella et al. (2000) performed better in Cuban oxisols (Medina et al., 2002) compared to temperate soil PTFs.

More recently, Sobieraj et al. (2001) argued that the hydrological behavior of tropical soils cannot be attributed to the mineralogical factors as it had been suggested by Tomasella and Hodnett (1996). Based on data from kaolinitic soils with hydrated vermiculite interlayers from western Amazonia (which are similar to ultisols found everywhere in the world, according to the authors), they concluded that the poor performance of temperate PTFs in predicting saturated hydraulic conductivity resulted from the lack of ability to reproduce the effects of macroporosity in tropical soils. This may indeed be the case, but it is likely that tropical soils typically have a greater macroporosity than temperate soils, and there is no reason why this should not be linked to their different mineralogy.

The aim of this work is to compare the performance of well-documented PTFs developed in tropical soils. As suggested by Pachepsky et al. (1999), the performance was examined in terms of their accuracy, reliability and utility. The potential effects of bulk density, selection of independent variables, and the methodology used for deriving the PTFs on their performance is discussed.

2. MATERIALS AND METHODS

The limited availability of detailed soil information in the tropics has generally precluded the development of PTFs able to provide hydraulic parameters in great detail. Most of the PTFs developed for tropical soils are limited to the prediction of the water content at a few water potentials, mostly at -10, -33 and -1500 kPa. Therefore, to compare the performance of PTFs developed for tropical soils on a common basis, it is necessary to constrain comparisons to a few number of measurements of water retention. It might be argued that modern soil moisture models demand the complete knowledge of the retention characteristics, rather than the discrete description provided by measurement points. However, Tomasella et al. (2003) recently showed that the calculation of soil water-retention characteristics based on the estimation of water content at selected water potentials using a PTF is, at least in the Brazilian soils, more accurate than the use of a PTF to estimate the retention parameters of the equation proposed by van Genuchten (1980). A possible explanation for this result, according to the authors, "might be related to the fact that soil moisture is controlled by different independent variables at different ranges of soil water potentials." This conclusion is in agreement with the results of van den Berg et al. (1997), who concluded that methods based on direct regressions are superior to methods based on parameter estimation.

A selection of PTFs developed using data of soils of the tropics is presented in Table 1. We selected PTFs that provide at least two points of the retention curve. Most of the PTFs have been developed for application within restricted geographical domains, and for a limited range of soil texture and soil types.

The performance of PTFs is usually characterized using statistics such as the mean squared error, mean error, etc., derived through the comparison between measured and estimated values of water contents. For many practical applications, these statistics are not sufficient, since they do not reflect the effect of the uncertainty in the estimated hydraulic parameter on the results of the applications (Pachepsky et al., 1999). One of the most sensitive soil parameters in many water balance models is the AWC. The AWC is generally determined as the difference between the water content at the conventional "field capacity" (-33 kPa) and a conventional "wilting point" (-1500 kPa). Crop yield models are quite sensitive to AWC (Pachepsky et al., 1999) since it represents an estimate of the water available for crops.

In order to make comparisons of the performance of PTFs developed for tropical soils on a common basis, we tested the PTFs shown in Table 1 for estimating the water content at water potential of -10, -33 and -1500 kPa, and the AWC between -10 and -1500 kPa and between -33 and -1500 kPa. The water content at -33 kPa as a measure of "field capacity" has often been considered to be too low for tropical soils (Pidgeon, 1972; Babalola, 1979; Lal, 1979; Reichardt, 1988; van den Berg et al., 1997). According to these authors, the water content at -10 kPa of water potential (or even as high as -6 kPa) provides a more accurate estimation of field capacity. For this reason, several PTFs developed for tropical soils include the estimation of water content at -10 kPa.

The performance of tropical PTFs was tested using the dataset of Hodnett and Tomasella (2002). The authors selected 771 horizons from 249 soil profiles of 22 countries from the IGBP-DIS soils database obtained from ISRIC. Each horizon had eight points of

Table 1
List of selected PTFs derived using data from tropical soils

Source	Outputs	Independent variables	Geographical domain
Aina and Periaswamy (1985)	$AWC_{33-1500\ kPa}$, $\theta_{33\ kPa}$, $\theta_{1500\ kPa}$	Sand, silt, clay, bulk density	Nigeria
Arruda et al. (1987)	$w_{33\ kPa}$, $w_{1500\ kPa}$	Silt and clay	South-East Brazil
Dijkerman (1988)	$w_{33\ kPa}$, $w_{1500\ kPa}$	Sand and clay	Sierra Leone
Hodnett and Tomasella (2002)	Van Genuchten retention parameters	Sand, silt, clay, CEC^a, OC^b, bulk density, pH	World tropical soils
Lal (1979)	$\theta_{10\ kPa}$, $\theta_{33\ kPa}$, $\theta_{1500\ kPa}$	Sand, clay	Nigeria
Oliveira et al. (2002)	$AWC_{33-1500\ kPa}$, $\theta_{33\ kPa}$, $\theta_{1500\ kPa}$	Sand, silt, clay, bulk density	North-East Brazil
Pidgeon (1972)	$w_{10\ kPa}$, $w_{33\ kPa}$, $w_{1500\ kPa}$	Sand, silt, clay, OC^b	Uganda
Tomasella et al. (2000)	Van Genuchten retention parameters	Coarse sand, fine sand, silt, clay, bulk density, OC^b, Me^c	Brazil
Tomasella et al. (2003)	Porosity, $\theta_{6\ kPa}$, $\theta_{10\ kPa}$, $\theta_{33\ kPa}$, $\theta_{100\ kPa}$, $\theta_{1500\ kPa}$	Coarse sand, fine sand, silt, clay, bulk density, OC^b, Me^c	Brazil
van den Berg et al. (1997)	Van Genuchten retention parameters, $AWC_{10-1500\ kPa}$, $\theta_{10\ kPa}$, $\theta_{1500\ kPa}$	Clay, silt, bulk density, OC^b, SS^d	World oxisols and related soils

θ = volumetric water content at a specific water potential; w = gravimetric water content at a specific water potential; AWC = Available water content within a range of water potential.
[a] Cation exchange capacity (cmol kg^{-1}).
[b] Organic carbon (g kg^{-1}).
[c] Moisture equivalent (g g^{-1}).
[d] Specific surface (m^2 g^{-1}).

the water-retention curve. This dataset was split randomly into two groups: (i) a calibration dataset, (492 curves); and (ii) a validation dataset (279 curves). The calibration dataset was used to derive the PTF of Hodnett and Tomasella (2002), and the validation dataset was used to test the PTF that was developed. Since the PTF proposed by Hodnett and Tomasella (2002) has been included in Table 1, all comparisons have been carried out using the validation dataset.

Since most of the tropical PTFs are site specific and have been developed for a limited range of soils, estimations of the output variables of Table 1 were restricted to the range of independent variables provided by the authors. Consequently, some PTFs were only applicable to a limited number of soil profiles. The results are therefore shown with an applicability index, defined as the ratio between the number of times that a particular PTF could be applied and the total number of data points (279), expressed in percentage. It should be noted that the application was not constrained by soil types, although some of the PTFs shown in Table 1 were developed for specific soils.

Several PTFs provide more then one equation depending on the basic soil data available. In all cases, except one, the variant that provided the best fitting to basic soil data (according to the original authors) was used. The only exception was the most accurate equation for estimating AWC suggested by van den Berg et al. (1997). This equation includes specific surface that was not known for soils in the validation dataset of Hodnett and Tomasella (2002). Therefore, the next more accurate PTF of van den Berg et al. (1997) was used to estimate AWC. In the case of the PTF of Lal (1979), the equations derived for group I were used, since they provided the best fittings. The PTF developed by Pidgeon (1972) estimates the "field capacity" equivalent to the soil moisture of a wetted plot after 48 h of free drainage. The author also provided equations that convert this value into water content of undisturbed cores at -10 and -33 kPa. These equations were used in this work for comparisons with other PTFs. The PTFs of Tomasella et al. (2000, 2003) use moisture equivalent as an input. This is basic information found in most Brazilian soil surveys, but it was not available in the dataset used by Hodnett and Tomasella (2002). For this purpose, it was estimated as the water content at -33 kPa expressed in gravimetric terms. It has been shown by Tomasella et al. (2000) that the gravimetric water content at -33 kPa provides an accurate estimation of the moisture equivalent in Brazilian soils. In the appendix, a complete list of the PTFs used in the comparisons is included.

The performance of different PTFs was evaluated using the determination coefficient, R^2; the mean error, ME, and the root mean squared error, RMSE, defined as:

$$\mathrm{ME} = \frac{\sum(\theta_m - \theta_p)}{\mathrm{NP}}$$

$$\mathrm{RMSE} = \sqrt{\frac{\sum(\theta_m - \theta_p)^2}{\mathrm{NP}}}$$

where the subscript p and m indicate predicted and measured values, and NP is the number of data points.

3. RESULTS

Table 2 presents the applicability index and statistics R^2, ME and RMSE resulting from the estimation of water content at -10, -33 and -1500 kPa. As expected, the PTF of Hodnett and Tomasella (2002) was applicable in all the samples, followed by the PTF proposed by Tomasella et al. (2000), which was applicable in 85% of the cases. The PTFs of Aina and Perisawamy (1985) and Lal (1979) were applicable to less than 20% of the dataset, while for the remainder of the PTFs the applicability varied from 60 to 81%.

In terms of ME, which is an indication of the bias of the estimation, the PTF of Hodnett and Tomasella (2002) presented the best performance for the water content at -10 and -1500 kPa. For the ME at -33 kPa and for the statistics RMSE and R^2 at all water potentials, the PTF of Tomasella et al. (2003) was superior to all of the PTFs tested. It is interesting to note that in terms of the RMSE and R^2, the PTF proposed by Tomasella et al. (2000) presented better performance than the PTF of Hodnett and Tomasella (2002), but slightly less performance for the water content at -1500 kPa. As general observation, the PTF of Tomasella et al. (2000) seems to underestimate values at -10 kPa in agreement with the conclusions of Medina et al. (2002).

Table 3 compares the performance of the selected PTFs for estimating $AWC_{10-1500\ kPa}$ and $AWC_{33-1500\ kPa}$. In terms of ME, the PTF of Hodnett and Tomasella (2002) presented the minimum biased in the range 10–1500 kPa, while the PTF proposed by Tomasella et al. (2003) was the least biased in the range 33–1500 kPa. In both ranges, PTF from Lal (1979) showed values of ME very close to the minimum.

The analysis of the values of RMSE revealed that the PTF of Lal (1979) had better performance in the range 10–1500 kPa, while the PTF of Tomasella et al. (2003) was slightly better in the range 33–1500 kPa. With the exception of the PTF of Lal (1979), all other PTFs tested in the range 10–1500 kPa showed similar performance in terms of the RMSE. Considering the values of R^2, the PTF proposed by Tomasella et al. (2003) was superior for both $AWC_{10-1500\ kPa}$ and $AWC_{33-1500\ kPa}$.

4. DISCUSSION

Table 2 demonstrated that the PTF proposed by Tomasella et al. (2003) was more accurate than the PTF of Hodnett and Tomasella (2002), at least for the water content at -10, -33 and -1500 kPa. This result was somewhat surprising, since Hodnett and Tomasella (2002) used IGBP data from 22 countries in the derivation, and the PTF was applicable to all of the soils in the dataset to which it was applied in this test. The PTF of Tomasella et al. (2003) was derived using only Brazilian soil data and was applicable to only 79% of the dataset, raising the possibility that the remaining datapoints could have introduced significant errors, which might have adversely affected the overall performance of the PTF. Toward that end, Hodnett and Tomasella (2002) showed that their PTF could not reproduce well the water release curves for soils with bulk density below 0.8 Mg m^{-3}. Their PTF was developed for soil with bulk densities greater than 0.28 Mg m^{-3}, in contrast with the PTF of Tomasella et al. (2003) which is only applicable for soils with bulk density greater than 0.72 Mg m^{-3}.

To investigate whether soils with bulk density less than 0.72 Mg m^{-3} affected the performance of the PTF of Hodnett and Tomasella (2002), the absolute error for -10 kPa

Table 2
Performance of different PTF at selected water potentials in terms of ME, RMSE, and the determination coefficient, R^2

PTF	Applicability (%)	−10 kPa			−33 kPa			−1500 kPa		
		ME ($m^3\ m^{-3}$)	RMSE ($m^3\ m^{-3}$)	R^2	ME ($m^3\ m^{-3}$)	RMSE ($m^3\ m^{-3}$)	R^2	ME ($m^3\ m^{-3}$)	RMSE ($m^3\ m^{-3}$)	R^2
Aina and Periaswamy (1985)	13	nd	nd	nd	−0.067	0.088	0.406	0.071	0.084	0.757
Arruda et al. (1987)	75	nd	nd	nd	0.070	0.155	0.027	0.115	0.148	0.106
Dijkerman (1988)	81	nd	nd	nd	0.072	0.138	0.199	0.047	0.098	0.290
Hodnett and Tomasella (2002)	100	−0.005	0.066	0.689	0.045	0.084	0.674	−0.004	0.065	0.675
Lal (1979)	20	0.082	0.097	0.417	0.085	0.102	0.339	0.079	0.092	0.466
Oliveira et al. (2002)	51	nd	nd	nd	0.056	0.101	0.443	0.046	0.070	0.693
Pidgeon (1972)	47	0.083	0.152	0.078	0.101	0.159	0.044	0.044	0.101	0.047
Tomasella et al. (2000)	85	−0.043	0.060	0.842	0.016	0.052	0.789	−0.008	0.066	0.602
Tomasella et al. (2003)	79	−0.032	0.037	0.953	0.011	0.017	0.993	0.007	0.044	0.801
van den Berg et al. (1997)	74	0.046	0.085	0.536	nd	nd	nd	0.033	0.073	0.488

Table 3
Performance of different PTFs for estimating water-holding capacity in terms of ME, RMSE, and the determination coefficient, R^2

PTF	Applicability (%)	$AWC_{10-1500\ kPa}$			$AWC_{33-1500\ kPa}$		
		ME ($m^3\ m^{-3}$)	RMSE ($m^3\ m^{-3}$)	R^2	ME ($m^3\ m^{-3}$)	RMSE ($m^3\ m^{-3}$)	R^2
Aina and Periaswamy (1985)	13	nd	nd	nd	−0.014	0.071	0.001
Arruda et al. (1987)	75	nd	nd	nd	−0.045	0.120	0.182
Dijkerman (1988)	81	nd	nd	nd	0.026	0.096	0.055
Hodnett and Tomasella (2002)	100	−0.002	0.067	0.374	0.048	0.081	0.313
Lal (1979)	20	0.003	0.043	0.033	0.006	0.044	0.001
Oliveira et al. (2002)	61	nd	nd	nd	0.006	0.072	0.000
Pidgeon (1972)	47	nd	nd	nd	0.013	0.068	0.102
Tomasella et al. (2000)	85	−0.035	0.065	0.297	0.023	0.055	0.388
Tomasella et al. (2003)	79	−0.038	0.061	0.514	0.004	0.040	0.578
van den Berg et al. (1997)	74	0.025	0.070	0.145	nd	nd	nd

of water potential was plotted as a function of the soil bulk density for both PTFs (Figure 1). The figure clearly shows that the PTF of Tomasella et al. (2003) consistently underestimated water content, particularly for low-density soils. Medina et al. (2002) concluded that the PTF of Tomasella et al. (2000), which was derived using the same Brazilian dataset, systematically underestimated water contents close to saturation. Therefore, a possible explanation for the underestimation of water content at −10 kPa of the PTF of Tomasella et al. (2003) could be a bias introduced by using only Brazilian dataset in PTF derivation.

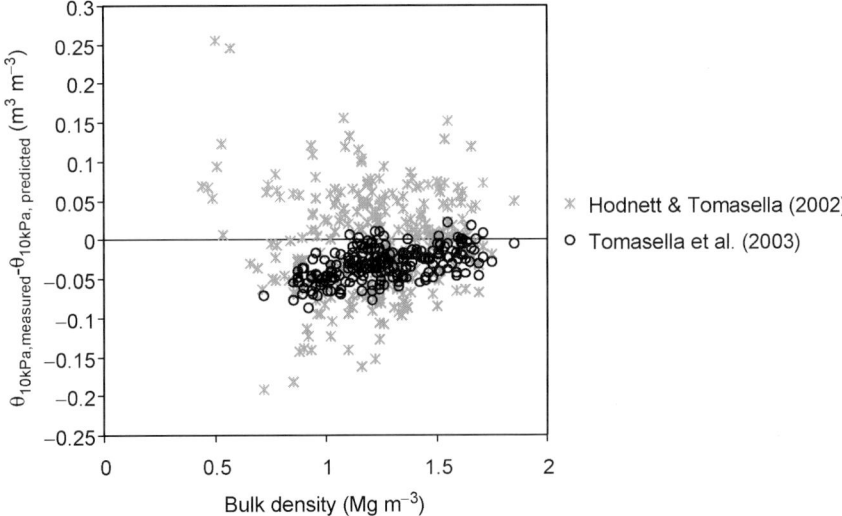

Figure 1. Differences between measured and predicted values of soil moisture at −10 kPa as a function of the bulk density for different PTFs.

Another conclusion from Figure 1 is that the PTF of Tomasella et al. (2003) shows much less scatter and smaller deviation from measurements compared with the PTF of Hodnett and Tomasella (2002). Although the discrepancies between measurements and estimations are greater in soils with low bulk density, the hypothesis that low-density soils are completely responsible for the inferior performance of Hodnett and Tomasella (2002) PTF must be disregarded.

Another argument could be related to the basic soil information used in the estimation: the PTF of Tomasella et al. (2003) uses moisture equivalent as an independent variable (which is very close to gravimetric water content at −33 kPa), which may be a better predictor than the independent variables used by Hodnett and Tomasella (2002). However, a close analysis of Table 2 shows that the PTF of Tomasella et al. (2003) performed better than the PTF of Tomasella et al. (2000), which used the same set of independent variables.

Since PTF of Hodnett and Tomasella (2002) is based on multiple regression, it is possible that the methodology used by Tomasella et al. (2003) for developing their PTF, namely the GMDH algorithm, could have some impact on the quality of the estimation. However, the most likely explanation for the better performance of the PTF of Tomasella et al. (2003) is that using PTFs to make direct estimates of water content at given potentials appears to be a more accurate procedure than estimating these water contents from the van Genuchten (1980) equation for water retention, after using a PTF to estimate the parameters of this equation.

The reason why the other PTFs in Table 1 were less successful in estimating water content could be related to their relatively simple formulations, with a very limited number of independent variables, which are soil and site specific. This would have constrained their ability to represent the variability encountered in the "global" IGBP soil database. Although the ranges of texture where those PTFs were developed are in some cases relatively wide (for instance, Arruda et al., 1987) and van den Berg et al. (1997) are applicable to 75% of the dataset), those formulations have developed for addressing specific soil types at local scale, and it is not expected that they could represent a great variety of soils.

In spite of its simplicity, the PTF of Lal (1979) showed a remarkable performance for estimating AWC in terms of the RMSE (Table 3). Since the application of this PTF was limited to 20% of the dataset, the range of estimated water content was narrow. This explains why the values of R^2 were relatively low for this PTF.

In general, the performance of tropical soil PTFs is comparable with those for temperate soils: Table 1 of Pachepsky et al. (1999) showed values of RMSE varying between 0.02 and 0.07 $m^3 m^{-3}$ for various PTFs developed in the US and Europe.

5. CONCLUSIONS

In general, the PTF of Tomasella et al. (2003) showed the best performance for estimating water content at −10, −33 and −1500 kPa. The reason why this PTF produced the best result might be related to the development based on the prediction of individual water-retention points rather the than the parameters of an analytical retention curve.

With regard to the estimation of $AWC_{10-1500\ kPa}$ and $AWC_{33-1500\ kPa}$, the PTF of Lal (1979) exhibited a good performance, although the range of applicability was limited to 20% of total data points. The PTF of Tomasella et al. (2003) performed better in the range between −33 and −1500 kPa, but was less accurate in the range between −10 and −1500 kPa.

The overall accuracy of PTFs for tropical soils seems to be affected by the presence of soils having low bulk density. Those soils should be treated in a separate group for which PTFs should be derived separately. The good performance of the PTF of Tomasella et al. (2003) encourages the development of PTF based on the estimation of individual water contents rather than the estimation of the parameters of the analytical equation, at least in tropical soils.

Considering the wide range of soils to which the PTFs were applied in this study, the general performance of PTFs developed for tropical soils is quite acceptable and certainly comparable to that of the PTFs for the soils of temperate regions.

ACKNOWLEDGEMENTS

The first author received financial support form the Fundação de Amparo a Pesquisa do Estado de São Paulo – Fapesp.

APPENDIX. PTFs USED TO ESTIMATE VOLUMETRIC WATER CONTENT AT SELECTED POTENTIALS AND AVAILABLE WATER CAPACITY

Symbols: CS, coarse sand (%); FS, fine sand (%); SA, sand (%); SI, silt (%); CL, clay (%); OC, organic carbon (g kg^{-1}); Db, bulk density (Mg m^{-3}); Me, moisture equivalent (g g^{-1}); CEC, cation exchange capacity (cmol kg^{-1}); $\theta_{10\ kPa}$, $\theta_{33\ kPa}$, $\theta_{1500\ kPa}$, volumetric water content at 10, 33 and 1500 kPa, respectively (m^3 m^{-3}); $w_{10\ kPa}$, $w_{33\ kPa}$, $w_{1500\ kPa}$, gravimetric water content at 10, 33 and 1500 kPa, respectively (g g^{-1}); AWC$_{33-1500\ kPa}$, available water capacity between 33 and 1500 kPa (m^3 m^{-3}); AWC$_{10-1500\ kPa}$, available water capacity between 10 and 1500 kPa (m^3 m^{-3}); α, n, θ_s, θ_r, van Genuchten retention parameters; x_{14-17}, z_{9-11}, auxiliary variables.

Aina and Periaswamy (1985)

$$\theta_{33\ kPa} = 0.6788 - 0.0055SA - 0.0013Db$$

$$\theta_{1500\ kPa} = 0.00213 + 0.0031CL$$

$$AWC_{33-1500\ kPa} = 01401 + 0.0003SICL - 0.0878Db$$

Arruda et al. (1987)

$$w_{33\ kPa} = 6.71198 \times 10^{-2} \exp[4.28 \times 10^{-2}(SI+CL) - 3.949 \times 10^{-4}(SI+CL)^2 + 6.68 \times 10^{-7}(SI+CL)^3]$$

$$w_{1500\ kPa} = 2.3662 \times 10^{-2}(SI+CL)^{1.20408 - 0.0872025\ \log(SI+CL)}$$

Dijkerman (1988)

$$w_{33\ kPa} = 0.3697 - 0.0035SA$$

$$w_{1500\ kPa} = 0.0074 + 0.0039CL$$

Hodnett and Tomasella (2002)

$$\ln \alpha = -0.02294 - 0.03526SI + 0.024OC - 7.6 \times 10^{-3}CEC - 0.11331pH$$

$$\ln n = 0.62986 - 0.00833CL - 0.00529OC + 0.00593pH + 7 \times 10^{-5}CL^2 - 1.4 \times 10^{-4}SASI$$

$$\theta_s = 0.81799 + 9.9 \times 10^{-4}\text{CL} - 0.3142\text{Db} + 1.8 \times 10^{-4}\text{CEC} + 0.00451\text{pH}$$
$$- 5 \times 10^{-6}\text{SACL}$$

$$\theta_r = 0.22733 - 0.00164\text{SA} + 0.00235\text{CEC} - 0.00831\text{pH} + 1.8 \times 10^{-5}\text{CL}^2$$
$$+ 2.6 \times 10^{-5}\text{SACL}$$

Lal (1979)

$$\theta_{10\ \text{kPa}} = 0.102 + 0.003\text{CL}$$

$$\theta_{33\ \text{kPa}} = 0.065 + 0.004\text{CL}$$

$$\theta_{1500\ \text{kPa}} = 0.006 + 0.003\text{CL}$$

Oliveira et al. (2002)

$$w_{33\ \text{kPa}} = 0.00333\text{SI} + 0.00387\text{CL}$$

$$w_{1500\ \text{kPa}} = 3.8 \times 10^{-4}\text{SA} + 0.00153\text{SI} + 0.00341\text{CL} + 0.030861\text{Db}$$

$$w_{33-1500\ \text{kPa}} = -2.1 \times 10^{-4}\text{SA} + 0.00203\text{SI} + 0.00054\text{CL} + 0.021656\text{Db}$$

Pidgeon (1972)

$$w_{\text{FC}} = 0.0738 + 0.0016\text{SI} + 0.003\text{CL} + 0.03\text{OC}$$

$$w_{10\ \text{kPa}} = \frac{100w_{\text{FC}} - 2.54}{91}$$

$$w_{33\ \text{kPa}} = \frac{100w_{\text{FC}} - 3.77}{95}$$

$$w_{1500\ \text{kPa}} = -0.0419 + 0.0019\text{SI} + 0.0039\text{CL} + 0.009\text{OC}$$

$$\text{AWC}_{33-1500\ \text{kPa}} = 0.1693 - 0.0015\text{CL} + 0.01218\text{OC}$$

Tomasella et al. (2000)

$$\ln \alpha = 2.6446 + 0.0121\text{CL} - 3.7861\text{Me} - 3.2834\text{Db} + 5.2 \times 10^{-5}\text{CSFS}$$
$$+ 9.63 \times 10^{-4}\text{FSCL} + 6.16 \times 10^{-4}\text{CS}^2$$

$$n = 2.1909 - 1.5296\text{Me} - 2.99 \times 10^{-4}\text{CSSI} - 3.45 \times 10^{-4}\text{FSCL} - 1.05 \times 10^{-4}\text{CS}^2$$
$$+ 2.5 \times 10^{-5}\text{FS}^2$$

$$\theta_s = 0.8219 - 1.77 \times 10^{-4}\text{SI} + 0.2324\text{Me} - 0.2867\text{Db} + 4.9 \times 10^{-5}\text{CSSI}$$
$$- 2.9 \times 10^{-5}\text{CSCL} + 2.7 \times 10^{-5}\text{FSCL} - 8 \times 10^{-6}\text{CS}^2$$

$$\theta_r = -0.1336 + 0.0025\text{SI} + 0.0034\text{CL} + 0.3991\text{Me} + 0.0768\text{Db} - 4.8 \times 10^{-5}\text{SI}^2$$
$$- 1.3 \times 10^{-5}\text{CL}^2$$

Tomasella et al. (2003)

$$x_{14} = -1.05501 + 0.0650857\text{SI}$$

$$x_{15} = -2.07588 + 0.0423954\text{CL}$$

$$x_{16} = -6.03402 + 4.80572\text{Db}$$

$$x_{17} = -2.18409 + 8.84963\text{Me}$$

$$z_9 = 0.175202 + 1.18513x_{17} - 0.0996042(x_{17})^2 + 0.327915x_{16} - 0.0758657(x_{16})^2$$

$$z_{10} = 0.929344z_9 + 0.132519x_{14}$$

$$\theta_{10\text{ kPa}} = 0.339255 + 0.112526z_{10}$$

$$z_{11} = 0.191452 + 1.25652x_{17} - 0.079098(x_{17})^2 + 0.393814x_{16} + 0.152095x_{17}x_{16}$$

$$\theta_{33\text{ kPa}} = 0.28951 + 0.103815z_{11}$$

$$z_{13} = 0.235084 + 0.33033x_{15} - 0.191838(x_{15})^2 + 0.0543679(x_{15})^3 + 0.977685x_{17}$$
$$+ 0.304174x_{15}x_{17} - 0.218857(x_{17})^2 - 0.164373x_{15}(x_{17})^2 + 0.0415057(x_{17})^3$$
$$+ 0.373361x_{16} + 0.0811861x_{17}x_{16} - 0.0768087x_{15}x_{17}x_{16}$$

$$\theta_{1500\text{ kPa}} = 0.214008 + 0.0862945z_{13}$$

van den Berg et al. (1997)

$$\theta_{10\text{ kPa}} = 0.1088 + 0.00347\text{CL} + 0.00211\text{SI} + 0.01756\text{OC}$$

$$\theta_{1500\text{ kPa}} = 0.00334\text{CLBd} + 0.00104\text{SIDb}$$

$$\text{AWC}_{10-1500\text{ kPa}} = 0.2817 - 0.1318\text{Db}$$

REFERENCES

Aina, P.O., Perisawamy, S.P., 1985. Estimating available water-holding capacity of western Nigerian soils from soil texture and bulk density, using core and sieved samples. Soil Sci. 140, 55-58.

Arruda, F.B., Zullo, J., Oliveira, J.B., 1987. Parâmetros de solo para cálculo de água disponível com base na textura do solo. R. Bras. Sci. Sol 11, 11-15.

Babalola, O., 1979. Spatial variability of soil water properties for a tropical soil of Nigeria. Soil Sci. 126, 269-279.

Carsel, R.F., Parrish, R.S., 1988. Developing joint probability distributions of soil water retention characteristics. Water Resour. Res. 24, 755-769.

Cassel, D.K., Lal, R., 1992. Soil physical properties of the tropics: common beliefs and management restraints. In: Lal, R., Sanches, P. (Eds.), Myths and Science of Soils of the Tropics, *Special Publication no. 29*. Soil Sci. Soc. Am., Madison, WI, pp. 61-88.

Chauvel, A., Grimaldi, M., Tessier, D., 1991. Changes in soil-pore space distribution following deforestation and revegetation: An example from the central Amazon. Brazil. For. Ecol. Manag. 38, 259-271.

Correa, J.C., 1984. Características físico-hídricas dos solos latossolo amarelo, podzólico vermelho-amarelo e podzol hidromórfico do estado do Amazonas. Pesq. Agropec. Bras. 19 (3), 347-360.

Demattê, J.L.I., 1988. Manejo de Solos ácidos dos Trópico úmidos – Região Amazônica. Fundação Cargill, Campinas, SP.

Dijkerman, J.C., 1988. An Ustult-Aguult-Tropept catena in Sierra Leone, West Africa. II. Land qualities and land evaluation. Geoderma 42, 29-49.

Hodnett, M.G., Tomasella, J., 2002. Marked differences between van Genuchten soil water-retention parameters for temperate and tropical soils: a new water-retention pedo-transfer function developed for tropical soils. Geoderma 108, 155-180.

Lal, R., 1979. Physical properties and moisture retention characteristics of some Nigerian soils. Geoderma 21, 209-223.

Medina, H., Mohamed, T., del Valle, A., Ruiz, M.E., 2002. Estimating water retention curve in rhodic ferralsols from basic soil data. Geoderma 108, 227-285.

Oliveira, L.B., Ribeiro, M.R., Jacomine, P.K.T., Rodrigues, J.J.V., Marques, F.A., 2002. Funções de pedotransferência para predição de umidade retida a potencias específicos em solos do estado de Pernambuco. Rev. Bras. Ci. Solo. 26, 315-323.

Pachepsky, Ya.A., Rawls, W.J., Timlin, D.J., 1999. The current status of pedotransfer functions: their accuracy, reliability, and utility in field- and regional-scale modeling. In: Corwin, D.L., Loague, K., Ellsworth, T.R. (Eds.), Assessment of Non-Point Source Pollution in the Vadose Zone, *Geophysical monograph 108*. American Geophysical Union, Washington, D.C., pp. 223-234.

Pidgeon, J.D., 1972. The measurement and prediction of available water capacity of ferralitic soils in Uganda. J. Soil Sci. 23, 431-441.

Rawls, W.J., Brakensiek, D.L., 1985. Prediction of soil water properties for hydrologic modeling. In: Jones, E.B., Ward, T.J. (Eds.), Proceedings of the Symposium on Watershed Management in the Eighties. April 30–May 1, (1985), Denver, CO. Am. Soc. Civil Eng., New York, NY, pp. 293-299.

Reichardt, K., 1988. Capacidade de campo. Rev. Bras. Ci. Solo 12, 211-216.

Sanchez, P.A., 1976. Properties and management of soils in the tropics. John Wiley, New York.

Sharma, M.L., Uehara, G., 1986. Influence of soil structure in water relations in a low humic latossol. I. Water retention. Soil Sci. Soc. Am. J. 32, 765-770.

Sobieraj, J., Elsenbeer, H., Vertessy, R.A., 2001. Pedotransfer functions for estimating hydraulic conductivity: implication for modelling storm flow generation. J. Hydrol. 251, 202-220.

Tomasella, J., Hodnett, M.G., 1996. Soil hydraulic properties and van Genuchten parameters for an oxisol under pasture in central Amazonia. In: Gash, J.H.C., Nobre, C.A., Roberts, J.M., Victoria, R.L. (Eds.), Amazonian Deforestation and Climate. John Wiley, Chichester, UK, pp. 101-124.

Tomasella, J., Hodnett, M.G., Rossato, L., 2000. Pedotransfer functions for the estimation of soil water retention in Brazilian soils. Soil Sci. Soc. Am. J. 64, 327-338.

Tomasella, J., Pachepsky, Y., Crestana, S., Rawls, W.J., 2003. Comparison of two techniques to develop pedotransfer functions for water retention. Soil. Sci. Soc. Am. J. 67, 1085-1092.

van den Berg, M., Klamt, E., van Reeuwijk, L.P., Sombroek, G., 1997. Pedotransfer functions for the estimation of moisture retention characteristics of Ferrasols and related soils. Geoderma 78, 161-180.

van Genuchten, M.Th., 1980. A closed form equation for predicting the hydraulic conductivity of unsaturated soils. Soil Sci. Soc. Am. J. 44, 892-989.

Vieira, L.S., Santos, P.C.T.C., 1987. Amazônia: Seus Solos e Outros Recursos Naturais. Agronômica Ceres, São Paulo.

Chapter 22

PEDOTRANSFER FUNCTIONS FOR EUROPE

J.H.M. Wösten[1,*] and A. Nemes[2]

[1]Alterra, Droevendaalsesteeg 3, 6700 AA Wageningen, The Netherlands

[2]USDA-ARS Hydrology & Remote Sensing Lab, Bldg. 007, Rm. 104, BARC-W, Beltsville, MD 20705-2350, USA

*Corresponding author: Tel. +31-317-474287

For the development of pedotransfer functions (PTFs) large, good quality data sets are required comprising measured hydraulic characteristics of a wide variety of soils. An example of such a large data set is the database of hydraulic properties of European soils (HYPRES). The HYPRES database contains information on a total of 5521 soil horizons (Wösten et al., 1999). Of these, 4030 horizons have soil hydraulic data that can be used for the development of PTFs. The soil information is donated by 20 institutions from 12 European countries.

Problems in constructing this large international database were two-fold: (i) various countries use different soil classification systems with the consequence that soil texture classes have different meanings in different countries, and (ii) the number of measured individual points along the various hydraulic characteristics varies considerably due to the application of different measurement techniques. Both problems had to be resolved to arrive at an international database that holds compatible, good quality data. It was decided to adhere to the FAO (1990) and the USDA (1951) particle-size class intervals. As a consequence, a new interpolation technique had to be developed to get an accurate estimation of missing particle-size fractions (Nemes et al., 1999). This interpolation technique was used to arrive at a standardized database. Even more important was the marked imbalance in the number of measured soil hydraulic data pairs for the different soil horizons due to use of different measurement techniques by the different national institutions. To avoid statistical bias, this imbalance was eliminated by approximating individual hydraulic characteristics with van Genuchten–Mualem equations (van Genuchten et al., 1991).

$$\theta(h) = \theta_r + \frac{\theta_s - \theta_r}{(1 + |\alpha h|^n)^{1-1/n}} \qquad (1)$$

$$K(h) = K_s \frac{((1 + |\alpha h|^n)^{1-1/n} - |\alpha h|^{n-1})^2}{(1 + |\alpha h|^n)^{(1-1/n)(L+2)}} \qquad (2)$$

As a result the measured hydraulic characteristics became available via optimised model parameters (i.e., θ_r, θ_s, K_s, α, L and n).

PTFs for the different texture classes were derived by firstly using the optimized parameters to determine water contents and hydraulic conductivities at 13 pressure heads: 0, 1, 2, 3, 5, 10, 25, 50, 100, 250, 500, 1000, 1600 kPa. Because the water retention ($\theta(h)$) and hydraulic conductivity ($K(h)$) values are log-normally distributed in nature, the geometric mean water contents and hydraulic conductivities at the 13 pressure heads were calculated. In addition to the geometric mean values, also the θ and K values plus and minus one standard deviation were calculated. These standard deviations gave an indication of the degree of variation of the individual curves around the geometric mean curve. Next Equations (1) and (2) were fitted to the geometric mean values at the 13 pressure heads to arrive at optimized model parameters for the mean characteristics. Since these parameters represent the mean hydraulic characteristics for a soil texture class, they are called class PTFs. In total, class PTFs for 11 texture classes have been established. The 11 texture classes consist of five FAO texture classes, each subdivided into topsoil and subsoil subclasses, plus the FAO class of organic soils used to prepare the European soil map on a scale 1:1,000,000. Figure 1 shows the calculated geometric mean water retention and hydraulic conductivity characteristic and the standard deviations for the texture class "Medium Fine Topsoils." The figure shows the characteristic shape of the hydraulic characteristics to be expected for this texture class. At the same time, the standard deviations demonstrate that there is considerable variability in hydraulic characteristics within the texture class. Table 1 gives the Mualem–van Genuchten parameters of the geometric-averaged curves for each texture class of the HYPRES database.

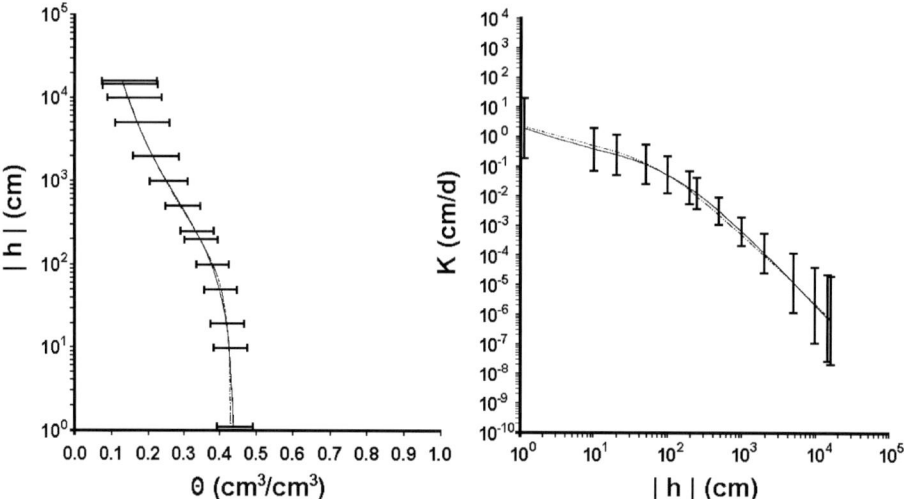

Figure 1. Geometric mean water retention (left graph) and hydraulic conductivity (right graph) characteristic (solid lines), standard deviations (bars) and Mualem–van Genuchten fits (dotted lines) for the texture class "Medium Fine Topsoil" (after Wösten et al., 1999).

In addition to the development of class PTFs, linear regression was used to investigate the dependency of each model parameter on more easily measured, basic soil properties. To comply with a number of physical boundary conditions, transformed parameters rather

Table 1
Mualem–van Genuchten parameters for geometric averaged water retention and hydraulic conductivity curves of the HYPRES database

		θ_r	θ_s	α	n	m	L	K_s
Topsoils	Coarse	0.025	0.403	0.0383	1.3774	0.2740	1.2500	60.000
	Medium	0.010	0.439	0.0314	1.1804	0.1528	−2.3421	12.061
	Medium fine	0.010	0.430	0.0083	1.2539	0.2025	−0.5884	2.272
	Fine	0.010	0.520	0.0367	1.1012	0.0919	−1.9772	24.800
	Very fine	0.010	0.614	0.0265	1.1033	0.0936	2.5000	15.000
Subsoils	Coarse	0.025	0.366	0.0430	1.5206	0.3424	1.2500	70.000
	Medium	0.010	0.392	0.0249	1.1689	0.1445	−0.7437	10.755
	Medium fine	0.010	0.412	0.0082	1.2179	0.1789	0.5000	4.000
	Fine	0.010	0.481	0.0198	1.0861	0.0793	−3.7124	8.500
	Very fine	0.010	0.538	0.0168	1.0730	0.0680	0.0001	8.235
	Organic[a]	0.010	0.766	0.0130	1.2039	0.1694	0.4000	8.000

[a]Within the organic soils no distinction is made in topsoils and subsoils.

than the original model parameters were used in the regression analysis. In this case, the imposed boundary conditions were: $K_s > 0$, $a > 0$, $n > 1$ and $-10 < L < +10$ (Wösten et al., 1999). The following basic soil properties were used as input variables: percentage clay, percentage silt, percentage organic matter; bulk density and also the qualitative variable topsoil or subsoil. The latter variable was entered in the regression equation by assigning the value 0 or 1. Linear, reciprocal, and exponential relationships of these basic soil properties were used in the regression analysis, and possible interactions were also investigated. As a consequence, the resulting regression model consists of various basic soil properties and the product of different basic soil properties, all of which contribute significantly to the description of the transformed model parameters. The models were selected using a subset selection method (Furnival and Wilson, 1974). Since these PTFs require basic soil properties for a specific point in the field instead of class average texture data, they are called continuous PTFs (Vereecken et al., 1992; Tietje and Tapkenhinrichs, 1993). The continuous PTFs developed using the HYPRES database, are presented in Table 2.

While class PTFs predict the hydraulic characteristics for rather broadly defined soil texture classes, and therefore cannot provide site-specific information, continuous PTFs can be applied in case of more site-specific applications. However, the R^2 values obtained indicate that the predictions of the hydraulic characteristics when using continuous PTFs are fairly inaccurate (Table 2). Subdividing the complete dataset into subsets of similar soil texture might improve these predictions.

The class PTFs for the 11 FAO texture classes are used to translate the representative profiles for the mapping units of the European soil map at a scale of 1:1,000,000 into soil hydraulic profiles. The result is a transformed European soil map that gives a spatial picture of the soil hydraulic properties of the unsaturated zone of European soils. Based on this map it was possible to prepare a map showing total available water on a European scale (Wösten et al., 1999). This map is just one example of the type of new spatial information that can be generated when PTFs are used in combination with other existing

Table 2
Continuous pedotransfer functions developed from the HYPRES database

$\theta_s = 0.7919 + 0.001691C - 0.29619D - 0.000001491S^2 + 0.0000821OM^2$
$\quad + 0.02427C^{-1} + 0.01113S^{-1} + 0.01472 \ln(S) - 0.0000733 \times OM \times C$
$\quad - 0.000619DC - 0.001183 \times D \times OM - 0.0001664 \times \text{topsoil} \times S \ (R^2 = 76\%)$

$\alpha^* = -14.96 + 0.03135C + 0.0351S + 0.646OM + 15.29D - 0.192\text{topsoil}$
$\quad - 4.671D^2 - 0.000781C^2 - 0.00687OM^2 + 0.0449OM^{-1} + 0.0663 \ln(S)$
$\quad + 0.1482 \ln(OM) - 0.04546DS - 0.4852 \times D \times OM + 0.00673 \times \text{topsoil} \times C$
$\quad (R^2 = 20\%)$

$n^* = -25.23 - 0.02195C + 0.0074S - 0.1940OM + 45.5D - 7.24D^2 + 0.0003658C^2$
$\quad + 0.002885OM^2 - 12.81D^{-1} - 0.1524S^{-1} - 0.01958OM^{-1} - 0.2876 \ln(S)$
$\quad - 0.0709 \ln(OM) - 44.6 \ln(D) - 0.02264DC + 0.0896 \times D \times OM + 0.00718$
$\quad \times \text{topsoil} \times C \ (R^2 = 54\%)$

$L^* = 0.0202 + 0.0006193C^2 - 0.001136OM^2 - 0.2316 \ln(OM) - 0.03544DC$
$\quad + 0.00283DS + 0.0488 \times D \times OM \ (R^2 = 12\%)$

$K_s^* = 7.755 + 0.0352S + 0.93\text{topsoil} - 0.967D^2 - 0.000484C^2 - 0.000322S^2$
$\quad + 0.001S^{-1} - 0.0748OM^{-1} - 0.643 \ln(S) - 0.01398DC - 0.1673 \times D \times OM$
$\quad + 0.02986 \times \text{topsoil} \times C - 0.03305 \times \text{topsoil} \times S \ (R^2 = 19\%)$

θ_s is a model parameter, α^*, n^*, L^* and K_s^* are transformed model parameters in the Mualem–van Genuchten equations; C, percent clay (i.e., percent < 2 μm); S, percent silt (i.e., percent between 2 and 50 μm); OM, percent organic matter; D, bulk density; topsoil and subsoil are qualitative variables having the value of 1 or 0; ln, natural logarithm.

spatial soil data. Other possible new products could be a travel time map for solutes or an infiltration rate map for erosion studies.

The HYPRES database served as input in a study by Nemes et al. (2003) in which the authors used national, continental and intercontinental scale data collections to derive PTFs for the estimation of soil water retention, using artificial neural networks. The goal of the study was to identify the relevance of PTF from international data collections for individual countries that contributed to the database or for countries located in areas with comparable climatic conditions. The same methodology and the same sets of predictors were used while the source database was changed. In order to study the influence of different combinations of predictors, 11 different PTFs were developed from each of the data sets. Larger residuals were found for soils that were underrepresented in the data sets. Water retention estimates were functionally tested. Soil moisture time series of seven soils were simulated, using water retention characteristics estimated by the different PTFs. Sum of square residuals were evaluated for simulated water contents and water contents observed in the field. Small differences were found for PTFs derived from different scale data collections. Moreover, PTF estimates resulted in simulations that were only marginally worse than simulations with estimates using laboratory measured water retention. In this case, differences in scales of PTF estimations contributed little to differences in simulated and observed water contents.

Application of HYPRES showed that it is possible to assign soil hydraulic characteristics to soils with a textural composition comparable to the soils for which the PTFs have been derived. However, care should be taken to use the functions for prediction of hydraulic characteristics of soils outside the range of the original database. Class PTFs give the mean hydraulic characteristics for rather broadly defined soil texture classes. As a consequence, these functions are generally better applicable, however, they can give limited site-specific information. In contrast, continuous PTFs can be more site-specific, but their general applicability is limited as they require more specific input data.

REFERENCES

FAO (Food and Agriculture Organisation), 1990. Guidelines for Soil Description, Third edition. FAO/ISRIC, Rome.
Furnival, G.M., Wilson, R.W., 1974. Regression by leaps and bounds. Technometrics 16, 499-511.
Nemes, A., Wösten, J.H.M., Lilly, A., Oude Voshaar, J.H., 1999. Evaluation of different procedures to interpolate particle-size distributions to achieve compatibility within soil databases. Geoderma 90, 187-202.
Nemes, A., Schaap, M.G., Wösten, J.H.M., 2003. Functional evaluation of pedo-transfer functions derived from different scales of data collection. Soil Sci. Soc. Am. J. 67, 1093-1102.
Tietje, O., Tapkenhinrichs, M., 1993. Evaluation of pedo-transfer functions. Soil Sci. Soc. Am. J. 57 (4), 1088-1095.
USDA (United States Department of Agriculture), 1951. Soil Survey Manual, U.S. Dept. Agriculture Handbook No. 18. USDA, Washington, DC.
van Genuchten, M.Th., Leij, F.J., Yates, S.R., 1991. The RETC code for quantifying the hydraulic functions of unsaturated soils. USDA, US Salinity Laboratory, Riverside, CA. United States Environmental Protection Agency, document EPA/600/2-91/065.
Vereecken, H., Diels, J., van Orshoven, J., Feyen, J., Bouma, J., 1992. Functional evaluation of pedotransfer functions for the estimation of soil hydraulic properties. Soil Sci. Soc. Am. J. 56, 1371-1378.
Wösten, J.H.M., Lilly, A., Nemes, A., Le Bas, C., 1999. Development and use of a database of hydraulic properties of European soils. Geoderma 90, 169-185.

Chapter 23

PEDOTRANSFER FUNCTIONS FOR THE UNITED STATES

W.J. Rawls

USDA-ARS Hydrology & Remote Sensing Lab, Bldg. 007, Rm. 104, BARC-W, Beltsville, MD 20705-2350, USA
Tel.: +1-301-504-8745

1. INTRODUCTION

Numerous pedotransfer functions have been developed in the United States; however, most have been developed using data sets that are representative of a region or smaller area. The pedotransfer functions presented in this chapter have been developed from national databases (Rawls et al., 1982; USDA, 1997; Rawls et al., 1998). The pedotransfer functions presented are for water retention and saturated hydraulic conductivity.

2. SOIL WATER RETENTION

Pedotransfer functions for water retention take one of two forms. First, pedotransfer functions that predict water retention at specific matric potentials; and second pedotransfer functions that predict the parameters of water retention models.

2.1. Pedotransfer functions for specific water potentials on the soil water retention curve

The two most frequently estimated water contents are those corresponding to soil water potentials of −33 and −1500 kPa, primarily because they are commonly measured and have commonly been referred to as field capacity and wilting point, respectively. Table 1 summarizes the soil water held at −33 and −1500 kPa for the USDA soil texture classes. Ahuja et al. (1985) showed that using a linear extrapolation between the water contents held at −33 and −1500 kPa on a log–log graph other water contents for specific water potentials could be adequately determined.

Table 2 summarizes equations developed from regression analysis that estimate soil water retention at specific water potentials using: (1) soil properties; (2) soil properties and water retained at −1500 kPa; and (3) soil properties and water retained at −33 and −1500 kPa (Rawls et al., 1982). As seen in Table 2, the accuracy of the regression equations increased when the water content held at −1500 kPa or both −33 and −1500 kPa were included with physical soil properties. Adding the water content held at −33 and −1500 kPa is more costly and time consuming to acquire; however, they increased the explained variation from 76 to 95%. In general, the water content held at −33 kPa was a more significant variable for estimating water retention at matric potentials

Table 1
Water retention properties classified by soil texture (Rawls et al., 1982)

Texture class	Sample size	Total porosity (φ) (cm³ cm⁻³)	Residual water content (θ_r) (cm³ cm⁻³)	Brooks–Corey parameters						Water retained at	
				Bubbling pressure (h_b)			Pore-size distribution (λ)			−33 kPa (cm³ cm⁻³)	−1500 kPa (cm³ cm⁻³)
				Arithmetic (cm)	Geometric[a] (cm)		Arithmetic	Geometric[a]			
Sand	762	0.437[b]	0.020	15.98	7.26		0.694	0.592		0.091	0.033
		(0.374–0.500)	(0.001–0.039)	(0.24–31.72)	(1.36–38.74)		(0.298–1.090)	(0.334–1.051)		(0.018–0.164)	(0.007–0.059)
Loamy sand	338	0.437	0.035	20.58	8.69		0.553	0.474		0.125	0.055
		(0.368–0.506)	(0.003–0.067)	(−4.04–45.20)	(1.80–41.85)		(0.234–0.872)	(0.271–0.827)		(0.060–0.190)	(0.019–0.091)
Sandy loam	666	0.453	0.041	30.20	14.66		0.378	0.322		0.207	0.095
		(0.351–0.555)	(−0.024–0.106)	(−3.61–64.01)	(3.45–62.24)		(0.140–0.616)	(0.186–0.558)		(0.126–0.288)	(0.031–0.159)
Loam	383	0.463	0.027	40.12	11.15		0.252	0.220		0.270	0.117
		(0.375–0.551)	(−0.020–0.074)	(−20.07–100.3)	(1.63–76.40)		(0.086–0.418)	(0.137–0.355)		(0.195–0.345)	(0.069–0.165)
Silt loam	1206	0.501	0.015	50.87	20.76		0.234	0.211		0.330	0.133
		(0.420–0.582)	(−0.028–0.058)	(−7.68–109.4)	(3.58–120.4)		(0.105–0.363)	(0.136–0.326)		(0.258–0.402)	(0.078–0.188)
Sandy clay loam	498	0.398	0.068	59.41	28.08		0.319	0.250		0.255	0.148
		(0.332–0.464)	(−0.001–0.137)	(−4.62–123.4)	(5.57–141.5)		(0.079–0.559)	(0.125–0.502)		(0.186–0.324)	(0.085–0.211)
Clay loam	366	0.464	0.075	56.43	25.89		0.242	0.194		0.318	0.197
		(0.409–0.519)	(−0.024–0.174)	(−11.44–124.3)	(5.80–115.7)		(0.070–0.414)	(0.100–0.377)		(0.250–0.386)	(0.115–0.279)
Silty clay loam	689	0.471	0.040	70.33	32.56		0.177	0.151		0.366	0.208
		(0.418–0.524)	(−0.038–0.118)	(−3.26–143.9)	(6.68–158.7)		(0.039–0.315)	(0.090–0.253)		(0.304–0.428)	(0.138–0.278)
Sandy clay	45	0.430	0.109	79.48	29.17		0.223	0.168		0.339	0.239
		(0.370–0.490)	(0.013–0.205)	(−20.15–179.1)	(4.96–171.6)		(0.048–0.398)	(0.078–0.364)		(0.245–0.433)	(0.162–0.316)
Silty clay	127	0.479	0.056	76.54	34.19		0.150	0.127		0.387	0.250
		(0.425–0.533)	(−0.024–0.136)	(−6.47–159.6)	(7.04–166.2)		(0.040–0.260)	(0.074–0.219)		(0.332–0.442)	(0.193–0.307)
Clay	291	0.475	0.090	85.60	37.30		0.165	0.131		0.396	0.272
		(0.427–0.523)	(−0.015–0.195)	(−4.92–176.1)	(7.43–187.2)		(0.037–0.293)	(0.068–0.253)		(0.326–0.466)	(0.208–0.336)

[a] Antilog of the log mean.
[b] First line is the mean value, the second line is ± one standard deviation about the mean.

Table 2
Linear regression equations for predicting soil water content at specific matric potentials (Rawls et al., 1982)

Matric potential (kPa)	Intercept	Sand (%)	Silt (%)	Clay (%)	Organic matter (%)	Bulk density (g cm^{-3})	−33 kPa water retention (cm^3 cm^{-3})	−1500 kPa water retention (cm^3 cm^{-3})	Correlation coefficient, R
	a	b	c	d	e	f	g	h	
Regression coefficients									
−4	0.7899	−0.0037			0.0100	−0.1315			0.58
	0.6275	−0.0041			0.0239		1.89	−0.08	0.57
	0.1829				−0.0246	−0.0376		−1.38	0.77
−7	0.7135	−0.0030		0.0017	0.0263	−0.1693			0.74
	0.4829	−0.0035					1.53	0.25	0.74
	0.8888	−0.0002			−0.0107			−0.81	0.91
−10	0.4118	−0.0030		0.0023	0.0317				0.81
	0.4103	0.0031			0.0260			0.41	0.81
	0.0619	−0.0002			−0.0067		1.34	−0.51	0.95
−20	0.3121	−0.0024		0.0032	0.0314				0.86
	0.3000	−0.0024			0.0235			0.61	0.89
	0.0319	−0.0002					1.01	−0.06	0.99
−33	0.2576	−0.0020		0.0036	0.0299				0.87
	0.2391	−0.0019			0.0210			0.72	0.92
−60	0.2065	−0.0016		0.0040	0.0275				0.87
	0.1814	−0.0015			0.0178			0.80	0.94
	0.0136					−0.0091	0.66	0.39	0.99
−100	0.0349		0.0014	0.0055	0.0251				0.87

Table 2. Continued

−0.0034	0.1417	−0.0012			0.0151		0.85	0.96
					0.0022		0.54	0.99
−200	0.0281		0.0011	0.0054	0.0200	0.52		0.86
	0.0986	0.0009			0.0116		0.90	0.97
	−0.0043				0.0026		0.69	0.99
−400	0.0238		0.0008	0.0052	0.0190			0.84
	0.0649	−0.0006			0.0085	0.36	0.93	0.98
	−0.0038				0.0026		0.79	0.99
−700	0.0216		0.0006	0.0050	0.0167			0.81
	0.0429	−0.0004			0.0062	0.24	0.94	0.98
	−0.0027				0.0024		0.86	0.99
−1000	0.0205		0.0005	0.0049	0.0154			0.81
	0.0309	−0.0003			0.0049	0.16	0.95	0.99
	−0.0019				0.0022		0.89	0.99
−1500	0.0260			0.0050	0.0158	0.11		0.80

Sand (%) + silt (%) + clay (%) = 100; Sand = 2.0–0.05 mm; Silt = 0.05–0.002 mm; Clay < 0.002 mm.
$\theta_x = a + b\text{sand}\,(\%) + c\text{silt}\,(\%) + d\text{clay}\,(\%) + e\text{organic matter}\,(\%) + f\text{bulk density}\,(\text{g cm}^{-3}) + g(-33\,\text{kPa moisture}\,(\text{cm}^3\,\text{cm}^{-3})) + h(-1500\,\text{kPa moisture}\,(\text{cm}^3\,\text{cm}^{-3}))$.
θ_x = predicted water retention (cm^3 cm^{-3}) for a given suction.
$a–h$ = regression coefficients.

between 0 and −33 kPa and the water content held at −1500 kPa was a more significant variable for estimating water retention at water potentials between −33 and −1500 kPa.

Ahuja et al. (1985) developed a method to estimate the soil water retention curve from soil bulk density, water content held at −33 kPa and a reference soil water retention curve for the soil texture class (Figure 1). The procedure is demonstrated in Figure 1 on page 73. The above procedures estimate points on the water retention curve to which water retention models such as those given in Table 3 can be fitted to describe the entire water retention curve.

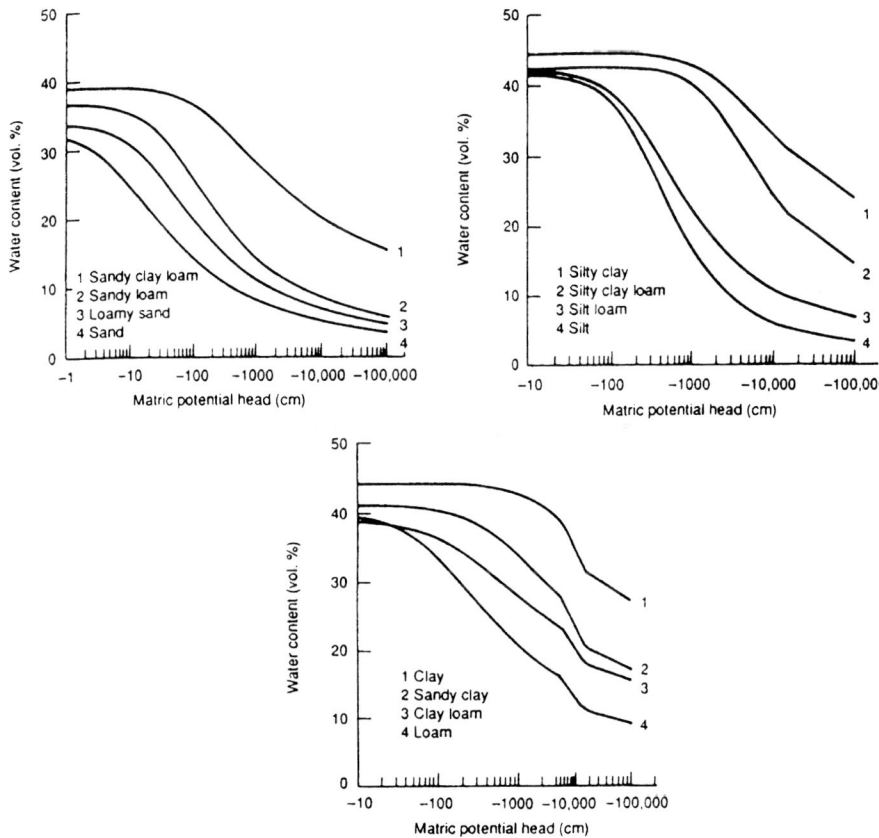

Figure 1. Representative water retention curves for USDA soil texture classes (Rawls et al., 1992).

2.2. Estimation of soil water retention model parameters

Table 4 summarizes Brooks–Corey (Brooks and Corey, 1964) parameters for the USDA texture classes. The model parameters subsequently related to physical soil properties using regression analysis (Table 4). Also included in Table 4 are independent equations for estimating the Campbell water retention model parameters (Campbell, 1974).

Table 3
Soil water retention and hydraulic conductivity relationships with parameter correspondence

Hydraulic soil characteristic	Parameters	Parameter correspondence
Brooks and Corey (1964)		
Soil water retention, $\dfrac{\theta - \theta_r}{\varphi - \theta_r} = \left(\dfrac{h_b}{h}\right)^\lambda$	λ = pore-size index h_b = bubbling capillary pressure θ_r = residual water content φ = porosity K_s = fully saturated conductivity ($\theta = \varphi$) $n = 3 + \dfrac{2}{\lambda}$	$\lambda = \lambda$ $h_b = h_b$ $\theta_r = \theta_r$ $\varphi = \varphi$ $K_s = K_s$
Hydraulic conductivity, $\dfrac{K(\theta)}{K_s} = \left(\dfrac{\theta - \theta_r}{\varphi - \theta_r}\right)^n = (S_e)^n$		
Campbell (1974)		
Soil water retention, $\dfrac{\theta}{\varphi} = \left(\dfrac{H_b}{h}\right)^{1/b}$	φ = porosity H_b = scaling parameter (length) b = constant	$\varphi = \varphi$ $H_b = h_b$ $b = \dfrac{1}{\lambda}$
Hydraulic conductivity, $\dfrac{K(\theta)}{K_s} = \left(\dfrac{\theta}{\varphi}\right)^n$	$n = 3 + 2b$	
van Genuchten (1980)		
Soil water retention, $\dfrac{\theta - \theta_r}{\varphi - \theta_r} = \left[\dfrac{1}{1+(\alpha h)^n}\right]^m$	φ = porosity θ_r = residual water content α = constant n = constant m = constant	$\varphi = \varphi$ $\theta_r = \theta_r$ $\alpha = (h_b)^{-1}$ $n = \lambda + 1$ $m = \dfrac{\lambda}{\lambda+1}$
Hydraulic conductivity, $\dfrac{K(\theta)}{K_s} = \left(\dfrac{\theta - \theta_r}{\varphi - \theta_r}\right)^{1/2} \left\{1 - \left[1 - \left(\dfrac{\theta - \theta_r}{\varphi - \theta_r}\right)^{1/m}\right]^m\right\}^2$		

θ = water content; h = capillary suction (cm); $K(\theta)$ = hydraulic conductivity for a given water content (cm h^{-1}).

Table 4
Estimation equations for the Brooks–Corey and Campbell water retention model parameters

Brooks–Corey parameters (Rawls and Brakensiek, 1985)
h_b – Brooks–Corey bubbling pressure (cm)
$h_b = e[5.340 + 0.185(C) - 2.484(\varphi) - 0.002(C)^2 - 0.044(S)(\varphi) - 0.617(C)(\varphi) + 0.001(S)^2(\varphi^2) - 0.009(C^2)(\varphi^2) - 0.00001(S^2)$
$\times (C) + 0.009(C^2)(\varphi) - 0.0007(S^2)(\varphi) + 0.000005(C^2)(S) + 0.500(\varphi^2)C]$
λ – Brooks–Corey pore size distribution index
$\lambda = e[-0.784 + 0.018(S) - 1.062(\varphi) - 0.00005(S^2) - 0.003(C^2) + 1.111(\varphi^2) - 0.031(S)(\varphi) + 0.0003(S^2)(\varphi^2) - 0.006(C^2)$
$\times (\varphi^2) - 0.0000002(S^2)(C) + 0.008(C^2)(\varphi) - 0.007)(\varphi^2)(C)]$
θ_r – Brooks–Corey residual water content (vol. fraction)
$\theta_r = -0.018 + 0.0009(S) + 0.005(C) + 0.029(\varphi) - 0.0002(C)^2 - 0.001(S)(\varphi) - 0.0002(C^2)(\varphi^2) + 0.0003(C^2)(\varphi) - 0.002(\varphi^2)(C)$

where:
$C = \%$ clay
$S = \%$ sand
$\varphi =$ porosity (vol. fraction)

Campbell parameters (Campbell, 1985)
Campbell air entry potential at standard bulk density (1.3 mg m^{-3}) meters:
$h_b = (-0.5(d_g)^{-\frac{1}{2}})(BD/1.3)(0.67b)$;

Campbell b:
$b = -2h_b + 0.2s_g$;

where:
BD = bulk density (mg m^{-3})
$\varphi =$ porosity (vol. fraction). If porosity (φ) is not known, estimate it from bulk density ($\varphi =$ air multiplier$(1 - BD/2.65)$) where air multiplier = 0.93 usually, must be (0.8, 0.98)
$d_g =$ geometric mean particle diameter (mm) = $\exp[-0.025 - 3.63\text{silt} - 6.88\text{clay}]$
$s_g =$ geometric standard deviation = $\exp[13.32\text{silt} + 47.7\text{clay} - \ln(d_g)\ln(d_g)]^{\frac{1}{2}}$

Table 5
Summary of soil water retention equation parameters derived by Saxton et al. (1986)

Applied tension range (−kPa)	Equation
>1500–10	$\Psi = A\theta^B$
	$A = \exp[a + b(\%C) + c(\%S)^2 + d(\%S)^2(\%C)]100$
	$B = e + f(\%C)^2 + g(\%S)^2(\%C)$
10–Ψ_e	$\Psi = 10.0 - (\theta - \theta_{10})(10.0 - \Psi_e)/(\theta_s - \theta_{10})$
	$\theta_{10} = \exp[(2.302 - \ln A)/B]$
	$\Psi_e = 100.0[m + n(\theta_s)]$
	$\theta_s = h + j(\%S) + k\log_{10}(\%C)$
Ψ_e–0.0	$\theta = \theta_s$
>1500–0.0	$K = 2.778 \times 10^{-6}\{\exp[p + q(\%S)$
	$+ [r + t(\%S) + u(\%C) + v(\%C)^2](1/\theta)]\}$

Coefficients

$a = -4.396$
$b = -0.0715$
$c = -4.880 \times 10^{-4}$
$d = -4.285 \times 10^{-5}$
$e = -3.140$
$f = -2.22 \times 10^{-3}$
$g = -3.484 \times 10^{-5}$
$h = 0.332$
$j = -7.251 \times 10^{-4}$
$k = 0.1276$
$m = -0.108$
$n = 0.341$
$p = 12.012$
$q = -7.55 \times 10^{-2}$
$r = -3.8950$
$t = 3.671 \times 10^{-2}$
$u = -0.1103$
$v = 8.7546 \times 10^{-4}$

Definitions

θ_{10} = water content at −10 kPa (m³ m⁻³)
K = water conductivity (m s⁻¹)

Ψ = water potential (−kPa)
Ψ_e = water potential at air entry (−kPa)
θ = water content (m³ m⁻³)
θ_s = water content at saturation (m³ m⁻³)

(%S) = percent sand (e.g., 40.0)
(%C) = percent clay (e.g., 30.0)

Table 6
Saturated hydraulic conductivity (K_s) classified by USDA soil texture classes and porosity (Rawls et al., 1998)

USDA Soil texture class	Geometric mean K_s^a (mm h^{-1})	Porosity (m^3 m^{-3})	Water retained at at −33 kPa (m^3 m^{-3})	Water retained at −1500 kPa (m^3 m^{-3})	Sand (%)	Clay (%)	Sample size
Sand	181.9 (266.8–96.5)	0.44	0.07	0.03	92	4	39
	91.4 (218.5–64.0)	0.39	0.09	0.02	91	4	30
Fine sand	141.3 (236.1–118.1)	0.49	0.07	0.03	89	3	14
	100.0 (219.8–68.1)	0.39	0.07	0.02	92	4	9
Loamy sand	123.0 (195.5–83.8)	0.45	0.09	0.04	82	6	19
	41.4 (77.6–30.5)	0.37	0.14	0.06	82	7	28
Loamy fine sand	62.2 (122.0–35.6)	0.46	0.11	0.06	82	6	18
	12.8 (116.0–6.8)	0.37	0.2	0.12	68	12	112
Sandy loam	55.8 (129.6–30.5)	0.47	0.23	0.1	65	11	75
	12.8 (31.3–5.1)	0.37	0.2	0.12	68	13	112
Fine sandy loam	22.4 (35.6–9.8)	0.45	0.24	0.1	70	14	24
	8.2 (17.0–3.4)	0.36	0.21	0.11	69	14	36
Loam	3.9 (28.4–1.6)	0.47	0.3	0.15	38	23	44
	6.2 (16.5–2.8)	0.39	0.28	0.13	43	22	65
Silt loam	14.4 (37.1–7.6)	0.49	0.34	0.14	18	19	61
	3.4 (9.9–1.0)	0.39	0.31	0.14	21	20	46
Sandy clay loam	7.7 (50.5–2.0)	0.44	0.31	0.2	56	26	20
	2.8 (10.9–1.0)	0.37	0.29	0.21	58	26	53
Clay loam	4.2 (13.1–2.2)	0.48	0.32	0.22	29	35	20
	0.7 (3.8–0.2)	0.4	0.34	0.25	35	35	53
Silty clay loam	3.7 (10.4–2.3)	0.50	0.37	0.23	10	34	26
	4.9 (14.0–2.3)	0.43	0.36	0.23	10	32	33
Sandy clay	0.9 (2.5–0.3)	0.39	0.3	0.22	51	36	14
Silty clay	1.8 (7.5–0.5)	0.53	0.41	0.27	4	49	10
Clay	2 (6.0–0.9)	0.48	0.4	0.31	18	53	20
	1.8 (6.9–0.3)	0.4	0.36	0.3	26	50	21

[a] K_s = saturated hydraulic conductivity; first line is mean value; in brackets are 25 and 75% percentile values.

Using the correspondence between model parameters given in Table 3, the equations in Table 4 can be used to apply the Brooks and Corey (1964), Campbell (1974) and van Genuchten (1980) water retention models. Saxton et al. (1986) developed pedotransfer functions for a modified model of Campbell (1974). A summary of the parameters is given in Table 5.

3. SATURATED HYDRAULIC CONDUCTIVITY

Rawls et al. (1998) assembled a national database of saturated hydraulic conductivity data from which the geometric means of the K_s, sorted according to the USDA soil texture classes and two bulk density classes, along with the 25 and 75% percentile values were developed and are given in Table 6.

Ahuja and associates (1984) proposed a generalized Kozeny–Carman equation (Carman, 1956) relating the saturated hydraulic conductivity to effective porosity in the following form:

$$K_s = C\phi_e^m \qquad (1)$$

where K_s is the saturated hydraulic conductivity (mm h^{-1}); ϕ_e, the effective porosity (m^3 m^{-3}) (total porosity, φ, minus water content at -33 kPa pressure head, $\theta_{1/3}$) and C and m are empirically derived constants. Rawls et al. (1998) also parameterized equation 1 by redefining the exponent m equal to 3 minus the Brooks–Corey pore size distribution index (λ) and C equal to 1930. The Brooks–Corey pore size distribution index (λ) was obtained by fitting a log–log plot of water content vs. pressure-head using only the -33 and -1500 kPa water contents.

REFERENCES

Ahuja, L.R., Naney, J.W., Green, R.E., Nielsen, D.R., 1984. Macroporosity to characterize spatial variability of hydraulic conductivity and effects of land management. Soil Sci. Soc. Am. J. 48, 699-702.

Ahuja, L.R., Naney, J.W., Williams, R.D., 1985. Estimating soil water characteristics from simpler properties or limited data. Soil Sci. Soc. Am. J. 49, 1100-1105.

Brooks, R.H., Corey, A.T., 1964. Hydraulic properties of porous media. Hydrology Paper No. 3, Colorado State University, Fort Collins, CO, 27 pp.

Campbell, G.S., 1974. A simple method for determining unsaturated conductivity from moisture retention data. Soil Sci. 117, 311-314.

Campbell, G.S., 1985. Soil Physics with BASIC: Transport Models for Soil–Plant Systems. Elsevier, New York, 150 pp.

Carman, P.C., 1956. Flow of Gases Through Porous Media. Academic Press Inc., New York.

Rawls, W.J., Brakensiek, D.L., 1985. Prediction of soil water properties for hydrologic modeling. *In*: Jones, E.B., Ward, T.J. (Eds.), Proceedings of the Symposium of

Watershed Management in the Eighties. April 30–May 1, 1985, Denver, CO. Am. Soc. Civil Eng., New York, NY, pp. 293-299.

Rawls, W.J., Brakensiek, D.L., Saxton, K.E., 1982. Estimation of soil water properties. Trans. ASAE 25 (5), 1316-1320, see also p. 1328.

Rawls, W.J., Ahuja, L.R., Brakensiek, D.L., Shirmohammadi, A., 1992. Infiltration and soil water movement. *In*: Maidment, D.R. (Ed.), Handbook of Hydrology. McGraw-Hill Inc., New York, Chapter 5.

Rawls, W.J., Gimènez, D., Grossman, R., 1998. Use of soil texture, bulk density, and the slope of the water retention curve to predict saturated hydraulic conductivity. Trans. ASAE 41 (4), 983-988.

Saxton, K.E., Rawls, W.J., Romberger, J.S., Papendick, R.I., 1986. Estimating generalized soil water characteristics for texture. Soil Sci. Soc. Am. J. 50, 1031-1036.

USDA, 1997. National Characterization Data. Soil Survey Laboratory, National Soil Survey Center, and Natural Resources Conservation Service, Lincoln, NE.

van Genuchten, M.Th., 1980. A closed-form equation for predicting the hydraulic conductivity of unsaturated soils. Soil Sci. Soc. Am. J. 44, 892-898.

Chapter 24

PEDOTRANSFER STUDIES IN POLAND

R. Walczak[*], B. Witkowska-Walczak and C. Sławiński

Institute of Agrophysics, Polish Academy of Sciences, Doswiadczalna 4, 20-290 Lublin 27, PO Box 201, Poland

[*]Corresponding author: Tel.: +48-81-7445061; E-mail: rwalczak@demeter.ipan.lublin.pl

1. WATER RETENTION

1.1. Importance of various soil solid phase parameters

The investigations of the impact of the soil physical and chemical parameters on the water retention curve and water conductivity curve were initiated in Poland in the 1970s. This research began from finding relations between the contents of granulometric fractions and the water contents at various values of the soil water potential. Turski et al. (1974, 1975) stated that the fraction of colloidal clay has a significant positive influence on the maximum hygroscopic water content and on the content of water unavailable for plants. Domżał (1979, 1983) described the impact of the contents of colloidal particles on the water retention curves. He emphasized that the impact of the soil compaction on the water retention depended on soil particle-size distribution, especially in the pF range from 2 to 2.7. Zawadzki (1970a) noted a considerable impact of the particle-size distribution on soil water retention curves. He (1970b) also proposed an equation to calculate the water content (W, vol. %,) at pF 2 as:

$$W_{pF2} = 0.667P + 7.54 \tag{1}$$

where P is the total porosity, and obtained correlation coefficient $r = 0.965$. Zawadzki et al. (1971) presented a formula to calculate the water content at pF 4,2 as:

$$W_{pF4.2} = 0.528S + 1.37 \tag{2}$$

where S is the specific soil surface area ($m^2 g^{-1}$). Trzecki (1974) reported equations to calculate the field capacity (WPP), the point of the beginning of the plant growth inhibition (WPHWR) and the permanent wilting point (WTWR) from the contents of six soil particle-size fractions and the organic matter content.

For the plow horizons the equations were:

$$\begin{aligned}
\text{WPP} &= 0.0188x_1 + 0.0879x_2 + 0.240x_3 + 0.296x_4 + 0.649x_5 + 0.316x_6 + 2.34x_7 \\
\text{WPHWR} &= -0.0213x_1 - 0.0338x_2 + 0.115x_3 + 0.451x_4 + 0.513x_5 + 0.323x_6 + 2.25x_7 \\
\text{WTWR} &= 0.00121x_1 - 0.00868x_2 + 0.0488x_3 + 0.0737x_4 + 0.0485x_5 + 0.142x_6 + 1.25x_7
\end{aligned} \tag{3}$$

In subsoil, the equations were:

$$WPP = 0.0157x_1 + 0.091x_2 + 0.284x_3 + 0.353x_4 + 0.105x_5 + 0.603x_6$$
$$WPHWR = -0.000227x_1 + 0.0205x_2 + 0.0395x_3 + + 0.303x_4 + 0.260x_5 + 0.524x_6 \quad (4)$$
$$WTWR = 0.00193x_1 + 0.243x_2 + 0.0111x_3 + + 0.0262x_4 + 0.193x_5 + 0.272x_6$$

where x_1, x_2, x_3, x_4, x_5 and x_6 are percentages fractions 1.0–0.1, 0.1–0.05, 0.05–0.02, 0.02–0.006, 0.006–0.002mm, and less than 0.002mm, respectively, x_7 being the content of organic matter (%).

The strong effect of compaction on soil water retention curves was revealed by numerous studies. It was shown that the increase in bulk density causes the decrease in the water retention in the range pF 0–pF 2. Compaction also leads to the increase in the water retained at pF 4,2 by 0.2–0.3vol.% in sandy soils and by 4–6vol.% in fine-textured loamy soils (Domżał, 1983).

Walczak (1977, 1984) studied soil compaction impact on soil water retention hysteresis in the pF range from 0 to 2.7 and stated that compaction strongly influenced the water retention hysteresis, especially at low bulk densities (Kaniewska and Walczak, 1974; Walczak and Kaniewska, 1975) as shown in Figure 1.

Soil water retention hysteresis can have marked effect on water content in soil profile if there are layers compacted to different extents. Consider two soil layers being brought in contact, one having low bulk density of $\rho = 1.175 \text{gcm}^{-3}$ and low moisture content, and another having high bulk density $\rho = 1.655 \text{gcm}^{-3}$ and high moisture content. As the equilibrium will be approached, the low-density layer will dry, while the high-density layer will become wetter. If, for example, the equilibrium will be achieved at pF 1, then as the data in Figure 1 show, the low-density layer will have water content of about 51vol.%, and the high-density layer will have water content of about $W = 39\text{vol}.\%$. Now let us take the opposite case, when the low-density layer is originally wet, the high-density layer is dry, and the equilibrium is again achieved at pF 1. The data in Figure 1 show that the compacted layer will have water content of about 39vol.%, whereas the low-density layer will have water content of 42vol.%, Hence, the distribution of the water in a soil profile can be substantially influenced by the differences in the compactions of the layers and by the effect of hysteresis. Pedotransfer functions should be developed to account for hysteresis, especially for small values of the soil bulk density.

Soil organic matter affects water retention because it can retain water by itself and it strongly affects soil microstructure. It was found difficult to establish direct relationships between the organic matter content and the water content at a given soil water potential. However, organic matter content has been frequently taken into account as a complementary parameter in soil hydraulic property estimations. Dobrzański et al. (1972) has found a good correlation between the soil surface area measured by water vapor adsorption and the organic matter content (Figure 2).

1.2. Pedotransfer functions for mineral soils

Walczak (1984) used data on water retention of disturbed soil samples from 39 genetic horizons of 14 different soils to establish pedotransfer functions that would predict both

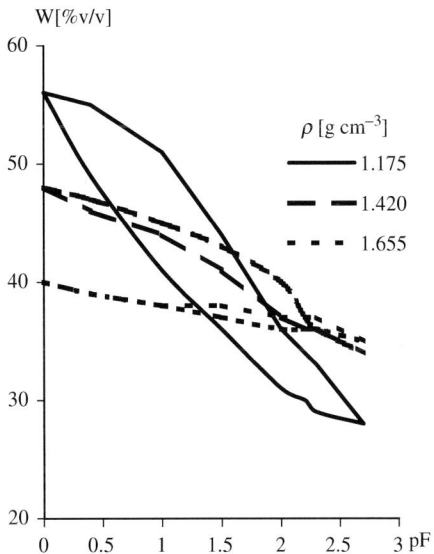

Figure 1. Hysteresis loops of soil water retention for maximum, average and minimum bulk density.

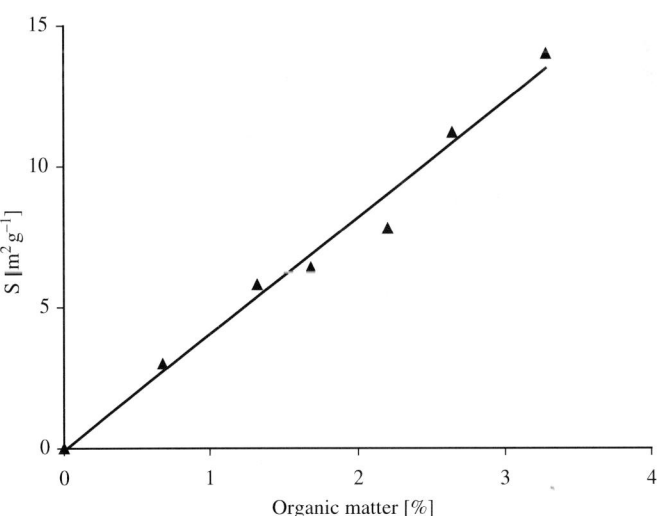

Figure 2. Relationship between the specific surface area measured by water vapor (S) and the content of organic matter.

drying and wetting water retention curves. Soil basic properties were soil specific surface, the particle-size distribution, soil bulk density, and soil organic matter content.

The multiple regression equations were developed as

$$W_p = b_0 + b_1 Y_1 + b_2 Y_2 + b_3 Y_3 \quad (5)$$

for capillary pressures in the range from pF 0 to pF 2.7 and

$$W_p = b_0 + b_1 Y_1 \quad (6)$$

for capillary pressures higher then pF 2.7. In equations (5) and (6), W_p is the predicted water content (gg^{-1}), Y_1 is the specific surface area ($m^2 g^{-1}$), Y_2 the weighed mean diameter of particles (mm), Y_3 the bulk density (gcm^{-3}) and the parameters b_0, b_1, b_2, b_3 are the regression coefficients. Regression coefficients for the drying and wetting water retention curves are given in Table 1. The correlation coefficients for these regressions were in the range from 0.94 to 0.98. The three arguments of the equation give a concise representation of main factors affecting water retention, i.e., mineralogical composition and nature of clay minerals, soil texture, and soil structure.

1.3. Comparison of selected pedotransfer function models

The structure of the Walczak's model (5) and (6) is similar to that of the Gupta–Larson (Gupta and Larson, 1979) and Rawls–Brakiensiek (Rawls and Brakensiek, 1982) models. These models are widely used for estimation of the water retention. The pedotransfer model of Gupta and Larson is based on the following multiple regression equation:

$$\theta_p = a_1 X_1 + a_2 X_2 + a_3 X_3 + a_4 X_4 + a_5 X_5 \quad (7)$$

where θ_p ($m^3 m^{-3}$) is the predicted water content, X_1 the percentage content of sandy fraction, X_2 the percentage content of silty fraction, X_3 the percentage content of clayey

Table 1
Parameters of the multiple linear regression to estimate drying and wetting water retention curves

pF	Drying water retention curve				Wetting water retention curve			
	b_0	b_1	b_2	b_3	b_0	b_1	b_2	b_3
0	91.60	0.06	−0.93	−43.77	nd	nd	nd	nd
1	79.79	0.07	−1.34	−37.36	64.46	0.09	−10.13	−28.06
1.5	73.66	0.08	−7.75	−33.40	55.53	0.09	−17.65	−22.21
1.6	71.37	0.09	−10.47	−31.83	53.33	0.10	−20.05	−20.82
2	58.86	0.09	−24.64	−23.02	44.80	0.10	−26.66	−15.63
2.2	49.46	0.10	−29.23	−17.52	nd	nd	nd	nd
2.3	49.93	0.10	−30.13	−15.69	38.23	0.10	−27.09	−12.44
2.7	31.07	0.11	−23.59	−9.72	nd	nd	nd	nd
3.7	−1.81	0.21	0.00	0.00	nd	nd	nd	nd
4.2	−1.74	0.18	0.00	0.00	nd	nd	nd	nd

fraction, X_4 the percentage content of the organic C, X_5 the bulk density (gcm^{-3}) and parameters a_1, a_2, a_3, a_4 and a_5 are the regression coefficients.

In the water retention model of Rawls and Brakensiek, the following equation of multiple regression is used:

$$\theta_p = a_0 + a_1 X_1 + a_2 X_2 + a_3 X_3 + a_4 X_4 + a_5 X_5 \tag{8}$$

where θ_p (m^{3}m^{-3}) is the predicted water content, X_1 the percentage content of sandy fraction, X_2 the percentage content of silt fraction, X_3 the percentage content of clayey fraction, X_4 the percentage content of organic carbon, X_5 the bulk density (gcm^{-3}) and parameters a_0, a_1, a_2, a_3, a_4 and a_5 are the regression coefficients.

We note that the use of contents of sand, silt and clay in the models of Gupta and Larson and Rawls and Brakensiek seems to be incorrect from a statistical point of view because these quantities are linearly dependent and their sum equals 100%.

Walczak et al. (2002) compared the above-mentioned models using data on 10 different soils. Figure 3 presents the results of the comparison. For each model, the regression "estimated θ_p vs. measured θ_m" water content was developed. The regression equations were:

$$\theta_{\text{Gupta}} = 0.11160 + 0.72197 \theta_{\text{measured}} \tag{9}$$

$$\theta_{\text{Rawls}} = 0.06762 + 0.88028 \theta_{\text{measured}} \tag{10}$$

$$\theta_{\text{Walczak}} = -0.0289 + 0.91 \theta_{\text{measured}}. \tag{11}$$

Statistics from the model comparison are presented in Table 2. Of the three compared models, the Walczak's model performed best (Table 2). That may be because properties of 10 text soils are closer to properties of the original soils used for the development of Walczak's pedotransfer function. It also may be that the Gupta–Larson and Rawls–Brakensiek models encompass very wide range of soils and give only approximate estimates for any subset of this multitude of soils.

1.4. Approach to pedotransfer functions for organic soils

Gnatowski (2001) studied statistical relations between the parameters of Mualem–van Genuchten equations describing soil water retention and the hydraulic conductivity coefficient for 87 peat soil samples, and basic soil physical properties (ρ, the bulk density (gcm^{-3}) and P, the ash content (%)). He used the following transformed variables:

$$\begin{aligned}
L^* &= \frac{L+10}{10-L} \\
\theta_r^* &= \frac{\theta_r + 0.05}{0.50 - \theta_r} \\
K_s^* &= \ln(K_s) \\
\alpha^* &= \ln(\alpha) \\
\theta_s^* &= \ln(\theta_s) \\
n^* &= \ln(n - 1.1)
\end{aligned} \tag{12}$$

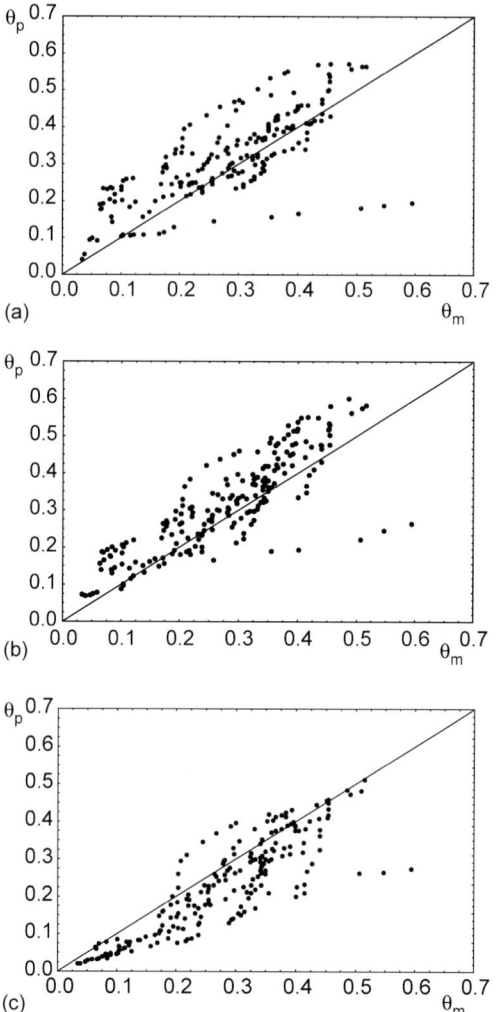

Figure 3. Performance of three pedotransfer functions: (a) Gupta and Larson model, (b) Rawls and Brakensiek model, (c) Walczak model.

where θ_s is the water content at saturation ($cm^3 cm^{-3}$), θ_r the residual water content ($cm^3 cm^{-3}$) and α (hPa^{-1}), n (–), L (–) are the model parameters. As a result of the performed statistical analysis, the following regression equations were obtained:

$$\theta_s^* = 0.138468 + 0.0108598 P - 0.762692 \rho - 0.111439 \ln(P)$$

$$\theta_r^* = 96.7386 - 436.177 \rho + 0.016193 P^2 + 867.459 \rho^2 + 27.6022 \ln(\rho)$$

$$\alpha^* = 114.697 - 0.659748 P - 238.113 \rho + 4.08889 \rho^{-1} + 3.63416 \ln(P) + 60.6568 \ln(\rho) + 2.14613 P \rho$$

Table 2
Correlation coefficients (R), standard errors of estimation (SEE), Snedecor coefficients (F) and (t) coefficients for the analyzed models

Model	R	SEE	F	t
Gupta and Larson	0.7352	0.0795	232	15.26
Rawls and Brakensiek	0.8239	0.0723	418	20.46
Walczak	0.8658	0.0633	593	24.35

$$n^* = 8.03053 + 0.650829P - 39.3579\rho + 134.476\rho^2 - 6.28299\ln(P)$$
$$K_s^* = 1179.23 - 0.462353P - 3324.32\rho + 22.454\rho^{-1} + 3648.58\rho^2 + 4.49555\ln(P) + 484.505\ln(\rho)$$
$$L^* = 4.29603 + 1.45987\ln(P) + 2.89744\ln(\rho) - 1.08517P\rho$$

(13)

For the above equations, these correlation coefficients were obtained: $R = 0.90$ for θ_s^*, $R = 0.82$ for θ_r^*, $R = 0.79$ for α^*, $R = 0.63$ for n^*, $R = 0.65$ for K_s^* and $R = 0.80$ for L^*.

2. HYDRAULIC CONDUCTIVITY

2.1. Saturated hydraulic conductivity

Investigations (Lipiec, 1983) that aimed at estimating soil hydraulic conductivity of undisturbed structure were performed. The research included 31 soil profiles studied at three depths of 0–30, 30–60 and 60–90cm. The determination of the coefficient of water conductivity K in saturated and unsaturated zones was performed with filtration method, using gypsum-sandy crusts. In this method, steel infiltrometers having a diameter of 28cm and a height of 15cm were applied in which soil samples were taken with undisturbed structure. On the base of the statistically significant correlation between soil water conductivity in saturated zone and the percentage content of sand, silt and clay fractions as well as the grain-size distribution index, the regression equations were determined to calculate the coefficient of water conductivity in the saturated zone. For particular layers, these equations have the following form, respectively:

$$K_s = 131.32 + 0.089\text{sand}^2 - 5.75\text{sand} \qquad R = 0.71(0-30\text{cm})$$

$$K_s = 118.56 + 0.091\text{sand}^2 - 5.87\text{sand} \qquad R = 0.90(30-60\text{cm})$$

$$K_s = 143.46 + 0.09 \times 5\text{sand}^2 - 6.32\text{sand} \qquad R = 0.87(60-90\text{cm})$$

$$K_s = 29.10 + \frac{1785.62}{\text{silt}} \qquad R = 0.70(0-30\text{cm})$$

$$K_s = 45.51 + \frac{1165.05}{\text{silt}} \quad R = 0.84(30-60\text{cm})$$

$$K_s = 87.45 + \frac{799.99}{\text{silt}} \quad R = 0.83(60-90\text{cm}) \tag{14}$$

$$K_s = 19.65 + \frac{1753.35}{\text{clay}} \quad R = 0.76(0-30\text{cm})$$

$$K_s = 24.66 + \frac{1312.45}{\text{clay}} \quad R = 0.89(30-60\text{cm})$$

$$K_s = 55.12 + \frac{1042.17}{\text{clay}} \quad R = 0.82(60-90\text{cm})$$

$$K_s = 50.94 + 318.15f \quad R = 0.49(0-30\text{cm})$$

$$K_s = 110.78 + 422.02f \quad R = 0.81(30-60\text{cm})$$

$$K_s = 75.51 + 390.0f \quad R = 0.80(60-90\text{cm})$$

where K_s is the coefficient of hydraulic conductivity in saturated zone (cmday^{-1}), sand, the percentage content of sand fraction, silt, the percentage content of silt fraction, clay, the percentage content of clay fraction and f is the grain size distribution index (Giesel et al., 1972) calculated as:

$$f = \frac{\sum_{i=1}^{k}(P_{i+1} - P_i)\frac{\log(P_{i+1}/P_i)}{\log(S_{i+1}/S_i)}}{P_k - P_i} \tag{15}$$

where P_i is the content of ith grain fraction, which is read from the cumulative grain distribution curve and the smaller grain fractions, S_i the maximal diameter of ith grain fraction and k the quantity of grain fractions. Index f is undimensional.

2.2. Unsaturated hydraulic conductivity

The impact of soil physical parameters on the unsaturated hydraulic conductivity at selected soil water potential values was investigated (Sławiński, 2003). The study was performed for 290 soil profiles selected from the Bank of Samples of the Mineral Arable Polish Soils. All the values of the water conductivity coefficient of the investigated soils were determined according to the same methodology (Sławiński, 2003; Sobczuk et al., 1992; Walczak et al., 1993). The instantaneous profiles method together with the measurement of the saturated hydraulic conductivity provided data to determine the relationship between the hydraulic conductivity and the soil water potential in the range of the capillary pressures from pF 0 to pF $\sim$ 3. The following values of the soil water potential and the respective pF values were selected for the statistical analysis: pF 0

(0.098kPa); pF 1 (0.98kPa); pF 1.5 (3.1kPa); pF 2 (9.8kPa); pF 2.2 (15.5kPa); pF 2.5 (31kPa); pF 2.7 (49kPa) and pF 3 (98kPa).

The following soil physical parameters were chosen as predictors: contents of sand, silt, and clay, F_{sand}, F_{silt}, and F_{clay} (%), the mean statistical diameter of particles, D_p (mm), the geometrical surface area of the soil particles, S_g (cm^{2}g^{-1}), the soil bulk density, ρ (gcm^{-3}), the specific surface area, S_{BET} (cm^{2}g^{-1}), the content of gravitational water, W_G (cm^{3}cm^{-3}) (the difference of water content between pF 0 and pF 2.2), the water content at pF 2.2 corresponding to the field capacity, FWC (cm^{3}cm^{-3}), the content of organic carbon, C_{org}, (%) and the total porosity P (cm^{3}cm^{-3}).

The mean statistical diameter of particles, D_p (mm) and the geometrical surface area of soil particles, (S_g) (cm^{2}g^{-1}) were calculated with assumption that soil particles have spherical shape (Walczak, 1984) with following equations:

$$D_p = \frac{\sum_{i=1}^{n} P_i \left(\frac{D_{imax} + D_{imin}}{2} \right)}{100\%} \qquad (16)$$

$$S_g = \frac{\sum_{i=1}^{n} P_i \frac{4\pi(\bar{D}_i/2)^2}{4/3\pi(\bar{D}_i/2)^3 \rho}}{100\%} \qquad (17)$$

where P_i is the percentage content of ith fraction (%), n the quantity of fractions, ρ the soil solid phase density (gcm^{-3}) and $\bar{D}_i = D_{i,max} + D_{i,min}/2$ is the mean diameter of ith fraction (mm), $D_{i,max}$ being the maximal diameter of ith fraction (mm) and $D_{i,min}$ the minimal diameter of ith fraction (mm).

For statistical analysis, 415 soil samples were chosen representing topsoils and subsoils. The partial correlation coefficients between the values of the hydraulic conductivity as well as its decimal logarithms and parameters of the soil solid phase were generally low. They did not exceed 0.25 for the value of the conductivity and 0.50 for its logarithm (Table 3). Logarithms of hydraulic conductivity coefficient were the subject of analysis.

Subgroups of soil solid phase parameters to develop PTFs were selected for each value of soil matric potential. The selection was based on two criteria: (a) a relatively high value of the partial correlation coefficient between a chosen parameter of a solid phase and the value of the logarithm of the hydraulic conductivity coefficient, and (b) lack of the functional or correlation dependence between the physical parameters of the soil, chosen to create the model. Various regression models were analyzed, i.e., the multiple linear regression, polynomial regression, factorial regression, and regression built on the base of the functions of physical parameters of the soil, e.g., logarithms, reciprocal functions, etc. Correlation coefficients of those regressions varied between 0.33 and 0.73.

The low accuracy of the developed regression lead to the assumptions that soils should be grouped prior to PTF development. The basic feature that makes it possible to distinguish the investigated soils from the point of view of their physical properties was the particle size distribution. The database was divided into four subsets according to four soil texture groups following the classification in Table 4. Sandy, loamy, silty, and clayey

Table 3
Correlation coefficients between values logarithms of hydraulic conductivity and analyzed parameters of soil structure (bold marks coefficients significantly different from zero)

Variable $\log_{10}K$	F_{clay}	F_{silt}	F_{sand}	D_p	S_g	P	ρ	W_G	FWC	S_{BET}	C_{org}
pF 0	−0.21	−0.11	0.19	0.20	−0.22	0.03	−0.12	0.29	−0.23	−0.12	0.07
pF 1	−0.45	−0.21	0.40	0.41	−0.37	0.05	−0.04	0.46	−0.36	−0.31	−0.13
pF 1.5	−0.46	−0.17	0.39	0.40	−0.40	0.01	0.03	0.43	−0.36	−0.38	−0.17
pF 2	−0.42	−0.07	0.31	0.31	−0.38	−0.01	0.09	0.34	−0.30	−0.41	−0.22
pF 2.2	−0.39	−0.01	0.26	0.26	−0.38	−0.05	0.14	0.27	−0.27	−0.44	−0.25
pF 2.5	−0.27	0.08	0.13	0.14	−0.29	−0.09	0.17	0.13	−0.18	−0.40	−0.25
pF 2.7	−0.16	0.14	0.02	0.03	−0.21	−0.11	0.18	0.02	−0.10	−0.35	−0.23
pF 3	−0.00	0.20	−0.12	−0.11	−0.08	−0.13	0.18	−0.15	0.03	−0.23	−0.17

Table 4
Soil classification according to granulometric group (Dobrzański and Zawadzki, 1995)

Granulometric group	Sand 1–0.1mm (%)	Silt 0.1–0.02mm (%)	Clay < 0.02mm (%)
Sandy soils	40–100	0–40	0–20
Loamy soils	10–79	0–40	21–90
Silty soils	0–59	41–100	0–50
Clay soils	0–9	0–49	51–100

sample groups contained 187, 115, 86, and 27 samples, respectively. The regression analyses were performed for each of these classes. The correlation coefficients remained relatively low, between 0.37 and 0.83. The data inspection showed that it would be beneficial to use a piece-wise linear regression in the form:

$$Y = A(b_{01} + b_{11}x_1 + b_{21}x_2 + \ldots + b_{m1}x_m) + B(b_{02} + b_{12}x_1 + b_{22}x_2 + \ldots + b_{m2}x_m) \quad (18)$$

where $A = 1$ and $B = 0$ for $Y \leq PP$ as well as $A = 0$ and $B = 1$ for $Y > PP$, where PP is a threshold value of the dependent variable. Correlation coefficients in the range from 0.81 to 0.85 were obtained with this regression equation with F_{clay}, F_{sand}, S_{BET}, C_{org}, W_G and FWC as input variables. The relatively high correlation coefficients suggested further applications of this model within soil texture groups. The correlation coefficient values within the range $0.86 \leq R \leq 0.96$ were obtained with F_{clay}, F_{sand}, S_{BET}, C_{org}, W_G and FWC as input variables. The general form of this model's equation is:

$$\log K = A(a_0 + a_1 F_{\text{clay}} + a_2 F_{\text{sand}} + a_3 S_{\text{BET}} + a_4 C_{\text{org}} + a_5 W_G + a_6 \text{FWC})$$
$$+ B(b_0 + b_1 F_{\text{clay}} + b_2 F_{\text{sand}} + b_3 S_{\text{BET}} + b_4 C_{\text{org}} + b_5 W_G + b_6 \text{FWC}) \quad (19)$$

where $A = 1$ and $B = 0$ for $\log K \leq PP$ as well as $A = 0$ and $B = 1$ for $\log K > PP$.

Tables of the coefficients and values of the breakpoints are available from authors upon request.

The PTF model was tested with the soil data that had not been used for its development. The verification was conducted for each textural group. For verification of the model, 145 soil samples were chosen, including 45 sandy soils, 32 loamy soils, 52 silty soils and 16 clayey soils. To verify the piece-wise linear model (18), it was necessary to determine according to which equation the coefficient of the soil water conductivity should be calculated for the given soil texture and the given value of the soil water potential. That is, it was necessary to determine whether the given value of the unsaturated hydraulic conductivity is larger than or less than the breakpoint value PP. To do that a first-guess PTFs for hydraulic conductivity was developed for each of the textural groups.

$$\log K = c_0 + c_1 F_{\text{clay}} + c_2 F_{\text{sand}} + c_3 S_{\text{BET}} + c_4 C_{\text{org}} + c_5 W_G + c_6 \text{FWC} \quad (20)$$

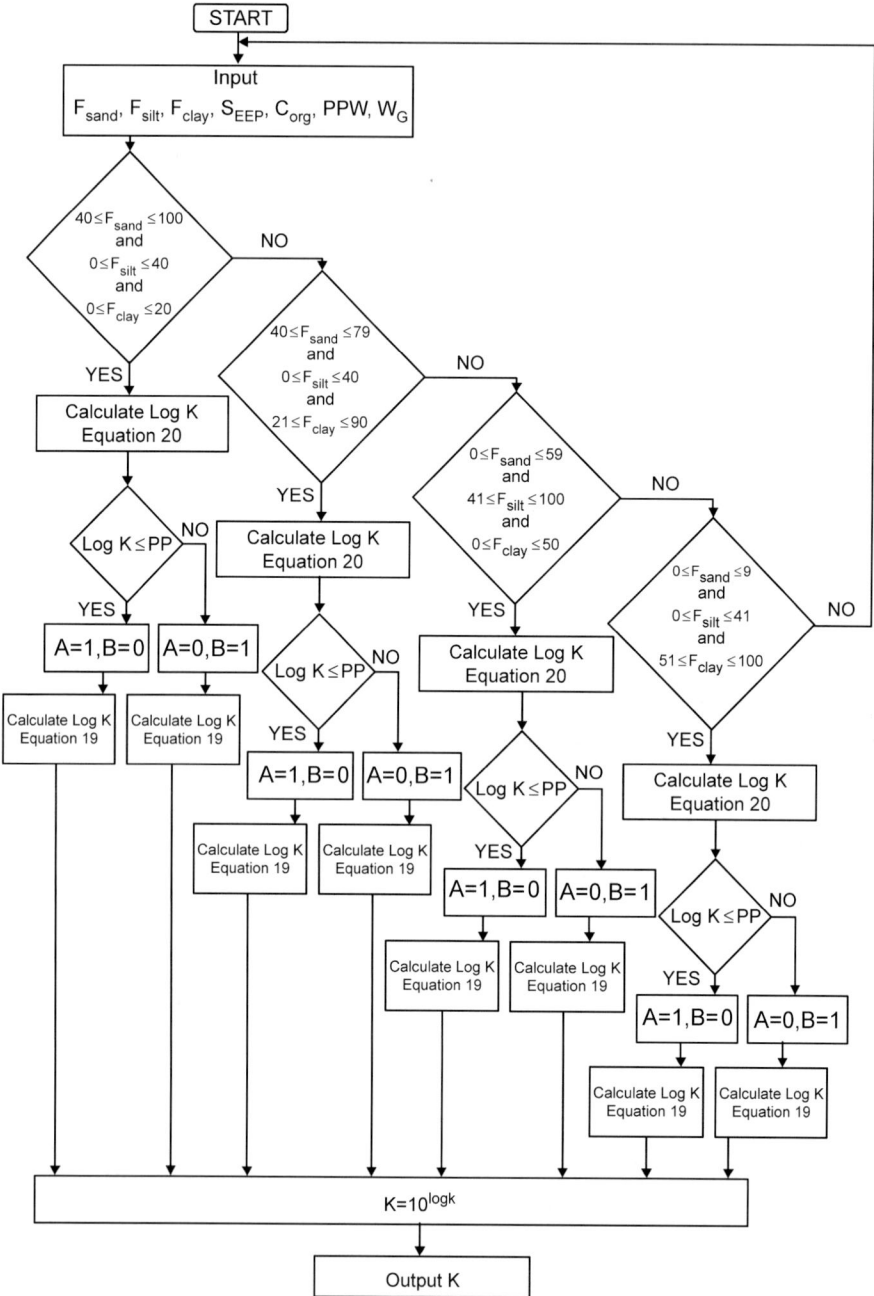

Figure 4. Algorithm of hydraulic conductivity coefficient calculation.

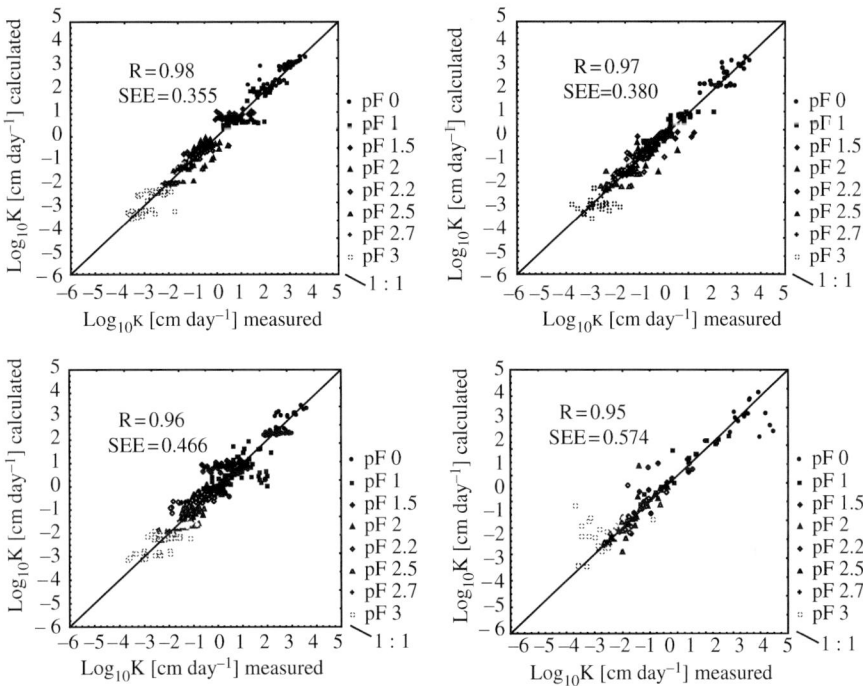

Figure 5. Measured and calculated values of the logarithm of the unsaturated hydraulic conductivity: (a) sandy soils, (b) loamy soils, (c) silty soils, (d) clay soils.

The coefficients of these equations together with correlation coefficients are available from the authors upon request. Using these equations, it is possible to obtain a rough estimate of the logarithm of the unsaturated conductivity coefficient for a given soil texture and a given soil water potential value. A value obtained in this way enables to determine which equation of the piece-wise regression should be used. The answer

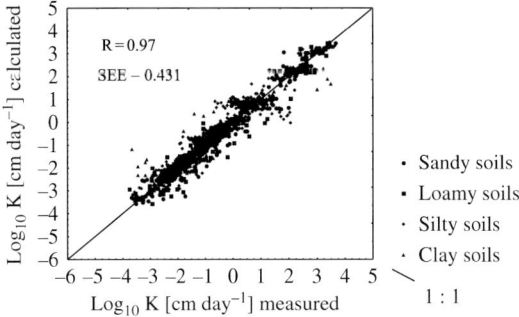

Figure 6. Measured and calculated values of the logarithm of the unsaturated hydraulic conductivity for the whole tested data set.

depends on whether the first-guess hydraulic conductivity is larger or less than the breakpoint value PP. The sequence of calculations is expressed in the form of an algorithm presented in Figure 4.

A comparison of the values of the logarithms of the soil water conductivity coefficients, calculated from the proposed model and measured for particular soil textures, is presented in Figure 5a–d. Figure 6 presents a comparison of calculated and measured values of the logarithms of the unsaturated hydraulic conductivity for the whole tested data set.

REFERENCES

Dobrzański, B., Dechnik, I., Stawiński, J., 1972. Correlation between the soil surface-area and humus compounds in the soil. Polish J. Soil Sci. 2, 99.
Domżał, H., 1979. Impact of soil compaction on water useful and unuseful for plants. Rocz. Gleboznawcze XXX (3), 45, (in Polish).
Domżał, H., 1983. Compaction of the solid phase and its role in the formation of the water–air properties of soil. Zesz. Problemowe Postępów Nauk Rolniczych 220/II, 137.
Giesel, W., Renger, M., Strebel, O., 1972. Berechnung des kapillaren Aufstiegs aus der Grundwasser in der Warzelraum unter stationaren Bedingungen. Z. Pflanzenern. U. Bodenkd. 132 (1), 9.
Gnatowski, T., 2001. Retention and hydraulic properties of peat and moorsh soils from Biebrza River Valley. Ph.D. Thesis, Agricultural Academy of Warsaw (in Polish).
Gupta, S.C., Larson, W.E., 1979. Estimating soil water retention characteristics from particle-size distribution, organic matter percent, and bulk density. Water Resour. Res. 15, 1633-1635.
Kaniewska, J., Walczak, R., 1974. A numerical interpretation of the hysteresis phenomenon in the drying and wetting processes of soil. Polish J. Soil Sci. VII, 63.
Lipiec, J., 1983. The estimation of hydraulic conductivity from soil physical properties. Problemy Agrofizyki 40, 5, (in Polish).
Rawls, W.J., Brakensiek, D.L., 1982. Estimation soil water retention from soil properties. J. Irrigation Drainage 108, 166.
Sławiński, C., 2003. The influence of soil physical parameters on values of hydraulic conductivity coefficient. Acta Agrophys. 90, 5, (in Polish).
Sobczuk, H.A., Plagge, R., Walczak, R.T., Roth, Ch.H., 1992. Laboratory equipment and calculation procedure to rapidly determine hysteresis of some soil hydrophysical properties under non-steady flow conditions. Pflanz. Bodenk. 155, 157.
Trzecki, S., 1974. Determination of water capacity of soils on the basis of their mechanical composition. Rocz. Gleboznawcze XXV, 33.
Turski, R., Domżał, H., Słowińska-Jurkiewicz, A., 1974. Impact of clay and humus content on maximal hygroscopity of loess soils. Rocz. Gleboznawcze XXV (3), 85, (in Polish).
Turski, R., Domżał, H., Słowińska-Jurkiewicz, A., Martyn, W., 1975. Influence of colloidal clay, $CaCO_3$ and humus content on water retention, plasticity and swelling of Polish rendzinas. Rocz. Gleboznawcze XXVI (3), 35.

Walczak, R., 1977. Model investigations of water binding energy in soils of different compaction. Zesz. Problemowe Postepów Nauk Rolniczych 197, 11.
Walczak, R., 1984. Relationship between water retention and parameters of solid phase of soils. Problemy Agrofizyki 41, 5, (in Polish).
Walczak, R., Kaniewska, J., 1975. A method for calculating energy effects and dissipation of energy in changes of soil moisture. Polish J. Soil Sci. VIII, 101.
Walczak, R.T., Sławiński, C., Malicki, M.A., Sobczuk, H., 1993. Measurement of water characteristics in soils using TDR technique, water characteristics of loess soil under different treatment. Int. Agrophys. 7, 175.
Walczak, R., Witkowska-Walczak, B., Sławiński, C., 2002. Comparison of correlation models for the estimation of the water retention characteristics of soils. Int. Agrophys. 16, 79.
Zawadzki, S., 1970a. The influence of mechanical composition of soils on soil moisture retention. Polish J. Soil Sci. III/1, 11.
Zawadzki, S., 1970b. An approximate method of determining water retention capacity of soils. Polish J. Soil Sci. III/2, 9.
Zawadzki, S., Michałowska, K., Stawiński, J., 1971. The application of surface area measurements of soils for determination of the content of water unavailable for plants. Polish J. Soil Sci. IV/2, 89.

Chapter 25

PEDOTRANSFER FUNCTIONS OF THE RYE ISLAND – SOUTHWEST SLOVAKIA

V. Štekauerová and J. Šútor

Institute of Hydrology, Slovak Academy of Sciences, Račianska 75 83102 Bratislava, Slovak Republic

Water retention is an important characteristic used in water movement models in vadose zone in order to evaluate such factors as the impact of anthropogenic activity on soil water balance factors, groundwater pollution, water storage in vadose zone, and global changes influence. However, water retention measurements require special experimental equipment and are time consuming, while project timing often does not allot enough time and resources to perform an exhaustive determination of soil water retention by traditional methods in the laboratory (Kutílek and Nielsen, 1994). To avoid this problem, a common methodology has been used in the literature where water retention is determined from available soil characteristics, e.g., grain size distribution, bulk density, organic C content, etc. That is, the water content in soil is assumed to depend on characteristics of its structure, bulk density, organic matter content or organic C content, and so on, via relationships described by what are known as pedotransfer functions (PTFs). PTFs not only forgo the difficulty of measuring water retention, but also allow us to use the pedological researches of an area that has been carried out in the past (Pachepsky et al., 1982; Rajkai et al., 1996).

Several methods have been devised in soil physics to determine soil hydraulic characteristics from basic soil parameters and to estimate functional forms of PTFs. Originally only the particle-size distribution was used (Brooks and Corey, 1964; Husz, 1967; Renger, 1971). Later dependence on the grain size distribution was expanded to include organic matters contents and bulk density (Gupta and Larson, 1979; Rawls et al., 1982). Three approaches have been developed to estimate water retention. The first approach is based on a regression predicting water retention at several soil water potentials (Cosby et al., 1984; Pachepsky et al., 1982; Puckett et al., 1985; Vereecken et al., 1989, 1990; Wősten et al., 1995; Wiliams et al., 1983; Bastet et al., 1998). The second approach is based on the following physical model of the soil–water system (Arya and Parys, 1981; Haverkamp and Parlange, 1986; Tietje and Tapkenhinrichs, 1993; Zeiliguer et al., 2000):

(a) size distribution of soil pores is determined from the grain size distribution or from water retention of separate textural fractions;
(b) water content in soil is calculated from the size distribution of soil pores;
(c) soil water potential is estimated as the moisture potential, which is in turn calculated from the size distribution of soil pores using a mathematical quantification of capillary phenomena in the soil–water system.

The third approach uses an analytical expression of the water retention curve, i.e., a dependence $\theta = F(\psi)$, e.g., according to Van Genuchten (1980). The individual parameters in the expression are obtained using a regression of the parameter values on basic soil characteristics such as the grain size distribution, porosity, organic carbon content, bulk and particle density.

One of the central questions in PTF development is searching for the PTF expression with general validity for all soils (Tietje and Tapkenhinrichs, 1993; Bastet et al., 1998; Van den Berg et al., 1997; Singh, 1998). However, in the present study, we concentrate on the soils of the Rye Island and develop PTFs using physical and hydrophysical characteristics of the soils of the Rye Island which we consider to be applicable with the presented accuracy only for this region.

1. AREA DESCRIPTION

The Rye Island region and its basic characteristics include the Danube river that flows through a relatively narrow valley along the Austrian–Slovak border. The Danube crosses a strip of mountains connecting the Alps and the Karpatians, passes Bratislava, and divides into two branches just below Slovak capital, the Danube and the Small Danube. The two branches of the Danube flow separately for approximately 100 km and then join each other again near the industrial town of Komarno. The area between the two branches of the Danube river is called Rye Island. Rye Island is part of the Danube Lowland and, because of its favorable climate, soil and morphological conditions, is one of the most productive agricultural areas of Slovakia. The average width of Rye Island is 20 km; its area is approximately 2000 km^2, which represents about 4% of the Slovak territory, but about 10% of the most productive arable land. Due to its extreme importance for Slovakia, preservation and improvement of the agricultural production potential of Rye Island is of great interest.

The Rye Island area forms a flat plain with only small differences in altitude. The altitude of Rye Island decreases in the South-East direction. Its altitude is between 135 and 136 m above sea level in the Bratislava area, decreasing downstream to 108 m (the town of Komarno). Rye Island is the result of sedimentation of the Danube, with sediments from upstream mountains being spread over its territory. During previous centuries (up to the 19th century), the river branched into multiple streams and frequently changed course within its own alluvial sediments. Remainders of old river branches can still be observed in the territory. The most important environmental factors that contributed to the formation of Rye Island were the Danube river and its water regime, climate changes and the Gabcikovo power station.

Soils in Rye Island area are a result of Danube river sedimentation, natural pedogenesis and antropogenic activities. Mollic Gleysols, Mollic Fluvisols and Calcaric Haplic Chernozems developed on alluvial sediments of Rye Island, especially in areas where the groundwater table is not in direct contact with the soil profile and where flooding does not occur regularly. Calcaric Fluvisols and Fluvic Gleysols evolved on young alluvial deposits of the Danubian Lowland. In the South-Eastern part of the Rye Island area many Solonchaks and Solonetz can be found. Due to different pedogenetic conditions across Rye Island, different soils can be found in the area. Close to the Danube and Small Danube rivers, young soils (fluvisols) with different degrees of gley processes can be identified.

Table 1
Statistical characteristics of the soil water content sets measured for different soil moisture potential. θ_{pF} – water content in soil water potential pF (vol.%)

Statistical characteristics	$\theta_{pF=0.3}$	$\theta_{pF=1.75}$	$\theta_{pF=2}$	$\theta_{pF=2.3}$	$\theta_{pF=2.74}$	$\theta_{pF=3}$	$\theta_{pF=3.48}$	$\theta_{pF=4.2}$	$\theta_{pF=4.78}$
Average	42.16	37.00	36.34	34.76	31.36	27.21	25.11	13.35	12.18
Standard deviation	0.36	0.42	0.61	0.45	0.47	0.55	0.38	0.54	0.37
Median	41.31	36.6	37.6	35.00	30.59	27.99	24.66	14.005	11.5
Modus	33.6	35.09	37.0	34.9	29.39	27.43	22.93	4.5	7.5
Standard error	5.35	5.35	7.50	6.38	5.98	8.19	4.85	6.10	5.61
Variance	23.58	28.58	56.21	40.65	35.77	67.09	23.50	37.24	31.47
Kurtosis	0.30	0.22	1.41	1.19	−0.04	0.20	0.51	−0.65	0.07
Skewness	0.40	0.27	−1.06	−0.44	0.36	−0.37	0.16	−0.09	0.57
Range	35.73	32.95	42.54	42.74	33.29	43.43	28.26	7.05	27.80
Minimum	25.80	21.22	9.66	10.60	16.32	5.79	11.07	0.40	2.0
Maximum	61.53	54.17	52.20	53.34	49.61	49.22	39.33	27.45	29.80
Sample size	221	159	151	199	159	224	159	128	235
Coefficient of variation	0.13	0.14	0.21	0.18	0.19	0.30	0.19	0.46	0.40

In depression areas continuously influenced by stagnant water, a specific catena of hydromorphic soils is found. Chernozems cover the major part of the Rye Island area. These soils have not been subjected to long-term effects by groundwater, and are among the most productive in Slovakia.

The soil cover of Rye Island area according to texture is as follows: course-textured soils (sandy soils) – 5%, medium-textured soils (loamy soils) – 66%, fine-textured soils (clayey soils) – 23%, and extremely fine-textured soils (clays) – 6%.

2. METHODS

Water retention curves of the Rye Island soils were determined under laboratory conditions using the pressure apparatus Soil Moisture Equipment, Santa Barbara, California. Soil samples were taken at selected localities of the Rye Island area during years 1990–1998 to represent relative occurrence of the soil kinds of the area. The data set contained data from 221 soil samples. Soil water potential values, or pF, of 0.3, 1.75, 2.0, 2.3, 2.74, 3.0, 3.48, and 4.2 were chosen to determine PTF from the above-mentioned data set. Water contents θ_{pF} for pF = 4.78 were obtained by an exicator method on 235 disturbed soil samples (Šútor and Komár, 1984; Šútor and Majerčák, 1988). Nine data subsets for PTF creation were obtained (one data subset for every pF point). Statistical characteristics of the subsets are presented in Table 1.

The PTF input parameters included the content (%) of the category I particles ($d <$ 0.01 mm), category II particles (0.01 mm $< d <$ 0.05 mm), category particles III (0.05 × mm $< d <$ 0.1 mm), and category IV particles (0.1 mm $< d <$ 2 mm) and the dry bulk density ρ_d (g cm^{-3}).

3. RESULTS AND DISCUSSION

Below we present regression equations for the water content in soil, θ_i, where i is the soil water potential pF:

$$\theta_{0.3} = 0.19396X_1 + 0.10889X_2 + 0.04911X_3 + 0.12254X_4 - 29.9873X_5 + 72.2029 \quad (1)$$

$$\theta_{1.75} = 0.07392X_1 + 0.16894X_2 - 0.21362X_3 + 0.03521X_4 - 21.3444X_5 + 62.0236 \quad (2)$$

$$\theta_{2.0} = -0.00095X_1 + 0.02568X_2 - 0.15271X_3 - 0.33344X_4 - 7.27798X_5 + 52.5034 \quad (3)$$

$$\theta_{2.3} = 0.0195X_1 + 0.00195X_2 - 0.20661X_3 - 0.4705X_4 - 21.8808X_5 + 70.3349 \quad (4)$$

$$\theta_{2.74} = 0.05301X_1 + 0.15938X_2 - 0.29316X_3 - 0.01822X_4 - 23.5785X_5 + 62.5331 \quad (5)$$

$$\theta_{3.0} = -0.06503X_1 - 0.09055X_2 - 0.59549X_3 - 0.3002X_4 - 12.687X_5 + 65.5509 \quad (6)$$

$$\theta_{3.48} = -0.07141X_1 - 0.18031X_2 - 0.4179X_3 - 0.2657X_4 - 15.5787X_5 + 66.4839 \quad (7)$$

$$\theta_{4.2} = 0.09661X_1 + 0.04884X_2 - 0.14197X_3 + 0.26499X_4 - 8.84463X_5 + 23.3886 \quad (8)$$

$$\theta_{4.78} = 0.229492 X_1 - 0.26074 \qquad (9)$$

where θ_i was measured in vol.%; X_1, X_2, X_3, and X_4 stand for the contents (%) of particles of the I, II, III, and IV categories; and X_5 is the dry bulk density ρ_d (g cm^{-3}). Statistics for individual pF values are shown in Table 2. The bulk density ρ_d was not used to develop equation (9) because the data obtained from disturbed soil samples were used.

Developed PTFs were tested with data on 24 soil samples from locations of the Rye Island that were not used when creating the PTFs. The accuracy of the calculated retention curve was quantified with the mean difference (MD) and the root of mean squared difference (RMSD). RMSD has been used for testing the closeness between measured and calculated water retention curves in works by Vereecken et al. (1989) and Tietje and Tapkenhinrichs (1993). MD and RMSD are used in the published work to directly compare the obtained PTFs with the data. The MD and the RMSD are calculated using the method of numerical quadrature within an interval of water potentials from -74130 to 0 cm:

$$\text{MD} = \frac{1}{b-a} \int_a^b (\theta_p - \theta_m) d\Psi \qquad (10)$$

$$\text{RMSD} = \left[\frac{1}{b-a} \int_a^b (\theta_p - \theta_m) d\Psi \right]^{1/2} \qquad (11)$$

where θ_m is the measured water content and θ_p is the water content calculated from PTF.

MD values can be positive as well as negative depending on whether the moisture values calculated from the PTF are over or below the measured moisture values. They are equal to zero in the case when the water retention curve from the measured data is identical to the water retention curve calculated using PTF. On the other hand, if MD $= 0$, it does not mean that the RMSD $= 0$. RMSD values determine the closeness between the measured values of the water retention curve and its values obtained using PTF. Tietje and Tapkenhinrichs (1993) present results from evaluating 13 PTFs using data from Lower Saxony, Germany. The best five PTFs for this database had the following values of MD and RMSD, respectively: 1.29 and 5.75 (Renger, 1971); 0.95 and 6.11 (Arya and Parys, 1981); -0.19 and 6.5 (Cosby et al., 1984); -5.27 and 7.51 (Rawls and Brakensiek, 1985); and -1.45 and 5.31 (Vereecken et al., 1990). From the equations (1)–(9) obtained using the test dataset in this work, the following average values were obtained: MD $= -2.5$ and RSMD $= 3.6$. Compared with the Tietje and Tarpenhinrich's evaluation, these values show a good applicability of equations (1)–(9).

Figure 1 shows the statistical distributions of MD and RMSD values. The distribution of MD is close to normal as the points in Figure 1 lie close to the straight line on the probability scale. The probability distribution of RMSD is far from normal. Small values of RMSD are observed much more frequently as compared to the large values. About 50% of the RMSD values are in the range between 0 and 2.5%, and another half of the points is in the range between 2.5 and 10%. We have also observed a strong relationship between MD and RMSD values as shown in Figure 2. This may indicate that using both MD and RMSD as PTF performance indicators may be excessive.

Table 2
Statistical characteristics of the representative soil physical properties for measured points of water retention curves

pF	Statistical characteristic	Content of the particles of the corresponding category (%)				Dry bulk density (g cm^{-3})
		I	II	III	IV	
0.3	Mean	37.36	36.52	17.91	7.96	1.44
	Min	16	12	2	0	0.91
	Max	66	52	39	42	1.72
	Sample size	221	221	221	221	221
1.76	Mean	35.69	38.65	17.62	7.87	1.44
	Min	16	22	2	0	0.91
	Max	60	52	39	41	1.72
	Sample size	159	159	159	159	159
2	Mean	28.00	35.75	27.82	8.27	1.38
	Min	4.2	8.51	2.71	0.04	0.97
	Max	77.81	76.2	82.32	59.1	1.7
	Sample size	151	151	151	151	151
2.3	Mean	37.70	34.63	22.44	4.95	1.44
	Min	4.48	8.51	2	0	0.91
	Max	77.81	52.78	82.32	41	1.72
	Sample size	199	199	199	199	199
2.74	Mean	35.69	38.65	17.62	7.87	1.44
	Min	16	22	2	0	0.91
	Max	60	52	39	41	1.72
	Sample size	159	159	159	159	159
3	Mean	29.10	40.52	20.08	10.26	1.40
	Min	4.2	12.3	2	0	0.91
	Max	60	76.2	75.13	59.1	1.72
	Sample size	224	224	224	224	224
3.48	Mean	35.69	38.65	17.62	7.87	1.44
	Min	16	22	2	0	0.91
	Max	60	52	39	41	1.72
	Sample size	159	159	159	159	159
4.2	Mean	37.99	31.84	25.60	4.13	1.44
	Min	4.48	8.51	2.71	0.04	1.00
	Max	77.81	52.78	82.32	41	1.70
	Sample size	128	128	128	128	128
4.78	Mean	12.18				
	Min	2				
	Max	29.80				
	Sample size	235				

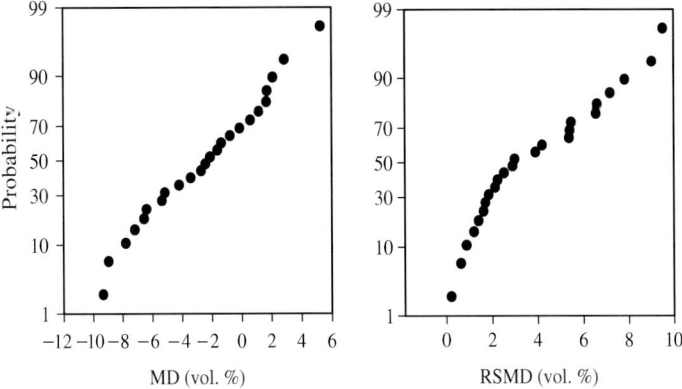

Figure 1. Probability distributions of the MD and RMSD between measured and PTF-estimated water retention in the test dataset.

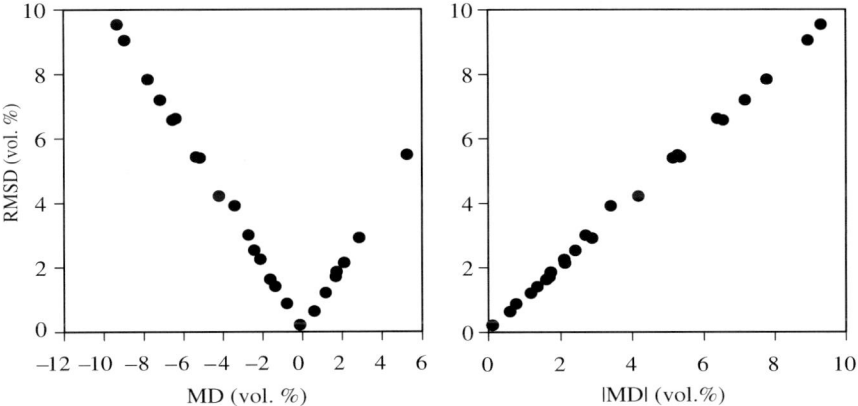

Figure 2. Relationship between the MD and RMSD obtained from the PTF testing.

4. CONCLUSION

This study presents a soil PTFs developed for the natural environment of the Rye Island. A data set of water retention curves determined under laboratory conditions was used. Multiple linear regression was used to express the dependency of water content in soil on the particle-size distribution in the ranges $d < 0.01$ mm, 0.01 mm $< d < 0.05 \times$ mm, 0.05 mm $< d < 0.1$ mm, and 0.1 mm $< d < 2$ mm and on dry bulk density. The obtained PTFs for 9 points of the water retention curve were tested with water retention curves from another data set, also from the Rye Island. Comparison of the water retention curves measured and calculated using PTFs shows a good agreement. We conclude that the developed PTFs may be successfully used as inputs into mathematical models of soil water regime for The Rye Island region.

ACKNOWLEDGEMENTS

The authors express their gratitude to the project VEGA 2/2003/2002 for the financial support.

REFERENCES

Arya, L.M., Parys, J.F., 1981. A physic-empirical model to predict the soil moisture characteristic from particle-size distribution and bulk density data. Soil Sci. Soc. Am. J. 45, 1023-1030.

Bastet, G., Bruand, A., Voltz, M., Bornard, M., Quentin, P., 1998. Prediction of water retention properties: performance of available pedotransfer functions and development of new approaches. *In*: Kutílek, M., Rieu, M. (Eds.), Proceedings of the XVI Congress of ISSS, Symposium 01. New concepts and theories in soil physics and their importance for studying changes induced by human activity, Montpellier, France, CD 8 pp.

Brooks, R.H., Corey, A.T., 1964. Hydraulic properties of porous media, Hydrol. Paper 3, Colorado State University, Fort Collins, CO, 27 pp.

Cosby, B.J., Hornberger, G.M., Clapp, R.B., Gin, T.R., 1984. A statistical exploration of the relationship of soil moisture characteristics to the physical properties of soils. Water Resour. Res. 20, 682-690.

Gupta, S.C., Larson, W.E., 1979. Estimating soil water characteristic from particle size distribution, organic matter percent, and bulk density. Water Resour. Res. 15, 1633-1635.

Haverkamp, R., Parlange, J.Y., 1986. Predicting the water retention curve from particle-size distribution: 1. Sandy soils without organic matter. Soil Sci. 142, 325-339.

Husz, G., 1967. The determination of pF-curves from texture using multiple regression. Z. Pflanzenernähr, Dung-Bodenkd. 116 (2), 23-29.

Kutílek, M., Nielsen, D.R., 1994. Soil Hydrology. Catena Verlag, Cremlingen-Destedt, Germany.

Pachepsky, Y., Shcherbakov, R.A., Varallyay, G., Rajkai, K., 1982. Statistical analysis of the correlation between the water retention and other physical properties of soils. Pochvovedenije 2, 42-52, (in Russian).

Puckett, W.E., Dane, J.H., Hajek, B.F., 1985. Physical and mineralogical data to determine soil hydraulic properties. Soil Sci. Soc. Am. J. 49, 831-836.

Rajkai, K., Kabos, S., Van Genuchten, M.Th., Jansson, P.E., 1996. Estimation of water-retention characteristics from the bulk density and particle-size distribution of Swedish soils. Soil Sci. 161, 832-845.

Rawls, W.J., Brakensiek, D.L., 1985. Prediction of soil water properties for hydrologic modeling. *In*: Jones, E., Ward, T.J. (Eds.), Watershed Manage, Eighties., Proceedings of the Symposium of ASCE, Denver, CO, New York.

Rawls, W.J., Brakensiek, D.L., Saxton, L.E., 1982. Estimation of soil water properties. Trans. ASAE 108, 1316-1320.

Renger, M., 1971. The estimation of pore size distribution from texture, organic matter content and bulk density. Z. Kluturtech. Flurbereinig 130, 53-67.

Singh, A.K., 1998. Use of pedotransfer functions in crop growth simulation. *In*: Clothier, B.E., Voltz, M.Y. (Eds.), Proceedings of the XVI Congress of ISSS, Symposium 03. Mass and energy transfers in soils, Montpellier, CD 6 pp.

Šútor, J., Komár, J., 1984. Vybrané hydrofyzikálne charakteristiky pôd Východoslovenskej nížiny. (Selected soil hydrophysical characteristics of East Slovakia Lowland). Proc. Sympozium Soil, Water, Plant KPVS Michalovce, Slovakia, pp. 0–10.

Šútor, J., Majerčák, J., 1988. Extrapolácia nameraných hodnôt hydrofyzikálnych charakteristík pôdy v rámci daného pôdneho druhu. (Extrapolation of measured values of soil hydrophysical characteristics in the frame of soil type). Vodohosp. Čas. (J. Hydrol. Hydromech.) 36, 639-654.

Tietje, O., Tapkenhinrichs, M., 1993. Evaluation of pedo-transfer functions. Soil Sci. Soc. Am. J. 57, 1088-1095.

Van den Berg, M., Klampt, E., Van Reeuwijk, L.P., Sombroek, W.G., 1997. Pedotransfer functions for the estimation of moisture retention characteristics of ferralsols and related soils. Geoderma 78, 161-180.

Van Genuchten, M.Th., 1980. A closed-form equation for predicting the hydraulic conductivity of unsaturated soils. Soil Sci. Soc. Am. J. 44, 987-996.

Vereecken, H.J., Maes, J., Feyen, J., Darius, P., 1989. Estimating the soil moisture retention characteristic from texture, bulk density and carbon content. Soil Sci. 148, 389-403.

Vereecken, H.J., Maes, J., Feyen, J., Darius, P., 1990. Estimating unsaturated hydraulic conductivity from easily measured soil properties. Soil Sci. 149, 1-12.

Wiliams, J., Prebble, R.E., Wiliams, W.T., Hignett, C.T., 1983. The influence of texture, structure and clay mineralogy on the soil moisture characteristic. Aust. J. Soil. Res. 21, 15-32.

Wősten, J.H., Finke, P.A., Jansen, M.J., 1995. Comparison of class and continuous pedotransfer functions to generate soil hydraulic characteristics. Geoderma 66, 227-237.

Zeiliguer, A.M., Pachepsky, Ya.A., Rawls, W.J., 2000. Estimating water retention of sandy soils using the additivity hypothesis. Soil Sci. 165, 373-383.

ADDITIONAL BIBLIOGRAPHY

Alessi, S., Prunty, L., Schuh, W.M., 1992. Infiltration simulations among five hydraulic property models. Soil Sci. Soc. Am. J. 56, 675-682.

Alexander, L., Skaggs, R.W., 1986. Predicting unsaturated hydraulic conductivity from the soil-water characteristic. Trans. ASAE 29, 176-184.

Alexander, L., Skaggs, R.W., 1987. Predicting unsaturated hydraulic conductivity from soil texture. J. Irrig. Drain. Engng 113, 184-197.

Allmaras, R.R., Rickman, R.W., Edin, L.G., Kimball, B.A., 1977. Chiseling influences on soil hydraulic properties. Soil Sci. Soc. Am. J. 41, 796-803.

Allamaras, R.R., Ward, K., Douglas, C.L., Ekin, L.G., 1982. Long-term cultivation effects on hydraulic properties of a Walla Walla silt-loam. Soil Tillage Res. 2, 265-279.

Anderson, S.H., Cassel, D.K., 1986. Statistical and autoregressive analysis of soil physical properties of Portsmouth sandy loam. Soil Sci. Soc. Am. J. 50, 1096-1104.

Anderson, M.G., Howes, S., Kneale, P.E., Shen, J.M., 1985. On soil retention curves and hydrological forecasting in ungauged catchments. Nordic Hydrol. 16, 11-32.

Andersson, S., 1953. Markfysikaliska undersökningar I odlad jord. II Om markens permeabilitet. Grundförbättring H.1, 28-45, (in Swedish).

Andersson, S., Wiklert, P., 1967a. Markfysikaliska undersökningar I odlad jord. XVIII. Om de vattenhållande egenskaperna hos rena system och blandsystem avand, lera och torv. Grundförbättring 20, 3-27.

Andersson, S., Wiklert, P., 1967b. On water-retaining properties of pure systems, and mixed systems of sand, clay and peat. Soil physical investigations in cultivated soil XVII. Grundförbättring 20, 1-27, (in Swedish).

Andersson, S., Wiklert, P., 1972. Markfysikaliska undersokningar I odlad jord. XXII. Om de vattenhollande egenskapema hos svenska jordar. Grundförbättring 25, 53-243.

Archer, J.R., Smith, P.D., 1972. The relation between bulk density, available water capacity, and air capacity of soils. J. Soil Sci. 23, 475-480.

Arshad, M.A., Franzluebbers, A.J., Azooz, R.H., 1999. Components of surface soil structure under conventional and no-tillage in northwestern Canada. Soil Tillage Res. 53, 41-47.

Arya, L.M., Parys, I.F., 1982. Reply to "comments on a physicoempirical model to predict the soil moisture characteristic from particle-size distribution and bulk density data". Soil Sci. Soc. Am. J. 46, 1348-1349.

Assouline, S., Rouault, Y., 1997. Modeling the relationship between particle and pore size distributions in multicomponent sphere packs: application to the water retention curve. Colloids Surf. A: Physicochem. Engng Asp. 127, 201-210.

Avissar, R., Pielke, R.A., 1989. A parameterization of heterogeneous land surface for atmospheric numerical models and its impact on regional hydrology. Mon. Weather Rev. 117, 2113-2136.

Bache, B.W., Frost, C.A., Inkson, R.H.E., 1981. Moisture release characteristics and porosity of twelve Scottish soil series and their variability. I. Soil Sci. 32, 505-520.

Bachmann, J., Harte, K.-H., 1992. Estimating soil water characteristics obtained by nasic soil data – a comparison of indirect methods. Z. Pflansenern. Bodenkd. 155, 109-114, (in German).

Banin, A., Amid, A., 1969. A correlative study of the chemical and physical properties of a group of natural soils of Israel. Geoderma 3, 185-198.

Bartelli, L.J., Peters, D.B., 1959. Integrating soil moisture characteristics with classification units of some Illinois soils. Soil Sci. Soc. Am. Proc. 23, 149-151.

Bauer, A., Black, A.L., 1992. Organic carbon effects on available water capacity of three soil textural groups. Soil Sci. Soc. Am. J. 56, 248-254.

Baumer, G.W., Brasher, B.R., 1982. Prediction of Water Contents at Selected Suctions, ASAE Paper No. 82-2590. American Society of Agricultural Engineering, St. Joseph, MI.

Becher, H.H., 1971a. Em verfahren zur messung der ungesättigten wasscrleitfähigkeit. Z. Pflanzenern. Bodenkd. 128, 1-12.

Becher, H.H., 1971b. Ergebnisse von wasserleitfähigkeismessungen im wasserungesättigten zustand. Z. Pflanzenern. Bodendkd. 128, 227-234.

Bejdonger, M.S., 1961. Relation between median grain size and permeability in the Arkansas River Valley, Arkansas. Professional paper. 424-C, US Geological Survey, C31–C32.

Bell, M.A., van Keulen, H., 1996. Effect of soil disturbance on pedotransfer function development for field capacity. Soil Technol. 8, 321-329.

Beskow, G., 1935. Tjälbildning och tällyftning med särskild hänsyn till vägar och järnvagar (with English summary), Sveriges Geologiska Undersökning, Ser. C, 375, Arshok 26(3), 242 pp.

Beverly, R.B., 1996. Evaluation of the moisture equivalent soil test for irrigation management. Commun. Soil Sci. Plant Anal. 27, 615-621.

Beverly, R.B., Tollner, E.W., Byous, A.W., Thain, S.M., 1994. Moisture equivalent as a routine soil physical test to guide irrigation management. Commun. Soil Sci. Plant Anal. 25, 1035-1043.

Bhatnagar, D., Nagarajarao, Y., Gupta, R.P., 1979. Influence of water content and soil properties on unsaturated hydraulic conductivity of some red and black soils. Z. Pflanzen. Bodenkd. 142, 99-108.

Bhushan, L.S., Varade, S.B., Gupta, C.P., 1973. Influence of tillage practices on cold size porosity and water retention. Indian J. Agric. Sci. 43, 466-471.

Bird, N.R.A., Bartoli, F., Dexter, A.R., 1996. Water retention models for fractal soil structures. Eur. J. Soil. Sci. 47, 1-6.

Bird, N.R.A., Perrier, E., Rieu, M., 2000. The water retention function for a model of soil structure with pore and solid fractal distributions. Eur. J. Soil Sci. 51, 55-63.

Biswas, T.D., Ali, M.H., 1969. Retention and availability of soil water as influenced by soil organic carbon. Indian J. Agric. Sci. 39, 582-583.

Bloemen, G.W., 1977a. Calculation of capillary conductivity and capillary rise from grain size distribution. I. Real and theoretical values of the exponent in a formula of Brooks and Corey for the calculation of hydraulic conductivities. ICW Nota No. 952, Wageningen, The Netherlands.

Bloemen, G.W., 1977b. Calculation of capillary conductivity and capillary rise from grain size distribution. II. Assessment of the values of the exponent in the formula of Brooks and Corey for the calculation of hydraulic conductivity from grain size distribution. ICW Nota No. 962, Wageningen, The Netherlands.

Bloemen, G.W., 1977c. Calculation of capillary conductivity and capillary rise from grain size distribution. III. Air entry pressure and saturated conductivity calculated from grain size distribution and median grain size. ICW Nota No. 990, Wageningen, The Netherlands.

Bloemen, G.W., 1977d. Calculation of capillary conductivity and capillary rise from grain size distribution. IV. Capillary rise in soil types and soil profiles. ICW Nota No. 1013, Wageningen, The Netherlands.

Bloemen, G.W., 1980. Calculation of hydraulic conductivities of soils from texture and organic matter content. ICW Technical Bulletin 120, Wageningen, The Netherlands.

Boisvert, J., Dyer, J.A., 1987. Le coefficient de sol dans les modèles empiriques de bilan hydrique. Can. Agric. Engng 29, 7-14.

Bond, W.J., Verburg, K., Snow, V.O., 1997. Heterogeneity and sensitivity – some implications for measuring and modeling flow and transport. In: Proceedings European Geophysical Society 22nd General Assembly, Vienna, April 1997, Ann. Geophys. 15(II), C 266.

Borg, H., 1982. Estimating soil hydraulic properties from textural data. Ph.D. Thesis, Washington State University, Pullman (Diss. Abstr. 83-03292).

Bork, H.-R., 1989. Regionalization of soil data measured at site scale. In: Bouma, J., Bregt, A.K. (Eds.), Land Qualities in Space and Time: Proceedings of a Symposium Organized by the International Society of Soil Science (ISSS), 22–26 August 1988, Wageningen, the Netherlands, pp. 289-298.

Bork, H.-R., Rohdenburg, H., 1986. Transferable parameterization methods for distributed hydrological and agroecological catchment models. Catena 13, 99-117.

Bouma, J., de Laat, P.J.M., 1981. Estimation of the moisture supply capacity of some swelling clay soils in the Netherlands. J. Hydrol. 49, 247-259.

Bouma, J., Droogers, P., 1999. Comparing different methods for estimating the soil moisture supply capacity of a soil series subjected to different types of management. Geoderma 92, 185-197.

Bouraoui, F., Haverkamp, R., Zammit, C., 1997. A physically based pedotransfer function for estimated water retention curve shape parameters, pp. 947–958 (van Genuchten, M.Th., Leij, F.J., Wu, L. (Eds.), c1999. Characterization and Measurement of the Hydraulic Properties of Unsaturated Porous Media: Proceedings of the International Workshop on Characterization and Measurement of the Hydraulic Properties of Unsaturated Porous Media: Riverside, California, October 22–24, 1997. U.S. Salinity Laboratory, Agricultural Research Service, U.S. Dept. of Agriculture/Dept. of Environmental Sciences, University of California, Riverside, CA).

Bouyoucos, G.J., 1935. A comparison between the suction method and the centrifuge method for determining the moisture equivalent of soils. Soil Sci. 40, 165-171.

Bowman, R.A., Reeder, J.D., Lober, R.W., 1990. Changes in soil properties in a Central Plains rangeland soil after 3, 20 and 60 years of cultivation. Soil Sci. 150, 851-857.

Brakensiek, D.L., 1977a. Estimating the effective capillary pressure in the Green and Ampt infiltration equation. Water Resour. Res. 13, 680-682.

Brakensiek, D.L., 1977b. Estimating the effective capillary pressure in the Green and Ampt infiltration equation. Water Resour. Res. 13, 680-682.

Brakensiek, D.L., 1979. Comments on "empirical equations for some soil hydraulic properties" by Roger B. Clapp and George M. Hornberger. Water Resour. Res. 15, 989-990.

Brakensiek, D.L., Engleman, R.L., Rawls, W.J., 1981a. Variation within texture classes of soil water parameters. Trans. ASAE 24, 335-339.

Brakensiek, D.L., Engleman, R.L., Rawls, W.J., 1981b. Variation within texture classes of soil water parameters. Trans. ASAE 24, 335-339.

Brakensiek, D.L., Rawls, W.J., Stephenson, G.R., 1984. Modifying SCS Hydrologic Soil Groups and Curve Numbers for Rangeland Soils, ASAE Paper. No. PNR-84-203. American Society of Agricultural Engineering, St. Joseph, MI.

Brakensiek, D.L., Rawls, W.J., Stephenson, G.R., 1986. Determining the saturated hydraulic conductivity of soil containing rock fragments. Soil Sci. Soc. Am. J. 50, 834-835.

Bregt, A.K., Beemster, J.G.R., 1989. Accuracy in predicting moisture deficit and changes in yield from soil maps. Geoderma 43, 301-310.

Bruand, A., 1990. Improved prediction of water-retention properties of clayey soils by pedological stratification. J. Soil Sci. 41, 491-497.

Bruce, R.R., 1972. Hydraulic conductivity evaluation of the soil profile from soil water relations. Soil Sci. Soc. Am. Proc. 36, 550-560.

Bruce, R.R., Dane, J.H., Quisenberry, V.L., Powell, N.L., Thomas, A.W., 1983. Physical characteristics of soils in the southern region: Cecil, Southern Cooperative Series Bull. 267. Georgia Agric. Exp. Station, Athens, GA.

Brust, K.J., van Bavel, C.H.M., Stirk, G.B., 1968. Hydraulic properties of clay loam soil and the field measurement of water uptake by roots. 3. Comparison of field and laboratory data on retention and of measured and calculated conductivities. Soil Sci. Soc. Am. Proc. 32, 322-326.

Brutsaert, W., 1966. Probability laws for pore size distributions. Soil Sci. 101, 85-92.

Brutsaert, W., 1967. Some methods of calculating unsaturated permeability. Trans. ASAE 10, 400-404.

Brutsaert, W., 1968. The permeability of porous medium determined from certain probability law for pore-size distribution. Water Resour. Res. 4, 425-434.

Burwell, R.E., Larson, W.E., 1969. Infiltration as influenced by tillage-induced random roughness and pore space. Soil Sci. Soc. Am. Proc. 33, 449-452.

Cameron, D.R., 1979. Variability of soil water retention curves and predicted hydraulic conductivities on a small plot. Soil Sci. 126, 364-371.

Camillo, P., Schmugge, T.J., 1983. Estimating soil moisture storage in the root zone from surface measurements. Soil Sci. 135, 245-264.

Canarache, A., 1993. Physical-technological maps – a possible product of soil survey for direct use in agriculture. Soil Technol. 6, 3-16.

Carlisle, V.W., Caldwell, R.E., Sodek, F., Hammond, L.C., Calhoun, F.G., Granger, M.A., Breland, H.L., 1978. Characterization Data for Selected Florida Soils, Soil Science Department Research Report 78-1. University of Florida, Gainesville, FL.

Carsel, R.F., Imhoff, J.C., Kittle, J.L. Jr., Hummel, P.R., 1991. Development of a database and model parameter analysis system for agricultural soils. J. Environ. Qual. 20, 642-647.

Carvallo, H.O., Cassel, D.K., Hammond, J., Bauer, A., 1976. Spatial variability of *in-situ* unsaturated hydraulic conductivity of Maddock sandy loam. Soil Sci. 121, 1-8.

Cassel, D.K., Bauer, A., 1975. Spatial variability in soils below depth of tillage. Bulk density and fifteen atmosphere percentage. Soil Sci. Soc. Am. Proc. 39, 247-250.

Cassel, D.K. (Ed.), 1985. Physical Characteristics of Soils of Southern Region: Summary of *In Situ* Unsaturated Hydraulic Conductivity, Southern Cooperative Series Bulletin 303. North Carolina State University, Raleigh, NC.

Cassel, D.K., Ratliff, L.F., Ritchie, J.T., 1983. Models for estimating *in situ* potential extractable water using soil physical and chemical properties. Soil Sci. Soc. Am. J. 47, 764-769.

Celia, M.A., Ferrand, L.A., 1992. A percolation-based model for the soil water retention function. pp. 71–80 (van Genuchten, M.Th., Leij, F.J., Lund, L.J. (Eds.), c1992. Indirect Methods for Estimating the Hydraulic Properties of Unsaturated Soils: Proceedings of the International Workshop on Indirect Methods for Estimating the Hydraulic Properties of Unsaturated Soils, Riverside, CA, October 11–13, 1989. U.S. Salinity Laboratory, Agricultural Research Service, U.S. Dept. of Agriculture/Dept. of Soil and Environmental Sciences, University of California, Riverside, CA).

Centurion, J.F., Moraes, M.H., Della Libera, C.L.F., 1997. Comparison of methods for determination of soil water retention curve. Rev. Bras. Cienc. Solo 21, 173-178.

Chatzis, I., Dullien, F.A.L., 1977. Modelling pore structure by 2-D and 3-D networks of capillary tubes to simulate the drainage capillary pressure curves of sandstones. J. Can. Petrol. Technol. 16, 97.

Chibber, R.K., 1964. Aggregate size distribution and water relationships amongst some typical Indian soils. Bull. Batn. Inst. Sci. India 26, 148-156.

Clapp, R.B., Hornberger, G.M., Cosby, B.J., 1983. Estimating spatial variability in soil moisture with a simplified dynamic model. Water Resour. Res. 19, 739-745.

Clausnitzer, V., Hopmans, J.W., Nielsen, D.R., 1992. Simultaneous scaling of soil water retention and hydraulic conductivity curves. Water Resour. Res. 28, 19-31.

Coelho, M.A., 1974. Spatial variability of water-related soil physical properties. Ph.D. Dissertation, Univ. of Arizona.

Coughian, K.J., Loch, R.J., Fox, W.E., 1978. Binary packing theory and the physical properties of aggregates. Aust. J. Soil Res. 16, 283-289.

Courtin, P., Feller, M.C., Klinka, K., 1983. Internal variability in some properties of disturbed soils in S.W. British Columbia, Canada. J. Soil Sci. 63, 529-539.

Crawford, J.W., 1994. The relationship between structure and the hydraulic conductivity of soil. Eur. J. Soil Sci. 45, 493-502.

Cresswell, H.P., Smiles, D.E., Williams, J., 1992. Soil structure, soil hydraulic properties and soil water balance. Aust. J. Soil Res. 30, 265-283.

Cresswell, H.P., McKenzie, N.J., Paydar, Z., 1997. A strategy for determination of hydraulic properties of Australian soils using direct measurement and pedotransfer functions. pp. 1143–1160 (van Genuchten, M.Th., Leij, F.J., Wu, L. (Eds.), c1999. Characterization and Measurement of the Hydraulic Properties of Unsaturated Porous Media: Proceedings of the International Workshop on Characterization and Measurement of the Hydraulic Properties of Unsaturated Porous Media: Riverside, CA, October 22–24, 1997. U.S. Salinity Laboratory, Agricultural Research Service, U.S. Dept. of Agriculture/Dept. of Environmental Sciences, University of California, Riverside, CA).

Croney, D., Coleman, J.D., 1954. Soil structure in relation to soil suction. J. Soil Sci. 5, 75-85.

Dane, J.H., Puckett, W.E., 1992. Field soil hydraulic properties based on physical and mineralogical information. pp. 389–404 (van Genuchten, M.Th., Leij, F.J., Lund, L.J. (Eds.), c1992. Indirect Methods for Estimating the Hydraulic Properties of Unsaturated Soils: Proceedings of the International Workshop on Indirect Methods for Estimating the Hydraulic Properties of Unsaturated Soils, Riverside, California, October 11–13, 1989. U.S. Salinity Laboratory, Agricultural Research Service, U.S. Dept. of Agriculture/Dept. of Soil and Environmental Sciences, University of California, Riverside, CA).

Dane, J.H., Cassel, D.K., Davidson, J.M., Pollans, W.L., Quinsberry, V.L., 1983. Physical Characteristics of Soils of the Southern Region – Troup and Lakeland Series, Southern Cooperative Series Bulletin 262. Alabama Agricultural Experimental Station, Auburn, AL.

Daniel, B.S., Rehfeldt, K.R., 1985. Evaluation of closed-form analytical models to calculate conductivity in a fine sand. Soil Sci. Soc. Am. J. 49, 12-19.

Daroussin, J., Kind, D., 1997. A pedotransfer rules database to interpret the soil geographical database of Europe for environmental purposes. pp. 25–40 (The use of pedotransfer in soil hydrology research in Europe. In: Bruand, A., Duval, O., Wösten, J.H.M., Lilly, A. (Eds.), c1997. Proceedings of the Second Workshop of the project "Using existing soil data to derive hydraulic parameters for simulation modelling in environmental studies and in land use planning" INRA Orléans, France, 10–12 October 1996. Orléans, France, INRA-Unité de Science du soil/EC – Joine research Centre, European Soil Bureau, Ispra, Italy/ECSC-EC-EAEC Brussels, Luxembourg).

Da Silva, A.P., Kay, B.D., Perfect, E., 1994. Characterization of the least limiting water range of soils. Soil Sci. Soc. Am. J. 58, 1775-1784.

De Jong, E., 1967. Moisture retention of selected Saskatchewan soils, Soil Plant Nutrients. Res. Rep. Dept. Soil Sci., Univ. of Sask., Saskatoon, Sask.

De Jong, R., 1982. Assessment of empirical parameters that describe soil water characteristics. Can. Agric. Engng 24, 65-70.

De Jong, R., Loebel, K., 1982. Empirical relations between soil components and water retention at 1/3 and 15 atmospheres. Can. J. Soil Sci. 62, 343-350.

De Jong, R., Campbell, C.A., Nicholaichuk, W., 1983. Water retention equations and their relationship to soil organic matter and particle size distribution for disturbed samples. Can. J. Soil Sci. 63, 291-302.

De Jong, R., Topp, G.C., Reynolds, W.D., 1992. The use of measured and estimated hydraulic properties in the simulation of soil water movement – a case study. pp. 569–584 (van Genuchten, M.Th., Leij, F.J., Lund, L.J. (Eds.), c1992. Indirect Methods for Estimating the Hydraulic Properties of Unsaturated Soils: Proceedings of the International Workshop on Indirect Methods for Estimating the Hydraulic Properties of Unsaturated Soils, Riverside, CA, October 11–13, 1989. U.S. Salinity Laboratory, Agricultural Research Service, U.S. Dept. of Agriculture/Dept. of Soil and Environmental Sciences, University of California, Riverside, CA).

De La Rosa, D., 1979. Relation of several pedological characteristics to engineering qualities of soil. J. Soil Sci. 30, 793-799.

De La Rosa, D., Cardona, D.F., Almorsa, J., 1981. Crop yield predictions based on properties of soils in Sevilla, Spain. Geoderma 25, 267-274.

Debano, L.F., 1971. The effect of hydrophobic substances on water movement in soil during infiltration. Soil Sci. Soc. Am. Proc. 35, 340-343.

Dechnik, I., Tarkiewich, S., 1976. Dynamics of total porosity of soils cultivated with the plough-miller. Pol. J. Soil Sci. 9, 71-77.

Denning, J.L., Bouma, J., Falayi, O., van Rooyen, D.J., 1974. Calculation of hydraulic conductivities of horizons in some major soils in Wisconsin. Geoderma 11, 1-16.

D'Hollander, E.H., 1979. Estimation of the pore size distribution from the moisture characteristic. Water Resour. Res. 15, 107-112.

Diaz, C.E., Chaizis, I., Dullien, F.A.L., 1987. Simulation of capillary pressure curves using bond correlated site percolation on a simple cubic network. Transp. Porous Media 2, 215.

Diekrügger, B., 1990. Calculation of soil properties for different scales. Trans. 14th Int. Cong. Soil. Sci. 1, 190-195.

Dombrowski, H.S., Brownell, L.E., 1954. Residual equilibrium saturation of porous media. Ind. Eng. Chem. 46, 1207-1219.

Dullien, F.A.L., Lai, F.S.Y., Macdonald, I.F., 1986. Hydraulic continuity of residual wetting phase in porous media. J. Colloid Interface Sci. 109, 201-218.

Dullien, F.A.L., Chatzis, I., Kantzas, A., Diaz, C.E., 1992. Simultaneous modeling of capillary pressure and two-phase relative permeability characteristics of porous media using network models. pp. 81–95 (van Genuchten, M.Th., Leij, F.J., Lund, L.J. (Eds.), c1992. Indirect Methods for Estimating the Hydraulic Properties of Unsaturated Soils: Proceedings of the International Workshop on Indirect Methods for Estimating the Hydraulic Properties of Unsaturated Soils, Riverside, California, October 11–13, 1989. U.S. Salinity Laboratory, Agricultural Research Service, U.S. Dept. of Agriculture/ Dept. of Soil and Environmental Sciences, University of California, Riverside, CA).

Dunne, K.A., Willmott, C.J., 1996. Global distribution of plant-extractable water capacity of soils. Int. J. Climatol. 16, 841-859.

Durner, W., 1994. Hydraulic conductivity estimation for soils with heterogeneous pore structure. Water Resour. Res. 30, 211-223.

El-Kadi, A.I., 1984. Automated estimation of the parameters of soil hydraulic properties. IGWAIC-GWAMI-12, International Groundwater Modeling Center, Holcomb Research Institute, Butler University, Indianapolis, IN.

El-Kadi, A.I., 1986. On estimating the hydraulic properties of soil, part 3. Parameters of the Philip infiltration equation. Adv. Water Resour. 9, 16-23.

Espino, A., Mallants, D., Vanclooster, M., Feyen, J., 1995. Cautionary notes on the use of pedotransfer functions for estimating soil hydraulic properties. Agric. Water Manag. 29, 235-253.

Farquharson, F.A.K., Mackney, D., Newson, M.D., Thomasson, A.J., 1978. Estimation of run-off potential of river catchments from soil surveys. Special Survey No. 11, Harpender, 29 pp.

Farrel, D.A., Larson, W.E., 1972. Modeling the pore structure of porous media. Water Resour. Res. 3, 699-706.

Fatt, I., 1956a. The network model of porous media. I. Capillary pressure characteristics. Trans. AIME Pet. Div. 207, 144-159.

Fatt, I., 1956b. The network model of porous media, II. Dynamic properties of a single size tube network. Trans. AIME Pet. Div. 207, 160-163.

Fatt, I., 1956c. The network model of porous media, III. Dynamic properties of networks with tube radius distribution. Trans. AIME Pet. Div. 207, 164-177.

Fayer, M.J., Simmons, C.S., 1995. Modified soil water retention functions for all matric suctions. Water Resour. Res. 31, 1233-1238.

Fayer, M.J., Gee, G.W., Jones, T.L., 1986. UNSAT-H version 1.0: Unsaturated flow code documentation and application for the Hanford Site. PNL-5899. Pacific Northwest Laboratories, Richland, WA.

Field, J.A., Parker, J.C., Powell, N.L., 1985. Comparison of field- and laboratory measured and predicted hydraulic properties of soil with macropores. Soil Sci. 138, 385-396.

Finke, P.A., Wösten, J.H.M., Kroes, J.G., 1996. Comparison of two approaches to characterize soil mapping unit behavior in solute transport studies. Soil Sci. Soc. Am. J. 60, 200-205.

Fisher, U., Celia, M.A., 1999. Prediction of relative and absolute permeabilites for gas and water from soil water retention curves using a pore-scale network model. Water Resour. Res. 35, 1089-1094.

Folk, R.L., Ward, W.C.A., 1957. A study on significance of grain size parameters. J. Sed. Pet. 27, 3-27.

Forrest, J.A., Beatty, J., Hignett, C.T., Pickering, J.H., Williams, R.G.P., 1985. A survey of the physical properties of wheatland soils in Eastern Australia, Div. Rep. No. 78. CSIRO Div. of Soils, Adelaide.

Free, G.R., Browning, G.M., Musgrave, G.W., 1940. Relative infiltration and related physical characteristics of certain soils, USDA Tech Bull. 729. US Government Printing Office, Washington, DC.

Frye, W.W., Ebelhard, S.A., Murdok, L.W., Blevins, R.L., 1982. Soil erosion effects on properties of two Kentucky soils. Soil Sci. Soc. Am. J. 46, 1051-1055.

Gardner, W.R., 1974. The permeability problem. Soil Sci. 117, 243-249.

Geeves, G.W., Cresswell, H.P., Murphy, B.W., Gessler, P.E., Chartres, C.J., Little, I.P., 1995. The physical, chemical, and morphological properties of soils in the wheat-belt of Southern NSW and Northern Victoria, NSW Department of Conservation and Land Management/CSIRO Div. Soils occasional rep. CSIRO, Australia, 178 pp.

Ghosh, R.K., 1976. Model of the soil-moisture characteristic. J. Ind. Soc. Soil Sci. 24, 353-355.

Ghosh, R.K., 1980. Estimation of soil-moisture characteristics from mechanical properties of soils. Soil Sci. 130, 60-63.

Gillham, R.W., Klute, A., Heerman, D.F., 1976. Hydraulic properties of a porous medium: measurement and empirical representation. Soil Sci. Soc. Am. J. 40, 203-207.

Giménez, D., Perfect, E., Rawls, W.J., Pachepsky, Ya.A., 1997. Fractal models for predicting soil hydraulic properties: a review. Engng Geol. 48, 161-183.

Gouweleeuw, B.T., van de Griend, A.A., 1996. Estimation of "effective" soil hydraulic properties by topsoil moisture and evaporation modeling applied to an arable site in central Spain. Water Resour. Res. 32, 1387-1392.

Gradwell, M.W., 1974. Measurements and predicted hydraulic conductivity for some New Zealand subsoils at water contents near field capacity. NZ J. Sci. 17, 463-473.

Greacen, E.L., 1981. Physical properties and water relations. In: Oades, J.M., Lewis, D.G., Norrish, K. (Eds.), Red-Brown Earths of Australia. Ch. 5. Waite Agricultural Research Institute/CSIRO Division of Soils, Adelaide.

Greacen, E.L., Ponsana, P., Forrest, J.A., 1974. The use of Marshall's model for describing the field hydraulic conductivity of a soil. Trans. 10th Int. Cong. Soil Sci. 1, 319-325.

Green, R.E., Corey, J.C., 1971. Calculation of hydraulic conductivity: a further evaluation of some predictive methods. Soil Sci. Soc. Am. Proc. 35, 3-8.

Green, R.E., Ahuja, L.R., Chong, S.K., Lau, L.S., 1982. Water Condition in Hawaii Oxic Soils, Tech Report 143. Water Resour. Res. Center, Univ. of Hawaii, 122 pp.

Green, T.R., Constanz, J.E., Freyberg, D.L., 1996. Upscaled soil-water retention using van Genuchten's function. J. Hydrol. Engng 1, 123-132.

Griffiths, E., 1991. Assessing permeability class from soil morphology. New Zealand DSIR Land Resources Technical Record No. 40.

Grismer, M.E., 1986. Pore-size distributions and infiltration. Soil Sci. 141, 249-260.

Gumma, G.A., 1978. Spatial variability of *in situ* available water. Ph.D. Thesis, Univ. of Arizona, Tucson, AZ.

Gummatov, N.G., Pachepsky, Ya.A., Shcherbakov, R.A., 1991. Spatial and temporal variability of water retention in Gray Forest soil. Pochvovedenie 9, 169-175, (in Russian).

Gumps, F.A., 1974. Comparison of laboratory and field determined saturated hydraulic conductivity and prediction from soil particle size. Trop. Agric. 51, 375-383.

Gupta, S.C., Larson, W.E., 1979. A model for predicting packing density of soils using particle-size distribution. Soil Sci. Soc. Am. J. 43, 758-764.

Gupta, S.C., Larson, W.E., 1982. Modeling soil mechanical behavior during tillage. In: Unger, P.W., van Doren, D.M. (Eds.), Predicting Tillage Effects on Soil Physical Properties and Processes, Special Publ. 44. ASA and SSSA, Madison, WI, pp. 151–178.

Haise, H.R., Haas, H.J., Jensen, L.R., 1955. Soil moisture of some Great Plains soils. II. Field capacity as related to 1/3-atmosphere percentage and "minimum point" as related to 15- and 26-atmosphere percentages. Soil Sci. Soc. Am. Proc. 19, 20-25.

Hamblin, A.P., Tennant, D., 1981. The influence of tillage on soil water behavior. Soil Sci. 132, 233-239.

Hantschel, R., 1987. Wasser- und elementbilanz von geschadigten, gedüngten fichtenökosystemen im fichtelgebirge unter berucksichtigung von physikalischer und chemischer bodenheterogenitat, bayreuther bodenkdl. Berichte 3, Univ. of Bayreuth.

Hantschel, R., Dumer, W., Horn, R., 1987. Die Bedeutung von Porenheterogenitaten fur die Ersteilung der pF/WG- und k/WG-Kurven. Deutsch. Bodenkdl. Mitt. 53, 403-409.

Hartge, K.H., 1965. Die Bestimmung von Porenvolumen und Porengrössenverteilung. Z. Für Kulturtech. Flurbereinig. 6, 193-206.

Haskett, J., Pachepsky, Ya.A., Acock, B., 1995a. Estimation of soybean yields at county and state level using GLYCIM: a case study for Iowa. Agron. J. 87, 926-931.

Haskett, J., Pachepsky, Ya.A., Acock, B., 1995b. Use of the beta-distribution for parameterizing variability of soil properties at the regional level for crop yield estimation. Agric. Syst. 48, 73-86.

Haverkamp, R., Parlange, J.-Y., 1982. Comments on "A physicoempirical model to predict the soil moisture characteristic from particle-size distribution and bulk density data". Soil Sci. Soc. Am. J. 46, 1348-1349.

Haverkamp, R., Parlange, J.-Y., Starr, J.L., Schmitz, G., Fuentes, C., 1990. Infiltration under ponded conditions. 3. A predictive equation based on physical parameters. Soil Sci. 149, 292-300.

Heiba, A.A., Sahimi, M., Scriven, L.E., Davis, H.T., 1982. Percolation theory of two phase relative permeability. SPE Paper No. 11015, 57th Annual Fall Technical Conference of SPE-AIME, New Orleans, LA, September 26–29.

Hennings, V., Müller, U., 1993. Überprüfung eines schätzverfahrens zur ermittlung von kennwerten der wasserbindung anhand der labordatenbank des niedersächsischen bodeninformationssystems. Z. Pflanzen. Bodenkd. 156, 67-73.

Hennings, V., Müller, U., 1997. Bewertung von pedotransferfunktionen zur schätzung der ungesättigten wasserleitfähigkeitsfunktion. Mitt. Dsch. Bodenkl. Ges. 85, 103-106.

Hill, J.N.S., Summer, M.E., 1967. Effect of bulk density on moisture characteristics of soils. Soil Sci. 103, 234-238.

Hill, R.L., Horton, R., Cruse, R.M., 1985. Tillage effects on soil water retention and pore size distribution of two Mollisols. Soil Sci. Soc. Am. J. 49, 1264-1270.

Hirschi, M.C., Moore, I.D., 1980. Estimating soil hydraulic properties from soil texture. Am. Soc. Agric. Engng Paper 80-2523.

Hollenbeck, K.J., Schmugge, T.J., Hornberger, G.M., Wang, J.R., 1996. Identifying soil hydraulic heterogeneity by detection of relative changes in passive microwave remote sensing observations. Water Resour. Res. 32, 139-148.

Hollis, J.M., 1989. A methodology for predicting Soil Wetness Class from soil and site properties. Soil Survey and Land Research Center report for MAFF. Silsoe, UK.

Holtan, H.N., England, C.B., Lawless, G.P., Schumaker, G.A., 1968. Moisture-tension data for selected soils on experimental watersheds, ARS 41-144. U.S. Dept of Agriculture, Beltsville, MD, 609 pp.

Hopmans, J.W., 1987a. A comparison of various methods to scale soil hydraulic properties. J. Hydrol. 93, 241-256.

Hopmans, J.W., 1987b. Presentation and application of an analytical model to describe soil hydraulic properties. J. Hydrol. 87, 135-143.

Hopmans, J.W., Dane, J.H., 1985. Effect of temperature dependence of soil hydraulic properties. Soil Sci. Soc. Am. J. 50, 4-9.

Hopmans, J.W., Dane, J.H., 1986. Temperature dependence of soil water retention curves. Soil Sci. Soc. Am. J. 50, 562-567.

Hubbard, R.K., Berdanier, C.R., Perkins, H.F., Leonard, R.A., 1985. Characteristics of selected upland soils of the Georgia Central Plain Pub 37. USDA-ARS, Washington, DC.

Ingevall, A., 1984. Estimation of clay content from water retention data, Report No. 140. Dept. of Soil Sci., Division of Agric. Hydrotechnics, Swedish Univ. of Agric. Sci. Uppsala, Sweden.

Jabro, J.D., 1992. Estimation of saturated hydraulic conductivity of soils from particle size distribution and bulk density data. Trans. ASAE 35, 557-560.

Jackson, R.D., 1972. On the calculation of hydraulic conductivity. Soil Sci. Soc. Am. Proc. 36, 380-382.

Jackson, T.J., 1980. Profile soil moisture from surface measurements. J. Irrig. Drain., Div. ASCE IR-2, 81-92.

Jacobsen, G.H., 1989. Umaettet hydraulisk ledningsevne I nogle danske jorde: metode of jordtypekarakterisering (Unsaturated hydraulic conductivity for some Danish soils; methods and characterization of soils; in Danish). Tidsskr. Planteavl. 93 (S-2030), 60.

Jakobsen, B.F., 1973. Interrelations of soil physical characteristics. Acta Agric. Scand. 23, 165-172.

Jamison, V.C., 1953. Changes in air–water relationships due to structural improvement of soils. Soil Sci. 76, 143-151.

Jaynes, D.B. 1992. Estimating hysteresis in the soil water retention function. pp. 219–232 (van Genuchten, M. Th., Leij, F.J., Lund, L.J. (Eds.), c1992. Indirect Methods for Estimating the Hydraulic Properties of Unsaturated Soils: Proceedings of the International Workshop on Indirect Methods for Estimating the Hydraulic Properties of Unsaturated Soils, Riverside, California, October 11–13, 1989. U.S. Salinity Laboratory, Agricultural Research Service, U.S. Dept. of Agriculture/Dept. of Soil and Environmental Sciences, University of California, Riverside, CA).

Johns, G.G., Smith, R.C.G., 1975. Accuracy of soil water budgets based on a range of relationships for the influence of soil water availability on actual water use. Aust. J. Agric. Res. 26, 871-883.

Jonasson, S.A., 1990. Estimation of soil water retention for natural sediments from grain size distribution and bulk density. Geologiska Institutionen CTH/GU, PubL A 62. Göteborg, Sweden.

Jones, S.B., Or, D., 1998. Design of porous media for optimal gas and liquid fluxes to plant roots. Soil Sci. Soc. Am. J. 62, 563-573.

Kabat, P., Hack ten Broeke, M.J.D., 1989. Input data for agrohydrological simulation models: some parameter estimation techniques. *In*: van Lanen, H.A.J., Bregt, A.K. (Eds.), Proc. EC Workshop on Application of Computerized EC Soil Maps and Climate Data, Wageningen, 15–16 November 1988, Report 9. DLO The Winard Staring Center for Integrated land, Soil and Water Research (SC-DLO), Wageningen.

Keisling, T.C., 1974. Precision with which selected physical properties of similar soils can be estimated. Ph.D. Thesis, Oklahoma State University, Stillwater, OK.

Kelly, G.E., 1975. Soil of the North Appalachian Experimental Watershed, Misc. Pub. 1296. USDA-ARS, Washington, DC.

Keng, J.C., Lin, C.S., 1982. A two-line approximation of hydraulic conductivity for structured soils. Can. Agric. Engng 24, 77-80.

Klavetter, E.A., Peters, R.R., 1986. Estimation of hydraulic properties of an unsaturated, fractured rock mass, SAND84-2642. Sandia National Laboratories, Albuquerque, NM.

Klute, A., Wilkinson, G.E., 1958. Some tests of the similar media concept of capillary flow: I. Reduced capillary conductivity and moisture characteristic data. Soil Sci. Soc. Am. Proc. 22, 278-281.

Kolterman, C.E., Gorelick, S.M., 1995. Fractional packing model for hydraulic conductivity derived from sediment mixtures. Water Resour. Res. 31, 3283-3297.

Koplik, J., Lisseter, T.J., 1985. Two-phase flow in random network models of porous media. Soc. Pet. Engng J. 25, 89.

Kosugi, K., 1994. Three-parameter lognormal distribution model for soil water retention. Water Resour. Res. 30, 891-901.

Kühnel, V., 1989. Scale Problems in Soil Moisture Flow. University College, Dept of Civil Engineering, Dublin, 225 pp.

Kuntze, H., 1966. Die messung des geschlossenen und offenen kapillarasaumes in natürlich gelagerten boden. Z. Für Pflanzener. Düngung Bodenkd. 111, 97-106.

Kunze, R.J., Uehara, G., Graham, K., 1968. Factors important in the calculation of hydraulic conductivity. Soil Sci Soc. Am. Proc. 32, 760-765.

Kutilek, M., 1973. The influence of clay minerals and exchangale cations on soil moisture potential. *In*: Hadas, A., Swartzendruber, D., Rijtema, P.E., Fuchs, M., Yaron, B. (Eds.), Ecological Studies. IV. Physical Aspects of Soil Water and Salts in Ecosystems. Chapman & Hall, London, pp. 153-160.

Labardi, P.C., Reinhardt, K., Nielsen, D.R., Biggar, J.W., 1980. Simple field methods for estimating soil hydraulic conductivity. Soil Sci. Soc. Am. J. 44, 3-6.

Laliberte, G.E., Brooks, R.H., 1967. Hydraulic properties of disturbed soil materials affected by porosity. Soil Sci. Soc. Am. Proc. 31, 451-454.

Lascano, R.I., van Bavel, C.H.M., 1982. Spatial variability of hydraulic and remotely sensed soil parameters. Soil Sci. Soc. Am. J. 46, 223-228.

Leenhardt, D., Voltz, M., Bornand, M., 1994a. Propagation of the error of spatial prediction of soil properties in simulating crop evapotranspiration. Eur. J. Soil Sci. 45, 303-310.

Leenhardt, D., Voltz, M., Bornand, M., 1994b. Evaluating soil maps for prediction of soil water properties. Eur. J. Soil Sci. 45, 293-301.

Lenhard, R.J., Parker, J.C., 1987. A model for hysteretic constitutive relations governing multiphase flow. 2. Permeability-saturation relations. Water Resour. Res. 23, 2197-2206.

Lenhard, R.J., Parker, J.C., Mishra, S., 1989. On the correspondence between Brooks–Corey and van Genuchten models. J. Irrig. Drain. Engng, Proc. ASCE 115, 744-751.

Leummens, H., Bouma, J., Bootlink, H.W.G., 1995. Interpreting differences among hydraulic parameters for different soil series by functional characterization. Soil Sci. Soc. Am. J. 59, 344-351.

Loague, K., 1992. Using soil texture to estimate saturated hydraulic conductivity and the impact on rainfall-runoff simulations. Water Resour. Res. 28, 687-693.

Logsdon, S.D., Allmaras, R.R., Wu, L., Swan, J.B., Randall, G.W., 1990. Macroporosity and its relation to saturated hydraulic conductivity under different tillage practices. Soil Sci. Soc. Am. J. 54, 1096-1101.

Long, F.L., Perkins, H.F., Carreker, J.R., Daniels, J.M., 1969. Morphological, chemical, and physical characteristics of eighteen representative soils of the Atlantic Coast Flatwoods, Res. Bull. No. 59. Univ. of Georgia, Athens.

Lund, Z.F., 1959. Available water-holding capacity of alluvial soils in Louisiana. Soil Sci Soc. Am. Proc. 23, 1-3.

Luxmoore, R.J., 1981. Micro-, meso-, and macroporosity of soil. Soil Sci. Soc. Am. J. 45, 671-672.

Luxmoore, R.J., 1982. Physical Characteristics of Soils of the Southern Region: Fullerton and Sequoia Series, ORNL-5868. Oak Ridge Nat. Lab., Oak Ridge, TN.

Madankumar, N., 1985. Prediction of soil moisture characteristics from mechanical analysis and bulk density data. Agric. Water Manag. 10, 305-312.

Manrique, L.A., Jones, C.A., 1991. Bulk density of soils in relation to soil physical and chemical properties. Soil Sci. Soc. Am. J. 55, 476-481.

Manrique, L.A., Jones, C.A., Ryke, P.T., 1990. Estimation of exchangeable bases and base saturation from soil physical and chemical data. Commun. Soil Sci. Plant Anal. 21, 2119-2134.

Manrique, L.A., Jones, C.A., Ryke, P.T., 1991. Predicting soil water retention characteristics from soil physical and chemical properties. Commun. Soil Sci. Plant Anal. 22, 1847-1860.

Masch, F.D., Denny, K.J., 1966. Grain size distribution and its effect on the permeability of the unconsolidated sands. Water Resour. Res. 2, 665-677.

McCuen, R.H., Rawls, W.J., Brakensiek, D.L., 1981. Statistical analysis of the Brooks–Corey and the Green–Ampt parameters across soil texture. Water Resour. Res. 17, 1005-1013.

McKenzie, N.J., Ryan, P.J., 1999. Spatial prediction of soil properties using environmental correlation. Geoderma 89, 67-94.

Mehuys, G.R., Stolzy, L.H., Leley, J., Weeks, L.V., 1975. Effect of stones on the hydraulic conductivity of relatively dry desert soils. Soil Sci. Soc. Am. Proc. 39, 37-42.

Meng, T.P., Taylor, H.M., Fryar, D.W., Gomez, J.F., 1987. Models to predict water retention in semiarid soils. Soil Sci. Soc. Am. J. 51, 1563-1565.

Messing, I., 1989. Estimation of saturated hydraulic conductivity in day from soil moisture retention data. Soil Sci. Soc. Am. J. 53, 665-668.

Milks, R.R., Fonteno, W.C., Larson, R.A., 1989. Hydrology of horizontal substrates. II. Predicting physical properties of media in containers. J. Am. Soc. Hortic. Sci. 114, 53-56.

Milly, P.C.D., 1987. Estimation of Brooks–Corey parameters from water retention data. Water Resour. Res. 23, 1085-1089.

Mishra, S., Parker, J.C., 1990. On the relation between saturated hydraulic conductivity and capillary retention characteristics. Ground Water 28, 775-777.

Mokma, D.L., 1966. Correlation of soil properties, percolation tests, and soil surveys in design of septic tanks, disposal fields in Eaton, Genese, Ingham, Macomb Counties, Michigan. Ph.D. Thesis, Michigan State Univ., East Lansing, MI.

Moolman, J.H., 1985. Spatial variability of two selected soil properties in a semiarid subcatchment of the Lower Great Fish River. S. Afr. J. Plant Soil 2, 72-78.

Moran, M.S., Clarke, T.R., Kustas, W.P., Weltz, M., Amer, S.A., 1994. Evaluation of hydraulic parameters in a semiarid rangeland using remotely sensed spectral data. Water Resour. Res. 30, 1287-1297.

Moreno, F., De La Rosa, D., Fernandez, J.E., Andreu, L., 1997. Preliminary analysis of soil properties variation for the development of pedotransfer functions. pp. 121–124 (The use of pedotransfer in soil hydrology research in Europe. In: Bruand, A., Duval, O., Wösten, J.H.M., Lilly, A. (Eds.), c1997. Proceedings of the Second Workshop of the project "Using existing soil data to derive hydraulic parameters for simulation modelling in environmental studies and in land use planning" INRA Orléans, France, 10–12 October 1996. Orléans, France, INRA-Unité de Science du soil/EC – Joine research Centre, European Soil Bureau, Ispra, Italy/ECSC-EC-EAEC Brussels, Luxembourg).

Mortland, M.M., 1954. Specific surface and its relationships to some physical and chemical properties of soil. Mich. Agric. Exp. Station J. Art. 1613, East Lansing, MI.

Mualem, Y., 1976. A new model for predicting the hydraulic conductivity of unsaturated porous media. Water Resour. Res. 12, 513-522.

Mualem, Y., 1977. Extension of the similarity hypothesis used for modeling the soil water characteristics. Water Resour. Res. 13, 773-780.

Mualem, Y., 1978. Hydraulic conductivity of unsaturated porous media: generalized macroscopic approach. Water Resour. Res. 14, 325-334.

Mungare, T.S., Shingte, A.K., Pharande, K.S., 1983. Relationship of available water capacity with some of the physical properties of soils. J. Maharashtra Agric. Univ. 8, 9-13.

Naney, J.W., Ahuja, L.P., Williams, R.D., Rawls, W.J., 1992. Estimating spacial distribution of hydraulic conductivity in a field using effective porosity data. pp. 515–528 (van Genuchten, M.Th., Leij, F.J., Lund, L.J. (Eds.), c1992. Indirect Methods for Estimating the Hydraulic Properties of Unsaturated Soils: Proceedings of the International Workshop on Indirect Methods for Estimating the Hydraulic Properties of Unsaturated Soils, Riverside, California, October 11–13, 1989. U.S. Salinity Laboratory, Agricultural Research Service, U.S. Dept. of Agriculture/Dept. of Soil and Environmental Sciences, University of California, Riverside, CA).

National Cooperative Soil Survey, 1974. Soil Survey Laboratory Data and Description for Some Soils of New Jersey. Soil Conservation Services, USDA and New Jersey Agric Exp. Sta., Rutgers Univ., New Brunswick, NJ.

Negm, M.A., Fanous, N.E., Mikhail, M.I., 1990. Some correlations between physical and chemical properties of Maryut desert soils. Egypt. J. Soil Sci. 30, 91-105.

Nielsen, D.R., Luckner, L., 1992. Theoretical aspects to estimate reasonable initial parameters and range limits in identification procedures for soil hydraulic properties. pp. 147–160 (van Genuchten, M.Th., Leij, F.J., Lund, L.J. (Eds.), c1992. Indirect Methods for Estimating the Hydraulic Properties of Unsaturated Soils: Proceedings of the International Workshop on Indirect Methods for Estimating the Hydraulic Properties of Unsaturated Soils, Riverside, California, October 11–13, 1989/U.S. Salinity Laboratory, Agricultural Research Service, U.S. Dept. of Agriculture/Dept. of Soil and Environmental Sciences, University of California, Riverside, CA).

Nielsen, D.R., Kirkham, D., 1960. Soil capillary conductivity; comparison of measured and calculated values. Soil Sci. Soc. Am. Proc. 24, 157-160.

Nikolaeva, S.A., Pachepsky, Ya.A., Shcherhakov, R.A., Shcheglov, A.I., Twetnova, O.B., Deryushchinskava, V.D., 1988. Modelling of moisture regime for calcareous chernozem. Pochvovedenie 1, 44-54, (in Russian).

Nikolaeva, S.A., Kuznetsov, Ya.M., Shcherbakov, R.A., Pachepskv, Ya.A., 1991. The effect of long-term irrigation on water retention in Caucasus piedmont chernozems. Vestnik MGU, Ser. Pochvovedenie 3, 71-74, (in Russian).

Nimmo, J.R., 1991. Comment on the treatment of residual water content in "a consistent set of parametric models for the two-phase flow of immiscible fluids in the subsurface" by L. Luckner et al. Water Resour. Res. 27, 661-662.

Nimmo, J.R., 2002. What measurable properties can predict preferential flow? Paper No. 324. Transactions of the 17th World Congress of Soil Science, August 14–21, Bangkok, Thailand.

Nimmo, J.R., Miller, E.E., 1986. The temperature dependence of isothermal moisture vs. potential characteristics of soils. Soil Sci. Soc. Am. J. 50, 1105-1113.

Nizeyimana, E., Olson, K.R., 1988. Chemical, mineralogical, and physical property differences between moderately and severely eroded Illinois soils. Soil Sci. Soc. Am. J. 52, 1740-1748.

Nofziger, D.L., Williams, I.R., Hornsby, A.G., Wood, A.L., 1983. Physical characteristics of soils in the southern region – Bethany, Konawa and Tipton series, Southern Cooperative Series Bull. 265. Oklahoma Agric. Exp. Station, Stillwater, OK.

Nordt, L.C., Jacob, J.S., Wilding, L.P., 1991. Quantifying map unit composition for quality control in soil survey. In: Mausbach, M.J., Wilding, L.P. (Eds.), Spatial Variabilities of Soils and Landforms, SSSA Spec. Publ., 28, pp. 183-197, Madison, WI.

Oosterveld, M., Chang, C., 1980. Empirical relations between laboratory determinations of soil texture and moisture characteristic. Can. J. Agric. Engng 22, 149-151.

Othmer, R., Diekkrüger, B., Kutilek, M., 1991. Bimodal porosity and unsaturated hydraulic conductivity. Soil Sci. 152, 139-149.

Pachepsky, Ya.A., Mironenko, E.V., Shcherbakov, R.A., 1992. Prediction and use of soil hydraulic properties (van Genuchten, M.Th., Leij, F.J., Lund, L.J. (Eds.), c1992. Indirect Methods for Estimating the Hydraulic Properties of Unsaturated Soils: Proceedings of the International Workshop on Indirect Methods for Estimating the Hydraulic Properties of Unsaturated Soils, Riverside, California, October 11–13, 1989. U.S. Salinity Laboratory, Agricultural Research Service, U.S. Dept. of Agriculture/Dept. of Soil and Environmental Sciences, University of California, Riverside, CA).

Pagliai, M., Guidi, G., La Marca, M., Giachetti, M., Lucamante, G., 1981. Effect of sewage sludges and composts on soil porosity and aggregation. J. Environ. Qual. 10, 556-561.

Paige, G.B., Hillel, D., 1998. Comparison of three methods for accessing soil hydraulic properties. Soil Sci. 155, 175-189.

Panian, T.F., 1987. Unsaturated Flow Properties Data Catalog, DOE/NV/10384-20, Vol. II. Desert Research Institute, Reno, NV, 493 pp.

Parkes, M.E., Waters, P.A., 1980. Comparison of measured and estimated unsaturated hydraulic conductivity. Water Resour. Res. 16, 749-754.

Perfect, E., McLaughlin, N.B., Kay, B.D., Topp, G.C., 1996. An improved fractal equations for the soil water retention curve. Water Resour. Res. 32, 281-287.

Perrier, E., Rieu, M., Sposito, G., de Marsily, G., 1996. Models of the water retention curve for soils with a fractal pore-size distribution. Water Resour. Res. 32, 3025-3031.

Pershinger, L.D., Yahner, J.E., 1970. Relation of percolation rates to soil texture on several Indiana soils. J. Soil Water Conserv. 25, 189-191.

Philip, J.R., 1980. Soil heterogeneity: some basic issues. Water Resour. Res. 16, 443-448.

Pluth, D.J., Adams, R.S., Rust, R.H., Peterson, J.R., 1970. Characteristics of selected horizons from 16 soils series in Minnesota. Univ. of Minnesota Agric. Exp. Stn. Tech. Bull. 272.

Pochatkova, T.N., 1975. Composition and properties of soil aggregates as a function of their dimensions. Ph.D. Thesis, Moscow, Russia, (in Russian).

Poelman, J.N.B., van Egmont, Th., 1979. Water Retention Curves for Sea- and River-Clay Soils Derived from Easily Measured Soil Properties, Report 1492. Winard Staring Center, Wageningen, The Netherlands, in Dutch.

Prebble, R.E., Stirk, G.B., 1959. Effect of free iron oxide on range of available water in soil. Soil Sci. 88, 213-217.

Quisenberry, V.L., Cassel, D.K., Dane, J.H., Parker, J.C., 1983. Physical characteristics of soils in the Southern Region-Norfolk, Dothan, Wagram and Goldsboro Series. Southern Cooperative Series Bull. 263. South Carolina Agric. Exp. St.

Raats, P.A.C., Gardner, W.R., 1971. A comparison of some empirical relationships between pressure head and hydraulic conductivity, and some observations on radially symmetric flow. Water Resour. Res. 7, 921-928.

Ragab, R., Cooper, J.D., 1990. Obtaining Soil Hydraulic Properties from Field, Laboratory and Predictive Methods. Institute of Hydrology, Wallingford, Oxforshire, UK.

Ragab, R., Feyen, J., Hillel, D., 1982. Effects of the method for determining pore size distribution on prediction of the hydraulic conductivity function and of infiltration. Soil Sci. 134, 141-145.

Rajkai, K., 1988. Relationships between water retention and different soil properties. Agrokémia es Talajtan 36–37, 15-30, (in Hungarian).

Rajkai, K., Várallyay, Gy., 1989. Estimative calculation of hydrophysical parameters from simply measurable soil properties. Agrokémia es Talajtan 38, 634-640.

Rajkai. K., Várallyay, Gy., 1992. Estimating soil water retention from simpler properties by regression techniques. pp. 417–426 (van Genuchten, M.Th., Leij, F.J., Lund, L.J. (Eds.), c1992. Indirect Methods for Estimating the Hydraulic Properties of Unsaturated Soils: Proceedings of the International Workshop on Indirect Methods for Estimating the Hydraulic Properties of Unsaturated Soils, Riverside, California, October 11–13, 1989. U.S. Salinity Laboratory, Agricultural Research Service, U.S. Dept. of Agriculture/Dept. of Soil and Environmental Sciences, University of California, Riverside, CA).

Rao, P.S.C., Jessup, R.E., Hornsby, A.C., Cassel, D.K., Pollans, W.A., 1983. Scaling soil microhydrological properties of Lakeland and Konowa soils using similar media concepts. Agric. Water Manag. 6, 277-290.

Rawls, W.J., 1983. Estimating soil bulk density from particle size analysis and organic matter content. Soil Sci. 135, 123-126.

Rawls, W.J., Brakensiek, D.L., 1983. A procedure to predict Green and Ampt infiltration parameters, Advances in Infiltration: Proceedings of the National Conference on Advances in Infiltration, December 12–13, 1983, Hyatt Regency Illinois Center, Chicago, IL. American Society of Agricultural Engineering, St. Joseph, MI, pp. 102–112.

Rawls, W., Yates, P., Asmussen, L., 1976. Calibration of Selected Infiltration Equations for the Georgia Coastal Plain, USDA-ARS-S-113, Washington, DC, 110 pp.

Rawls, W.J., Brakensiek, D.L., Logsdon, S.D., 1993. Predicting saturated hydraulic conductivity utilizing fractal principles. Soil Sci. Soc. Am. J. 57, 1193-1197.

Reignier, D., 1986. Estimation de la Conductivité Hydraulique. Institut National Agronomique Paris, INRA. Science du Sol, Centre de Recherches d'Avignon, Montfavet, France.

Rice, J., Baumer, O.W., 1986. Computation of soil water hydraulic parameters. Technical Paper. Lincoln, Nebraska. USDA-SCS, Midwest National Technical Center.

Richards, S.J., Weeks, L.V., 1953. Capillary conductivity values from moisture yield and tension measurements on soil columns. Soil. Sci. Soc. Am. Proc. 17, 206-209.

Rivière, L.M., Dartigues, A., Lemaire, F., 1984. Some properties of French peats for use in ornamental horticulture. Proceedings of 7th International Peat Congress 4, 21–35, Dublin, Ireland.

Robbins, C.W., 1977. Hydraulic Conductivity and Moisture Retention Characteristics of Southern Idaho's Silt Loam Soils, Res. Bull 99. Agric. Exp. Sta., Univ. of Idaho, Moscow, ID.

Rockhold, M.L., Rossi, R.E., Hills, R.G., 1996. Application of similar media scaling and conditional simulation for modeling water flow and tritium transport at the Las Cruces trench site. Water Resour. Res. 32 (3), 595-609.

Rogowski, A.S., 1972a. Watershed physics: model of the soil moisture characteristic. Water Resour. Res. 7, 1575-1582.

Rogowski, A.S., 1972b. Estimation of the soil moisture characteristic and hydraulic conductivity: comparison of models. Soil Sci. 123, 423-429.

Rogowski, A.S., 1972c. Watershed physics: soil variability criteria. Water Resour. Res. 7, 1015-1023.

Rolston, D.E., 1976. Evaluation of field method for measuring or predicting soil water properties. J. Indian Soc. Soil Sci. 24, 101-113.

Romkens, M.J.M., Selim, H.M., Scot, H.D., Phillips, R.E., Whisler, F.D., 1986. Physical Characteristics of Soils in the Southern Region, Southern Cooperative Series Bull. 266. Mississippi Agric. and Forestry Exp. Sta., Oxford, MS.

Rossi, C., Nimmo, J.R., 1994. Modeling of soil water retention from saturation to oven dryness. Water Resour. Res. 30 (3), 701-708.

Rourke, R.V., Beck, C., 1968. Soil-Water, Chemical and Physical Characteristics of Eight Soil Series in Maine, Tech. Bull. 29. Maine Agric. Exp. Sta., Univ. of Maine, Orono, Maine.

Römkens, M.J.M., Selim, H.M., Scott, H.D., Phiullips, R.E., Whisler, F.D., 1986. Physical Characteristics of Soils in the Southern Region: Captina, Gigger, Grenada, Loring, Oliver, and Sharkey Series, Southern Cooperative. Series Bull. 264. Mississippi Agric. and Forestry Exp. Sta., Oxford, MS, 180 pp.

Russel, E.R., Mickle, J.L., 1971. Correlation of suction curves with the plasticity index of soils. J. Mater. 6, 320-331.

Russo, D., Bresler, E., 1980. Soil-water-suction relationships as affected by soil solution composition and concentration. In: Banin, A., Kafkafi, U. (Eds.), Agrochemicals in Soils. Pergamon Press, Elmsford, NY, pp. 287-296.

Salter, P.J., Williams, J.B., 1965. The Influence of Texture on the Moisture Characteristics of Soils. II. Available Water Capacity and Moisture Release Characteristics. J. Soil Sci. 16, 310-317.

Salter, P.J., Berry, G., Williams, J.B., 1966. The influence of texture on the moisture characteristics of soils, III. Quantitative relationships between particle size, composition, and available-water capacity. J. Soil Sci. 17, 93-98.

Satterwhite, M.B., 1980. Evaluating soil moisture and textural relationships using regression analysis. Report ETL-0226, Engineering Topographic Laboratories, U.S. Army Corps of Engineers, Fort Belvoir, VA, 31 pp.

Schaap, M.G., 1996. The role of soil organic matter in the hydrology of forests on dry sandy soils. Ph.D. Thesis, University of Amsterdam, the Netherlands, 150 pp.

Scheinost, A.C., Auerswald, K., 1995. Pedotransfer-funktionen zur erzeugung von wasseretentionskurven unter berücksichtigung bimodaler porensystem. Mitt. Deutsch. Bod. Ges. 76, 141-144.

Schindler, U., Bohne, K., Sauerbrey, R., 1985. Comparison of different measuring and calculating methods to quantify the hydraulic conductivity of unsaturated soil. Z. Planzenern. Bodenk 148, 607-617.

Schjonning, P., 1985. Porositetsforhold i landbrugsjord. I. Modeller og jordtypeforskelle (Soil pore characteristics, I. Models and soil type differences). Tidsskr. Planteavl. 89, 411-423, (in Danish).

Schoeneberger, P.J., 1990. Physical properties and water movement through a soil and saprolite at different geomorphic positions. Ph.D. Thesis, North Carolina State University, Raleigh, NC.

Schuh, W.M., 1992. Calibration of soil hydraulic parameters through separation of subpopulations in reference to soil texture. pp. 489–498 (van Genuchten, M.Th., Leij, F.J., Lund, L.J. (Eds.), c1992. Indirect Methods for Estimating the Hydraulic Properties of Unsaturated Soils: Proceedings of the International Workshop on Indirect Methods for Estimating the Hydraulic Properties of Unsaturated Soils, Riverside, California, October 11–13, 1989. U.S. Salinity Laboratory, Agricultural Research Service, U.S. Dept. of Agriculture/Dept. of Soil and Environmental Sciences, University of California, Riverside, CA).

Schuh, W.M., Cline, R.L., 1990. Effect of soil properties on unsaturated hydraulic conductivity pore interaction factors. Soil Sci. Soc. Am. J. 54, 1509-1519.

Scotter, D.R., Clothier, B.E., Sauer, I.J., 1988. A critical assessment of the role of measured hydraulic properties in the simulation of absorption, infiltration and redistribution of soil water. Agric. Water Manag. 15, 73-86.

Setiawan, B.I., Nakano, M., 1993. On the determination of unsaturated hydraulic conductivity from soil moisture profiles and from water retention curves. Soil Sci. 156, 389-395.

Shainberg, I., Rhoades, J.D., Prather, R.J., 1981. Effect of low electrolyte concentration on clay dispersion and hydraulic conductivity of a sodic soil. Soil Sci. Soc. Am. J. 45, 273-277.

Sharma, M.L., 1966. Influence of soil structure on water retention, water movement and thermodynamic properties of adsorbed water. Ph.D. Thesis, Univ. of Hawaii, 190 pp.

Shaykewich, C.F., 1970. Hydraulic properties of disturbed and undisturbed soils. Can. J. Soil Sci. 50, 431-437.

Shaykewich, C.F., Zwarich, M.A., 1968. Relationships between soil physical constants and soil physical components of some Manitoba soils. Can. J. Soil Sci. 48, 199-204.

Shuh, W.M., Cline, R.D., Sweeney, M.D., 1988. Comparison of a laboratory procedure and a textural model for predicting *in situ* water retention. Soil Sci. Soc. Am. J. 52, 1218-1227.

Simota, C., Mayr, T., 1996a. Predicting the soil water retention curve from readily-available data obtained during soil surveys. Int. Agrophys. 10, 185-188.

Simota, C., Mayr, T., 1996b. Pedotransfer functions. *In*: Loveland, P.J., Rousvelt, M.D.A. (Eds.), Agroclimatic Change and European Soil Suitability. Cranfield University.

Sinclair, L.R., 1981. Improved conductivities from desorption data. ASAE Paper No. 81-2027.

Skaggs, R.W., Wells, L.G., Ghate, S.R., 1978. Predicted and measured drainable porosities for field soils. Trans. ASAE 21, 522-528.

Smettem, K.R.J., Ross, P.J., 1992. Measurement and prediction of water movement in a field soil: the matrix-macropore dichotomy. Hydrol. Proc. 6, 1-10.

Snyder, V.A., 1996. Statistical hydraulic conductivity models and scaling of capillary phenomena in porous media. Soil Sci. Soc. Am. J. 60, 771-774.

Sodek, F., Carlise, V.W., Collins, M.E., Hammond, L.C., Harris, W.G., 1990. Characterization Data for Selected Florida Soils, Soil Sci. Res. Rept. 90-1. Univ. of Florida, Institute of Food and Agric. Sci., Gainesville, FL.

Southard, R.J., Buol, S.W., 1988. Subsoil saturated hydraulic conductivity in relation to soil properties in the North Carolina Coastal Plain. Soil Sci. Soc. Am. J. 52, 1091-1094.

Spomer, L.A., 1980. Prediction and control of porosity and water characteristic in sand-soil mixtures for drained turf sites. Agron. J. 72, 361-362.

Sposito, G., Jury, W.A., 1990. Miller similitude and generalized scaling phenomena. *In*: Hillel, D., Elrick, D. (Eds.), Scaling in Soil Physics, SSSA Spec. Publ. 25. SSSA, Madison, WI, pp. 13-22.

Springer, E.P., Cundy, T.W., 1987. Field-scale evaluation of infiltration parameters from soil texture for hydraulic analysis. Water Resour. Res. 23, 325-334.

Stephens, D.B., 1992. A comparison of calculated and measured unsaturated hydraulic conductivity of two uniform soils in New Mexico. pp. 249–263 (van Genuchten, M.Th., Leij, F.J., Lund, L.J. (Eds.), c1992. Indirect Methods for Estimating the Hydraulic Properties of Unsaturated Soils: Proceedings of the International Workshop on Indirect Methods for Estimating the Hydraulic Properties of Unsaturated Soils, Riverside, California, October 11–13, 1989. U.S. Salinity Laboratory, Agricultural Research Service, U.S. Dept. of Agriculture/Dept. of Soil and Environmental Sciences, University of California, Riverside, CA).

Stephens, D.B., Rehfeldt, D.R., 1985. Evaluation of closed-form analytical models to calculate conductivity in a fine sand. Soil Sci. Soc. Am. J. 49, 12-19.

Stephens, D.B., Lambert, K., Watson, D., 1987. Regression models for hydraulic conductivity and field test of the borehole permeameter. Water Resour. Res. 23, 2207-2214.

Stevenson, D.S., 1974. Influence of peat moss on soil water retention for plants. Can. J. Soil Sci. 54, 109-110.

Stewart, V.I., Adams, W.A., Abdulla, H.H., 1970. Quantitative pedological studies on soils derived from Silurian mudstones. II. The relationship between stone content and the apparent density of the fine earth. J. Soil Sci. 21, 248-255.

Strait, S., Saxton, K.E., Papendick, R.I., 1979. Pressure and hydraulic conductivity curves for various soil textures. Internal Report, USDA-ARS, Washington State University, Pullman.

Su, C., Brooks, R.H., 1976. Hydraulic Functions of Soils from Physical Experiments and Their Applications, WRRI-41. Water Resour. Res. Inst., Oregon State Univ., Corvallis, OR.

Talsma, T., 1985. Prediction of hydraulic conductivity from soil water retention data. Soil Sci. 140, 184-188.

Talsma, F., Flint, S.E., 1958. Some factors determining the hydraulic conductivity of subsoils with special reference to tile drainage problems. Soil Sci. 85, 198-206.

Tamari, S., 1994. Relation between pore space and hydraulic properties in compacted beds of silty loam aggregates. Soil Technol. 7, 57-73.

Tester, C.F., 1990. Organic amendment effects on physical and chemical properties of a sandy soil. Soil Sci. Soc. Am. J. 54, 827-831.

Thomasson, A.J., Carter, A.D., 1992. Current and future uses of the UK soil water retention dataset. pp. 355–358 (van Genuchten, M.Th., Leij, F.J., Lund, L.J. (Eds.), c1992. Indirect Methods for Estimating the Hydraulic Properties of Unsaturated Soils: Proceedings of the International Workshop on Indirect Methods for Estimating the Hydraulic Properties of Unsaturated Soils, Riverside, California, October 11–13, 1989/U.S. Salinity Laboratory, Agricultural Research Service, U.S. Dept. of

Agriculture/Dept. of Soil and Environmental Sciences, University of California, Riverside, CA).

Toledo, P.G., Novy, R.A., Davis, H.T., Scriven, L.E., 1990. Hydraulic conductivity of porous media at low water content. Soil Sci. Soc. Am. J. 54, 673-679.

Tripathi, R.P., Ghildyal, B.P., 1975. Evaluation of pore size distribution methods for calculating the hydraulic conductivity of soil. J. Ind. Soil Sci. 23, 273-279.

Tyler, S.W., Wheatcraft, S.W., 1992. Fractal scaling of soil particle size distributions: analysis and limitations. Soil Sci. Soc. Am. J. 56, 362-369.

Van Brakel, J., 1975. Pore space models for transport phenomena in porous media, review and evaluation with special emphasis on capillary liquid transport. Powder Technol. 11, 205.

Van de Griend, A.A., O'Neil, P.E., 1986. Discrimination of soil hydraulic properties by combined thermal infrared and microwave remote sensing. Proceedings of IGARSS '86 Symposium, ESA SP-254. Eur. Space Agency, Neuilly, France, pp. 839–845.

van Genuchten, M.Th., Nielsen, D.R., 1985. On describing and predicting the hydraulic properties of unsaturated soils. Ann. Geophys. 3, 615-628.

van Genuchten, M.Th., Kaveh, F., Russell, W.B., Yates, S.R., 1989. Direct and indirect methods for estimating the hydraulic properties of unsaturated soils. In: Bouma, J., Bregt, A.K. (Eds.), Land Qualities in Space and Time: Proceedings of a Symposium Organized by the International Society of Soil Science (ISSS), 22–26 August 1988, Wageningen, the Netherlands, pp. 61-72.

van Ommen, H.C., Hopmans, J.W., van der Zee, S.E.A.T.M., 1989. Prediction of solute breakthrough from scaled soil physical properties. J. Hydrol. 105, 263-273.

Vereecken, H., Diels, J., Feyen, J., 1990. Functional Validation of Pedotransfer Functions for Soil Hydraulic Properties, Transactions of the 14th International Congress of Soil Science, Vol. V. Kyoto, Japan, August 12–18 1990, pp. 533–534.

Vetterlein, E., 1990. Estimation of saturated hydraulic conductivity from soil texture and structure. Arch. Agron. Soil Sci. 34, 435-441.

Visser, W.C., 1969. The relation between lithological properties and the shape of desorption curve. Water in Unsaturated Zone, Proceedings of UNESCO/IASH Symposium, Wageningen, Netherlands, pp. 305–311.

Vogel, T., Císlerová, M., 1988. On the reliability of unsaturated hydraulic conductivity calculated from the moisture retention curve. Transp. Porous Media 3, 1-15.

Voltz, M., Goulard, M., 1994. Spatial interpolation of soil moisture retention curves. Geoderma 62, 109-123.

Voltz, M., Lagacherie, P., Louchart, X., 1997. Prediction soil properties over a region using sample information from a mapped reference area. Eur. J. Soil Sci. 48, 19-30.

Vukovic, M., Soro, A., 1992. Determination of Hydraulic Conductivity of Porous Media from Grain-Size Composition, Water Resources Publication 91-067280, Littleton, CO, 82 pp.

Wagenet, R.J., 1990. Estimation and application of soil hydrological properties for land evaluation and environmental protection, Transactions of International Congress on Soil Science 24th, Kyoto, Japan. 12–18 August, 1990, Vol. 1, p. 49.

Wagenet, R.J., Knighton, R.E., Bresler, E., 1984. Soil chemical and physical effects on spatial variability of hydraulic conductivity. Soil Sci. 137, 252-262.

Wagenet, R.J., Bouma, J., Grossman, R.B., 1991. Minimum data sets for use of soil survey information in soil interpretive models. In: Musbauch, Wilding (Eds.), Spatial

Variabilities of Soil and Landform, SSSA Special Publication No. 28, pp. 161-182, Madison, WI.

Walter, C., Curmi, P., Gascuel-Odoux, C., 1996. Pertinence du découpage pédologique pour l'estimation spatiale des propriétés physique du sol. Validation à l'échelle d'un bassin versant. *In*: Christophe, C., Lardon, S., Monestiez, P. (Eds.), Etudes des Phénomènes Spatiaux en Agriculture, La Rochelle, 6–8 Dec. 1995, Les Colloques no. 78. INRA, Paris, pp. 97-110.

Wang, J.S.Y., Narasimhan, T.N., 1989. Processes, mechanisms, parameters and modeling approaches for partially saturated flow in soil and rock media. SAND88-70S4. Sandia National Laboratories, Albuquerque, NM, and LBL-26224, Berkeley Laboratory, Berkeley, CA.

Wang, J.S.Y., Narasimhan, T.N., 1992. Distributions and correlations of hydraulic parameters of rocks and soils (van Genuchten, M.Th., Leij, F.J., Lund, L.J. (Eds.), c1992. Indirect Methods for Estimating the Hydraulic Properties of Unsaturated Soils: Proceedings of the International Workshop on Indirect Methods for Estimating the Hydraulic Properties of Unsaturated Soils, Riverside, California, October 11–13, 1989/U.S. Salinity Laboratory, Agricultural Research Service, U.S. Dept. of Agriculture/Dept. of Soil and Environmental Sciences, University of California, Riverside, CA).

Ward, A., Wells, L.G., Phillips, R.E., 1983. Characterizing unsaturated hydraulic conductivity of western Kentucky surface mine spoils and soils. Soil Sci. Soc. Am. J. 47, 847-854.

White, I., Sully, M.J., 1992. On the variability and use of the hydraulic conductivity alpha parameter in stochastic treatments of unsaturated flow. Water Resour. Res. 28, 209-213.

Widiatmaka, Curmi, P., 1994. Soil horizons hydrodynamic characteristics of an acid soil system. Interest of their grouping according to functional properties for spatial transposition, 15th World Congress of Soil Sci., Acapulco, Mexico, July 10–16, 1994. Transaction, Vol. 2b, pp. 151–152.

Wierenga, P., Hudson, J.D., Winson, J., Nash, M., Toorman, A., Hills, R.G., 1989. Soil Physical Properties at the I-as Cruces Trench Site, NUREG/CR-5441. U.S. Nuclear Regulatory Commission, Washington, DC.

Wiersma, D., 1984. Soil Water Characteristic Data from Some Indiana Soils, Purdue Univ. Agric. Exp Stn. Bull. 542, West Lafayette, IN.

Wilcox, J.D., Spilsbury, R.H., 1941. Soil moisture studies. II. Some relationships between moisture measurements and mechanical analysis. Sci. Agric. 21, 159 172.

Williams, J.R., Prebble, E., Williams, W.T., Hignett, C.T., 1983. The influence of texture, structure and clay mineralogy on the soil moisture characteristic. Aust. J. Soil Res. 21, 15-31.

Wu, L., Allmaras, R.R., Lamb, J.B., Johnsen, K.E., 1996. Model sensitivity to measured and estimated hydraulic properties of a Zimmerman fine sand. Soil Sci. Soc. Am. J. 60, 1283-1290.

Yoshida, I., Kuona, H., Chikushi, J., 1985. A study on the prediction of a soil moisture characteristic curve from particle-size distribution. J. Fac. Agric., Tottori Univ. 20, 45-54.

Young, K.K., Dixon, J.D., 1966. Overestimation of water content at field capacity from sieved sample data. Soil Sci. 101, 104-107.

Zeiliguer, A.M., 1992. A hierarchical system to model the pore structure of soils: indirect methods for evaluating the hydraulic properties. pp. 499–514 (van Genuchten, M.Th., Leij, F.J., Lund, L.J. (Eds.), c1992. Indirect Methods for Estimating the Hydraulic Properties of Unsaturated Soils: Proceedings of the International Workshop on Indirect Methods for Estimating the Hydraulic Properties of Unsaturated Soils, Riverside, California, October 11–13, 1989. U.S. Salinity Laboratory, Agricultural Research Service, U.S. Dept. of Agriculture/Dept. of Soil and Environmental Sciences, University of California, Riverside, CA).

Zobeck, T.M., Fausey, N.R., Al-Hamden, N.S., 1985. Effect of sample cross-section area on saturated hydraulic conductivity in two structured soils. Trans. ASAE 28, 791-794.

Index

A

Abductive networks 25
Absorption 156
Accuracy indices 6, 363
Accuracy of PTF 4, 6, 35, 123, 132, 171, 359, 392, 393, 424
Activation function 22, 23
Adsorption coefficient 197, 198, 216
Adsorption isoterm models 197
Adsorption isoterm 196
Advanced Very High Resolution Radiometer (AVHRR) 289
Advective dispersion equation 205
Aggregate size distribution 136, 144–146, 149, 150
Aggregate stability 117
Aggregation factor 27, 368
Aggregation of modules 368
Agricultural modeling 7
Airborne radiometry 265
Air entry point 232, 260, 374
Air-entry potential 78, 79, 259
Air-filled porosity 203, 330, 337, 338
Alfisols 9, 64, 98, 99, 102, 103, 105, 106
Alluvial sediments 466
Analysis of variance 12, 391, 393
Ancillary data 5, 264, 275, 288
Ancillary information 274, 275, 284, 287, 288, 289, 297, 313, 335, 336
Andosols 99
ANSWERS model 267
Antecedent soil moisture 180
Antropogenic activities 466
Aquic Ustoll 162, 168, 171
Arable land 144, 466
Aridisols 99, 102, 103, 105
Artificial neural networks 2, 21, 25, 29, 30, 64, 99, 130, 136, 171, 172, 248, 267, 282, 286, 303, 354, 408, 434

Arya–Paris model 51, 325
Aspect 276
Attenuation coefficient 264
Australian National and State soil database 257
Available water capacity 49, 53, 82, 83, 111, 120, 136, 161, 162, 163, 164, 172, 231, 330, 336, 338, 415, 425

B

Backpropagation algorithm 23
Backward elimination 9
Bank of samples of the Mineral Arable Polish Soils 456
Barron criteria 25
Bayesian analyses 40
Bayesian probabilities 265
Bethany soil 87
Biased regression estimators 12
Bimodal porous medium-, micro- and macro-pore systems 225
Biopores 116, 118, 119, 124, 135
Black-box models 23
Bootstrap method 24, 40, 41, 313, 349, 351, 353, 354, 358, 377
Boundary conditions 210, 256, 288, 295, 297, 432, 433
Brooks and Corey water retention parameters 362
Brooks–Corey equation 78, 79, 80, 82, 89, 90, 156, 161, 164
Brooks–Corey retention function parameters 372
Brownian motion 214
Brutsaert's equation 359
Bubbling capillary pressure 442
Bubbling pressure 7, 156, 438, 443
Bucket-type water balance model 255

Buckingham solute diffusion model 204
Bulk density 3, 4, 5, 9, 10, 11, 12, 17, 18, 41, 51, 64, 71, 72, 73, 74, 77, 78, 85, 90, 91, 95, 96, 97, 98, 99, 104, 109, 110, 111, 123, 127, 129, 131, 132, 150, 155, 156, 160, 162, 164, 165, 166, 167, 168, 169, 171, 172, 186, 197, 203, 225, 235, 237, 238, 256, 258, 260, 264, 267, 273, 283, 284, 287, 289, 290, 298, 313, 314, 323, 324, 326, 328, 329, 330, 334, 338, 340, 342, 345, 352, 354, 359, 360, 362, 366, 369, 372, 373, 374, 375, 377, 383, 388, 406, 408, 409, 415, 416, 418, 420, 423, 424, 425, 433, 435, 439, 441, 443, 446, 447, 450, 451, 452, 453, 457, 462, 465, 468, 469, 470, 472

C

Cadmiun breakthrough 391
Calcaric Haplic Chernozems 391
Calcic Ustochrept 303
Calcisols 144
Calibration data set 34, 40, 354
Campbell retention function parameters 372
Campbell water retention model 75, 263, 441, 443
Campbell's equation 203, 329, 334
Capillary potential 87, 183, 243
Capillary pressure 241, 243, 244, 442, 452, 456
Catchment 5, 18, 121, 125, 126, 135, 136, 191, 217, 253, 255, 257, 260, 263, 265, 266, 267, 276, 279, 280, 284, 289, 290, 295, 297, 303, 316, 318
Categorical soil data 2
Catena of hydromorphic soils 468
Cation exchange capacity (CEC) 198
Cecil soil 87
Centroid calculation 402
CERES-Maize model 267
Characteristic points 4, 238, 241, 245, 278
Chemical, Runoff and Erosion from Agricultural Management Systems, CREAMS erosion model 181

Chernozems 9, 64, 144, 466, 468
Chi-square statistics 4, 5, 27, 309, 397
Chloride breakthrough 391
Cholesky decomposition 391
Chromic Calciustert 303
Class definitions 48
Class pedotransfer function 136, 172
Clay accumulation intensity 161
Clay bridges 116
Clay coating 29, 115, 116, 161, 233, 449
Clay content 17, 47, 48, 49, 50, 51, 52, 53, 54, 60, 61, 62, 63, 102, 103, 104, 105, 107, 108, 109, 121, 122, 146, 153, 154, 155, 156, 162, 165, 167, 168, 182, 183, 184, 198, 199, 209, 212, 213, 215, 216, 217, 256, 258, 260, 264, 266, 267, 283, 287, 304, 313, 314, 323, 327, 328, 329, 336, 338, 340, 343, 362, 373, 376, 415, 416
Clay particles 150, 153, 156, 227
Clay plasma 233
Climatic region 160, 161
Coarse-textured soils 37, 64, 100, 101, 103, 106, 107, 109, 111, 116, 130, 201, 248
Coefficient of correlation 366
Coefficient of determination 4, 13, 17, 18, 85, 122, 180, 190, 359, 397
Coefficient of linear extensibility 4, 233
Coefficient of variation 198, 306, 307, 340, 397, 467
Co-kriging 281
Colloidal particles 449
Combinatorial algorithms 25
Comparison of grouping 168
Complete membership 369, 400
Complex system 2, 21
Component loading matrix 8
Compound topographical variables 314
Concentration of solute in soil solution 197
Conditional probability approach 282, 290
Confidence intervals 9, 10, 15, 18, 39, 40, 41, 260
Consecutive grouping 167

Conservation practice factor 178
Constant slope impedance factor (CSIF) model 202
Content of gravitational water 457
Content of water unavailable for plants 449, 462
Continuity equation 205, 254
Continuous classification 282, 290
Continuous pedotransfer function 172, 318, 354, 409, 435, 472
Contour-based networks 277
CONUS-SOIL 257
Convection dispersion equation 204
Convective flux 205, 206
Convective-dispersive solute transport 204
Convergence of lateral fluxes 276
Correlation coefficient 12, 36, 130, 359, 366, 399, 449, 452, 455, 457, 458, 459, 461
Correlation-based statistics 398
Cost function 377, 388, 397
Cost-complexity factor 57
Covariance matrix 328, 391
Kozeny–Carman equation 84, 85, 161, 323, 346
Critical ODR 330, 337
Critical shear stress 185, 186, 188, 189
Critical values 5
Crop rotation 177, 178
Cropping system 178, 238, 406, 409
Cross-validation procedures 29
Crossing scales 230
Crust factor 184, 191
Crust infiltrometer 370
CSIF solute transport model 195
Cumulative size distribution curve 456

D

Darcy flux 205, 207
Data analysis 2, 3, 4, 9, 18, 30, 64, 111, 144, 318, 345, 409
Data exploration 21
Data mining 2, 6, 21, 29, 345
Data variability 2, 409
Days with good workability 391

Days with sufficient aeration 391
Decision rules 363, 365, 366, 368, 369, 400, 401, 403, 404
Decision support 1, 238, 406
Decomposition 274, 391
Defuzzification 402
Degree of erosion 178
Degree of saturation 72, 207, 257
Depositional landforms 281
Descriptive statistics 4, 305
Deviance 27, 28, 190
Diameter of grain fraction 456
Dielectric constant 340, 342, 343
Difference-based statistics 394
Diffusion–dispersion coefficient 205, 207, 208, 209, 220
Diffusive flux 205, 206
Diffusive solute transport 200
Digital Elevation Model (DEM) 266, 291
Digital Terrain Model or DTM 277
Digitized contour lines 275
Direct solar radiation index 276
Direction of lateral fluxes 276
Disconnectivity parameter 203
Dispersion coefficient 205, 209, 217
Dispersivity predictor 4
Dispersivity PTF 333
Distribution coefficient 197, 199, 217, 340
Double cross-validation 17, 18
Downstream acceleration of lateral fluxes 276
Drained upper limit 256
DRAINMOD 248, 392
Dry consistency class 131
Drying water retention curve 452

E

Effective hydraulic conductivity 183, 184, 189, 190, 191
Effective porosity 84, 85, 91, 110, 136, 330, 409, 446
Effective rainfall duration 185
Effective rainfall intensity 185
Effective runoff duration 185
Eigenvalue problem 8

Electric conductivity meter 7
Electric properties of the soil 340
Elementary soil particles 160
Elevation data on a regular grid 275
Elevation relief ratio 314
Empirical models 54, 187
Entisols 99, 102, 103, 105, 106
Environmental impact of pesticides 363
Environmental modeling 1, 7, 257
Environmental monitoring 195
Environmental variables 2, 5, 274, 282, 287, 393
EPIC erosion model 373, 181
EPIC-ASW Utilities 373
Equipotential lines 277
Erosion model pedotransfer functions 186
Erosivity factors 181
European database HYPRES 162, 163, 165
European Soil Erosion Model (EUROSEM) 192
Evaporation 4, 9, 111
Evapotranspiration 255, 274, 276, 287, 288, 318, 392, 406
Exchange reactions 205
Expert system 122, 324, 345, 363, 400, 405, 406, 409
Expert weight 366, 368, 369, 401, 402, 404
Extractable phosphorus 282
Extragrade membership 328, 329
Extrapolation 121, 264, 296, 299, 327, 328, 440

F

Factorial regression 457
FAO triangle 164, 165, 167
Favorable, F, membership 329, 369, 364, 401
Feed-forward back-propagation 22, 313
Ferrasols 425
F-based pattern index 399
F-ratios 13
F-test 13, 14, 305, 399
F-value 17
Field bulk density 330

Field capacity 4, 8, 49, 97, 98, 103, 120, 121, 155, 161, 163, 164, 225, 230, 232, 233, 235, 236, 238, 241, 243, 244, 248, 283, 295, 296, 328, 330, 335, 336, 337, 339, 340, 349, 372, 417, 440, 449, 457
Fine-textured soils 35, 37, 39, 98, 103, 107, 109, 130, 207, 209, 215, 468
Finite-difference schemes 277
First-order Taylor analysis 323, 358
First, second, third, and fourth moments 323
Flow discharge 188
Flow shear stress 185, 188, 189, 190
Flow stream power 188
Flow velocity 187, 188
Fluvisols 144, 466
Flux density 201
Forward selection 9
Fractal dimension 51, 54, 196, 214, 217, 330, 334
Fractal geometry 64, 196, 213, 214, 217, 267
Fraction of colloidal clay 449
Fraction of oxygen in air 330
Fractional bias 397
Fractional variance 396
Freundlich adsorption isotherm 219
Freundlich parameter 198, 217
Fully saturated conductivity 442
Functional criteria 6, 392
Functional evaluation of PTF 6, 390, 393
Functional forms of PTFs 465
Functional soil behavior 390
Functional validation 18
Funnel effect 388
Fuzziness exponent 327
Fuzzy based logic rules 364
Fuzzy c-means classification procedure 282
Fuzzy expert systems 400, 406
Fuzzy k-means with extragrades 327
Fuzzy logic methods 5, 6
Fuzzy set theory 363, 400, 406
Fuzzy solution set 402, 404

G

Gamma radiometric technique 264
Gas diffusion coefficient 327, 337, 345
Gas diffusivity 201, 203, 204, 217, 330, 337
Gas retention 4
General circulation model 1, 415
Generalized linear models 282
Generalized parameters 78, 80
Genetic grouping 160
Geochemical processes 274
Geographic resources analysis support system GRASS 313
Geo-referenced soil database 351
Geometric mean diameter 283
Geometric standard deviation 54, 55, 211, 360, 374
Geometry of the pore network 160
Geomorphological processes 274
Geomorphometry 5
Geophysical techniques 7
Geostatistical interpolation 273
GIS-based regional modeling 7
Gleysols 144, 466
Global output 402, 404
Goodness-of-fit 9, 12, 13, 18, 129
Grain size distribution index 456
Grain size distribution 64, 456, 465, 466
Granulometric composition of materials 8
Granulometric fractions 449
Gravimetric water content 4, 109, 144, 155, 243, 324, 419, 423, 425
Gravitational potential 254
Gravity drainage 205, 318
Green-Ampt-Mein-Larson, GAML procedure 183
Gregson-Hector-McCowan (GHM) model xxii, 73
Grid networks 277
Groundwater pollution 465
Ground-penetrating radar 7
Group method of data handling 2, 24, 25, 30, 55, 58, 64, 99, 104, 105, 130, 408

Grouping based on structure and bulk density 165
Grouping strategy 159
Guelph permeameter 119, 120, 370
GUEST (Griffith University Erosion System Template) 181
Gupta-Larson 452, 453

H

Hairsine-Rose model 181
Halloysite content 153
Heavy metals 196, 197, 198, 199, 200
Heterotopic co-kriging 281, 282
Hierarchical polynomial regressions 57, 99
High resolution airborne radiometric systems 265
Histogram 6, 15, 307
Horizon-based grouping 161
Horizonation 274
HUNSODA database 55
Hutson-Cass retention function parameters 372
Hydraulic conductivity curve 195, 325, 435
Hydraulic conductivity "matching point" xxvi, 263
Hydraulic conductivity 3, 6, 8, 18, 30, 33, 34, 41, 50, 51, 52, 53, 64, 71, 83, 84, 85, 86, 88, 90, 91, 95, 110, 111, 116, 117, 118, 119, 120, 121, 123, 126, 127, 129, 136, 150, 156, 159, 161, 163, 164, 166, 168, 172, 183, 184, 189, 190, 191, 195, 205, 206, 210, 213, 225, 248, 254, 255, 256, 263, 264, 267, 273, 281, 283, 284, 286, 287, 289, 290, 295, 296, 298, 300, 301, 302, 303, 304, 307, 310, 312, 316, 318, 323, 325, 327, 339, 345, 351, 352, 354, 358, 369, 370, 388, 393, 406, 409, 416, 425, 432, 435, 440, 442, 445, 446, 447, 453, 455, 456, 457, 458, 459, 460, 461, 462, 472
Hydraulic functions 33, 34, 41, 91, 136, 267, 285, 296, 298, 315, 318, 409, 435
Hydraulic radius of the rill flow 189

HYDRO-GEN code 303
Hydrophobic 481
HYPRES database 9, 369, 376, 431, 432, 433, 434, 435
Hysteresis loop 451

I

Ideal gas equation 337
IGBP-DIS soils database 42, 154
Illite content 153
Impedance factor at saturation 202
Impedance factor 202, 217
Improvement of prediction 168
Incremental sum of squares 160
Index of agreement 397
Individual aggregate fractions ix, 143
Infiltration depth 183
Inflection point 150, 238, 330, 334, 335
Input-output relationship 132
Instantaneous profile method 256
Integrated indices 6, 357, 363, 367, 370, 377, 383, 388, 390
Inter-aggregate macro-porosity 233
International Unsaturated Soil Database, UNSODA 55, 164
Interped pore structure 24, 119, 180
Interpedal pores 116, 230
Interpolating soil properties 280, 282
Interpolation 3, 38, 39, 53, 54, 55, 61, 64, 77, 121, 273, 274, 275, 277, 280, 281, 288, 289, 290, 298, 299, 373, 393, 431
Interrill erodibility 185, 186, 188
Interstratified material 153
Intrapedal pores 116
Intrinsically stationary process 299
Inverse of the bubbling pressure 7
Inverse optimization scheme 256
Iron oxides 153, 154, 198
Irregularly spaced (x, y, z) data 275
Irrigation 134, 136, 185, 209, 255, 298, 317, 330, 339, 462
Isoproturon breakthrough 391, 392
Iteration 2, 18, 24, 25, 100, 132, 336
Iterative calibration 21, 23

J

Jackknife techniques 29

K

Kaolinite content 153
Key water contents 4, 241, 242, 243, 245, 246
K-fold 29
Kinematic wave models 213
Kolmogorov-Smirnov statistics 5
Konawa soil 87, 488
Kozeny-Carman equation 323
Kriging with external drift 281, 282
Kriging with uncertain data 282
Kriging 274, 277, 281, 282
Kurtosis 6, 7, 304, 307, 467

L

Lack of fit 14, 15, 17
Lakeland series 480
Land surface shape 5
Land use 6, 9, 115, 130, 159, 177, 303, 311, 316, 415, 416
Landform attributes 276, 290
Landscape description 275
Landscape position 5, 287, 290
Landscape 5, 7, 41, 91, 117, 121, 254, 265, 266, 273, 275, 276, 285, 287, 290, 298
Lateral distribution of the flow 277
Lateral flux 276, 278
Lateral subsurface flow vi
Latin-Hypercube technique 274
LEACHM, Leaching estimation and chemistry model 407
Leaching 64, 199, 213, 217, 267, 274, 297, 317, 406, 415
Learning networks 25
Leave-one-out approach 29
Levenberg-Marquardt algorithm 23, 313
Linear adsorption 197, 205, 208
Liquid limit 4, 234, 243, 248
Lithic Haplustoll 303
Litter surface cover 185

503

Loam 10, 49, 51, 64, 75, 76, 81, 82, 97, 98, 99, 100, 101, 102, 106, 111, 118, 128, 144, 145, 146, 147, 148, 149, 150, 155, 156, 159, 164, 165, 167, 168, 172, 207, 235, 236, 244, 248, 267, 336, 369, 416, 438, 445, 450, 457, 459, 461
Local climatic gradient 276
Local flow accumulation 276
Local terrain attributes 277

M

Macroflow region 161
Macro-pore specific volume of pedostructure 238
Macromorphological attributes 127, 130, 134
Macropore 84, 119, 122, 127, 128, 129, 133, 135, 136, 166, 196, 274, 287
Mahalanobis distance 328
Map variability 391
Marshall solute diffusion model 204
Mass fractal dimension 214, 330, 334
"Matching point" in the Brooks-Corey $K(h)$ or $K(\theta)$ relation xxvi, 263
Matrix pore space 391
Maximum adsorbed water content 241, 243
Maximum capillary-sorptive water content 241, 243
Maximum hydroscopic water content 449
Maximum molecular water content 241
Mean absolute error 17, 359, 397
Mean bias error 261, 394
Mean geometric diameter 330
Mean square error 17, 88
Mean statistical diameter of particles 457
Mechanical composition of soil 49, 462
Mechanical dispersion of solute molecules 205
Mechanical resistance 337
Median absolute error 397
Medium Fine Topsoils 432
Membership class 363, 365, 368, 369, 400, 404, 405
Mesopore 127, 129
Microflow region 161

Micro-pore specific volume of pedostructure 238
Micromorphological data 23
Micropore 127, 129, 133
Microscopic capillary length 345
Millingtom-Quirk solute diffusivity model 204
Mineral bulk density 96
Mineral grains 96, 274, 450
Mineral soils 133, 136, 450
Mineral weathering 274
Mineralogical composition of clay fraction 153
Mixing cell models 213
mixture model of De Vries 340
Mobile water content 210, 212, 213, 217, 339
Mobile-immobile model, MIM xxv, 196, 207
Model adequacy 3
Model calibration 8, 267
Model EUR-M3 376
Model fit 12, 363
Model sensitivity 495
Model specification 3, 13
Model validation 17, 57, 60, 217, 357, 405, 409
Modeling applications 257, 399
Modeling efficiency, EF 359
Modeling objective 4
Moist consistence 134
Moisture equivalent 64, 248, 418, 419, 423, 425
Moisture supply capacity of soil 391
Molecular diffusion 205, 206, 211, 217
Mollic Fluvisols 466
Mollic Gleysols 466
Mollisols ix, 98, 99, 102, 103, 105, 106, 484
Monte Carlo analysis 40, 41
Montmorillonite content 153
Morphological quantification system 127
Morphometric indices 127, 129, 130
Mottles 115, 134
MRE statistics 309

Mualem-Van Genuchten parameters 122
Multicollinearity 2, 10, 12, 18
Multilayered iterational algorithms, MIA 25
Multiple linear regression 4, 161, 198, 452, 457, 469
Multiple-flow-direction, MFD, algorithms 278
Multivariate methods 7

N
Natural pedogenesis 466
Net acid addition rate 330, 343
Net infiltration 255
Neuro Multistep software 325
Neuro θ, Theta 325
Neuropack 352, 353
Neurons 22, 23, 376
New Zealand regional datasets 258
Nodes 22, 28, 100, 146, 283, 286, 313, 376
nonlimiting water range (NLWR) 327
Normalized average error 397
normalized difference vegetation index (NDVI) 289
Nugget 299, 302, 309, 310, 311, 312
Number of predictors 167
Numerical soil data 8

O
Olsen-Kemper solute diffusion model 204
One-dimensional flow 254, 255
Optimal water content for tillage 327, 336
Optimization criteria 2, 33, 36
Organic carbon content 3, 35, 95, 97, 98, 99, 100, 101, 102, 103, 104, 105, 106, 107, 108, 109, 110, 111, 127, 129, 154, 162, 198, 199, 273, 284, 313, 314, 326, 329, 338, 343, 359, 366, 367, 377, 466
Organic carbon partition coefficient 340
Organic matter content 5, 7, 34, 41, 64, 71, 95, 111, 123, 164, 171, 198, 199, 209, 216, 243, 256, 282, 289, 298, 327, 345, 349, 360, 362, 369, 377, 449, 450, 452, 465, 472

Organic soils 6, 135, 163, 432, 435, 453
Outlier detection 15, 18
Oven-dry bulk density 273, 283, 284, 287, 289, 313, 314
Oxysols 98

P
P buffering capacity 327
Papendick and Campbell solute diffusivity model 204, 327
Parameter estimation 3, 12, 35, 53, 54, 241, 254, 267, 358, 397, 409, 417
Parametric methods 34, 35, 71
Parent material grouping 167
Parent material 4, 117, 118, 121, 122, 156, 161, 167, 168, 171, 265, 274, 279
Partial or fuzzy membership 364, 400
Particle density 51, 97, 330, 466
Particle-size distribution 47, 50, 52, 55, 64, 127, 160, 211, 260, 267, 283, 289, 304, 323, 328, 345, 435, 449, 452, 462, 465, 469, 472
PAWC model 256
Peak runoff rate 185
Pearson correlation coefficient 130, 399
Peclet number 208
Ped shape 124, 128, 129
Ped size 117, 118, 119, 121, 124, 128, 129
Pedality 116, 127, 128, 129, 130, 131, 160
Pedogentic variable 274
Pedohydral parameters 232
Pedological similarities 161
Pedology 317
Pedon-SEI 325
Pedostructure interpedal water pool 238
Pedostructure pore specific volume 238
Pedostructure specific volume 238
Pedostructure water content 238
Pedostructure 226, 227, 229, 230, 231, 232, 233, 234, 235, 236, 238
Pedotransfer function validation 362
Pedotransfer function 1, 3, 4, 5, 6, 9, 10, 17, 18, 21, 30, 33, 41, 47, 49, 50, 52, 64, 71, 72, 87, 91, 95, 111, 115, 135, 136, 150, 156, 172, 177, 178, 180, 186, 188, 189,

190, 195, 213, 217, 225, 229, 235, 241, 248, 254, 255, 257, 263, 267, 273, 290, 295, 318, 323, 324, 326, 341, 344, 345, 349, 354, 357, 358, 359, 362, 363, 364, 365, 366, 367, 368, 370, 371, 377, 379, 380, 381, 382, 383, 384, 388, 389, 390, 391, 393, 406, 408, 409, 411, 415, 425, 431, 435, 440, 446, 450, 452, 453, 454, 465, 472
Pedotransfer models 106
Pedotransfer rules 3, 8, 122, 125, 136
PEf statistics 314
Penetrometer 7
Penman solute diffusion model 204
Percolation theory 213
Performance of the model 4
Permanent wilting point 49, 120, 121, 155, 161, 164, 231, 233, 235, 236, 449
Permittivity 330, 340, 342
Pesticides 196, 198, 217, 318, 363
pH buffering capacity 330, 343
Phosphate absorption capacity 5
Physical process-based computer simulation model 181
Physical-empirical model 256
Piece-wise linear model 459
Pima clay 82
Pipette method 144, 260
Plan curvatures 276, 277, 279, 284, 287, 293, 314
Planar-void pore-interaction method 126
Planosols 144
Plastic limit 4, 243, 248, 335
Plasticity class 131
Plasticity 131, 232, 462
PMQ solute transport model 204
Podzoluvisols 144
Point-based soil water balance 255
Point PTF 363
Point-scale system 127
Polynomial networks 25
Pore connectivity 209
Pore volume number 208
Pore water velocity 205, 209, 213
Pore-size distribution index 80, 85, 110

Pore-size distribution 4, 51, 79, 80, 85, 90, 110, 203, 210, 216, 300, 438
Pore-size index 442
Porosity 34, 35, 50, 64, 72, 80, 81, 84, 85, 90, 91, 95, 97, 110, 115, 117, 118, 131, 133, 135, 136, 153, 156, 166, 168, 172, 201, 202, 203, 204, 216, 227, 230, 238, 267, 273, 287, 330, 337, 338, 345, 360, 372, 375, 409, 418, 438, 443, 445, 446, 449, 457, 466
Porous media 41, 91, 111, 126, 150, 156, 172, 195, 196, 201, 206, 210, 214, 217, 267, 290, 295, 318, 345, 405, 406, 446, 472
Potential erosion 276
Precision agriculture 1, 297, 317, 345
Precision farming 273, 295
Prediction interval 15
Predictor 2, 3, 4, 9, 10, 11, 12, 14, 15, 17, 18, 21, 24, 25, 26, 30, 34, 35, 51, 52, 55, 57, 60, 63, 64, 84, 98, 99, 100, 102, 103, 104, 109, 111, 121, 129, 131, 132, 136, 153, 154, 155, 164, 166, 167, 168, 198, 199, 200, 209, 216, 275, 282, 283, 285, 288, 313, 314, 315, 326, 327, 340, 362, 365, 380, 392, 393, 423, 434, 457
Preferential flow 119, 123, 126, 217, 225, 267, 289, 409
PREFLO model 392
Preliminary analysis of soil data 4
Pressure head 7, 23, 30, 36, 38, 39, 51, 72, 110, 153, 162, 163, 164, 166, 167, 254, 263, 284, 285, 286, 287, 289, 295, 296, 300, 304, 305, 306, 307, 308, 309, 310, 311, 313, 314, 315, 334, 359, 360, 362, 365, 432, 446
Primary peds water content 238
Primary peds 226, 227, 229, 230, 232, 233, 235, 238
Primary soil structure 119
Primary terrain attributes 276, 279
Principal factor analysis 8
Principal component analysis 7, 8, 9, 18
Principal component regression 12
Profile composition 390

Profile curvatures 159, 390
PTF reliability 6, 7, 8
PTF software 5, 325
PTF uncertainty 2, 8, 390
PTF-HYPRES 304, 305, 306, 307, 308, 309, 311, 312, 315

R

Radial basis function 22
Radiometric data 265, 266, 267
Rainstorm event 183
Random error 37, 38, 299
Range 299, 312
Range based pattern index 399
Range pressure chamber 370
rate of erosion process 276
rate of lateral accumulation 276
Rate of lateral fluxes 276
Ratio of scatter 397
Rawls-Brakensiek model 452
reduced major axis method 394, 399
Reductionist paradigm 255
Regional PTF development 6
Regression equation 1, 2, 4, 9, 12, 21, 24, 28, 71, 83, 87, 95, 121, 162, 207, 242, 248, 258, 275, 303, 304, 305, 433, 439, 440, 452, 453, 454, 455, 459, 468
Regression tree modeling 2, 26, 99, 145
Relative complex dielectric constant of soil-water mixture 342
Relative intensity of the solar radiation 276
Relative mean bias 397
Relative saturation 207, 208
Relative variability of evapotranspiration rate 276
Remote sensing 5, 7, 9, 47, 91, 95, 145, 265, 297, 406, 431, 440
Renflow sandy loam soil 144
Representative soil databases 36
Required aerobic depth 330
Residual water content 7, 74, 122, 144, 156, 254, 375, 443, 454
Retardation coefficient 197, 205

Retention 1, 3, 4, 6, 7, 8, 9, 10, 18, 30, 33, 34, 35, 37, 38, 39, 40, 41, 47, 50, 51, 52, 53, 55, 57, 60, 61, 64, 71, 72, 73, 74, 75, 76, 80, 82, 84, 89, 90, 91, 95, 97, 98, 99, 100, 101, 102, 103, 104, 105, 106, 107, 108, 109, 110, 111, 120, 121, 122, 123, 126, 131, 132, 135, 136, 144, 145, 148, 149, 150, 153, 154, 155, 156, 159, 160, 161, 162, 163, 164, 165, 166, 167, 171, 172, 195, 198, 202, 203, 206, 207, 209, 210, 211, 213, 215, 216, 217, 225, 230, 233, 241, 242, 243, 244, 245, 246, 248, 254, 255, 256, 257, 258, 260, 262, 263, 267, 273, 283, 284, 285, 286, 287, 289, 290, 295, 296, 298, 299, 300, 303, 304, 305, 306, 307, 308, 309, 310, 311, 312, 313, 314, 315, 318, 323, 325, 327, 329, 330, 334, 335, 336, 339, 345, 350, 351, 352, 353, 354, 359, 360, 362, 363, 364, 367, 368, 369, 370, 372, 373, 374, 375, 376, 388, 392, 393, 405, 406, 408, 409, 416, 417, 418, 424, 425, 432, 434, 435, 438, 439, 440, 441, 442, 443, 444, 446, 447, 449, 450, 451, 452, 453, 462, 465, 466, 468, 469, 470, 471, 472
Richards' equation 253
Ridge regression 12
Rill erodibility 185, 186, 188, 189
Rill erosion parameters 188
Ring infiltrometer 120, 370
Risk for soil acidification 327
risk-analysis software 327
River sedimentation 466
Root mean square deviation, RMSD 359
Root mean square error 17, 36, 38, 78, 306, 307, 354, 359, 397
Root mean square residuals (RMSR) 52
Root mean square variation 397
Root Zone Water Quality Model (RZWQM) 89, 297
Rosetta 41, 52, 53, 64, 325, 345, 351, 352, 354, 408
RUSLE erosion model 24

Runoff 125, 177, 178, 179, 181, 183, 185, 187, 188, 190, 191, 253, 255, 267, 287, 290, 317, 318, 392, 409

S
slope orientation (Aspect) 276
Sadeghi-Kissel-Cabrera solute diffusivity model 204
Sand content 5, 12, 55, 103, 107, 109, 146, 148, 160, 184, 185, 211, 235, 282, 304, 313, 326, 328, 338
Saturated hydraulic conductivity xii, xxii, xxiii, xxviii, 31, 50, 51, 66, 67, 83, 84, 88, 90, 92, 110, 111, 113, 116, 119, 136, 137, 138, 139, 140, 156, 163, 168, 173, 183, 195, 205, 206, 263, 264, 269, 270, 284, 287, 289, 293, 300, 303, 304, 323, 325, 339, 347, 351, 352, 356, 358, 369, 370, 388, 406, 409, 416, 437, 445, 446, 447, 455, 478, 483, 484, 486, 487, 490, 493, 494, 496
Scale effects 4, 84, 393
Scale-invariant 215
Scale of application x
Scale 1, 4, 5, 6, 8, 9, 10, 41, 64, 72, 73, 81, 82, 84, 87, 90, 91, 125, 126, 127, 134, 135, 136, 160, 162, 195, 206, 213, 214, 215, 217, 230, 231, 238, 248, 253, 254, 255, 256, 258, 264, 267, 273, 274, 275, 278, 279, 280, 281, 282, 288, 289, 290, 295, 296, 297, 298, 299, 300, 301, 304, 309, 311, 312, 315, 316, 317, 318, 334, 345, 368, 392, 393, 394, 408, 415, 424, 432, 433, 434, 435, 469
Scaling laws 3
Scaling method 72, 73, 74, 89, 90, 91, 217
Secondary soil structure 119
Secondary terrain attributes 276, 278, 280
Sediment discharge rate 188
Sediment particle composition 24
Sediment particle fractions 181
Sediment transport capacity indices 288, 314

Sediment transport index 276
Sediment 39, 64, 154, 156, 167, 177, 181, 182, 185, 187, 188, 190, 191, 274, 276, 288, 314, 466
Sedimentation 64, 274, 288, 466
Semiempirical dielectrical mixing model 342
Semi-physical model 23
Semivariance 299, 300, 309, 310, 311, 312
Semivariogram 274, 281, 282, 298, 299, 300, 301, 302, 303, 309, 310, 311, 312, 314, 315
Sensitivity analysis 5, 100, 109, 191, 267, 362, 363, 409
Shallow landslide hazard 275
Shape of the distribution 6
Shapiro-Wilk statistics 5
SH-Pro 323, 325, 345
Shrinkage curve 4, 225, 227, 237, 238
Shrinkage limit 232
Shrinkage parameters 225
Shrink-swell dynamics x
Shrink-swell properties 225
S-shaped membership function 401
Sigmoidal function 23, 51
Sill 299, 311, 312, 317
Silt coating 116
Silt content 60, 61, 62, 117, 162, 172, 304, 313
Simplex method 56, 64, 304
Simplified predictive methods 298
Simulation bias 397
Simulation model 6, 9, 18, 87, 90, 91, 123, 136, 181, 267, 297, 298, 315, 362, 391, 392, 393, 405, 406, 411
soil inference system SINFERS 328
Single-flow-direction, SFD, algorithms 278
Sink strength 254
Skewness of the distribution 6
Slickensides 116
Slope angle 276
Slope at inflection point 330
Slope length 177, 178, 179

Slope orientation (aspect), 276
Slope steepness 177, 178
Snedecor coefficient, F 455
Snowmelt 255
Soddy-podzolic soil 245, 246, 247
Sodium absorption ratio, SAR 156
Soil aggregate composition 3, 145, 148
Soil alkalinity 9
Soil bulk density 3, 4, 72, 85, 95, 96, 97, 111, 186, 197, 203, 235, 441, 450, 452, 457
Soil chemical properties 155
Soil classification system 136, 431
Soil classification 64, 115, 136, 180, 303, 326, 345, 431, 459
Soil clay films 115
Soil compaction 267, 350, 449, 450, 462
Soil concretions 115
Soil consistence 30, 64, 111, 136
Soil cover 6, 7, 468
Soil database 30, 52, 64, 136, 144, 150, 164, 233, 257, 264, 273, 275, 326, 345, 351, 424, 435
Soil deposition 274
Soil detachment 185
Soil erodibility factor 179
Soil erosion modeling 4
Soil erosion 4, 7, 71, 177, 178, 180, 181, 186, 190, 191
Soil European database 162
Soil formation 265, 274, 275, 288
Soil horizon designations v
Soil horizon 87, 88, 96, 99, 116, 117, 118, 119, 122, 123, 124, 128, 130, 172, 226, 230, 231, 238, 286, 287, 313, 409, 431
Soil hydraulic conductivity 3, 30, 111, 136, 156, 267, 289, 296, 310, 318, 354, 409, 455
Soil hydraulic database in the Moscow State University 243
Soil hydraulic properties 3, 4, 5, 6, 7, 8, 9, 10, 11, 18, 30, 33, 47, 50, 51, 52, 55, 57, 60, 61, 62, 63, 64, 71, 88, 89, 90, 91, 95, 99, 111, 115, 116, 117, 123, 126, 127, 130, 133, 134, 135, 136, 153, 155, 156, 172, 195, 213, 215, 217, 225, 267, 273, 274, 275, 280, 283, 284, 287, 288, 289, 290, 295, 296, 297, 298, 299, 300, 303, 312, 313, 315, 316, 318, 323, 345, 362, 408, 409, 425, 433, 435, 472
Soil hydraulic property variations 298
Soil hydrology 3, 9, 123, 136, 256, 317, 318, 405, 472
Soil inference system 5, 9, 136, 323, 324, 326, 328, 330, 344, 345, 406
Soil infiltration parameters 4, 190
Soil mapping technique 265
Soil matric potential 4, 8, 126, 225, 241, 242, 248, 388, 457
Soil matrix 119, 231, 274, 289
Soil mineralogical properties 9
Soil moisture deficit 183
Soil morphological attributes 3, 115, 116, 120, 121, 122, 123, 127, 131, 132, 136
Soil morphology 3, 115, 117, 118, 120, 122, 125, 126, 127, 133, 134, 135, 136, 290
Soil organic carbon composition 9, 340
Soil organic carbon content 3, 97, 198, 367, 377
Soil parameters 1, 4, 130, 183, 217, 318, 417, 465
Soil particles median diameter 378, 381, 386
Soil pH 4, 9, 10, 41, 64, 91, 111, 127, 136, 155, 156, 189, 197, 198, 216, 217, 232, 248, 257, 267, 273, 275, 280, 283, 284, 286, 289, 290, 296, 313, 318, 324, 326, 327, 328, 329, 334, 343, 345, 350, 391, 405, 406, 408, 425, 446, 449, 453, 456, 457, 462, 465, 470, 472
Soil Plant Atmosphere Water Field and Pond Hydrology, SPAW, package 349
Soil profile 85, 98, 115, 117, 118, 125, 136, 159, 161, 231, 255, 256, 274, 288, 295, 296, 388, 391, 393, 415, 417, 419, 450, 455, 456, 466
Soil rheology 4

Soil salinity 4, 156
Soil solid phase density 457
Soil spatial characterization 276
Soil stability factor 184
Soil strength 327, 337
Soil structure code 180
Soil structure 3, 7, 18, 52, 63, 97, 111, 115, 116, 117, 118, 119, 120, 121, 122, 123, 124, 126, 127, 129, 131, 132, 133, 134, 135, 136, 145, 153, 166, 168, 172, 180, 195, 196, 217, 225, 257, 267, 283, 289, 339, 425, 452, 458
Soil surface O_2 concentration 337
Soil survey 8, 30, 47, 48, 49, 53, 54, 64, 83, 91, 95, 98, 110, 111, 115, 116, 117, 120, 121, 122, 123, 127, 130, 131, 133, 136, 156, 159, 163, 172, 188, 217, 225, 255, 257, 262, 267, 290, 303, 317, 318, 324, 345, 384, 419, 435, 447
Soil taxonomic order 100, 101, 102
Soil textural triangle 164
Soil texture class system solid phase parameters 449
Soil texture class system 47
Soil texture class 47, 48, 49, 50, 52, 53, 79, 82, 89, 100, 101, 102, 104, 121, 124, 133, 304, 431, 432, 433, 435, 440, 441, 445, 446
Soil texture diagrams 48
Soil thermal conductivity 340, 341
Soil thermal properties 323
Soil type 34, 73, 74, 76, 80, 84, 86, 122, 123, 135, 136, 161, 197, 198, 202, 203, 207, 211, 214, 216, 236, 256, 275, 285, 326, 336, 345, 417, 419, 424, 472
Soil water capacity 22
Soil Water Characteristics from Texture, SWCT, package 349
Soil water content at field capacity 238, 372
Soil water content at wilting point 337, 372
Soil water content 4, 56, 71, 72, 74, 77, 81, 82, 84, 85, 88, 89, 90, 91, 100, 202, 203, 210, 216, 217, 238, 241, 243, 245, 254, 283, 285, 286, 289, 290, 295, 296, 300, 306, 311, 313, 315, 318, 337, 343, 345, 364, 365, 367, 370, 372, 374, 378, 381, 388, 393, 416, 439, 467
Soil water retention hysteresis 6, 450
Soil water retention 3, 4, 6, 8, 9, 10, 18, 30, 41, 47, 50, 51, 52, 64, 71, 72, 73, 74, 76, 82, 89, 91, 95, 97, 98, 100, 101, 102, 103, 110, 111, 132, 136, 144, 145, 149, 150, 154, 156, 171, 172, 195, 203, 217, 241, 243, 245, 248, 255, 256, 267, 273, 283, 284, 286, 289, 290, 295, 296, 298, 299, 300, 304, 305, 306, 309, 310, 311, 312, 313, 318, 323, 345, 354, 362, 363, 367, 368, 369, 388, 392, 406, 408, 409, 425, 434, 440, 441, 442, 444, 449, 450, 451, 453, 462, 465
Soil water tension 386, 388
Soil-landscape analysis 275, 276, 280
Soil-Loss Estimation Model for South Africa (SLEMSA) 180
Soil-water internal configuration 225
Soil–water medium functional model xv, 226
Soil-water medium hierarchy and functionality xv, 226
SOILOSS computer program 181
program SOILPAR 323
SOILPROP software 325
Solids specific volume 238
Solute adsorption parameters 196
Solute degradation 196
Solute dispersivity 4, 196, 217
Solute mineralization 196
Solute retention 195
Solute transformation 196
Solute transport in soils 18, 217
Solute transport models 195, 213, 217, 406
Solute transport parameters 195, 196, 197, 207, 217, 327
Solute velocity 217
Sorptivity 136, 267, 330, 338, 339
Source strength 254
Spatial coverage 7, 253
Spatial discretization 277
Spatial heterogeneity 253

Spatial organization 281, 290
Spatial resolution 265, 275, 279, 290
Spatial variability of soil moisture 280
Spatial variation 5, 195, 273, 297, 299, 303, 305, 312, 313, 316
Spatially dense physical information 7
Spearman coefficient 313, 399
Spearman ranking 313
Specific catchment area 276
Specific goal of model use 357
Specific heat capacity 340
Specific matrix potential 391
Specific surface area 203, 451, 452, 457
Spectral stripping algorithm 266
Spline interpolation 55, 277
Spodosols 98, 99, 102, 103
Stagewise method 9, 18
Stagnant water 207, 213, 217, 466
Standard deviation 6, 28, 37, 51, 54, 55, 56, 58, 60, 63, 73, 84, 85, 146, 148, 171, 210, 211, 283, 300, 305, 306, 307, 314, 327, 328, 329, 359, 360, 373, 374, 394, 432, 435, 438, 443, 467
Standard error of estimation (SEE) 455
Standard normal deviates 15
Standardization 120, 133
Standardized database 431
Standardized scores 8, 9
Statistical regression 1, 2, 3, 5, 17, 21, 28, 29, 132, 188, 303
Statistical validation 17, 18, 406
Statistically representative set 390
STATSGO database 257
Steady-state runoff rate 188
Stepwise method 9
Stochastic simulation 273, 274, 307
Stream–power indices 276
Stream tubes 277
Stress surfaces 116
Structural cracks 116
Structural shrinkage 232
Structure grade 131, 161
Structure of spatial variability 299
Structure strength 180
Structural surface features 116

Structure 2, 3, 7, 18, 21, 26, 30, 50, 51, 52, 63, 64, 97, 99, 103, 111, 115, 116, 117, 118, 119, 120, 121, 122, 123, 124, 126, 127, 128, 129, 131, 132, 133, 134, 135, 136, 145, 149, 153, 156, 159, 160, 161, 162, 165, 166, 167, 168, 171, 172, 179, 180, 195, 196, 210, 217, 225, 234, 238, 257, 258, 267, 274, 277, 278, 281, 282, 283, 287, 289, 290, 295, 297, 299, 300, 302, 303, 304, 311, 315, 316, 318, 339, 345, 353, 367, 388, 400, 415, 425, 452, 455, 458, 465, 472
Student t-test 15, 305, 314
Studentized residuals 14, 15
Subsoils 49, 64, 134, 154, 162, 163, 165, 167, 435, 457
Sugeno's inference method 401, 402
Sum of square error 395
Sustainable land use system vi
SWAP solute transport model 206
Swelling potential of soil 232
SWIM solute transport model 206
Systematic error 37, 38, 39, 289, 296, 312
Systematic pattern in residuals 363

T

Tactile qualities of the soil 115
Takagi-Sugeno-Kang method of fuzzy inference 363
Tangential curvatures 276, 277, 313
Teller soil 82
Temperate regions 161, 415, 424
Temperature indices 276
Temporal heterogeneity 253
Temporal resolution 256
Temporal variation of soil properties 7
Terminal node 28, 100, 146, 286
Terrain analysis 273, 275, 277, 284, 288, 290, 297, 318
Terrain attributes 5, 136, 275, 276, 277, 278, 279, 280, 281, 282, 283, 284, 285, 287, 288, 289, 290, 309, 313, 314, 318, 408
Terrain characterization index 280
Terrain indices 275, 278, 290

Terrain variables 5, 283, 288, 314
Textural class 3, 10, 40, 41, 47, 49, 50, 54, 73, 74, 75, 78, 79, 80, 81, 82, 88, 89, 90, 100, 103, 104, 111, 122, 127, 128, 129, 131, 132, 144, 145, 161, 163, 164, 165, 168, 171, 172, 215, 216, 267, 281, 290, 349, 359, 360, 362, 409
Textural composition 103, 104, 145, 435
Textural parameters in PTF 3
Texture classification 50, 54, 55, 150, 159
Texture grouping 77, 90, 163, 164
Texture 4, 5, 10, 12, 18, 30, 35, 39, 40, 41, 47, 48, 49, 50, 51, 52, 53, 54, 55, 64, 71, 74, 75, 76, 77, 78, 79, 80, 81, 82, 88, 89, 90, 91, 95, 97, 98, 99, 100, 101, 102, 103, 104, 106, 109, 110, 111, 115, 116, 117, 118, 119, 120, 121, 122, 123, 124, 127, 128, 130, 131, 132, 133, 135, 136, 144, 145, 146, 148, 149, 150, 156, 159, 160, 161, 162, 163, 164, 165, 166, 167, 168, 169, 171, 172, 199, 202, 203, 207, 217, 225, 230, 235, 237, 238, 243, 257, 258, 264, 265, 267, 273, 283, 286, 287, 290, 298, 304, 323, 326, 336, 344, 345, 349, 350, 354, 362, 369, 390, 409, 416, 417, 424, 425, 431, 432, 433, 435, 438, 440, 441, 445, 446, 447, 452, 457, 459, 461, 462, 468, 472
Thermal capacity 330
Thermal conductivity 330, 340, 341
Thermal diffusivity 330, 340, 341
Threshold water content 202, 203
Tillage limit 330, 334, 335
Tillage practice 7, 298, 311
Time to ponding 183, 330
Time-domain reflectometry 89
Topographic attributes 5, 135, 280, 284, 288, 290
Topographic information 7, 91, 275, 285
Topographic wetness indices 276
Topsoil 162, 163, 165, 166, 167, 217, 238, 371, 372, 376, 384, 432, 433, 435, 457

Tortuosity parameter 202
Tortuous pore 51
Total effective surface cover 184
Total salt content 155
Total shrinkage of primary peds 232
Total sum of squares 12
Training set 5, 327, 328, 338
Transfer function models 213
Transfer function 8, 213, 217, 267, 313
Transpedal pores 116
Transport of water and chemicals in soil 1
Tree pruning 28
Triangulated Irregular Networks (TINs) 277
Troup soil 245, 247
Typic Haplaquods 162
Typic Haploboroll 162
Typic Ustifluvent 303

U

Ultisols 99, 102, 103, 105, 106, 167, 416
Uncertainty in model inputs parameters 390
Uncertainty in model input 390
Unfavorable, U, membership 364
Unified diffusivity model (UDM) 203
Universal function approximators 21, 23
Universal Soil Loss Equation (USLE) 179
Unsaturated hydraulic conductivity 6, 18, 41, 51, 53, 64, 85, 86, 90, 111, 119, 156, 159, 164, 172, 213, 225, 255, 263, 267, 289, 301, 304, 307, 310, 312, 318, 325, 345, 351, 354, 456, 459, 461, 462, 472
Unsaturated zone 217, 255, 274, 318, 390, 433, 455
UNSODA database vii, 173
Upper quartile absolute error 397
Upscaling of PTF estimates 8
US National Soil Characterization Database 3

USDA-SCS National Soil Survey
 Laboratory Pedon Database 262
User-friendly interface 181
User-oriented software 27
User-oriented techniques 5
USLE database 4, 190, 191

V

Vadoze zone transport processes 255
Vadose zone xv, xvi, 92, 140, 206, 217, 218, 219, 239, 248, 253, 255, 270, 296, 318, 408, 428, 465
Validation dataset 100, 354, 419
Van Genuchten equations 3
Van Genuchten parameters, xv, 12, 57, 64, 131, 145, 148, 149, 161, 165, 245, 334, 425
Van Genuchten retention function parameters 372
Vapor equilibrium technique 370
Variability in modeling results 390, 391
Variability 2, 5, 8, 9, 12, 13, 18, 33, 41, 73, 91, 111, 119, 159, 161, 167, 171, 209, 214, 217, 255, 267, 274, 275, 276, 279, 280, 281, 282, 283, 284, 288, 289, 290, 295, 296, 297, 298, 299, 300, 302, 306, 311, 312, 313, 316, 317, 318, 358, 359, 390, 391, 392, 397, 399, 409, 424, 425, 432, 446
Variance ratio 397
Variance 5, 8, 12, 15, 25, 36, 37, 40, 54, 159, 180, 286, 300, 301, 305, 310, 312, 317, 324, 326, 328, 344, 358, 359, 363, 377, 391, 393, 394, 397, 399, 467

Variance-covariance matrix 8, 12, 15
Vegetation patterns 276
Velocity of lateral fluxes 276
Vermiculite content 153
Vertisols 99, 102, 103, 105, 106, 109, 136
Visual qualities of soil 115
Volumetric heat capacity 340
Volumetric soil water content 56, 84, 295, 364, 370, 378, 381

W

Walczak's model 452, 453
Water balance equation 255
Water content at wilting point 97, 337, 372
Water erosion prediction project (WEPP) 181
Water management 6, 156, 345, 409
Water storage in vadose zone 465
Water table depth 134
Water table 117, 134, 136, 159
Water-holding capacity of soil 316
Watershed modeling 71
Weighted-average predictions 8
WEPP erodibility 185
WEPP infiltration 183
Wetting water retention curve 452
Wilting coefficient 9, 49, 64, 290
Wilting point 4, 49, 97, 98, 103, 120, 121, 144, 155, 161, 164, 225, 230, 231, 233, 235, 236, 238, 244, 296, 335, 336, 337, 339, 349, 372, 440, 449